Coarse Grained Simulation and Turbulent Mixing

Small-scale turbulent flow dynamics is traditionally viewed as universal and as enslaved to that of larger scales. In coarse grained simulation (CGS), large energy-containing structures are resolved, smaller structures are spatially filtered out, and unresolved subgrid scale effects are modeled. *Coarse Grained Simulation and Turbulent Mixing* reviews our understanding of CGS. Beginning with an introduction to the fundamental theory, the discussion then moves to the crucial challenges of predictability. Next, it addresses verification and validation, the primary means of assessing accuracy and reliability of numerical simulation. The final part reports on the progress made in addressing difficult nonequilibrium applications of timely current interest involving variable density turbulent mixing.

The book will be of fundamental interest to graduate students, research scientists, and professionals involved in the design and analysis of complex turbulent flows.

Fernando F. Grinstein is a scientist at the X-Computational Physics Division of the Los Alamos National Laboratory. He is a world leader in issues of large eddy simulation (LES) of turbulent material mixing physics in complex multidisciplinary applications. He has led integration efforts of the pioneers of the implicit LES techniques in workshops and special meetings worldwide, and in the first comprehensive description of the methodology, *Implicit LES: Computing Turbulent Flow Dynamics*, written with Len Margolin and William Rider.

Coarse Grained Simulation and Turbulent Mixing

FERNANDO F. GRINSTEIN

Los Alamos National Laboratory

CAMBRIDGE
UNIVERSITY PRESS

One Liberty Plaza, 20th Floor, New York, NY 10006, USA

Cambridge University Press is part of the University of Cambridge.

It furthers the University's mission by disseminating knowledge in the pursuit
of education, learning, and research at the highest international levels of excellence.

www.cambridge.org
Information on this title: www.cambridge.org/9781107137042

First published 2016

Printed in the United Kingdom by TJ International Ltd. Padstow Cornwall

A catalog record for this publication is available from the British Library.

Library of Congress Cataloging in Publication Data
Names: Grinstein, Fenando F., editor.
Title: Coarse grained simulation and turbulent mixing / edited by Fernando F. Grinstein,
Los Alamos National Laboratory.
Description: New York, NY : Cambridge University Press, [2016] | ©2016 |
Includes bibliographical references and index.
Identifiers: LCCN 2015042013| ISBN 9781107137042 (Hardback ; alk. paper) |
ISBN 1107137047 (Hardback; alk. paper)
Subjects: LCSH: Turbulence. | Fluid dynamics.
Classification: LCC TA357.5.T87 C53 2016 | DDC 532/.0527–dc23 LC
record available at http://lccn.loc.gov/2015042013

ISBN 978-1-107-13704-2 Hardback

To
Julia and Frederic,
and to the many contributors to this volume.

Contents

Color plate section between pages 212 and 213

Contributors

Malcolm J. Andrews
Los Alamos National Laboratory

Nicholas A. Denissen
Los Alamos National Laboratory

Ekaterina Fedina
Swedish Defense Research Agency

James R. Fincke
Los Alamos National Laboratory

Christer Fureby
Swedish Defense Research Agency

Kalyana C. Gottiparthi
School of Aerospace Engineering, Georgia Institute of Technology

Akshay A. Gowardhan
Lawrence Livermore National Laboratory

Fernando F. Grinstein
Los Alamos National Laboratory

Brian M. Haines
Los Alamos National Laboratory

James R. Kamm
Los Alamos National Laboratory

Robert J. Kares
Los Alamos National Laboratory

Len G. Margolin
Los Alamos National Laboratory

Suresh Menon
School of Aerospace Engineering, Georgia Institute of Technology

Reetesh Ranjan
School of Aerospace Engineering, Georgia Institute of Technology

Jon Reisner
Los Alamos National Laboratory

William J. Rider
Sandia National Laboratories

J. Raymond Ristorcelli
Los Alamos National Laboratory

Bertrand Rollin
Department of Aerospace Engineering, University of Florida

Vincent A. Thomas
Los Alamos National Laboratory

Adam J. Wachtor
Los Alamos National Laboratory

V. Gregory Weirs
Sandia National Laboratories

Leslie Welser-Sherrill
Los Alamos National Laboratory

Ye Zhou
Lawrence Livermore National Laboratory

Preface

The small scale turbulent flow dynamics is traditionally viewed as universal and enslaved to that of larger scales. In coarse grained simulation (CGS) large energy containing structures are resolved, smaller structures are spatially filtered out, and unresolved subgrid scale (SGS) effects are modeled. CGS includes classical large eddy simulation (LES) strategies focusing on explicit SGS models, implicit LES (ILES) relying on SGS modeling and filtering provided by physics capturing numerical algorithms, and, more generally, LES combining mixed explicit/implicit SGS modeling. The CGS strategy of separating resolved/unresolved physics constitutes the viable approach to address complex transition, unsteady flow, and multiphysics in practical geometries.

The validity of the scale separation assumptions in CGS needs to be carefully tested when potentially important SGS flow physics is involved, specifically, for turbulent material mixing – the underlying focus of the book. Fundamental CGS issues receiving special dedicated attention, include: (1) coupling convectively driven flow with relevant other physics – for example, with material mixing and combustion; (2) inherent sensitivities of turbulent flow to initial conditions; and (3) capturing complex turbulent mixing consequences. The book reviews our understanding of CGS, its theoretical basis, verification, validation, predictability aspects, and reports progress in difficult nonequilibrium applications of timely current interest involving variable density turbulent mixing.

The research surveyed here was supported by many sponsors. It is a pleasure to specially acknowledge the support from the Los Alamos National Laboratory Laboratory Directed Research and Development (LDRD) Program on "Turbulence by Design," and the LDRD Exploratory Research Program on "LES Modeling for Predictive Simulations of Material Mixing."

Fernando F. Grinstein

Prologue
Introduction to Coarse Grained Simulation

Fernando F. Grinstein

Mixing of materials by small scale turbulent motion is a critical element of many flow systems of interest in engineering, geophysics, and astrophysics. Numerical simulation plays a crucial role and turbulent mixing predictability is a major concern. Small scale resolution requirements focus typically on those of continuum fluid mechanics described by Navier–Stokes (NS) equations; different requirements are involved depending on the regime considered and on the relative importance of coupled physics such as multispecies diffusion and combustion as determined by Reynolds number (Re), Knudsen, Schmidt, Damköhler, and other characteristic nondimensional numbers. Direct numerical simulation (DNS), resolving *all relevant* physical space/time scales, is prohibitively expensive in the foreseeable future for most practical flows and regimes of interest at moderate-to-high Re. On the other end of the simulation spectrum are the Reynolds-averaged Navier–Stokes (RANS) approaches, which focus on statistical moments for an ensemble of realizations and model the turbulent effects.

Small scale turbulent flow dynamics is traditionally viewed as universal and enslaved to that of larger scales (Fig. P.1). In *coarse grained* simulation (CGS) large energy containing structures are resolved, smaller structures are spatially filtered out, and unresolved subgrid scale (SGS) effects are modeled. CGS includes classical large eddy simulation (LES) strategies [1] focusing on explicit SGS models, implicit LES (ILES) [2] relying on SGS modeling and filtering provided by *physics capturing* numerical algorithms, and more general LES using suitably mixed explicit/implicit SGS modeling. Transition to turbulence involves unsteady large scale dynamics, which can be captured by CGS but not by single-point closures typical in RANS [3]. Our fundamental views of the so-called "spectral gap" between large and small scales have significantly evolved over the past decade to provide a solid basis for the ideas of enslavement in turbulence as they relate to CGS [4]. *The CGS strategy of separating resolved/unresolved physics constitutes a viable intermediate approach between DNS and RANS to address practical geometries and multiphysics.*

As complex turbulent flow applications typically involve underresolved simulations, robustness of CGS predictions becomes the unsettled issue. If the information contained in the filtered-out smaller and SGS spatial scales can significantly alter the evolution of the larger scales of motion and practical integral measures, then the utility of CGS is questionable. The validity of the scale separation assumptions in CGS needs to be carefully tested when potentially important SGS flow physics is involved, specifically,

Figure P.1 Paintings by Hokusai (top) and Da Vinci (bottom) depicting universality and enslavement of small scale flow dynamics to that of large scales.

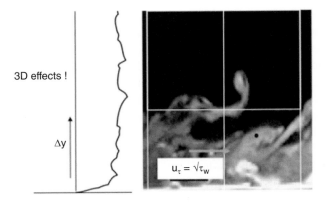

3D effects !

Δy

$u_\tau = \sqrt{\tau_w}$

Figure P.2 Subgrid near wall features in turbulent channel flow; Δy denotes a typical CGS grid size in the direction normal to the wall, and u_τ is the inner friction velocity.

for turbulent wall bounded flows (Fig. P.2) and for turbulent material mixing (Fig. P.3) – the main focus of the book.

The book reviews our understanding of CGS of turbulent mixing, its theoretical basis, verification, validation, predictability aspects, and progress addressing difficult open issues in nonequilibrium applications involving single- as well as multiphase turbulent flow (surveyed in [5]).

grid size

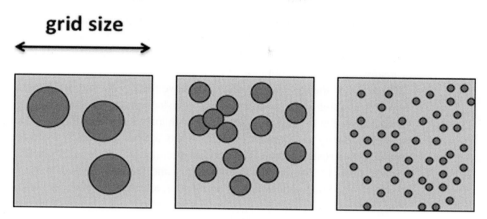

Figure P.3 Different realizations of subgrid material mixing having the same volume fractions.

P.1   CGS Fundamentals

We first address the fundamentals of the CGS paradigm: *small scales are enslaved to the dynamics of the largest*, or, put in other words, the spectral cascade rate of energy (the rate limiting step) is determined by the *initial and boundary condition constrained large scale dynamics*. The physics, computation, and metrics of enslavement and mixing for dissipative forward cascade dominated flow are introduced, and the evolution of our understanding of the role of the *spectral gap* between large (resolved) and small (modeled) scales is examined in this context.

The sensitivity of turbulent flows to initial conditions (IC) is now well recognized. Far field (late time) turbulent flows remember their particular near field (initial) features [6–9], and the mechanism by which they transition from IC to asymptotic flow occurs involves unsteady large scale coherent structure dynamics – which can be captured by CGS but not by single point closure turbulence modeling [3, 4]. The long standing view that an IC independent universal turbulence state is eventually achieved in the far field (or at late time) – for example, [10] – has been replaced by the recognition that different self-similar end states are possible depending on IC [7].

Extensive evaluations of CGS in fundamental forced and decaying isotropic homogeneous turbulence have been reported based on both Euler and NS equations [11–13]. Small scale enslavement ideas are relevant in addressing *underresolved* simulation of high-*Re* scalar mixing driven by *underresolved* velocity field in (equilibrium) isotropic forced turbulence [14, 15]. Chapter 1 shows that properly constructed CGS can be used to accurately capture [15] the dominant aspects of the mixing transition [16, 17] and established turbulent scalar mixing characteristics predicted by theory [18] and DNS [19, 20].

Chapter 2 examines fundamental ways to ensure that enough scale separation is involved in the CGS by addressing the following practical question [17]: at what flow conditions can researchers be sure that their numerical simulations have reproduced the most influential physics of the flows and fields of practical interest? Our current understanding of the energy transfer process of the turbulent flows, which forms the foundation of our scale separation arguments, is surveyed. A metric is developed using suitably available resolved tools to indicate whether the necessary physics of the flows of interest have been captured.

Finite scale theory [21], nonlinear enslavement [22], and ILES [2] all represent "top down" philosophies of modeling physical processes that rationalize a coarse grained approach to numerical simulation. Chapter 3 explores the overlap among these theories to justify and perhaps establish limits to coarse grained modeling of turbulent mixing. Finite scale equations are a coarse graining of the NS equations to describe high *Re* fluid flow. In this chapter the derivation and the properties of these equations are reviewed. The finite scale coarse graining as an extension of moment methods of kinetic theory and how the finite scale equations can be considered a hydrodynamic theory at second order are discussed. Finally an alternate explanation is proposed for physical viscosity, potentially extending the regime of the finite scale equations to length scales of the order of the molecular mean free path.

Chapter 4 examines the basics of material conservation of passive scalar mixing in finite scale NS (FSNS) fluid turbulence. Fluid motion of a NS fluid when observed at finite length and time scales is described by the equations of a second-order fluid that we call the FSNS equations. FSNS equations can also be viewed as an analytic model for ILES equations. The Reynolds averaging procedure is applied to understand the statistical fluid physics of FSNS fluid turbulence. The model problem investigated is the mixing of a passive scalar by a homogenous stationary isotropic turbulence in the presence of a mean scalar gradient. Passive scalar mixing by FSNS turbulence is shown to be mathematically consistent with the mixing fluid physics in NS turbulence, both satisfying a scalar variance conserved invariant property not exhibited by eddy viscosity models.

P.2   Challenges to Predictability

We next discuss the crucial challenges to predictability. In addition to the basic concerns regarding convergence of predictions as function of resolution, we must address the fundamental aspects of characterizing and modeling flow conditions at the SGS level – *within a computational cell or instrumentation resolution*, and at the supergrid (SPG) scale – *at initialization and boundaries of the (numerical or laboratory) flow experiments*. An overview of SGS and SPG issues and CGS strategies is presented, and capturing the dominant features of transition to turbulence is a particular focus.

Accurate and reliable characterization and modeling of material interfaces is an essential aspect in simulating the material mixing dynamics. Interfaces between fluids can be miscible or immiscible, with their *dynamic* character changing during evolution (and eventual phase change) of the materials. Because of SGS and SPG issues, *observations* based on numerical *or* laboratory experiments are inherently intrusive, and convergence metrics and *specific frameworks* for model verification, validation, and uncertainty quantification (VVUQ) designed *for the problems at hand* are needed to establish predictability.

Chapter 5 addresses the SGS and SPG issues. CGS strategies based on the augmented Euler or NS equations, involve the classical LES and ILES approaches. Classical LES using explicit SGS models includes many proposals ranging from simple eddy viscosity formulations to more sophisticated and accurate dynamic and mixed models [1]. Relying only on the SGS modeling and filtering provided by the numerics has been denoted numerical LES (e.g., [23]). Good or bad SGS physics can be built into the simulation model depending on the choice of numerics and its particular implementation. By sheer serendipity, physical features designed over decades into popular finite volume (FV) nonoscillatory numerical schemes to simulate shocks also provide implicit SGS models suitable for turbulent flow simulation in the ILES framework [2]. Depending on the regime of interest and grid resolution, CGS generally will involve a mixed explicit/implicit SGS strategy. Cascade pathways driven by vortex instabilities, self-induced deformations, stretching, and reconnections have been demonstrated [24, 25], and the global dynamics of transition to turbulence has been examined in terms of well-defined prototypes such as the Taylor–Green vortex [26, 27]. These processes are fundamental building blocks to be captured by CGS. Inherent to any CGS approach is the fact that the smallest characteristic resolved turbulence scale will be determined by the resolution cutoff wavelength prescribed by an explicit or implicit spatial filtering process, and practical convergence metrics must be formulated in terms of a suitable effective *Re*.

Chapter 6 examines predictability issues in multimaterial reactive flow modeling. Clouds are formed via the conversion of water vapor into cloud water or ice. This process is known as condensation, with its inverse being evaporation and the two sided process being labeled in this chapter as latent heat release. While both the time and spatial scales for latent heat release are extremely short, the integrated impact of this energy release is what drives systems such as hurricanes. Hence, when simulating systems such as hurricanes, small scales should be either resolved or represented by a SGS model; if these scales are improperly modeled, resulting predictions of intensity and track could be significantly impacted. To illustrate the impact of these numerical errors on the ability of a model to predict a system such as a hurricane, in this chapter a range of problems are presented ranging from simple advection of an isolated cloud to slightly more complex simulations of stratus clouds and finally to relatively complex simulations of hurricanes. For the two latter phenomena some of the best data sets ever obtained are compared against simulations using either traditional Eulerian or recently formulated Lagrangian cloud modeling approaches, with these comparisons revealing how differing numerical errors present in either formulation impact model predictability. Further, an aspect stressed in this chapter is how small scale processes such as evaporation can significantly influence the larger scale circulation and how numerical errors, especially those found in traditional Eulerian cloud models, make simulating this upscale energy exchange exceedingly difficult.

Chapter 7 addresses the VVUQ issues. Verification and validation are the primary means by which to assess the accuracy and reliability of numerical simulation [28].

Verification is the process of assuring that the code is solving the equations correctly, and it is usually addressed through convergence studies and analytic test problems. Validation is the process of demonstrating that one is solving the appropriate equations with the relevant SPG conditions, and is usually addressed through comparisons with available laboratory data and theoretical studies. A fundamental aspect in this context is that of assessing uncertainties in the numerical and laboratory experiments. The overall conduct of VVUQ is discussed through the construction of a relevant workflow, which by necessity is complex and hierarchical in nature. The particular characteristics of VVUQ elements depend on where the VVUQ activity takes place in the overall hierarchy of physics and models. In this chapter, the focus is on the differences between and interplay among validation, calibration, and uncertainty quantification (UQ), as well as the difference between UQ and sensitivity analysis. The complementary approach of best estimate plus reduced uncertainties approach is also discussed in some detail. The discussion is at a relatively high level; it explains the key issues associated with the overall conduct of VVUQ and offers guidance on conducting VVUQ analyses toward truly predictive turbulent mixing calculations.

P.3   Complex Mixing Consequences

Difficult challenges simulating nonequilibrium complex mixing consequences in variable density turbulent flow problems of current interest are demonstrated. For the flows discussed in this section, well-characterized whole scale laboratory studies are impossible or very difficult. Deterministic simulation studies are expensive and critically constrained by limitations in characterizing, modeling, and validating all the relevant physical subprocesses, and acquiring all the necessary and relevant SGS and SPG information. Here, we can only extrapolate from our understanding and established analysis of CGS performance in equilibrium turbulent flows, and we rely on the simulation model confidence developed from building block VVUQ and testing.

The challenging problem of underresolved mixing of material scalars promoted by underresolved velocity *and underresolved IC* in shock driven turbulent flows is examined in Chapter 8. In many areas of interest, such as inertial confinement fusion (ICF), understanding the collapse of the outer cores of supernovas, and supersonic combustion engines, vorticity is introduced at material interfaces by the impulsive loading of shock waves, and turbulence is generated via Richtmyer–Meshkov (RM) instabilities [29]. The complexity of shock waves and other compressibility effects add to the physics of material mixing, and to difficult issues of characterization and modeling of the initial and dynamic material interfaces. The inherent difficulties with the open problem of predictability of material stirring and mixing by underresolved multiscale turbulent velocity fields are now compounded with the inherent sensitivity of turbulent flows to IC [3]. The extensive recent RM simulation work [30–33] has demonstrated ILES to be an effective CGS *strategy* in this context, because of its unique combination of shock and turbulence emulation capabilities. We focus on effects of initial spectral content and interfacial morphology on transitional and late time turbulent mixing in fundamental shocktube experiments, and examine practical challenges encountered in CGS predictability evaluations in complex configurations [33] for which state of the art laboratory data and diagnostics are available.

In many noted areas of interest, such as ICF, understanding the collapse of the outer cores of supernovas, and supersonic combustion engines, vorticity is driven by the impulsive loading of shock waves and shear velocity gradients at material interfaces. In Chapter 9 we report simulations of laser driven reshock, shear, and ICF capsule experiments at the Omega laser facility [34] in the strong shock high energy density regime [35]. Validation of the simulations is based on direct comparison with radiographic data. Simulations are also compared with available DNS and theory of homogeneous isotropic turbulence. Despite the fact that the flow is neither homogeneous, isotropic, nor fully turbulent, there are local regions in which the flow demonstrates characteristics of homogeneous isotropic turbulence. We identify and isolate these regions by the presence of high levels of turbulent kinetic energy and vorticity. Our results show that in laser (shock) driven transitional flows, turbulent features such as self-similarity and isotropy only fully develop once decorrelation, characteristic vorticity distributions, and integrated turbulent kinetic energy (TKE) have decayed significantly.

Next, in Chapter 10, drive asymmetry, convergence, and the origin of turbulence in ICF implosions are examined. Uniform, spherically symmetric laser illumination of an ICF target is critical to achieving the long sought goal of high yield thermonuclear burn and gain in laser driven fusion. Unfortunately, highly symmetric laser drive of the target is extremely difficult to achieve in practice and some degree of drive asymmetry seems unavoidable. In this article we use very high resolution two- and three-dimensional numerical simulations with the LANL's RAGE code to investigate the connection between drive asymmetry and the generation of turbulence in the deuterium-tritium (DT) fuel in a simplified ICF implosion [36]. Long wavelength deviations from spherical symmetry in the pressure drive lead to the generation of coherent vortical structures in the DT gas and it is the three-dimensional instability of these structures that in turn leads to turbulence and mix. The simulations suggest that this mechanism may be an additional important source of mix in ICF implosions and may play a role in the problems being encountered at the National Ignition Facility in achieving ignition of a laser driven fusion target.

Modeling turbulent Rayleigh–Taylor driven mixing with dynamic interfaces is examined in Chapter 11. The behavior of turbulence in the presence of strong density gradients and compressibility is fundamental in applications ranging from ICF and supernovae to environmental flows. The dominant physical mechanisms at work in variable density turbulence include Kelvin–Helmholtz, Rayleigh–Taylor, and RM instabilities. All three mechanisms must be accounted for in a unified way in multiphysics simulations. Even in simplified test problems, such unstably stratified, unsteady, inhomogeneous flows pose a challenge to turbulence modeling. Rayleigh–Taylor mixing – where heavy fluid is (unstably) placed over a light fluid under the influence of gravity, is the particular focus of this chapter. Unsteady RANS have been proposed to model the mixing process across a range of complexity. Here the focus is on modeling multimaterial mixing and a dynamic interface in the tilted rig experiments, which have been studied experimentally [37] and computationally using LES [38].

Chapter 12 addresses spray combustion issues and simulation strategies in swirling flow. Nearly all flows in nature and engineering applications are turbulent and many involve complex chemical reactions. This complexity is aggravated if the flow field includes more than one phase (gas plus a liquid/solid phase) such as those encountered in gas turbine, liquid fueled rocket, and diesel engines. Combustion in such two phase flows is predominantly non-premixed, in which the heat release is controlled by the fuel–air mixing rate. Since turbulence plays an important role in increasing scalar gradients, and thus increasing mixing rates, turbulence/chemistry interaction is much more stronger in non-premixed flames compared with premixed flames. In two phase flows, the introduction of spray droplets adds several other processes with numerous time and length scales. Time scales are especially important since the liquid droplet has to vaporize and the vaporized fuel need to mix with the ambient air efficiently for combustion to occur. In order to achieve this efficiently most gas turbine combustors employ swirl as a means for aerodynamic control of spray mixing and flame holding. Prediction of spray breakdown and fuel–air mixing is critical to predict combustion efficiency in these complex systems. This chapter discusses the physics of liquid jet injection, spray formation physics, and its interaction in a swirling flow environment that is typically employed in both propulsion and power generation systems. The numerical methodologies applied to these systems are discussed along with identification of key physics critical for accurate predictions. Comparison with experimental data is used to highlight strengths and limitations of both the computational and the experimental strategies [39, 40].

Last but not least, Chapter 13 examines combustion in afterburning. Our knowledge of condensed phase explosions is limited, although explosives have existed since the gunpowder was discovered in China in the ninth century. The physical complexity of condensed phase explosions, involving extremely high pressures and temperatures, phase transitions, turbulence, shocks, mixing, instabilities, chemical reactions, and shock turbulence interactions puts very high demands on the physical modeling and the numerical simulation techniques. CGS is a cost effective approach to design full scale experiments. In enhanced blast explosives, metal particles – usually aluminum – are added to the explosive compound in order to increase the afterburning energy release by allowing the metal particles and detonation products combust with air. This presents another modeling challenge since the combustion becomes multiphased. Recent work is reviewed in the field of CGS of afterburning, including afterburning of trinitrotoluene (TNT) in unconfined air and varying the height of blast (HoB) [41], as well as multiphase afterburning of TNT/aluminum and nitromethane/steel in a spherical sector domain [42]. The main objective is to examine the use of combustion CGS to elucidate the physical processes involved in unconfined air and near ground air blasts to demonstrate effects of the HoB on afterburning and how metal particles affect combustion. The aim is to capture the most significant stages of turbulent mixing, involving the initial blast wave, secondary and reflected shocks, possible implosions, and the constant mixing stage(s), providing further knowledge of the complicated processes that occur during condensed phase explosion events.

P.4 Summary

Our present focus is on complex turbulent mixing consequences driven by variable density and compressible flow instabilities. Throughout our presentation, we emphasize the inherently intrusive nature of *coarse grained observations in computational and laboratory experiments*, intimately linked to their SGS and SPG specifics. Difficult challenges are then related to characterizing and modeling the unresolved SGS and SPG aspects and assessing uncertainties associated with CGS predictions and laboratory measurements. VVUQ provides the rational basis to decide when the CGS modeling is *good enough for its intended purpose*. Ensemble averaged CGS over a *suitably complete* set of realizations covering relevant IC variability is envisioned as the simulation strategy of choice to address complex flow dynamics in practical geometries.

References

1. Sagaut P., *Large Eddy Simulation for Incompressible Flows*, 3rd edition, Springer, New York, 2006.
2. Grinstein, F.F., Margolin, L.G., and Rider, W.J., editors, *Implicit Large Eddy Simulation: Computing Turbulent Flow Dynamics*, Cambridge University Press, New York, 2007; 2010.
3. George, W.K. and Davidson, L., "Role of initial conditions in establishing asymptotic flow behavior," *AIAA Journal*, 42, 438–446, 2004.
4. George, W.K. and Tutkun, M., "Mind the gap: A guideline for large eddy simulation," *Phil. Trans. R. Soc. A*, 367(1899), 2839–2847, 2009.
5. Balachandar, S. and Eaton, J.K., "Turbulent dispersed multiphase flow," *Annu. Rev. Fluid Mech.*, 42, 111–133, 2010.
6. Wygnanski, I., Champagne, F., and Marasli, B., "On the large-scale structures in two-dimensional, small-deficit, turbulent wakes," *J. Fluid Mech.*, 168, 31–71, 1986.
7. George, W.K., "The self-preservation of turbulent flows and its relation to initial conditions and coherent structures," in *Advances in Turbulence*, edited by W.K. George and R.E.A. Arndt, Hemisphere, New York, 1989.
8. Slessor, M.D., Bond, C.L., and Dimotakis, P.E., "Turbulent shear-layer mixing at high Reynolds numbers: Effects of inflow conditions," *J. Fluid Mech.*, 376, 115–38, 1998.
9. Ramaprabhu, P., Dimonte, G., and Andrews, M. J., "A numerical study of the influence of initial perturbations on the turbulent Rayleigh–Taylor instability," *J. Fluid Mech.*, 536, 285–319, 2005.
10. Townsend, A.A., *Structure of Turbulent Shear Flow*, Cambridge University Press, Cambridge, 1976.
11. Fureby, C. and Grinstein, F.F., "Monotonically integrated large eddy simulation of free shear flows," *AIAA Journal*, 37, 544–56, 1999.

12. Fureby, C. and Grinstein, F.F., "Large eddy simulation of high Reynolds-number free and wall bounded flows," *J. Comput. Phys.*, 181, 68–97, 2002.

13. Domaradzki, J.A., Xiao, Z., and Smolarkiewicz, P., "Effective eddy viscosities in implicit large eddy simulations of turbulent flows," *Phys. Fluids*, 15, 3890–3893, 2003.

14. Pullin, D.I., "A vortex-based model for the subgrid flux of a passive scalar," *Phys. Fluids*, 12, 2311, 2000.

15. Wachtor, A.J., Grinstein, F.F., Devore, C.R., Ristorcelli, J.R., and Margolin, L.G., "Implicit large-eddy simulations of passive scalar mixing in statistically stationary isotropic turbulence," *Physics of Fluids*, 25, 025101, 2013.

16. Dimotakis, P.E., "The mixing transition in turbulent flows," *J. Fluid Mech.*, 409, 69–98, 2000.

17. Zhou, Y., "Unification and extension of the similarity scaling criteria and mixing transition for studying astrophysics using high energy density laboratory experiments or numerical simulations," *Physics of Plasmas*, 14, 082701, 2007.

18. Ristorcelli, J.R., "Passive scalar mixing: Analytic study of time scale ratio, variance, and mix rate," *Phys. Fluids*, 18, 1–17, 2006.

19. Overholt, M.R. and Pope, S.B., "Direct numerical simulation of a passive scalar with imposed mean gradient in isotropic turbulence," *Phys. Fluids*, 8, 3128–3148, 1996.

20. Gotoh, T., Watanabe, T., and Suzuki, Y., "Universality and anisotropy in passive scalar fluctuations in turbulence with uniform mean gradient," *Journal of Turbulence*, 12(48), 1–27, 2011.

21. Margolin, L.G., "Finite-scale equations for compressible fluid flow," *Phil. Trans. R. Soc. A*, 367(1899), 2861–2871, 2009.

22. Jones, D.A., Margolin, L.G., and Poje, A.C., "Enslaved finite difference schemes for nonlinear dissipative PDEs," *Numerical Methods for Partial Differential Equations*, 12(1), 13–40, 1996.

23. Pope, S.B., "Ten questions concerning the large eddy simulation of turbulent flows," *New J. Phys.*, 6(35), 2004.

24. Grinstein, F.F., "Self-induced vortex ring dynamics in subsonic rectangular jets," *Phys. Fluids*, 7, 2519–2521, 1995.

25. Grinstein, F.F., "Vortex dynamics and entrainment in regular free jets," *J. Fluid Mech.*, 437, 69–101, 2001.

26. Frisch, U., *Turbulence*, Cambridge University Press, New York, 1995.

27. Drikakis, D., Fureby, C., Grinstein, F.F., and Youngs, D., "Simulation of transition and turbulence decay in the Taylor-Green vortex," *Journal of Turbulence*, 8(020), 1–12, 2007.

28. Oberkampf, W.L. and Roy, C. J., *Verification and Validation in Scientific Computing*, Cambridge University Press, New York, 2010.

29. Brouillette, M., "The Ritchmyer-Meshkov instability," *Annu. Rev. Fluid Mech.*, 34, 445–468, 2002.

30. Leinov, E., Malamud, G., Elbaz, Y., Levin, A., Ben-dor, G., Shvarts, D., and Sadot, O., "Experimental and numerical investigation of the Richtmyer–Meshkov instability under reshock conditions," *J. Fluid Mech.*, 626, 449–475, 2009.

31. Thornber, B., Drikakis, D., Williams R.J.R., and Youngs, D.L., "The influence of initial conditions on turbulent mixing due to Richtmyer-Meshkov instability," *J. Fluid Mech.* 654, 99–139, 2010.

32. Gowardhan, A.A., Ristorcelli, J.R., and Grinstein, F.F., "The bipolar behavior of the Richtmyer–Meshkov instability," *Physics of Fluids* 23 (*Letters*), 071701, 2011.

33. Gowardhan, A.A., and Grinstein, F.F., "Numerical simulation of Richtmyer–Meshkov instabilities in shocked gas curtains," *Journal of Turbulence*, 12(43), 1–24, 2011.

34. Welser-Sherrill, L., Fincke, J., Doss, F., Loomis, E., Flippo, K., Offermann, D., Keiter, P., Haines, B.M., and Grinstein, F.F. , "Two laser-driven mix experiments to study reshock and shear," *High Energy Density Physics*, 9, 496–499, 2013.

35. Haines, B.M., Grinstein, F.F., Welser-Sherrill, L., and Fincke, J., "Simulations of material mixing in laser–driven reshock dxperiments," *Phys. Plasmas*, 20(022309), 2013.

36. Thomas, V.A. and Kares, R.J., "Drive asymmetry and the origin of turbulence in an ICF implosion," *Phys. Rev. Lett.* 109(075004), 2012.

37. Andrews, M. and Spalding, D., "A simple experiment to investigate two-dimensional mixing by Rayleigh-Taylor instability," *Physics of Fluids*, A2, 922–927, 1990.

38. Holford, J., Dalziel, S., and Youngs, D., "Rayleigh–Taylor instability at a tilted interface in laboratory experiments and numerical simulations," *Laser and Particle Beams* 21, 419–423, 2003.

39. Patel, N. and Menon, S., "Simulation of spray-turbulence-flame interactions in a lean direct injection combustor," *Combustion & Flame*, 153, 228–257, 2008.

40. Srinivasan, S., Smith, A., and Menon, S., "Accuracy, reliability and performance of spray combustion models in large-eddy simulations," in *Quality and Reliability of Large-Eddy Simulations II*, ed. by Salvetti et al., 211–220, Springer Verlag, 2011.

41. Fedina, E. and Fureby, C., "Investigating ground effects on mixing and afterburning during a TNT explosion," *Shock Waves*, 23, 251–261, 2013.

42. Gottiparthi, K. and Menon, S., "A study of interaction of clouds of inert particles with detonation in gases," *Comb. Sci. and Tech.*, 184, 406–433, 2012.

Part I

Fundamentals

1 Proof of Concept
Enslaved Turbulent Mixing

Fernando F. Grinstein and Adam J. Wachtor

1.1 Introduction

In practical turbulent flow applications exhibiting extreme geometrical complexity and a broad range of length and time scales, direct numerical simulation (DNS) is prohibitively expensive, and dependable large scale predictions of highly nonlinear processes must be typically achieved with underresolved computer simulation models. In coarse grained simulation (CGS) large energy containing structures are resolved, smaller structures are spatially filtered out, and unresolved subgrid scale (SGS) effects are modeled. CGS includes classical large eddy simulation (LES) [1] using explicit SGS models, and implicit LES (ILES) relying on the SGS modeling and filtering implicitly provided by *physics capturing* numerical algorithms [2].

At moderately high Reynolds numbers (Re) – when convective time scales are much smaller than those associated with molecular diffusion – the primary concern of the numerical simulation is the convectively driven interpenetration mixing processes (entrainment and stirring due to velocity gradient fluctuations), which can be captured with sufficiently resolved ILES [2]. Assessing predictability of *underresolved scalar mixing by an underresolved turbulent velocity field* in the ILES framework is the particular focus of this chapter, where we revisit recent ILES studies of turbulent mixing of a passive scalar in the framework of forced isotropic turbulence [3].

The mixing of a passive scalar by a fluctuating flow field is a classical problem in turbulence and is relevant in many industrial flow applications. Overholt and Pope [4] conducted DNS of the mixing of a passive scalar in the presence of a mean scalar gradient, by forced, spatially periodic, isotropic turbulence. In this flow, a statistically steady state scalar variance is achieved by balancing the scalar variance production and dissipation. This problem was first investigated in the context of LES by Pullin [5], who proposed a model for the flux of a passive scalar by the SGS motions. The LES results predicted that the normalized scalar variance asymptotically approaches a nearly constant value for large Taylor Re (Re_λ), consistent with laboratory experiments [6] also indicating essentially constant scalar variance as a function of Re_λ for Schmidt number $Sc \approx 1$. The velocity-to-scalar dissipation time scale ratio was also reported to be asymptotically constant in [5], but comparisons with the DNS [4] available at the time were inconclusive as to whether such a result captured physical behavior. Subsequent theoretical [7] and higher Re DNS [8] studies have shown that the time scale ratio should exhibit continued growth with increasing Re – also consistent with the trends of the earlier DNS [4]. The asymptotic behaviors of the scalar variance and squared velocity-to-scalar Taylor microscales ratio are specifically used to assess the performance of ILES against the available theoretical, laboratory, and computational studies.

The present ILES strategy uses a multidimensional compressible flux corrected transport (FCT) algorithm, with low wavenumber momentum forcing imposed separately for the solenoidal and dilatational velocity components. Effects of grid resolution on the flow and scalar mixing are investigated at turbulent Mach numbers 0.13 and 0.27. Turbulence metrics are used to show that ILES can accurately capture the mixing transition and asymptotic self-similar behaviors as functions of an effective Reynolds number determined by grid resolution. The results are used to demonstrate the feasibility of predictive underresolved simulations of high Reynolds number turbulent scalar mixing using ILES.

1.2   Forced Isotropic Turbulence

Positive evaluations of ILES of forced and decaying isotropic turbulence have been reported [9–16]. Comparisons of instantaneous probability distribution functions (PDFs) of explicit (LES) and implicit (ILES) SGS viscosities for forced isotropic turbulence in [10] showed similar behaviors sensitive to the actual SGS models involved (Fig. 1.1), and cumulative distribution functions of the vorticity and strain rate magnitudes from ILES and LES were found to be in good agreement with those of DNS of isotropic turbulence. Domaradzki et al. [12] and Thornber et al. [15] reported well-behaved ILES spectral eddy viscosities in agreement with theory. In contrast, poor performances in this fundamental context were reported by Garnier et al. [16] using finite difference discretizations of popular shock capturing schemes; the latter failings have been attributed to the inherent inadequacy of such implementations for ILES [17].

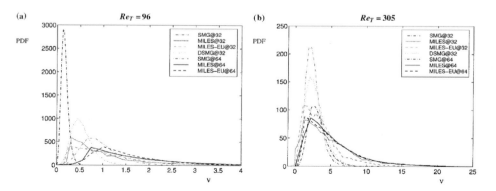

Figure 1.1 PDFs of the explicit and implicit SGS viscosities from ILES and various LES models [29] used in simulations of forced homogeneous isotropic turbulence, for, (a) Re=96, and, (b) Re=305. Effective implicit ILES SGS viscosity is computed based on Eq. (1.5) below.

ILES is based on solving the compressible conservation equations for mass, momentum, energy, and a passive scalar, with a well-established multidimensional fourth order FCT algorithm (e.g., [9]) suitably extended for the forced isotropic turbulence studies [3]. The fluid motion is described in a nominally inviscid framework:

$$\rho,_t + (\rho\, u_j),_j = 0 \tag{1.1}$$

$$(\rho\, u_i),_t + (\rho\, u_i u_j),_j + p,_i = f_i \tag{1.2}$$

$$E,_t + \left[(E+p)\, u_j \right],_j = 0 \tag{1.3}$$

$$(\rho\theta),_t + (\rho\, \theta\, u_j),_j = f_\theta \tag{1.4}$$

where ρ is the mass density, u_i the velocity, p the pressure, f_i the momentum forcing, $E = \rho u_j u_j / 2 - p/(\gamma - 1)$ the total energy density, f_θ the scalar forcing, $(),_i \equiv \partial()/\partial x_i$, and summation over repeated *roman* indices is implied. The passive scalar θ is forced by imposing a uniform mean scalar gradient, α_i, prescribed along the i direction ($\alpha_i = \alpha_1 \delta_{1i}$ chosen here), such that $f_\theta = -\rho\, \alpha_i u_i$.

An isotropic turbulence simulation strategy [18] is implemented by which low wave number forcing can be enforced separately for solenoidal and dilatational components on the momentum equations. The configuration studied assumes triply periodic boundary conditions on a cubical domain with unit box length and uniformly spaced 32^3, 64^3, 128^3, and 256^3 grids. The simulated flow is characterized by volume–time-averaged quantities: *rms*-velocity-fluctuation u', velocity and scalar Taylor microscales λ and λ_θ,

$$\lambda_\theta = \frac{1}{3} \sum_{\beta=1}^{3} \sqrt{\langle \theta^2 \rangle / \langle (\theta,_\beta)^2 \rangle}, \quad \lambda = \frac{1}{3} \sum_{\beta=1}^{3} \sqrt{\langle u_\beta^2 \rangle / \langle (u_{\beta,\beta})^2 \rangle}.$$

On dimensional grounds, an effective viscosity ν_{eff} can be computed as ratio of forced dissipation and mean squared strain rate magnitude [10],

$$\nu_{eff} = \varepsilon_s / \left[2\langle s_{ij}s_{ij} \rangle_T \right], \tag{1.5}$$

where ε_s denotes the dissipation imposed by the forcing scheme, $<>_T$ denotes volume–time average, and $S_{ij} = (u_{i,j} + u_{j,i})/2$ is the strain-tensor. For a given grid resolution, and u' based turbulence Mach numbers $Ma = 0.13, 0.29$, effective $Re_{eff} = u'\lambda/\nu_{eff}$ can be thus directly evaluated from the developed (raw) simulation data.

Figure 1.2 exemplifies the developed vorticity field for the 128^3 and 256^3 resolutions, depicting flow dominated by elongated structures characteristic of high Re isotropic turbulence; Figure 1.3 shows typical scalar visualizations for the 256^3 resolution case. Figure 1.4 demonstrates scalar and velocity spectra as functions of grid resolution. The scalar spectra exhibit longer inertial ranges and more pronounced spectral bumps than their velocity counterparts at corresponding resolutions – consistent with [8]; the latter results directly reflect on the $\lambda^2/\lambda_\theta^2$ results discussed in the following.

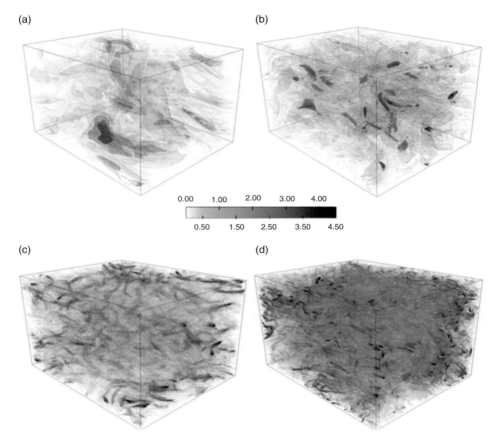

Figure 1.2 Instantaneous visualizations of the vorticity magnitude for (a) 32^3; (b) 64^3; (c) 128^3; (d) 256^3 resolutions.

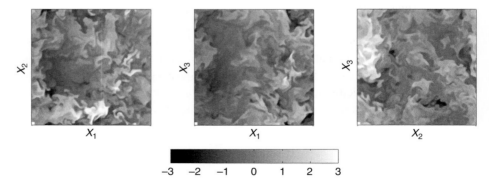

Figure 1.3 Colormaps of fluctuating scalar field scaled by the rms of the fluctuating scalar field in mid-planes of the domain for 32^3 grid resolution; for reference, superimposed on the lower left of each frame is a box with side length equal to the scalar Taylor microscale.

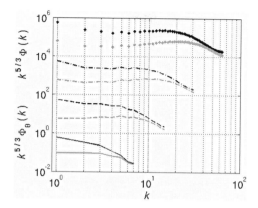

Figure 1.4 Compensated velocity (black) and scalar variance spectra (gray). Solid line: 32^3; dashed line: 64^3; dash-dot line: 128^3; diamond: 256^3; $Ma = 0.27$ case.

1.3   Turbulence Statistics versus Effective Reynolds Number

PDF analysis associated with isotropic turbulence has been reported for DNS (e.g., [19, 20]) and is extensively used as validation tool for ILES (e.g., [3, 10, 11, 21]). The basic ideas are illustrated in Figure 1.5 in terms of the PDFs of characteristic velocity function measures. PDF trends with increasing ILES resolution are the same as for the increasing Re of the DNS data, suggesting an effective Re_{eff} for ILES determined by grid resolution. Of considerable interest in the context of assessing resolution as relevant to ILES *convergence* is the fact that PDFs associated with the DNS data approach what appears to be a high Re limit above the mixing transition threshold $Re_\lambda \approx 100$ [22] (e.g., Fig. 1.5(d)). As Re_{eff} is increased, the SGS modeling implicitly provided by a well-designed ILES numerics consistently captures expected asymptotic turbulence characteristics such as the Gaussian behavior of the simulated fluctuations (Figs. 1.5(b) and 1.6(a)) and the non-Gaussian PDF tails of the derivatives (Figs. 1.5(a), (e) and 1.6(b), (c)). The skewness, Sk, and kurtosis, K, defined as follows for both the fluctuating scalar and velocity fields below and listed in Table 1.1, quantify the close agreement with pure Gaussian features (for which $Sk = 0$ and $K = 3$ are expected). The tails of the velocity PDF tend to be narrower than the Gaussian at all resolutions, but trend closer to Gaussian as resolution increases. Additional characteristic turbulent metrics are exemplified in Table 1.1; more detailed analysis of the velocity and scalar simulation data is reported in [3].

$$Sk_\theta = \langle \theta^3 \rangle / \langle \theta^2 \rangle^{3/2}, \; Sk = \frac{1}{3} \sum_{\beta=1}^{3} \langle u_\beta^3 \rangle / \langle u_\beta^2 \rangle^{3/2}$$

$$K_\theta = \langle \theta^4 \rangle / \langle \theta^2 \rangle^2, \; K = \frac{1}{3} \sum_{\beta=1}^{3} \langle u_\beta^4 \rangle / \langle u_\beta^2 \rangle^2$$

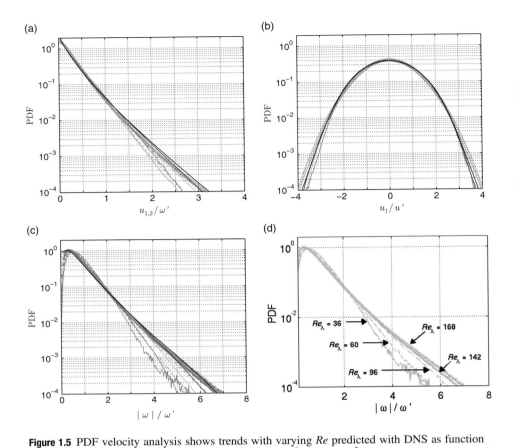

Figure 1.5 PDF velocity analysis shows trends with varying Re predicted with DNS as function of grid resolution. ILES – blue: 32^3; red: 64^3; green: 128^3; black: 256^3. DNS by Jimenez et al. [19] (gray) – solid line: $Re_\lambda = 36$; dashed line: $Re_\lambda = 60$; dash-dot line: $Re_\lambda = 96$; diamond: $Re_\lambda = 142$; X: $Re_\lambda = 168$; $W' = W_{rms}$ and results are for the $Ma = 0.27$ case.
(a) Transverse velocity derivatives; (b) velocity fluctuations; (c, d) vorticity magnitude; (e, f) longitudinal velocity derivatives. PDFs of longitudinal velocity derivatives are based on the full velocity field in (e), whereas the corresponding figure (f) is based only on its solenoidal part; the solenoidal velocity is extracted by inverse-Fourier transformation after a Helmholtz decomposition of the velocity field in Fourier space. (g, h) Vortex stretching scaled with the strain rate magnitude (based on full and solenoidal velocity, respectively). ILES – blue: 32^3; red: 64^3; green: 128^3; black: 256^3. DNS (gray) – solid line: $Re_\lambda = 36$; dashed line: $Re_\lambda = 60$; dash-dot line: $Re_\lambda = 96$; diamond: $Re_\lambda = 142$; X: $Re_\lambda = 168$. A black and white version of this figure will appear in some formats. For the color version, please refer to the plate section.

The established negative bias in the PDF of the longitudinal derivatives [19, 20] is also captured by ILES (Fig. 1.5(e), (f)); this bias is recognized as due to turbulent intermittency and self-amplification of longitudinal velocity increments in isotropic turbulence [19]. Figure 1.5(e) shows that the tails of the PDFs for the ILES based longitudinal derivatives are wider than for the DNS (with ILES trending closer to incompressible DNS as Ma decreases [3]). Figure 1.5(f) shows significantly improved agreement between ILES and DNS when the PDF analysis is based only on the

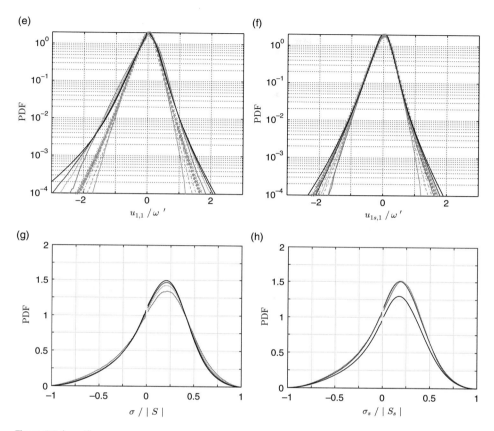

Figure 1.5 (*cont.*)

solenoidal portion of the velocity field. Compressibility mainly affects functions of longitudinal derivatives, whereas PDFs of transverse derivatives and functions thereof (e.g., vorticity $\omega = \nabla \times u$, evaluated based on the simulated velocity u data) are less affected by the dilatational velocity (Fig. 1.5(a), (c)). Finally, trends for the PDF of the vortex stretching $\sigma = \omega_i S_{ij} \omega_j / \omega^2$ (Fig. 1.5(g), (h)) are also consistent with those seen in the DNS [19], giving strong support that ILES is able to capture not only the appropriate amplitudes of quantities such as vorticity and strain [3], but also their characteristic alignment [23].

The ratio of momentum diffusivity to scalar diffusivity defines the Schmidt number *Sc*. Conventional wisdom is to expect $Sc \sim O(1)$ for ILES when momentum and scalar transport equations use the same numerical scheme. An effective diffusivity, D_{eff}, may be determined by:

$$D_{eff} = \varepsilon_\theta \Big/ \Big\langle (\theta_{,i})^2 \Big\rangle_{\mathrm{T}} \tag{1.6}$$

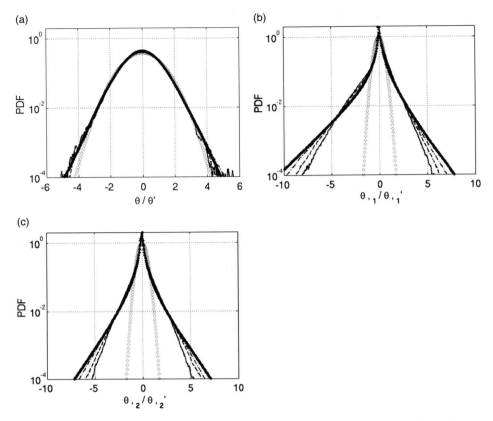

Figure 1.6 PDFs of (a) fluctuations of the passive scalar scaled with *rms* of the passive scalar fluctuations; (b) magnitude of fluctuating scalar derivative in mean scalar gradient direction scaled with its *rms* magnitude; and (c) magnitude of fluctuating scalar derivative in isotropic directions scaled with its *rms* magnitude; Case 1. Solid line: 32^3; dashed line: 64^3; dash-dot line: 128^3; diamond: 256^3; gray open circle: Gaussian.

in terms of the forced scalar dissipation $\varepsilon_\theta = \alpha_1 \langle u_1 \theta \rangle_T$. An effective Sc can then be found from the ratio of effective viscosity and effective diffusivity:

$$Sc = v_{eff}/D_{eff}. \tag{1.7}$$

The values for effective Sc (see Table 1.1) consistently approach $Sc \sim 0.7$ [24].

Further (positive) ILES performance aspects are revealed in terms of joint PDFs (JPDFs) of vorticity magnitude ω and strain rate magnitude $s_{ij}s_{ij}$, vorticity magnitude and vortex stretching σ, and strain rate magnitude and vortex stretching (Fig. 1.7). These two dimensional statistics depict a weak correlation between strong vortices and high strain, while high vortex stretching is associated more with moderate levels of vorticity magnitude, not the intense vortices that create worm like structures. JPDF shapes, trends as function of grid resolution, and depicted weak correspondence

Table 1.1 Time-Averaged Flow Statistics and Parameters for $Ma \sim 0.27$

Resolution	32^3	64^3	128^3	256^3
λ/L	0.15	0.099	0.066	0.043
λ_θ/L	0.10	0.059	0.037	0.023
$v^*/v^*_{32^3}$	1.0	0.44	0.19	0.08
Re_λ	67	93	145	225
Sc	0.61	0.68	0.69	not available
Ma	0.26	0.27	0.27	0.27
$u'\tau/L$	0.69	0.71	0.71	0.71
$\omega'L/u'$	20.93	32.24	50.28	78.80
$a_1 L/\theta'$	2.62	3.05	3.09	3.21
$\langle\theta u_1\rangle/(\theta' u')$	−0.521	−0.457	−0.497	−0.470
$\langle\theta u_2\rangle/(\theta' u')$	−0.007	−0.003	−0.0008	0.019
$\langle\theta u_3\rangle/(\theta' u')$	−0.007	−0.006	−0.005	0.005
$\theta,_1{}'/a_1$	3.87	5.66	8.84	13.46
$\theta,_2{}'/a_1$	3.85	5.53	8.77	13.29
TKE_d/TKE_s	0.42	0.40	0.38	0.39
Sk, Sk_θ	0.05, −0.21	0.01, −0.05	−0.01, 0.04	0.02, −0.003
K, K_θ	2.79, 5.59	2.81, 4.18	2.73, 3.61	2.77, 3.54
Sk_2	−0.69	−0.77	−0.80	−0.99
Sk_{2_s}	−0.34	−0.42	−0.49	−0.57
$Sk_{2_{\theta,1}}$	−0.41	−0.41	−0.41	−0.41
$Sk_{2_{\theta,2}}$	−0.20	−0.27	−0.30	−0.33
$\dfrac{\langle\theta,_i s_{ij}\theta,_j\rangle_s}{\langle\theta,_i s_{ij}\theta,_j\rangle}$	0.83	0.80	0.77	0.76

between strain rate magnitude and vortex stretching are in close agreement with those in [19] as function of Re; a somewhat wider strain rate data spread than for the incompressible DNS in Figure 1.7(a) is associated with the noted effects of compressibility on the distributions of longitudinal velocity derivatives (affecting also the strain rate [3]).

JPDFs for the scalar and scalar gradient with vorticity and vortex stretching are shown in Figure 1.8. The JPDF of the scalar fluctuations and the vortex stretching reveals that scalar fluctuations skewed toward regions of vortex stretching ($\sigma > 0$), rather than those of vortex compression ($\sigma < 0$). In Figure 1.8(b), it is quite evident that large scalar gradients are associated with regions of low vorticity on the order of the *rms* vorticity (vortex sheets) and strong vortices (worm like structures) are correlated with small scalar gradients. This result shows that convective mixing due to the fluctuating vorticity field primarily occurs in the vortex sheets and that high vorticity levels will not directly correspond to increased mixing. The asymmetry in the JPDF of the scalar gradient and the vortex stretching shows that vortex stretching and vortex compression act differently on the mixing process. Whereas vortex compression is correlated with a wide range of scales for the scalar gradient, vortex compression is seen to result in areas where the scalar gradient is small.

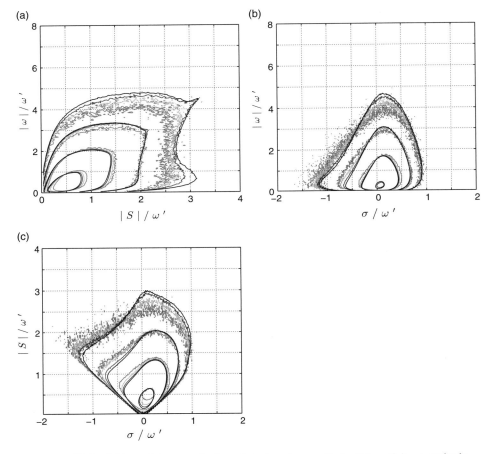

Figure 1.7 JPDFs of (a) vorticity magnitude and strain rate magnitude; (b) vorticity magnitude and vortex stretching; and (c) strain rate magnitude and vortex stretching. All axes are scaled with the *rms* vorticity ω'. Blue: 32^3; red: 64^3; green: 128^3; black: 256^3. Contour levels are plotted at 10^3, 10^{-4}, 10^{-5}, and 10^{-6}. A black and white version of this figure will appear in some formats. For the color version, please refer to the plate section.

1.4   Asymptotic Analysis

The nondimensional scalar variance $\langle \theta^2 \rangle_{\rm T}/(\alpha_1 L_\varepsilon)^2$, where $L_\varepsilon = (u')^3/\varepsilon_s$, is compared with predictions of state-of-the-art mixing LES using a (stretched vortex) SGS stress model [5]. Figure 1.9 shows that the ILES scalar variance results asymptotically reach nearly constant behavior above the cited mixing transition threshold for $\mathrm{Re}_\lambda = \mathrm{Re}_{mix} \sim 100$ [22], in close agreement with predictions of laboratory experiments [6] and previous LES [5] – when a scalar SGS model is used. The unsuccessful results of "LES without scalar SGS model" in Figure 1.9 show that the implicit SGS scalar modeling provided by the particular (dispersive) numerics used in [5] is inadequate by itself to predict the expected asymptotic behavior. Lower scalar variance predictions with present ILES are mainly attributed to differences in the forcing schemes, and

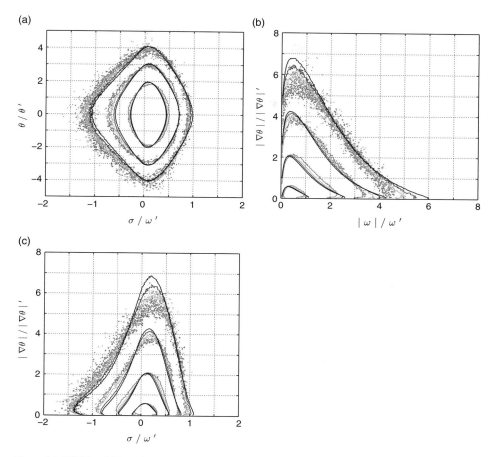

Figure 1.8 JPDFs of (a) scalar fluctuation scaled by the *rms* of the scalar fluctuation and vortex stretching scaled by the *rms* vorticity; (b) magnitude of the scalar gradient scaled with the *rms* of the scalar gradient and vorticity scaled with the *rms* vorticity; and (c) magnitude of the scalar gradient scaled with the *rms* of the scalar gradient and vortex stretching scaled with the *rms* vorticity. Blue: 32^3; red: 64^3; green: 128^3; black: 256^3. Contour levels are plotted at 10^{-3}, 10^{-4}, 10^{-5}, and 10^{-6}. A black and white version of this figure will appear in some formats. For the color version, please refer to the plate section.

somewhat less to compressibility effects – since the solenoidal velocity component is responsible for the generation of a passive scalar flux [25] and subsequent scalar stirring, $\langle \theta,_i s_{ij} \theta,_j \rangle$ (Table 1.1). Scatter of previous and current results around presumably constant asymptotic scalar variance behaviors suggests comparable overall uncertainties.

The ratio $\lambda^2/\lambda_\theta^2$ – directly proportional to the velocity-to-scalar dissipation timescale ratio, r, through $\lambda^2/\lambda_\theta^2 = Sc(5/3)r$ – is plotted in Figure 1.10 for the current ILES, the $(Sc = 0.7)$ LES of [5], DNS [4, 8], for which $Sc = 0.7$ and $Sc = 1.0$, respectively. Following the theory in [7], using small scale isotropy assumptions, and rearranging the stationary forced equations for the scalar variance and scalar dissipation, we can derive an expression for $\lambda^2/\lambda_\theta^2$ [3], which exhibits its leading linear dependence with

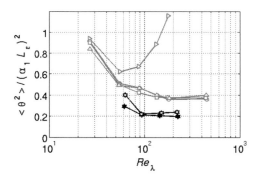

Figure 1.9 Nondimensional scalar variance as a function of Re_λ. Gray open square/triangle/circle: LES [5]; Gray open right-triangle: LES without scalar SGS model [5]; Black full star: ILES Ma = 0.27; Black open star: ILES Ma = 0.13, vs. Re_λ

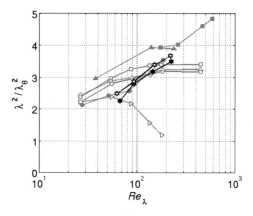

Figure 1.10 Ratio $(\lambda / \lambda_\theta)^2$ as a function of Re_λ. Gray full circle – line: DNS [4]; Gray full square – line: DNS [8]; Gray open square/triangle/circle: LES [5] with scalar SGS model; Gray open right-triangle: LES [5] without scalar SGS model; full star: ILES Ma = 0.27; open star: ILES Ma = 0.13; full black-line: theory [3].

Re_λ for $Sc = 1.0$ and high Re – consistent with trends in Figure 1.10. This result is noteworthy on several accounts. For one it shows that the squared ratio of Taylor microscales $\lambda^2/\lambda_\theta^2$ is a function of Re in developed turbulence. That the eddy turnover rate of turbulent kinetic energy and scalar variance are proportional to each other and not a function of Re has been a fundamental assumption underlying many models in turbulence phenomenology (e.g., as noted in [5]).

Early DNS results [4] showed continued growth of $\lambda^2/\lambda_\theta^2$ with increasing Re_λ, while the LES results using an explicit scalar SGS model [5] are asymptotically constant. Moreover, when the explicit scalar SGS model of [4] is turned off, the ratio rapidly decreases with Re_λ. However, ILES (also without an explicit scalar SGS term) exhibits neither decreasing nor asymptotically constant behavior, but shows continued growth over the simulated range of effective Re_λ very consistent with the early DNS [4], the theory [7], and the trends suggested by the more recent DNS [8]. We interpret the latter results as pointing at required mixing realizability constraints for SGS mix modeling, which appear consistently built into a (numerically) well-designed ILES, but not so in the mainstream classical SGS modeling strategies. This issue is revisited in Chapter 4.

1.5   Summary

Sufficiently resolved ILES can capture the fundamental aspects of the stirring driven mixing transition and characteristics of developed isotropic turbulence for high (finite) Re and $Sc \approx 1$. Detailed analysis presented was based on statistical turbulence metrics and PDFs of velocity and scalar functions, as function of Ma and effective Re measures determined by grid resolution.

Computation of the ILES v_{eff} through Eq. (1.5) based on the computed ratio of dissipation and mean squared strain-rate is distinct from the local residual numerical viscosity associated with the algorithm specifics (as Newtonian viscosity) through modified equation analysis [26]. For high-Re inertially dominated flow, the computed v_{eff} characterizes the simulated dissipation in the inertial subrange, whereas the local numerical viscosity directly affecting the (grid level) smallest scales of the simulated flow characterizes the simulated viscous dissipation subrange (see Fig. 1.11). As the

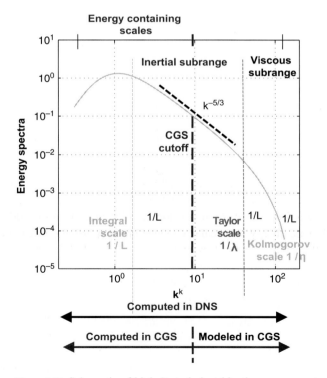

Figure 1.11 Schematic of high Re turbulent kinetic energy spectra.

ν_{eff} based Re is increased, the SGS scalar mixing model implicitly provided by a well-designed ILES numerics is found to be adequate to consistently capture well-established mixing characteristics from the reported studies, namely, the Gaussian behavior of fluctuating velocity and scalar PDFs, non-Gaussian (and appropriately biased – when applicable) PDF tails of their derivatives, asymptotically constant nondimensional scalar variance, and increasing squared ratio of the Taylor micro-scales with Re.

The results are regarded as clear demonstration of the feasibility of predictive under-resolved simulations of stirring driven high Re turbulent scalar mixing with ILES. They strongly suggest enslavement of the small scale mixing dynamics to that of the larger scales of the flow for sufficiently large Re.

1.6   Acknowledgments

Los Alamos National Laboratory (LANL) is operated by the Los Alamos National Security, LLC for the U.S. Department of Energy NNSA under Contract No. DE-AC52-06NA25396. This work was made possible by funding from the LANL LDRD-ER on "LES Modeling for Predictive Simulations of Material Mixing," through project number 20100441ER.

References

1. Sagaut, P. 2006, *Large Eddy Simulation for Incompressible Flows*, 3rd ed., Springer.
2. Grinstein, F.F., Margolin, L.G., and Rider, W.J. 2010, *Implicit Large Eddy Simulation: Computing Turbulent Flow Dynamics*, 2nd printing, Cambridge University Press.
3. Wachtor, A.J., Grinstein, F.F., DeVore, C.R., Ristorcelli, J.R., and Margolin, L.G. 2013, "Implicit large-eddy simulation of passive scalar mixing in statistically stationary isotropic turbulence," *Phys. Fluids*, 25(2), 1–19.
4. Overholt, M.R. and Pope, S.B. 1996, "Direct numerical simulation of a passive scalar with imposed mean gradient in isotropic turbulence," *Phys. Fluids*, 8(11), 3128–3148.
5. Pullin, D.I. 2000, "A vortex-based model for the subgrid flux of a passive scalar," *Phys. Fluids*, 12(9), 2311–2319.
6. Dowling, D.R. and Dimotakis, P.E. 1990, "Similarity of concentration field of gas-phase turbulent jets," *J. Fluid Mech.*, 218, 109–141.
7. Ristorcelli, J.R. 2006, "Passive scalar mixing: Analytic study of time scale ratio, variance, and mix rate," *Phys. Fluids*, 18(7), 1–17.
8. Gotoh, T., Watanabe, T., and Suzuki, Y. 2011, "Universality and anisotropy in passive scalar fluctuations in turbulence with uniform mean gradient," *J. Turb.*, 12(48), 1–27.
9. Grinstein, F.F. 2010, "Vortex dynamics and transition to turbulence in free shear flows," ch. 8 in *Implicit Large Eddy Simulation: Computing Turbulent Flow Dynamics*, 2nd printing, Cambridge University Press.
10. Fureby, C. and Grinstein, F.F. 1999, "Monotonically integrated large eddy simulation of free shear flows," *AIAA Journal*, 37(5), 544–556.

11. Porter, D.H., Woodward, P.R., and Pouquet, A. 1998, "Inertial range structures in decaying turbulent flows," *Phys. of Fluids* 10, 237–245.

12. Fureby C. and Grinstein F.F. 2002, "Large eddy simulation of high Reynolds-number free and wall bounded flows," *J. Comput. Phys.*, 181, 68.

13. Domaradzki, J.A., Xiao, Z., and Smolarkiewicz, P. 2003, "Effective eddy viscosities in implicit large eddy simulations of turbulent flows," *Phys. Fluids*, 15, 3890–3893.

14. Margolin, L.G., Rider W.J., and Grinstein, F.F. 2006, "Modeling turbulent flow with implicit LES," *Journal of Turbulence*, 7, N15, 1–27.

15. Thornber, B., Mosedale, A., and Drikakis, D. 2007, "On the implicit large eddy simulations of homogeneous decaying turbulence," *Journal of Computational Physics*, 226, 1902–1929.

16. Garnier, E., Mossi, M., Sagaut, P., Comte, P., and Deville, M. 1999, "On the use of shock-capturing schemes for large eddy simulation," *J. Comput. Phys.* 153, 273–311.

17. Rider, W.J. and Margolin, L.G. 2010, "Numerical regularization: The numerical analysis of implicit subgrid models," ch. 5 in *Implicit Large Eddy Simulation: Computing Turbulent Fluid Dynamics*, ed. by F.F. Grinstein, L.G. Margolin, and W.J. Rider, 2nd printing, Cambridge University Press.

18. Petersen, M.R. and Livescu, D. 2010, "Forcing for statistically stationary compressible isotropic turbulence," *Phys. Fluids*, 22(11), 1–11.

19. Jimenez, J., Wray, A.A., Saffman, P.G., and Rogallo, R.S. 1993, "The structure of intense vorticity in isotropic turbulence," *J. Fluid Mech.* 255, 65–90 832.

20. Li, Y. and Meneveau, C. 2005, "Origin of non-Gaussian statistics in hydrodynamic turbulence," *Phys. Rev. Letters*, 95, 1–4.

21. Li, Y., Perlman, E., Wan, M., Yang, Y., Burns, R., Meneveau, C., Burns, R., Chen, S., Szalay, A., and Eyink, G. 2008, "A public turbulence database cluster and applications to study Lagrangian evolution of velocity increments in turbulence," *J. Turbulence* 9, N31.

22. Dimotakis, P.E. 2000, "The mixing transition in turbulent flows," *J. Fluid Mech.*, 409, 69–98.

23. Jiménez, J. 1992, "Kinematic alignment effects in turbulent flows," *Phys. Fluids A*, 4, 652–654.

24. Wachtor, A.J. 2012, "On the predictability of turbulent mixing with implicit large-eddy simulation," PhD Thesis Dissertation, University of California, Irvine.

25. Blaisdell, G.A., Mansour, N.N., and Reynolds, W.C. 1994, "Compressibility effects on the passive scalar flux within homogeneous turbulence," *Phys. Fluids* 6, 3498–3500.

26. Hirt, C.W. 1969, "Computer studies of time-dependent turbulent flows," *Phys. Fluids* 12, II–219.

2 A Minimum Turbulence State for Coarse Grained Simulation

Ye Zhou

2.1 Introduction and Summary

In many problems in fundamental physics, we must simultaneously deal with uncertainty in the underlying equations of motion and with uncertainty in our ability to solve them. In turbulence, we have only the latter [1]. As noted by renowned physicist Richard Feynman, the turbulence problem is still referred to as the last unresolved classical physics problem [2].

The presence of strong nonlinear interactions makes turbulence a truly multiple scale problem. The challenge in direct numerical simulations (DNS) of a very high Reynolds number (Re) flow is to account for all the scales, starting from the largest where the energy injection occurs to scales that are roughly two times the Kolmogorov dissipation wavenumber. The Re is around 10^8 for airplane wing and fuselage [3] and even higher for turbulent flows in space and astrophysical setting [4]. Specifically, the Re, which is the ratio between the inertial forces to the viscous actions, is defined as

$$Re = \mathrm{U}L/v, \tag{2.1}$$

where U and L are the characteristic velocity and length scales and v is the kinetic viscosity [5]. Figure 2.1 shows the distinctive spectral scales from a large collection of the wind tunnel and geophysical experiments (reproduced from [6]).

In such a computationally intensive field, we have witnessed an unprecedented advancement of the capabilities of the supercomputers. For grid generated turbulence or turbulent flows in a periodic box, brute force DNS has already matched or surpassed experiments [7]. Pope [8] even suggested that we have entered an era of sufficient computer power.

In this chapter, the classical "turbulence problem" is narrowed down and redefined for scientific and engineering applications. From an application perspective, accurate computation of large scale transport of the turbulent flows is needed. Here, an analysis that allows for the large scales of very high Re turbulent flows to be handled by the available supercomputers is inspected. Current understanding of the energy transfer process of the turbulent flows, which forms the foundation of our argument, is discussed. Two distinctive interactions, namely, the distance and near grid interactions, are inspected for large scale simulations. The distant interactions in the subgrid scales in an inertial range can be effectively modeled by an eddy damping. The near grid interactions must be carefully incorporated. The data redundancy in the inertial range is demonstrated.

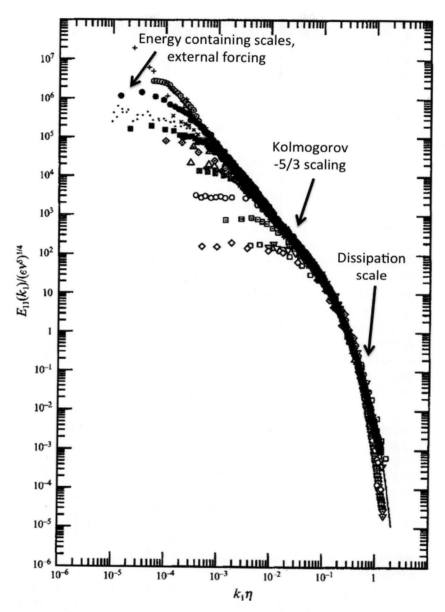

Figure 2.1 The compensated longitudinal energy spectra are plotted against the longitudinal wavenumbers. The data are obtained from a collection of the wind tunnel and geophysical flows (reproduced from [6], with permission from Cambridge University Press). The values of the Reynolds numbers are based on the Taylor microscales.

Furthermore, the minimum state is defined as the turbulent flow that has the lowest Re that captures the energy containing scales of the astrophysics problems in a laboratory or simulation setting. The transition criterion for the time dependent flows is presented.

Two types of applications are briefly summarized. First, the spatial and temporal criteria have found applications to a range of high energy density physics experiments. Second, the procedure for estimating the numerical viscosity for implicit large eddy simulation (ILES) has been advanced.

2.2   "Turbulence Problem" Redefined

Most important properties of a high Re turbulent flow are determined by the transport dynamics of the large scales. Therefore, it makes sense to focus computing resources on capturing these scales accurately. From an application perspective, accurate and time dependent three-dimensional computations of the large scales may be all that is needed. It is therefore necessary to choose the grid size in a uniform fashion, using the boundary between the large scales and inertial range.

In support of this argument, it will be illustrated that the self-similarity properties lead to the data redundancy; an advantage that should be fully exploited. The universality of the inertial range is indeed remarkable. At high Re, this universality can be observed by the establishment of an extended inertial range (Fig. 2.1) or the asymptotic behavior of the normalized energy dissipation rate. Indeed, the characteristic time scale of the energy containing eddies, U/L, should be in the same order of magnitude as the time scale of the energy dissipation rate ε/U^2. Here, the energy dissipation rate is denoted as ε. Therefore, a dimensionless ratio, D, can be introduced as

$$D = \varepsilon L / U. \qquad (2.2)$$

Recently, a significant amount of work has been devoted to investigating the behavior of D as a function of Re ([9–13], see Figs. 2.2 and 2.3).

This article notes that (1) two distinct interactions have been identified, (2) a model that incorporates both interactions already exists, (3) a refined boundary between the

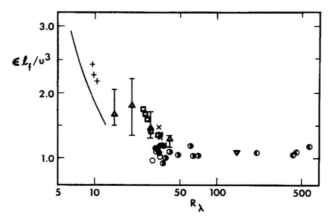

Figure 2.2 Nondimensional parameters: $\varepsilon\, l_f/u^3$ based on laboratory data, with l_f denoting a longitudinal integral scale – from [11], with L denoting the integral scale.

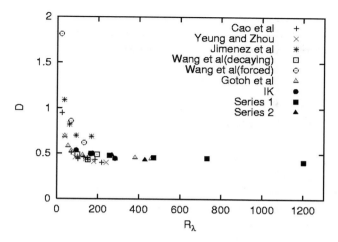

Figure 2.3 Nondimensional parameters: $D = \varepsilon L/u^3$ based on computational simulation data – from [12], with L denoting the integral scale.

energy containing and inertial ranges can be located, and (4) the self-similarity in the inertial range has been demonstrated. As a result, a scaling argument to scale the extremely high Re flows to high, but a manageable Re in order to fit into the existing supercomputers is proposed.

2.2.1   Filter Classification and Their Spectral Support

The objective of the filters is to separate the large scales as faithfully as possible. Therefore, the filtering operation, which divides the flow into the subgrid and resolvable scales, should not adversely affect the large scale properties.

Zhou et al. [14] pointed out that the resolvable scale interactions are affected when the filters with same spectral support are utilized. The so-called "Type A category" filters include familiar Gaussian [15] and exponential filters [16]. The subgrid scale field, as well as the subgrid stresses, can be directly evaluated from the resolvable scale field (see, for example, [17–19]).

However, in LES implementation, the maximum wavenumber is determined by the grid size. For wavenumbers up the cutoff, k_C, all functions, the original, resolvable, and subgrid, are known. Nothing is known for $k > k_C$. The sharp cutoff filter (the "Type B category") has distinctive spectral support ([14]).

2.2.2   Near Grid and Distant Interactions

As noted already, from an application perspective, the time dependent, three-dimensional computations of the energy containing scales may be all that is needed. Using a sharp cutoff filter, the subgrid of a given problem can be subdivided into two distinctive areas. The resolvable scale wavenumber is denoted $\mathbf{k}$, which satisfies a triad $\mathbf{k=p+q}$. The subgrid region for *the near grid interactions*, where one of the wavenumber is greater

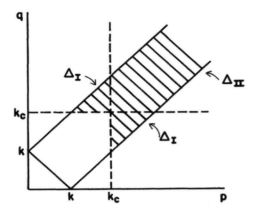

Figure 2.4 Region of integration for the sharp-cut filter. From [20].

and the other is less than the cutoff wavenumber, k_C, is denoted as Δ_I (Fig. 2.4). The subgrid region where both wavenumbers are greater than k_C (*the distant interaction region*) is denoted as Δ_{II}. Detailed studies of subgrid models in the energy transfer and momentum equations reveal the relationship between the eddy damping, backscatter, and the Reynolds and cross stresses [20]. It is convenient to introduce the superscript notation $>$ for subgrid quantities and $<$ for resolvable scale quantities. Hence, the measurements resulted from "subgrid–subgrid" and "subgrid-resolvable" couplings would be denoted with the superscript notation $>>$ and $><$, respectively.

The near grid and distant interactions play distinctive roles in the energy transfer process [21]. The eddy viscosity $v^{>>}(k)$, resulting from distant interactions, behaves in the same manner as the molecular viscosity. Therefore, an eddy viscosity model is acceptable. The eddy viscosity $v^{><}(k)$, resulting from the near grid interactions, is responsible for the cusp like behavior of the spectral eddy viscosity first identified by Kraichnan [22] (Fig. 2.5).

To confirm the importance of the near grid interaction dynamics, the model equation should be accurately resolved [23]. In an illustrative example, a fictitious cutoff wavenumber is introduced in a DNS and the near grid interactions between the resolvable and subgrid scales are kept. Again, the results demonstrated that the near grid interactions are critical for faithful computation of the large scale evolutions.

2.2.3   Resolvable Scale Model Equation

The first order of business is to derive a resolvable scale equation using the method of recursive renormalization (r-RG) group theory. This methodology was first proposed by Rose [24] for a model problem of passive scalar advection and was extended to the Navier–Stokes equation [25–26]. Starting from the Kolmogorov dissipation wavenumber, the inertial range is divided into multiple shells, with their length as thin as possible. The first resolvable scale equation can be written symbolically as P (which denotes the projection operator)

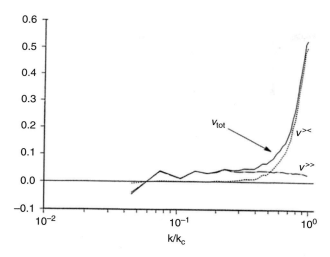

Figure 2.5 Spectral eddy viscosity and the individual contributions from both the near grid and distant interactions. From [21].

$$\partial u^< / \partial t + v_0\, k^2\, u^< = P\,[u^< u^< +\; 2\, u^< u^> +\; u^> u^>]. \qquad (2.3)$$

After removing the first subgrid shell, two types of subgrid interactions will make their distinctive contributions, and hence, they must be considered individually. First, the distant interactions (the last term on the right hand of Eq. 2.3) will result in an enhanced eddy viscosity, v_1. Second, the near grid interaction (the second term on the right hand of Eq. 2.3) should be either computed directly or approximated by an expression in the resolvable scale field. This process is repeated to remove the remaining subgrid scales shells.

The resulting recursion relation for these subgrid distant interactions lead to a fixed point – the eddy viscosity. In the suggested model for the resolvable scale equation, the near grid interactions are considered explicitly

$$\partial u^< / \partial t + v_{eff}\; k^2 u^< = P\,[u^< u^< +\; 2\, u^< u^>]. \qquad (2.4)$$

The wavenumber domain for the left hand side of Eq. (2.4) and the first term on the right hand side is $[0, k^*]$, while that for the second term is $[0, 2k^*]$. Here, k^* is used to denote the boundary between the energy containing and inertial ranges, which will be defined more precisely in the next section. The eddy viscosity, v_{eff}, has the contributions from the molecular viscosity and the subgrid–subgrid interactions.

2.3   Coarse Grained Simulations

2.3.1   Determination of the Boundary between the Energy Containing and Inertial Ranges

The traditional definition of the inertial range is the existence of a scale that is free from the large scale forcing and small scale viscous dissipation. A more precise definition can be introduced, where the upper and lower boundaries of the inertial range depend on the outer scale (δ) and Re [27–30]:

$$\text{Lower bound}: \mathrm{L}\,v \approx 50\,\mathrm{Re}^{-3/4}\delta, \tag{2.5}$$

$$\text{Upper bound}: \mathrm{L}_{\text{L-T}} \approx 5\,\mathrm{Re}^{-1/2}\delta. \tag{2.6}$$

The well-known estimation that $Re > 10^4$ (Hinze, [31]), (or 100 when the Taylor microscale is used) is needed for an inertial range.

The upper bound of the inertial range in this model for resolvable scale equation is our minimum grid size, which can be chosen as the grid size in physical space or cutoff wavenumber in spectral space.

2.3.2   Data Redundant in the Inertial Range

How can we compute the large scale of these higher Re turbulent flows found in astrophysical or geophysical flows? The answer lies in the universality of the inertial range, which we should exploit and utilize.

In fact, the universality in the inertial range can be demonstrated by computing the triadic energy transfer functions (for definition, see, for example, Domaradzki and Rogallo [32]). In Zhou [33–35], these triadic interactions are selected such that they satisfy the self-similarity scaling laws of Kraichnan [36]. The reconstructed energy transfer function, T(k), based on calculations for several grid sizes, has been shown to differ only in its range (or extent) in the spectral domain. With a given energy input, this ideal Kolmogorov inertial range is essentially a pipe without leak. The different length of the pipe only reflects the different resolutions (or, in other words, different Re) of the flows (Fig. 2.6).

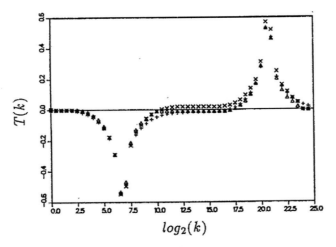

Figure 2.6 The energy transfer function of different grid sides for idealized Kolmogorov inertial range wavenumber. Rescaled to illustrate the "pipe without leak" analogy. From [33, 34].

Based on this understanding, one can rescale these energy transfer functions without affecting the large scale. This is clear evidence of data redundancy in the inertial range.

2.3.3 Local and Nonlocal Interactions in High Reynolds Number Flows

What would be the minimum resolution requirement (minimum model) for a faithful model calculation of the large scale of a flow? The answer is that the near grid interactions must be in the inertial range. The condition for this requirement can be found by demanding that the upper wavenumber of the near grid scale, $2k_c$, be equal to the lower boundary of the inertial range inner viscous scale).

In order to separate the local and nonlocal interactions, the so-called "disparity" parameter [33–35],

$$s(k, p, q) = \max(k, p, q)/\min(k, p, q) \tag{2.7}$$

is employed the extend of the interacting scales. In spectral space, the wavenumber $\mathbf{k}$ satisfies a triad $\mathbf{k=p+q}$. This parameter has been used to classify the interactions as local (~ 2) and nonlocal ($s > 2$) by Lesieur [37].

In the inertial range, the fractional energy flux for a given wavenumber scales versus the scale disparity parameter has been determined to follow a $-4/3$ power law ([33–35]; see Fig. 2.7)

$$\Pi(k, s)/\Pi(k) \sim s^{-4/3}. \tag{2.8}$$

The work by Gotoh and Watanable [38] also provided support to this scaling. These authors used both direct numerical simulations and a statistical closure theory, the Lagrangian renormalized approximation (LRA) (Kaneda [39]).

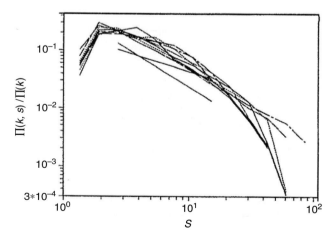

Figure 2.7 Fractional contribution $\Pi(k,s)/\Pi(k)$ to the energy flux in the inertial range LES. The various curves are for $k=2^{n/2}$, $6 \leq n \leq 14$. The straight lines indicate $s^{-2/3}$ and $s^{-4/3}$ behaviors [33–34].

The peak of a normalized flux function can be found along with the value of s at which it occurs, s_p. Let s_h be the scale disparity parameter where the normalized flux reduces to half of its peak value [28, 40–41], that is, $\Pi(k, s_h)=(1/2) \Pi(k, s_p)$. The s^{-M} scaling properties lead to

$$S_h/S_p = 2^{(1/M)}. \tag{2.9}$$

The range of interactions one must keep is $[k_{L\text{-}T}, 2\ k_{L\text{-}T}]$, as a result of $s_h/s_p = 2$.

Finally, several published numerical works by Domaradzki et al. [42–43] suggested the specific length of the inertial range that one must keep. These authors have numerically investigated isotropic and anisotropic channel flows. In both the homogeneous and inhomogeneous directions, they found that the modes that are smaller than a fictitious cutoff wave number k_C will not extend their interactions beyond $2k_C$.

2.4   The Minimum State

The mixing transition concept for *stationary* fluid flows refers to the transition to a turbulent state in which the flow drives rapid mixing at the molecular scale. This turbulent state leads to rapid dissipation of momentum and of concentration fluctuations (mixing). The extent of the effective inertial range could be narrowed to [27, 28]

$$\lambda_K < \lambda_v << \lambda << \lambda_{L-T} < \delta. \tag{2.10}$$

Here the lower limit of the inertial range is the inner viscous scale $\lambda_v = 50\lambda_K$, where the Kolmogorov microscale can be rewritten as $\lambda_K = \delta Re^{-3/4}$. The upper limit of the inertial range is the Liepmann–Taylor scale $\lambda_{L-T} = 2.17\lambda_T$, where $\lambda_T = 2.3\delta Re^{-1/2}$ is the well-known Taylor correlation microscale [5, 27].

To ensure the integrity of the physics of the large scale dynamics of the flows of practical interest, the corresponding large scale modes computed or measured in a simulation or a laboratory setting should not be contaminated because of their interaction with the dissipation range, which is not universal (Martinez et al., [44]). This requirement can be satisfied by maintaining a sufficiently broad inertial range. The required length of the inertial range needed can be deduced, for example, from our understanding of the interacting scales discussed in the previous section.

The *minimum state* is defined as the turbulent flow which has the *lowest Re* that captures the energy containing scales of the astrophysics problems in a laboratory or simulation setting (Zhou, [28, 40, 41]). The *minimum state* is therefore the lowest *Re* turbulent flow where all the modes in the energy containing scales will only interact with modes in the same spectral range or those within the inertial range. Obviously, this requirement is introduced to take full advantage of the universality of the inertial range.

The *minimum state*, according to the foregoing analysis, is the turbulent flow that takes the value of $k^*_z \equiv 2\ k^*_{L\text{-}T}$ equal to the inner viscous wavenumber, k_v (the end of the inertial range; see Fig. 2.8). Using the definition of $k_{L\text{-}T}$ and k_v, one finds that the *critical Re* of the *minimum state* is

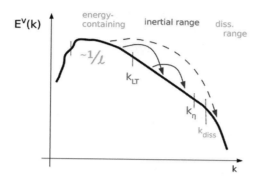

Figure 2.8 Sketch of a kinetic energy spectrum indicating the energy containing, inertial, and dissipation ranges and their wavenumber boundaries. The idea behind the minimum state is that the inertial range should be long enough that direct interactions between modes in the energy containing and dissipation ranges are energetically weak, indicated by the dashed (green) arrow. Some strong interactions are indicated via the solid (green) arrows. From [40]. A black and white version of this figure will appear in some formats. For the color version, please refer to the plate section.

$$Re^* = 1.6 \times 10^5. \tag{2.11}$$

The outer scale Re is approximately related to the Taylor microscale Re by $R_\lambda = (20/3)^{1/2}$ $Re^{1/2}$ in isotropic flow. As a result, the corresponding *critical* Taylor microscale Re of the *minimum state* is $R_\lambda {}^* \approx 1.4 \times Re^{1/2}$ ($R_\lambda {}^* \approx 560$).

Figure 2.9 provides another way to illustrate this argument using the physical length scales variation with the Re. The Reynolds numbers where an inertial range first occurs and the "minimum state" are marked short and long dash lines, respectively. The shaded area represents the redundant data of the inertial range. As a result, any high Re flow can be easily scaled down to a very high, but computationally achievable, Re flow.

2.4.1 Temporal Criterion

Many experiments have been conducted in classical fluid dynamics facilities, shock tubes, and laser facilities (such as the OMEGA laser and National Ignition Facility) to understand the complex flows induced by various instabilities. For example, the published examples were specialized to the Rayleigh–Taylor [45, 46], Richtmyer–Meshkov [47, 48], and Kelvin–Helmholtz flows [49]. Due to diagnostic limitations, typical measurements consist solely of the growth of the mixing zone width. While these widths (of the bubble and spike fronts, individually or combined) are usually measured, it is difficult to know whether or not a particular experiment has reached mixing transition.

We have extended the stationary mixing transition to time dependent flows and applied it to a wide range of experiments [29, 30]. The Liepmann–Taylor scale essentially describes the internal laminar vorticity growth layer generated by viscous shear along the boundaries of a large scale feature of size δ. The temporal development of such a laminar viscous layer is well known to go as $(vt)^{1/2}$ (Stokes [50], Lamb [51]).

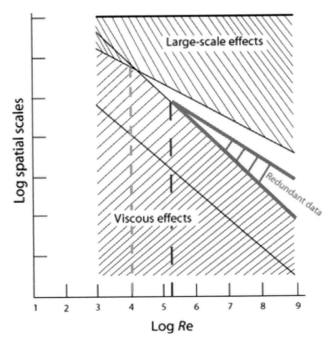

Figure 2.9 An illustration of how an arbitrarily high *Re* flow can be scaled down to a manageable one (shown in green), but still captures the physics of the large scale. Based on a figure in Dimotakis with added marks in color. From [40]. A black and white version of this figure will appear in some formats. For the color version, please refer to the plate section.

$$\lambda_D = C(vt)^{1/2} \tag{2.12}$$

Hence, the upper bound of the developing inertial range is the smaller of the Liepmann–Taylor scale, λ_{L-T} and λ_D. Here the coefficient of the diffusion layer, C, was suggested as $C \approx \sqrt{15}$ both for isotropic, homogeneous turbulence [52] and for steady parallel flows, and as $C \equiv 5$ for laminar boundary layer flows.

The inertial range is presumed to be established when the evolution of the large scale, $\min\{\lambda_D, \lambda_{L-T}\}$, is decoupled from the inner viscous scale, λ_v. For time dependent flows, the mixing transition is achieved when a range of scales exists such that the temporally evolving upper bound $[\min\{\lambda_v, \lambda_{L-T}\}]$ is significantly larger than the temporally evolving lower bound, λ_v. Thus, the mixing transition occurs if and when the inequality

$$\min\{\lambda_{L-T}(t), \lambda_D(t)\} > \lambda_v(t) \equiv 50\lambda_k(t) \tag{2.13}$$

is satisfied [28, 40, 41].

In designing and interpreting experiments for time dependent flows, the time dependent mixing transition provided the desirable ability to estimate the time required to achieve the mixing transition state.

2.5 Applications

2.5.1 High Energy Density Physics Experiments

Achieving the minimum state should be a goal for high energy density physics (HEDP) experimentalists when designing and developing their investigations [53–64]. The *Re* of the minimum state may be achievable already for some of the most advanced experimental platforms by optimizing the maximum potential of existing experimental facilities.

The theoretical analysis for a HEDP experiment to achieve the *Re* at the *minimum state* or beyond can be summarized. The viscosity is evaluated, the outer scale or the characteristic velocity is computed, and the *Re* is determined. For time dependent flows, several length scales should be evaluated in order to inspect whether both Eq. (2.11) and Eq. (2.13) are satisfied.

2.5.2 ILES Effective Reynolds Number

The utility of ILES for practical scientific and engineering simulations is clear. The broadly defined ILES [65] generally uses high resolution nonoscillatory FV (NFV) algorithms to solve unfiltered Euler or Navier–Stokes equations. Recently, an unique issue has been advanced for ILES: how one can estimate the values of the relevant effective viscosity (v_{eff}) produced by the numerical method [66]. It is critical to make progress on this matter, and further formalizing the perceived abilities of ILES in computing high *Re* flow would indeed offer significant values.

Since ILES is applied to a broad range of engineering and scientific applications, a scheme should be developed in physical space. In this fashion, v_{eff} can be written as

$$v_{eff} = \varepsilon/\Omega, \tag{2.14}$$

where Ω is the enstrophy $\Omega = \omega^2/2$ and the vorticity, $\omega = \nabla \times u$, can be readily evaluated based on the simulated velocity u data. As noted in [66], the key to accurately estimating v_{eff} is to find a way to do so away from the smallest scales of the simulated flow, where numerically controlled dissipation takes place. Once the dissipation rate is obtained, v_{eff} can be written down based on a dimensional argument.

The turbulent flows of practical interest are usually both high *Re* and complex. Yet, as a first cut, an interesting scheme can be developed from the following procedure:

- For a highly resolved ILES flow, the compensated inertial range occurs at lower spectral ranges. The bottleneck occurs as a plateau at higher wave numbers [67–71].
- The flux is still the only link between the energetic and dissipative scales of motion.
- The effective dissipations of the computations can be obtained directly from the inflow profiles of the inertial range energy transfer.

Hence, a methodology should be utilized to estimate the "input" of the energy into the energy transfer from the large scales. The key is to carry out this estimation around the energy containing scales so that the small scale information, where the dissipation

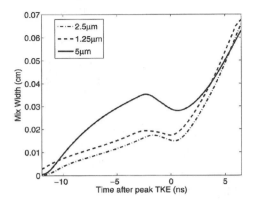

Figure 2.10 Asymptotic measures of the mixing width.

is controlled by numerical methods, is not needed. Fortunately, a large body of experimentally and numerically generated data suggested that a nondimensional parameter such as D approaches a constant when Re becomes sufficiently large (e.g., Figs. 2.2 and 2.3). We note that the outer scale (L) based Re is approximately related to the Taylor microscale Re_λ, by $Re_\lambda \approx (20/3)^{1/2}\sqrt{Re_L}$ in isotropic flow, and by $Re_\lambda \approx 1.4\sqrt{Re_L}$ for turbulence in the far field of a jet. The literature shows internal consistency between the establishment of an inertial range and the asymptotic behavior of $D \longrightarrow D_\infty$ = constant. The laboratory data compiled by Sreenivasan [10, 11] exhibited constancy for $Re_\lambda > 100$.

The collection of the computational data by Kaneda [12] and Sreenivasan [10, 11] indicated that somewhat higher Re_λ is needed, perhaps around 200, for which the inertial range is about one decade; observed data scatter for lower Re (e.g., in Figs. 2.2 and 2.3) has been attributed to differences among forcing schemes, forced long wavelength ranges, and box sizes. At high Re limit, the energy flux can be estimated directly

$$\varepsilon = D_\infty U^3/L. \tag{2.15}$$

Recent high resolution ILES [72–75], show a limited inertial range of the kinetic energy spectrum as well as distinct flow features associated with high Re flows. The viscosity independent dissipation rate may have been thus captured for these high resolution ILES flows, and a dimensional estimation such as $v_{eff} = \varepsilon/\Omega$ – through Eqs. (2.14) and (2.15) – may provide a reasonable v_{eff} in this context.

The laboratory reshock experiments [55] were performed using the University of Rochester's OMEGA laser. The laboratory target consists of a cylindrical beryllium (Be) tube ≈ 1.4 mm in length and ≈ 0.5 mm in diameter with a ≈ 100 μm wall thickness. The target is successively hit from both sides by two laser driven shocks. The first, ≈ 5 kJ, at t = 0 ns impacts the plastic ablator on the left, driving a Mach ≈ 5 shock through the 20 μm aluminum (Al) tracer disk adjoining the ablator. The tracer disk is thus propelled to the right down the center of the cylinder, which is filled with a low density (60mg/cc) CH foam. The second shock, ≈ 4 kJ at 5 ns, impacts a plastic ablator

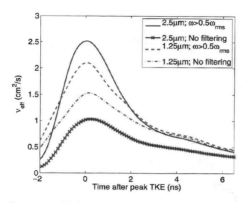

Figure 2.11 Estimated effective eddy viscosity v_{eff}.

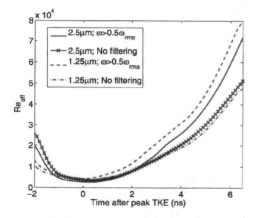

Figure 2.12 Estimated outer scale Re_{eff}.

at the right end of the tube. The shocks collide at approximately 8 ns to the right of the mixing layer and reshock the mixing layer at approximately 10 ns, causing it to compress until approximately 13 ns. At approximately 17 ns, the second shock exits the mixing layer.

Asymptotic estimates for an outer scale Re_{eff} can be also generated. Following the discussion, we assume that high Re isotropic turbulence regimes for which $D = \varepsilon L/U^3 \approx \frac{1}{2}$ have been achieved,[1] and define a time dependent L to be the outer scale prescribed by a mixing width δ_{MZ} (Fig. 2.10). This mixing measure is designed to yield $\delta_{MZ} = h$ for $\psi(y) = 1 + \tanh(2(y - y_c)/h)]$, where y_c defines the center of the mixing layer. We thus formulate an *asymptotic* model for the dissipation, $\varepsilon = DU^3/(2L)$; in this context we can now evaluate outer scale (asymptotic) measures of $v_{eff} = \varepsilon/[2\langle s_{ij}s_{ij}\rangle]$ (Fig. 2.11) and $Re_{eff} = UL/v_{eff}$ (Fig. 2.12).

---

[1] Following Kaneda et al. [12] (Fig. 3), we expect $D \sim 0.5 - 2$, with $D_\infty \approx 0.5$.

2.6   Acknowledgments

This work was performed under the auspices of the Lawrence Livermore National Security, LLC under Contract No. DE-AC52-07NA27344.

BIBLIOGRAPHY

[1] Nelkin, M., "Universality and scaling in fully developed turbulence," *Adv. in Phys.*, 43, 143–181, 1994.

[2] Zhou, Y., and Speziale, C.G., "Advances in the fundamental aspects of turbulence, energy transfer, interacting scales, and self-preservation in isotropic decay," *Appl. Mech. Rev.*, 51, 267, 1998.

[3] Moin, P., and Kim, J., "Tracking turbulence with supercomputers," *Scientific American*, 276, 62–68, 1997.

[4] Zhou, Y., Matthaeus, W.H., and Dmitruk, P., "Magnetohydrodynamic turbulence and time scales in astrophysical and space plasma," *Rev. Mod. Phys.*, 76, 1015–1035, 2004.

[5] Batchelor, G.K., *Theory of Homogeneous Turbulence*, Cambridge University Press, Cambridge, 1953.

[6] Saddoughi, S.G. and Veeravalli, S.V., "Local isotropy in turbulent boundary layers at high Reynolds number," *J. Fluid Mech.*, 268, 333, 1994.

[7] Jiménez, J., "Computing high-Reynolds-number turbulence: Will simulation ever replace experiments?," *J. Turbulence*, 4(022), 1–14, 2003.

[8] Pope, S.B., "Ten questions concerning the large-eddy simulation of turbulent flows," *New J. Phys.*, 6(35), 1–24, 2004.

[9] Burattini, P., Lavole, P., and Antonia, R.A., "On the normalized energy dissipation rate," *Phys. Fluids*, 17(098103), 1–4, 2005.

[10] Sreenivasan, K.R., "On the scaling of the turbulence energy dissipation rate," *Phys. Fluids*, 27, 1048–1051, 1984.

[11] Sreenivasan, K.R., "An update on the energy dissipation rate in isotropic turbulence," *Phys. Fluids*, 10, 528–529, 1998.

[12] Keneda, Y., Ishihara, T., Yokokawa, M., Itakura, K., and Uno, A., "Energy dissipation rate and energy spectrum in high resolution direct numerical simulations of turbulence in a periodic box," *Phys. Fluids*, 15, L21–24, 2003.

[13] Pearson, B.R., Krogstad, P.-A., and van de Water, W., "Measurements of the turbulent energy dissipation rate," *Phys. Fluids*, 14, 1288–1290, 2002.

[14] Zhou, Y., Hossain, M., and Vahala, G., "A critical look at the use of filters in large eddy simulation," *Phys. Lett. A*, 139, 330–332, 1989.

[15] Leonard, A., "Energy cascade in large-eddy simulations of turbulent fluid flows," *Adv. Geophy. S.*, 18A, 237–248, 1974.

[16] Germano, M., "Differential filters for the large eddy simulation of turbulent flows," *Phys. Fluids*, 29, 1755–1757, 1986.

[17] Langford, J.A. and Moser, R.D., "Optimal LES formulations for isotropic turbulence," *J. Fluid Mech.*, 398, 321–346, 1999.

[18] Domaradzki, J.A. and Loh, K.-C., "The subgrid-scale estimation model in the physical space representation," *Phys. Fluids*, 11, 2330–2342, 1999.

[19] Domaradzki, J.A. and Adams, N.A., "Direct modelling of subgrid scales of turbulence in large eddy simulations," *J. of Turbulence*, 3(024), 1–19, 2002.

[20] Zhou, Y., "Eddy damping, backscatter, and subgrid stresses in subgrid modeling of turbulence, "*Phys. Rev. A*, 43, 7049–7052, 1991.

[21] Zhou, Y. and Vahala, G., "Reformulation of recursive-renormalization-group based subgrid modeling of turbulence," *Phys. Rev. E*, 47, 2503–2519, 1993.

[22] Kraichnan, R.H., "Eddy viscosity in two and three dimensions," *J. Atmos. Sci.*, 33, 1521–1536, 1976.

[23] Dubois, T., Jauberteau, F., and Zhou, Y., "Influences of subgrid scale dynamics on resolvable scale statistics in large-eddy simulations," *Physica D*, 100, 390–406, 1997.

[24] Rose, H.A., "Eddy diffusivity, eddy noise, and subgrid-scale modelling," *J. Fluid Mech.*, 81, 719–734, 1977.

[25] Zhou, Y., Vahala, G., and Hossain, M., "Renormalization-group theory for the eddy viscosity in subgrid modeling," *Phys. Rev. A*, 37, 2590–2598, 1988.

[26] Zhou, Y., Vahala, G., and Hossain, M., "Renormalization eddy viscosity and Kolmogorov's constant in forced Navier-Stokes turbulence," *Phys. Rev. A.*, 40, 5865–5874, 1989.

[27] Dimotakis, P.E., "The mixing transition in turbulent flows," *J. Fluid Mech.*, 409, 69–98, 2000.

[28] Zhou, Y., "Unification and extension of the concepts of similarity criteria and mixing transition for studying astrophysics using high energy density laboratory experiments or numerical simulations," *Phys. Plasmas*, 14, 082701, 2007.

[29] Zhou, Y., Robey, H.F., and Buckingham, A.C., "Onset of turbulence in accelerated high-Reynolds-number flow," *Phys. Rev. E*, 67, 056305, 1–11, 2003.

[30] Zhou, Y., Remington, B.A., Robey, H.F., et al., "Progress in understanding turbulent mixing induced by Rayleigh-Taylor and Richtmyer-Meshkov instability," *Phys. Plasma*, 10, 1883–1896, 2003.

[31] Hinze, J.O., *Turbulence*, 2nd edition, McGraw-Hill, New York, 1975.

[32] Domaradzki, J.A. and Rogallo, R.S., "Local energy transfer and non-local interactions in homogeneous, isotropic turbulence," *Phys. Fluids A*, 2, 413–426, 1990.

[33] Zhou, Y., "Degree of locality of energy transfer in the inertial range," *Phys. Fluids A*, 5, 1092–1094, 1993.

[34] Zhou, Y., "Interacting scales and energy transfer in isotropic turbulence," *Phys. Fluids A*, 5, 2511–2524, 1993.

[35] Zhou, Y., "Renormalization group theory for fluid and plasma turbulence," *Phys. Report*, 488, 1, 2010.

[36] Kraichnan, R.H., "Inertial range transfer in two- and three-dimensional turbulence," *J. Fluid Mech.*, 47, 525–535, 1971.

[37] Lesieur, M., *Turbulence in Fluids*, Martinus Nijhoff, Dordrecht, 1990.

[38] Gotoh, T. and Watanabe, T., "Statistics of transfer fluxes of the kinetic energy and scalar variance," *J. of Turbulence*, 6(33), 1–18, 2005.

[39] Kaneda, Y., "Renormalization expansions in the theory of turbulence with the use of the Lagrangian position function," *J. Fluid Mech*, 107, 131, 1981.

[40] Zhou, Y. and Oughton, S., "Nonlocality and the critical Reynolds numbers of the minimum state magnetohydrodynamic turbulence," *Phys. Plasma*, 18, 072304, 2011.

[41] Zhou, Y., Buckingham, A.C., Bataille, F., and Mathelin,L., "Manimum state for high Reynolds and Peclet number turbulent flows," *Phys. Lett. A.*, 373, 2746, 2009.

[42] Domaradzki, J.A., Liu, W., and Brachet, M. E., "An analysis of subgrid-scale interactions in numerically simulated isotropic turbulence," *Phys. Fluids A* 5, 1747, 1993.

[43] Domaradzki, J.A., Liu, W., Hartel, C., and Kleiser, L., "Energy transfer in numerically simulated wall-bounded turbulent flows," *Phys. Fluids*, 6, 1583, 1994.

[44] Martinez, D.O., Chen, S., Dolen, G.D., Kraichnan, R.H., Wang, L.P., and Zhou, Y., "Energy spectrum in the dissipation range of fluid turbulence," *J. Plasma Phys.*, 57, 195, 1997.

[45] Lord Rayleigh, *Scientific Papers*, Dover, New York, 1965.

[46] Taylor, G. I., "The instability of liquid surfaces when accelerated in a direction perpendicular to their plane," *Proc. R. Soc. London, Ser. A*, 201, 192, 1950.

[47] Richtmyer, R. D., "Taylor instability in shock acceleration of compressible fluids," *Commun. Pure Appl. Math.*, 13, 297, 1960.

[48] Meshkov, E. E., "Instability of the interface of two gases accelerated by a shock wave," *Izv., Acad. Sci., USSR Fluid Dyn.*, 4, 101, 1969.

[49] Chandrasekhar, S., *Hydrodynamic and Hydromagnetic Stability*, Dover, New York, 1961.

[50] Stokes, G.G., "On the effect of the internal friction of fluids on the motion of pendulums," *Trans. Cambridge Philos. Soc.*, 9, 8, 1901.

[51] Lamb, H., *Hydrodynamics*, Cambridge University Press, London, 1932.

[52] Tennekes, H. and Lumley, J.L., *First Course in Turbulence*, MIT Press, Cambridge, 1972.

[53] Hurricane, O.A., Hansen, J.F., Robey, H.F., et al., "A high energy density shock driven Kelvin–Helmholtz shear layer experiment," *Phys. Plasma*, 16(5), p056305, 2008.

[54] Haines, B.M., Grinstein, F.F., Welser-Sherrill, L., et al., "Analysis of the effects of energy deposition on shock-driven turbulent mixing," *Phys. Plasma*, 20(7), 072306, 2013.

[55] Welser-Sherrill, L., Fincke, J., Doss, F., et al., "Two laser-driven mix experiments to study reshock and shear," *High Energy Density Phys.*, 9(3), 496, 2013.

[56] Haines, B.M., Grinstein, F.F., Welser-Sherrill, L., et al., "Simulation ensemble for a laser-driven shear experiment," *Phys. Plasma*, 20(9), 092301, 2013.

[57] Grinstein, F.F., Gowandhan, A.A., et al., "On coarse-grained simulations of turbulent material mixing," *Physica Scripta*, 86(5), 058203, 2012.

[58] Smalyk, V.A., Hansen, J.F., and Hurricane, O.A., "Experimental observations of turbulent mixing due to Kelvin–Helmholtz instability on the OMEGA Laser Facility," *Phys. Plasma*, 19(9), 092702, 2012.

[59] Gowandhan, A.A. and Grinstein, F.F., "Numerical simulation of Richtmyer–Meshkov instabilities in shocked gas curtains," *J. Turbulence*, 12, pN43, 2012.

[60] Hurricane, O.A., Smalyk, V.A., Raman, K., et al., "Validation of a turbulent Kelvin–Helmholtz shear layer model using a high-energy-density OMEGA laser experiment," *Phys. Rev. Lett.*, 109(15), 2012.

[61] Di Stefano, C.A., Kuranz, C.C., Keiter, P.A., et al., "Late-time breakup of laser-driven hydrodynamics experiments," *High Energy Density Phys.*, 8(4), 360, 2012.

[62] Hurricane, O.A., Smalyk, V.A., Hansen, J.F., et al., "Measurements of turbulent mixing due to Kelvin–Helmholtz instability in high-energy-density plasmas," *High Energy Density Phys.*, 9(1), 47, 2013.

[63] Doss, F.W., Loomis, E.N., Welser-Sherrill, L., et al., "Instability, mixing, and transition to turbulence in a laser-driven counterflowing shear experiment," *Phys. Plasma*, 20(1), 012707, 2013.

[64] Drake, R.P., Harding, E.C., and Kuranz, C.C., "Approaches to turbulence in high-energy-density experiments," *Physica Scripta*, 2008, p014022, 2008.

[65] Grinstein, F.F., Margolin, L.G., and Rider, W.G., editors, *Implicit Large Eddy Simulation: Computing Turbulent Flow Dynamics*, Cambridge University Press, New York, 2010.

[66] Zhou, Y., Grinstein, F.F., Wachtor, A.J., and Haines, B.M., "Estimating the effective Reynolds number in implicit large eddy simulation," *Phys. Rev. E.*, 89, 013303, 2014.

[67] Falkovich, G., "Bottleneck phenomenon in developed turbulence," *Phys. Fluids*, 6, 1411, 1994.

[68] Yeung, P.K. and Zhou, Y., "Universality of the Kolmogorov constant in numerical simulations of turbulence," *Phys. Rev. E.*, 56, 1746, 1997.

[69] She, Z.S., Chen, S., Doolen, G., Kraichnan, R.H., and Orszag, S.A., "Reynolds number dependence of isotropic Navier-Stokes turbulence," *Phys. Rev. Lett.*, 70, 3251, 1993.

[70] Cao, N., Chen, S., and Doolen, G.D., "Statistics and structures of pressure in isotropic turbulence," *Phys. Fluids*, 11, 2235, 1999.

[71] Wang, L.P., Chen, S., Brasseur, J.G., and Wyngaard, J.C., "Examination of hypotheses in the Kolmogorov refined turbulence theory through high- resolution simulations," *J. Fluid Mech.*, 309, 113, 1996.

[72] Grinstein, F.F., Gowardhan, A.A., and Wachtor, A.J., 2011, "Simulations of Richtmyer–Meshkov instabilities in planar shock-tube experiments," *Physics of Fluids*, 23, 034106.

[73] Fureby, C. and Grinstein, F.F., "Monotonically integrated large eddy simulation of free shear flows," *AIAA Journal*, 37, 544–556, 1999.

[74] Porter, D.H., Woodward, P.R., and Pouquet, A., "Inertial range structures in decaying turbulent flows," *Phys. of Fluids*, 10, 237–245, 1998.

[75] Fureby, C. and Grinstein, F.F., "Large eddy simulation of high Reynolds number free and wall bounded flows," *J. Comput. Phys.*, 181, 68, 2002.

3 Finite Scale Theory
Compressible Hydrodynamics at Second Order

Len G. Margolin

3.1 Introduction

"All our knowledge has its origin in our perceptions."

Leonardo da Vinci

"What we observe is not nature itself, but nature exposed to our method of questioning."

Werner Heisenberg

Finite scale equations were derived as a coarse graining of the Navier–Stokes equations to describe high Reynolds number fluid flow. In this chapter I will review the motivation, the derivation, and the properties of these equations. I will discuss finite scale coarse graining as an extension of moment methods of kinetic theory and show how the finite scale equations should be considered a hydrodynamic theory at second order. I will present new material that describes the extension of finite scale theory to small length scales of the order of the molecular mean free path where Navier–Stokes theory is known to fail.

Finite scale equations describe the evolution of finite volumes of fluid in time. The opening quotations of this chapter are meant to motivate interest in deriving and analyzing the properties of those equations and their solutions. Physical experiments are done with instruments that have finite size and hence finite precision. Similarly, numerical calculations are carried out at finite scale represented by the discretization parameters of the simulation, that is, the computational cell size Δx and the time step Δt. So the finite scale equations represent a coarse graining of the theoretical fluid flow equations as well as a practical model for numerical simulation. At least one important class of numerical methods, the nonoscillatory finite volume (NFV) schemes, have been shown a posteriori to be solving finite scale equations [45].

"Using a term like nonlinear science is like referring to the bulk of zoology as the study of non-elephant animals" (Stan Ulam). It is only relatively recently that physicists have begun to realize that the nonlinearity of the laws that govern well-known physical systems can give rise to complex and even chaotic behavior [23]. Further, new techniques have been developed so that nonlinearity could be embraced rather than avoided. Two of these techniques that have influenced the development of finite scale theory are nonlinear enslavement, which began in the mathematical community [18], and NFV differencing [7], which arose in the computational physics community.

The combination of those ideas originating in those two very different communities led to the closure theorem [45], which is discussed in Section 3.3.3 and is the central element of the derivation of the finite scale equations.

This article is primarily concerned with modeling high Reynolds number flows, that is, unsmooth flows with turbulence and/or shocks. As will be discussed in Section 3.2, these are flows where Navier–Stokes equations, supported by their derivation from kinetic theory, should be an appropriate model. However, when discretized and applied as the underlying model for numerical simulation, Navier–Stokes equations typically fail to provide accurate or even stable results for high Reynolds number flows. This failure is essentially due to lack of sufficient dissipation, or equivalently lack of sufficient entropy production. For over sixty years, numerical modelers have supplemented the Navier–Stokes equations with additional dissipative fluxes that are generically (and perhaps pejoratively) termed artificial viscosity. The origin and nature of these flux terms are empirical and are described in Section 3.5.

Over the history of computational fluid dynamics, code developers have carefully distinguished between physics models and discretization error. Using the technique of modified equation analysis [29], one can expand the discrete equations into equivalent partial differential equations (PDEs). Terms depending on the discretization parameters, for example, the cell size and the time step, are termed truncation error. Artificial viscosity, which depends on the square cell size [79], would appear to be a deliberate introduction of a second-order error into the discrete equations. Similarly, subgrid scale models for unresolved turbulence depend also on the square of the cell size [68].

My subtitle, "Compressible Hydrodynamics at Second Order," is designed to imply that some truncation terms belong to the physics models and are not discretization errors. It is no coincidence that the additional terms that appear from coarse graining the Navier–Stokes equations in Section 3.3.2 also depend on the square of the averaging length. Inspection of the proof of the closure theorem in Section 3.3.3 as well as other coarse graining techniques such as the Reynolds decomposition discussed in Section 3.6.1 shows that first-order terms vanish in general due to the isotropy of space. Similarly, long experience with first-order numerical methods has shown them too diffusive for practical use. Both in the finite scale equations and in numerical methods, the order parameter is the macroscopic length Δx. More precisely, it is the macroscopic gradients of density, velocity, and so on, measured over distances of $\mathcal{O}(\Delta x)$ and properly normalized that are small so that the new flux terms represent small departures from thermodynamic equilibrium.

The material in this chapter consists of two related topics: (1) a review of the finite scale theory at macroscopic length scales and (2) a new extension of the theory to microscopic length scales of the order of the molecular mean free path. The first theme seems to qualify as a coarse graining technique in that it upscales the Navier–Stokes equations using renormalization techniques. The second theme tells a different story, bypassing the Navier–Stokes equations to more directly construct a *hydrodynamic* theory from kinetic theory. I would term this "ensemble averaging" rather than "coarse graining," that is, finding a smooth description of what is essentially a stochastic process. Indeed, the macroscopic theory in its application as a turbulence model also represents ensemble averaging.

A second distinction between finite scale theory and coarse graining is that the new terms of the finite scale equations do not depend on fundamental fluid parameters at smaller scales, but rather on the resolved flow. More generally, I would say that the finite scale description is independent of unresolved details. I have related this independence to modern theories of nonlinear enslavement and inviscid dissipation. However, I would remind that the very successful theory of thermodynamics is also based on the idea that (often) the details don't matter.

Finally, I would mention that form invariance, the central concept in the proof of the closure theorem, is itself at odds with the task of creating coarse grained models. The remarkable result of Section 3.7 is that the equations of a microscopic finite scale theory have the same form as the macroscopic equations, illustrating this contradiction. It is the goal of this chapter to provide sufficient physical reasoning and mathematical detail for the reader to consider the utility of finite scale theory.

An outline of this chapter is as follows. Sections 3.3 through 3.5 constitute a review of previously published material. Sections 3.2 and 3.6 through 3.8 are new material. I will briefly describe the derivation of Navier–Stokes from kinetic theory in Section 3.2. This section is meant to show the how macroscopic length scales should naturally enter a hydrodynamic theory and to emphasize that the form of the balance laws (i.e., conservation equations) does not depend on Chapman–Enskog (CE) approximation. Section 3.3 summarizes the derivation of the finite scale theory and contains a brief description of the closure theorem. The role of unresolved kinetic energy is also introduced in this section and the connection to other hydrodynamic models, Brenner's theory of volume transport and Holms's alpha models, is discussed. Section 3.4 discusses the properties of the finite scale equations and relates those properties to nonlinear enslavement and inviscid diffusion.

The practical issues of numerical stability and convergence provide powerful constraints on computational models. An early history of artificial viscosity, finite volume methods, and nonoscillatory advection is presented in Section 3.5.1 and that of implicit large eddy simulation (ILES) in Section 3.5.2. It was the close connection between artificial viscosity and nonoscillatory methods, exposed by modified equation analysis, that first led me to a rationale for ILES [45] and the beginning of finite scale theory. Also, the proof of the closure theorem in Section 3.3.3 is rooted in my earlier applications of inertial manifold theory (e.g., nonlinear enslavement) to numerical algorithms [34].

In Section 3.2, I criticize the CE derivation of physical viscosity and heat conduction as inappropriate for high Reynolds number flows. Nevertheless, those terms are well founded in empirical studies that have been conducted over more than one hundred years. The material in Sections 3.6, 3.7, and 3.8 proposes an alternate explanation that accounts for those well-verified transport terms on macroscopic scales, but differs on small length scales of the order of the molecular mean free path. This potentially leads to hydrodynamic theories and numerical programs that are simpler and more economical than schemes based directly on kinetic theory but still appropriate for such small scales.

The finite scale theory reviewed in Sections 3.3 and 3.4 begins with the Navier–Stokes equations and so contains all the limitations of that theory. The essence of a hydrodynamic theory lies in the approximations made to design an approximate solution to the Boltzmann equation. In particular, Navier–Stokes theory is based on the idea of local thermodynamic equilibrium (LTE). The velocity distribution functions do not depend on system size Δx and so the constitutive relations (e.g., pressure) derived from them are also independent of Δx. In Section 3.7, I generalize finite scale ideas and form invariance first to show that LTE is not a consistent second-order approximation and then to derive a distribution function that does depend consistently on Δx^2.

In Section 3.8, I apply these new results to the important problem of calculating the profile of a shock. This is a problem where Navier–Stokes theory fails both quantitatively in predicting the width of the shock and qualitatively in predicting the dependence of width on incident Mach number. That more fundamental calculations based directly on the Boltzmann equation do accurately reproduce the experimental measurements of shocks [1] would seem to support my criticisms of CE theory. My preliminary results indicate that finite scale theory will predict wider shocks.

3.2   Hydrodynamics from Kinetic Theory

> The fundamental task of nonequilibrium statistical mechanics is to deduce the evolution
> of the macroscopic state of physical systems from the knowledge of their underlying
> microscopic dynamics. [66]

Hydrodynamics is a macroscopic description of dynamics in a fluid state of matter. The usual path in deriving Navier–Stokes equations from kinetic theory starts with the Boltzmann equation and uses CE successive approximation in a small parameter (Knudsen number) to calculate departures of the probability distribution function (PDF) from equilibrium. It is not always appreciated that much of Navier–Stokes theory is actually independent of the form of the distribution function. The objectives of this section are first to isolate the elements of the Navier–Stokes equations that depend on CE, next to contrast the regime of high Reynolds number flows with those of rarefied gases [9], and finally to raise some issues of CE approximation in the high Reynolds number regime.

3.2.1   Balance Equations

A derivation of hydrodynamic equations from the viewpoint of statistical physics begins with the Boltzmann equation, which describes the time evolution of the PDF for molecular velocity $f(x_i, c_i, t)$ in the six-dimensional phase space of position x_i and velocity c_i:

$$\frac{\partial f}{\partial t} + c_i \frac{\partial f}{\partial x_i} = \mathcal{C}_i. \tag{3.1}$$

Here, C_i is the Boltzmann collision integral. The PDF f has the units of a number of molecules per volume of (physical) space per volume of (velocity) space. There are many detailed treatments of the Boltzmann equation and of the collision integral; see, for example, [28, 35].

The hydrodynamic description is a coarse graining of the Boltzmann dynamics in terms of the lowest order moments of the PDF f, thermodynamic equilibrium (LTE), density, macroscopic momentum and internal energy. Assume a monatomic gas of identical molecules of individual mass m, linear momentum mc_i, and energy $\frac{1}{2}mc^2$. For any PDF f, one defines the macroscopic hydrodynamic variables density (ρ), momentum density (ρv_i), and total energy density (ρE) *at a point* x_i in terms of the integrals over the velocity phase space

$$\rho \equiv m \int f(x_i, c_i, t)\, d^3 c \tag{3.2}$$

$$\rho v_i \equiv m \int c_i f(x_i, c_i, t)\, d^3 c \tag{3.3}$$

$$\rho E \equiv \frac{m}{2} \int c^2 f(x_i, c_i, t)\, d^3 c \tag{3.4}$$

where all integrals extend from $(-\infty, \infty)$. Further, define the relative (sometimes termed the peculiar) velocity $C_i = c_i - v_i$. Then one can define the microscopic internal energy density $(\mathcal{I})$

$$\rho \mathcal{I} \equiv \frac{m}{2} \int C^2 f(x_i, c_i, t)\, d^3 c \tag{3.5}$$

from which it follows

$$\rho E = \frac{1}{2}\rho v^2 + \rho \mathcal{I}. \tag{3.6}$$

That is, the total macroscopic energy density is the sum of the kinetic energy density and the internal energy density.

The derivation of the balance laws now follows easily from the definitions given here and from the conservation of mass, momentum, and energy. For example, multiplying the Boltzmann equation (3.1) by molecular mass m and integrating over all components of the velocity c_i leads to the continuity equation:

$$\frac{\partial \rho}{\partial t} + \frac{\partial \rho v_i}{\partial x_i} = 0 \tag{3.7}$$

where the integral over the collision integral has vanished due to mass conservation. Similarly, multiplying Eq. (3.1) by the momentum mc_i and integrating over all components of the velocity leads to the equation for conservation of momentum:

$$\frac{\partial \rho v_i}{\partial t} + \frac{\partial \rho v_i v_j}{\partial x_i} = -\frac{\partial \sigma_{ij}}{\partial x_i}. \tag{3.8}$$

Finally, multiplication of Eq. (3.1) by $\frac{m}{2}C^2$ and integrating over all components of velocity leads to the equation for conservation of total energy:

$$\frac{\partial}{\partial t}\left[\rho\left(\mathcal{I}+\frac{1}{2}v^2\right)\right]+\frac{\partial}{\partial x_i}\left[\rho\left(\mathcal{I}+\frac{1}{2}v^2\right)\right]=-\frac{\partial}{\partial x_i}\left(\sigma_{ij}v_j+q_i\right). \tag{3.9}$$

Equations (3.6)–(3.9) are valid for any PDF $f(x_i, c_i, t)$. However, they are not complete, as constitutive relations are needed for the higher-order moments that appear, namely the pressure tensor σ_{ij}:

$$\sigma_{ij}=m\int C_i C_j f(x_i, c_i, t)\, d^3 c \tag{3.10}$$

and the heat flux q_i:

$$q_i \equiv \frac{m}{2}\int C^2 C_i f(x_i, c_i, t)\, d^3 c. \tag{3.11}$$

3.2.2 Brief Remarks on Chapman–Enskog Theory

The pressure tensor and heat flux do depend on the particular form of the PDF $f(x_i, c_i, t)$. In the conventional CE successive approximation [10], constitutive relations follow from approximating the fluxes in terms of the basic state variables and their spatial and temporal derivatives. These relations are derived from perturbation solutions of the Boltzmann equation for the PDF f.

In the case of a (classical) system in equilibrium, that is, in the absence of macroscopic gradients, the equilibrium PDF is the Maxwell–Boltzmann (Gaussian) distribution. In three dimensions,

$$f^{eq}=N\left(\frac{m}{2\pi kT}\right)^{\frac{3}{2}}\exp-\left(\frac{mC^2}{2kT}\right). \tag{3.12}$$

Here N is the total number of particles per unit volume, T is temperature,[1] and k is Boltzmann's constant. In this case, we derive the equation of state

$$p=\frac{2}{3}\rho\mathcal{I}=n\rho RT \tag{3.13}$$

where n is the amount of gas measured in moles and R is the ideal gas constant equal to the product of the Boltzmann constant and Avogadro's number. In equilibrium, the deviator components of the pressure tensor and the heat flux vanish, leading to the Euler equations, which are the lowest (equilibrium) order.

In the presence of macroscopic gradients, it is much more difficult to derive constitutive relations. Independently and more than forty-five years after the introduction of

---

[1] For a simple gas in Navier–Stokes theory, temperature and internal energy are equivalent. But in the finite scale theory, the situation is complicated by the presence of unresolved kinetic energy; see Sections 3.3.2 and 3.7.3.

the Boltzmann equation, Chapman and then Enskog published results for the transport coefficients of viscosity and heat conductivity. A subsequent and somewhat simpler derivation is due to Grad [25], who assumed an expansion of the PDF $f(x_i, c_i, t)$ in terms of Hermite polynomials. CE theory [10] is based on perturbation solutions of the Boltzmann equation. The perturbation parameter is the Knudsen number, essentially the ratio of molecular scattering lengths to macroscopic scales of interest, which in the language of finite scale theory I have previously denoted Δx. Through a very complicated calculation, first-order corrections to the pressure tensor and the heat flux are derived; these corrections respectively have the form of Newtonian viscosity, $\frac{1}{2}\rho\eta\left(\frac{\partial v_i}{\partial x_j} + \frac{\partial v_j}{\partial x_i}\right)$, and Fourier's Law, $-\rho\kappa\frac{\partial \mathcal{T}}{\partial x_i}$. The transport coefficients for the viscosity, η, and for heat conduction, κ, are predicted by CE theory based on microscopic physical parameters. In conjunction with Eqs. (3.7)–(3.9), a first-order CE theory constitutes a statistical physics derivation of the Navier–Stokes equations.

Higher-order expansions beyond Navier-Stokes theory have been carried out. Higher-order terms represent corrections for larger Knudsen numbers; such corrections are necessary for studies of rarified gas dynamics, but not for simulations of high Reynolds number flows where macroscopic length scales far exceed viscous length scales. Indeed, high Reynolds number is the limit where the Knudsen number tends toward zero. I will be concerned with other issues of CE theory in this limit.

3.2.3   Hydrodynamics at Second Order

The vast majority of practical applications can be satisfactorily studied with the Navier–Stokes equations, especially for single phase low speed Newtonian flows. After all, Newtonian viscosity, Fourier's law of heat conduction, and the Navier–Stokes equations were all known empirically long before kinetic theory was developed. However, as a model for computational fluid dynamics, Navier–Stokes equations fall noticeably short in the regime of high Reynolds number flows. This important regime includes turbulent flows and flows with shocks.

In principle, the Reynolds number represents the inverse of the ratio of the molecular mean free path to the macroscopic scale length, that is, the inverse Knudsen number; high Reynolds numbers should represent a regime where CE theory works well. The issue is that on the discrete mesh, velocities and velocity gradients are smoothed to the point that viscosity is unable to provide sufficient dissipation. Practically speaking, when the Navier–Stokes equations are used as a model for high Reynolds number flows, kinetic energy piles up at the smallest scales of the problem (the computational cell size), destroying the accuracy and even the stability of the numerical result.

This result has led to well-known modifications of the discrete models used for high Reynolds number flows; additional terms generically described as *artificial viscosity* [79] and subgrid scale models [63] are added to the Euler equations. A fuller description of the origin and nature of these computational terms will be presented in Section 3.5.

Finite scale theory was originally developed to provide a theoretical rationale for the numerical technique of ILES [45]. The underlying idea – that the spatial averages of the conserved densities obey Navier–Stokes equations emended by additional terms arising from averaging the nonlinearity of advection – addresses a potential shortcoming of the CE process. If CE represents a perturbation solution of the Boltzmann equation for a small Knudsen number, why does the Knudsen number itself not appear in the equations? More particularly, why are the Navier–Stokes equations independent of Δx, the macroscopic length scale that makes the Knudsen number small?

Recall that the balance equations, (3.7)–(3.9), are the result of integrating the Boltzmann equation over the velocity phase space. Finite scale theory further integrates the equations over the coordinate space; the macroscopic length scales then explicitly enter the equations as the limits of the spatial integration; compare this to Eq. (3.14) in Section 3.3.1. We term this "macroscopic" or "averaging" length Δx in deference to its first origin in numerical methods. The identification of Δx with the computational cell size gives a precise quantification of the otherwise vague idea of a macroscopic length scale. In the case of experimental physics, Δx can be associated with the uncertainty of measurement devices, making a classical connection with the quantum emphasis on the role of *the observer* [44].

The spatial integration of the balance laws will be discussed in the following section. However, note that finite scale fluxes in Eqs. (3.17) and (3.20) at lowest order are proportional to Δx^2; the linear terms vanish due to parity, the left–right symmetry of space. This is the meaning of my subtitle: Compressible Hydrodynamics at Second Order.

The dependence of the finite scale fluxes on Δx is not unexpected, but if the Navier–Stokes equations were truly an expansion in Knudsen number, one should expect to see Δx appear *in the denominator* of the fluxes. Instead, Eqs. (3.19) and (3.20) illustrate that the small quantities that allow the expansion are the higher-order gradients of the macroscopic variables. This is consistent with the intent of Navier–Stokes to describe small departures from equilibrium, but not with CE theory where it is Knudsen number that is assumed to be small. In other words, CE theory predicts that the Navier–Stokes equations become more accurate as Δx gets large, in contrast to the finite scale equations, which predict the Navier–Stokes equations will become less accurate. The latter result is more consistent with numerical experience in simulating shocks and turbulence, where artificial viscosity and subgrid scale models also increase in magnitude with Δx.

Finally, I note that these second-order terms in Eqs. (3.17) and (3.20) originate from averaging the advective terms of the balance equations. The possibility that the PDF f itself depends on Δx will be addressed in Section 3.6.

3.3   Finite Scale Equations

In this section, I will outline the derivation of the finite scale equations from the one-dimensional Navier–Stokes compressible equations. Technically speaking, I should be referring to finite scale Navier–Stokes, since the coarse graining methodology can be applied to many other evolutionary equations. I will eschew this additional qualification from this point on.

3.3.1   Length Scales

The essence of finite scale theory is the existence of a length scale that is not inherent in the physical process being modeled,[2] but that arises from the question being asked. In the case of numerical methods, this is the computational cell size Δx. In the case of experiments, this might be the measurement uncertainty. In the case of theoretical development, this is the macroscale referred to in the previous section. In general, I will refer to this in words as *the observer* and notationally[3] as Δx. I will also refer to "physical scales of fluid processes," by which I mean the length scales at which physical viscosity and heat conductivity are significant. In high Reynolds number flows, to be defined in Eq. (3.30), this length scale might represent the size of the smallest eddies (Kolmogorov scale) or the width of the shock profile. In general, this is the length scale contained in the physical viscosity that is the order of the molecular mean free path. In this chapter, I am primarily concerned with regimes where the observer is much larger than the physical scales of the process. In terms of the previous section, this is the regime of small Knudsen numbers, where Navier–Stokes is justifiable on the basis of kinetic theory and in particular of CE expansions.

I will refer to the dependent variables of the finite scale equations either as coarse grained or as macroscopic variables (indistinguishably) and designate these variables with a hat; for example, the macroscopic density in one spatial dimension (1D) is:

$$\widehat{\rho}(x) \equiv \frac{1}{\Delta x} \int_{x-\frac{1}{2}\Delta x}^{x+\frac{1}{2}\Delta x} \rho(x') \, dx'. \tag{3.14}$$

I will use a tilde to indicate mass (Favre) averaged variables; for example, the mass averaged velocity in 1D is:

$$\widetilde{u}(x) \equiv \frac{1}{\widehat{\rho}\,\Delta x} \int_{x-\frac{1}{2}\Delta x}^{x+\frac{1}{2}\Delta x} \rho(x') u(x') \, dx'. \tag{3.15}$$

3.3.2   Presenting the Finite Scale Equations

The derivation of the finite scale equations involves applying the averaging operator of Eq. (3.14) to the one-dimensional version of the Navier–Stokes equations that were presented in the previous section, Eqs. (3.7)–(3.9). Then the closure theorem (3.29) is applied to the nonlinear terms that appear in the equations representing advection and work. The details of this process are found in [43] and are briefly described in Section 3.3.3. I will begin by first presenting the finite scale equations and will point out some new features of those equations with respect to Navier–Stokes.

Choose as the primary finite scale variables averaged density $\widehat{\rho}$, the momentum density $\widehat{\mathcal{M}} = \widehat{\rho u}$, energy density $\widehat{\mathcal{E}} = \widehat{\rho E}$, and pressure $\widehat{p}$. Then the finite scale Navier–Stokes equations for a compressible fluid in 1D derived in [43] are:

---

[2] I mean to include time scales as well, but will assume this is understood in general.
[3] In previous works, the observer length scale is sometimes denoted by "L."

$$\frac{\partial \widehat{\rho}}{\partial t} = -\frac{\partial}{\partial x}(\widehat{\rho}\widetilde{u}), \tag{3.16}$$

$$\frac{\partial \widehat{\mathcal{M}}}{\partial t} = -\frac{\partial}{\partial x}\left\{\widehat{\mathcal{M}}\widetilde{u} + \widehat{p} + \mathcal{P}\right\}, \tag{3.17}$$

$$\frac{\partial \widehat{\mathcal{E}}}{\partial t} = -\frac{\partial}{\partial x}\left\{\widehat{\mathcal{E}}\widetilde{u} + (\widehat{p} + \mathcal{P})\widetilde{u} + \mathcal{Q}\right\}. \tag{3.18}$$

Here, $\mathcal{P}$ is the momentum flux and $\mathcal{Q}$ is the heat flux. Typically, the pressure $\widehat{p}$ will depend on the specific internal energy $\widetilde{\mathcal{I}}$ and the density $\widehat{\rho}$.

The momentum flux $\mathcal{P}$ consists of the classical Newtonian viscosity, where η is the kinematic viscosity, and a new term of order $\mathcal{O}(\Delta x^2)$ that arises from the averaging of the momentum advection:

$$\mathcal{P} \equiv \frac{1}{3}\left(\frac{\Delta x}{2}\right)^2 \widehat{\rho}\widetilde{u}_x^2 - \widehat{\rho}\eta\widetilde{u}_x. \tag{3.19}$$

Similarly, the heat flux $\mathcal{Q}$ consists of the classical Fourier heat flux, where κ is the thermal conductivity, and two new terms. The first new term arises from averaging the nonlinear energy advection and is also $\mathcal{O}(\Delta x^2)$. The second new term arises from the averaging of the nonlinear work term. In [43], it is shown to have the same form as the first term in the case of a perfect gas. More generally, this term will depend on the form of the equation of state, but will also be $\mathcal{O}(\Delta x^2)$.

$$\mathcal{Q} \equiv \frac{1}{3}\left(\frac{\Delta x}{2}\right)^2 \widehat{\rho}\widetilde{u}_x\widetilde{\mathcal{I}}_x + (\widehat{pu} - \widehat{p}\widetilde{u}) - \widehat{\rho}\kappa\widetilde{\mathcal{I}}_x. \tag{3.20}$$

Let us begin a discussion of these equations by noting they are written in terms of the average *momentum velocity*, $\widetilde{u}$, which is not equal to the average velocity $\widehat{u}$ in a compressible fluid:

$$\widetilde{u} \equiv \frac{\widehat{M}}{\widehat{\rho}} \neq \widehat{u}. \tag{3.21}$$

Similarly, the macroscopic specific internal energy is

$$\widetilde{\mathcal{I}} \equiv \frac{\widehat{\rho\mathcal{I}}}{\widehat{\rho}} \neq \widehat{\mathcal{I}}. \tag{3.22}$$

In compressible turbulence theory, this is termed Favre averaging [17]. This choice is necessary as the equation set is not closed in terms of $\widehat{u}$ nor $\widehat{\mathcal{I}}$.

The idea that there is more than one fluid velocity required to describe fluid flow has appeared recently in other research, most notably in Brenner's theory of volume transport [8] and in Holm's development of the α–model [19].[4] In particular, compare the constitutive relationship shown in eq. 3.2 of [43]

---

[4] Brenner does not derive his theory in the context of coarse graining. The α–model is a coarse graining along Lagrangian trajectories.

$$\tilde{u} \equiv \hat{u} + \frac{1}{3}\left(\frac{\Delta x}{2}\right)^2 \frac{\widehat{\rho_x \hat{u}_x}}{\hat{\rho}} \tag{3.23}$$

with eq. 4.6 of Greenshields and Reese's description of Brenner's theory applied to shock profiles [26]

$$u_v = u_m + \alpha \frac{1}{\rho} \nabla \rho, \tag{3.24}$$

which implies a particular (flow dependent) form for the postulated constant α of Brenner's volume transport.

Next, note that the specific internal energy is defined as the difference of total specific energy and total specific kinetic energy. That is, averaging Eq. (3.6) and applying the closure theorem to the product $\tilde{\rho} u^2$ yields

$$\hat{\rho} \tilde{\mathcal{I}} = \hat{\tilde{\mathcal{E}}} - \frac{1}{2}\hat{\rho} \tilde{u}2 - \mathcal{S}, \tag{3.25}$$

where

$$\mathcal{S} \equiv \frac{1}{2}\left(\widehat{\rho u^2} - \hat{\rho}\tilde{u}^2\right) \approx \frac{1}{6}\left(\frac{\Delta x}{2}\right)^2 \hat{\rho}\tilde{u}_x^2. \tag{3.26}$$

In words, $\mathcal{S}$ represents *unresolved* kinetic energy, which turns out to be $\mathcal{O}(\Delta x^2)$. I note parenthetically that in numerical codes, it is the quantity

$$\mathcal{U} \equiv \mathcal{I} + \frac{\mathcal{S}}{\hat{\rho}} \tag{3.27}$$

that is calculated and used in the equation of state. I will postpone sorting out the effects of this inaccuracy to Section 3.7.

In the derivation of the α–model [19], the kinetic energy (see Eq. 2.21) is postulated to have the form

$$E_\alpha = \int \left[\frac{1}{2}|u|^2 + \frac{\alpha^2}{2}\nabla u^2\right] d^3 x. \tag{3.28}$$

Here, "u is the spatially filtered Eulerian fluid velocity" [19]. Comparison with Eq. (3.26) then allows a tentative relation between the length scales α and Δx.

From a macroscopic point of view (i.e., from the point of view of an observer whose instruments can only resolve motion on scales greater than Δx), the cascade of fluid motions to scales smaller than Δx in the finite scale equtions appears to dissipate macroscopic kinetic energy. However, from the continuum viewpoint, this unresolved kinetic energy must continue its cascade down in length scale until it is small enough to be dissipated by viscosity, and so does not affect the thermodynamic pressure immediately. This situation has been long recognized in the turbulence simulation community where an additional partition of the total energy, termed turbulence kinetic energy or TKE, is recognized, and a transport equation for turbulent kinetic energy is often used to represent its evolution; see, for example, the discussion in [58] (124–128).

More generally in numerical simulations, the relation of internal energy and temperature must be treated carefully, especially when reactions and other processes that are thermally activated are involved.

The dependence of the finite scale fluxes on Δx is not unexpected, but if the Navier–Stokes equations were truly an expansion in Knudsen number, one should expect to see Δx appear *in the denominator* of the fluxes. Instead, Eqs. (3.19) and (3.20) show that the small quantities that allow the expansion are the gradients of the macroscopic variables. This is consistent with the intent of Navier–Stokes to describe small departures from equilibrium, but not with CE theory.

On the other hand, the dependence of the lowest order finite scale fluxes on Δx^2 rather than Δx results from the closure theorem (3.29). Inspection of the proof of the closure theorem [45] in turn shows it is a simple consequence of the isotropy of space. In the derivation of the finite scale equations (3.16)–(3.20) terms of order $\mathcal{O}(\Delta x^4)$ and higher would appear, but have been truncated. In addition, as shown in [45], terms of order $\mathcal{O}(\Delta t^2)$ and higher may be expected due to averaging over the observer time scale Δt. In the next section, I will briefly describe the basis of the closure theorem; a detailed derivation can be found in the cited references.

3.3.3 Closure Theorem

The proof in [45] focuses on Burgers's equation and so treats only fluid velocity. However, the new terms in the earlier equations arise from the nonlinearity of the underlying equations, and in particular of the advective terms that are similar for incompressible and compressible Navier–Stokes. The compressible finite scale equations as derived in [43] and more briefly described earlier follow from the following general *closure* theorem, presented in 1D as follows, but readily extended to three spatial dimensions and to the time domain.

Closure Theorem: *For any (continuum) fields A and B that are sufficiently smooth at small scales,*

$$\widehat{AB} = \widehat{A}\,\widehat{B} + \frac{1}{3}\left(\frac{\Delta x}{2}\right)^2 \widehat{A_x}\widehat{B_x} + HOT, \qquad (3.29)$$

where $A_x \equiv \frac{\partial A}{\partial x}$, HOT are higher-order terms, for example, $\mathcal{O}(\Delta x^4)$, and averaging (overbars) are defined as in Eq. (3.14).

The derivation begins with the assumption that at sufficiently small length scales, physical viscosity ensures that the flow field may be expanded in a convergent Taylor series. In this case, the nonlinear advective terms may be explicitly evaluated as a series in even powers of Δx. Equation (3.29) is readily verified for Δx in this regime.

For larger values of Δx, the proof proceeds with a combination of induction and renormalization. That is, I assume that Eq. (3.29) holds for some large value of Δx and prove that the same form holds for $2\Delta x$. In this stage, it is important to recognize that the meaning of the overbar depends on the averaging length scale, and so one is effectively changing variables in the renormalization. The invariance of the parametric dependence of the equations is termed form invariance. Details of the proof can be found in [45] where the time domain is considered and in [46, 27] where multiple space dimensions are considered.

The step-by-step process of renormalization has an interesting physical interpretation. In Section 3.2.1 the mean and the relative (peculiar) velocities were introduced. The total fluxes, of mass, momentum, and energy, at a point can be decomposed into two terms, the advective flux due to the mean velocity and the diffusive flux due to the relative velocity. In the process of renormalization, the mean velocity changes, that is, it is smoothed because the renormalized velocity at each stage is the average of two velocities at the smaller scale. The conservation laws ensure that the *sum* of the advective and diffusive fluxes at each stage remains unchanged.

In the conventional view, the advective fluxes are purely flow dependent whereas the diffusive fluxes depend on a material dependent diffusion coefficient. In the finite scale view, as one progresses to larger averaging scales, the diffusive flux "increases" at the expense of the advective flux; this is reflected in the form of the finite scale fluxes in Eqs. (3.19) and (3.20) where the diffusion constants are the velocity gradients – that is, are not material dependent.

From a numerical point of view I note that similar arguments led Richtmyer to the analogous form of artificial viscosity in [62]. In this unpublished report, Richtmyer reasoned that all discrete shocks should have the same width as measured in the number of computational cells and was led to the quadratic form later published in [76]. The equivalence of advection and diffusion was also exploited by Smolarkiewicz in the construction of the NFV methodology MPDATA [70]. These and other connections of finite scale theory to numerics are discussed in Section 3.5.

From a historical point of view, the proof of the closure theorem follows earlier work on enslaved difference schemes. In [34], Don Jones, Andrew Poje, and I posed the question: "How can a coarsely resolved calculation of Burgers's equation most closely reproduce a more finely resolved calculation?" Our answer used a numerical renormalization that, in the sense of modified equation analysis, is the discrete analog of (3.29).

3.4   Properties of Finite Scale Equations

Truth is in the eye of the observer.

(With apologies to David Hume.)

In this section, I will discuss some properties of the finite scale equations from the viewpoints of fluxes and energy dissipation. One immediate observation is that since the finite scale (3.16)–(3.18), depend explicitly on the observer Δx, so must their solutions. This simple observation has several interesting consequences. In regimes where the observer length scale is much larger than the physical length scales of the process, the observer will see fluxes, energy dissipation, fluctuations, and so on that are independent both of the material properties of the fluid and of the small scale details of the flow, yet which may dominate the flow. This shows that the finite scale equations have properties previously discussed in the literature such as enslavement [18] and inviscid dissipation [15]. In addition, the finite scale equations require a generalization of the ideas of Lagrangian volumes and of local thermodynamic equilibrium for finite-sized systems.

3.4.1 Fluxes

I will start by offering a physical explanation for the origin of the finite scale fluxes. The finite scale equation for density, Eq. (3.16), is a gradient equation without sources. This implies that any finite volume of fluid whose boundary is moved with the momentum velocity $\tilde{u}$ has a constant mass. This does not preclude the exchange of mass with neighbors, for it is only the net flux of mass integrated around the boundary of the volume that vanishes. Indeed, the velocities of individual parcels[5] contribute to the average momentum velocity but each is, in general, different. Note, however, that each parcel of fluid carries its own momentum and energy, and the surface integrals of these fluxes do not in general vanish, leading to the finite scale fluxes.

From the form of the finite scale fluxes in Eqs. (3.19) and (3.20), several points should be noted. First, there is a kind of Onsager relation between the coefficients of finite scale momentum diffusion and energy diffusion. This reflects that the collection of parcels responsible for transfering momentum in and out of a finite volume is the same collection that transfers energy.

Second, both the finite viscosity and thermal conductivity are proportional to density as distinct from Newtonian viscosity or Fourier heat conduction. For Newtonian viscosity, a kinetic argument that dates back to Maxwell (see [60] for a standard exposition) shows the viscous coefficient is proportional to the product of density and molecular mean free path; as the latter is inversely proportional to density, the coefficient is independent of density. In the case of the finite scale viscosity, mean free path is replaced by the observer who is independent of the flow.

Third, the finite scale fluxes allow countergradient diffusion of both momentum and energy locally, unlike Newtonian viscosity or Fourier heat conduction. Since the finite scale equations contain dissipation on the scale of the observer, it is appropriate that they generate fluctuations on that scale as well. Do hydrodynamic fluctuations play a role in the simulation of macroscopic fluid flows? Deardorff [14] notes the importance of this countergradient flux in modeling the turbulent transport in the planetary boundary layer. Countergradient diffusion also appears in Stull's theory of *transilient turbulence* [71] which attempts to account for nonlocal mixing in turbulent atmospheres due to transport by large eddies. A related idea (in spectral space) is stochastic backscatter, that is, the inverse cascade of energy from small wavenumbers to larger. Backscatter is especially important in geostrophic (two-dimensional) turbulence where

---

[5] I use the term "parcel" to indicate a volume of fluid much smaller than the finite volume under consideration, but still large enough to be considered a continuum.

the net energy dissipation may be the small difference of the forward and inverse cascade of energy [20]. As backscatter does not appear explicitly in LES models, it must be added as part of the explicit subgrid scale model [39].

The ratio of the finite scale flux of momentum to the classical flux is a dimensionless indicator of the importance of finite scale effects. From Eq. (3.17), the ratio

$$\widetilde{R} \equiv \frac{\Delta x^2}{12\eta} \frac{\partial u}{\partial x} \qquad (3.30)$$

is a Reynolds number. Large values of $\widetilde{R}$ imply the dominance of the finite scale fluxes. The presence of Δx implies that turbulence (a phenomenon associated with high Reynolds number) is not simply a property of a flow, but rather it is a property of the flow and the observer. This has numerical ramifications. There have been several recent attempts to extract an "effective Reynolds number" from ILES simulations using the Euler equations where physical viscosity is omitted [3, 82]. From the point of view of finite scale theory, it is true that different grid resolutions of the same problem represent different physical problems. This is illustrated in [51] where the energy dissipation histories of one particular problem simulated at different resolutions are compared for a variety of Reynolds numbers. It is shown there that the dissipation histories for small $\widetilde{R}$ (where physical viscosity is important) converge as a function of resolution. Similarly, it is shown that dissipation histories for large $\widetilde{R}$ also converge as a function of resolution. This emphasizes that ILES and the finite scale equations not only are valid for turbulent flows, but are valid for all fluid flows.

But now consider a different question. Given a particular high Reynolds problem to be simulated at some chosen resolution, how does the problem setup depend on the Reynolds number? The answer is that if the resolution is not sufficient to resolve viscous effects, the problem setup *does not depend* on Reynolds number, and in that case neither does the simulation. Perhaps more to the point, one needs to know the magnitude of physical viscosity a priori to decide whether it is relevant for a particular problem.

One final numerical note. The fact that the solutions of the finite scale equations, Eqs. (3.16)–(3.18), depend explicitly on Δx has implications for convergence testing, a popular method for estimating the order of the error in numerical algorithms as a function of cell size. Most often, error is evaluated in terms of the difference of an analytic solution and the numerical solution at a point. However, in the case of finite volume schemes, it is more appropriate to compare the integral of the analytic solution over the cell with the numerical solution. In problems with steep gradients, for example, with shocks, the difference can be significant. Issues associated with convergence testing are discussed in more detail in [49].

3.4.2   Energetics

From Eq. (3.18) the global rate of energy dissipation by the finite scale fluxes in a one-dimensional domain $\mathcal{D}$ is

$$\frac{dE_I}{dt} = -\frac{1}{2} \int_{\mathcal{D}} \left[\frac{1}{3} \left(\frac{\Delta x}{2} \right)^2 \widetilde{u} \left(\overline{\rho} \, \widetilde{u}_x \right)^2 - \overline{\rho} \, \eta \, \widetilde{u}_x^2 \right] dx. \qquad (3.31)$$

Integrating by parts and neglecting surface terms (work done by external forces) yields

$$\frac{dE_I}{dt} = \frac{1}{6} \left(\frac{\Delta x}{2} \right)^2 \langle \overline{\rho} \widetilde{u}_x^3 \rangle - \eta \langle \overline{\rho} \widetilde{u}_x^2 \rangle. \tag{3.32}$$

Here the brackets indicate spatial integration over the entire domain.

The length scale associated with viscous dissipation, usually termed the Kolmorogov microscale, is defined by

$$L_K \equiv \left(\frac{\eta^3}{\varepsilon} \right)^{1/4} \tag{3.33}$$

where $\varepsilon = \frac{\partial}{\partial t} \left(\overline{u}^2 \right)$ is the average rate of dissipation of kinetic energy per unit mass. In situations where $\Delta x \gg L_K$, the first term in Eq. (3.32) dominates and essentially all the energy dissipation is *inviscid*, that is, independent of η. In fact, the observed rate of energy dissipation in turbulent flows is found to be independent of the molecular viscosity. Kolmogorov's 1941 theory contains the basic assumption that there is a finite value of ε in the limit as $\eta \to 0$. These ideas are discussed in a mathematical context by Eyink in [15] where it is suggested that the "velocity field at small length scales must be sufficiently irregular to support a finite energy dissipation: in particular, it must be non-differentiable." In both Kolmogorov's theory (4/5ths law) and Eyink's later results, $\varepsilon \sim \Delta u^3$, similar to the finite scale results in Eq. (3.32) in the limit of vanishing viscosity [16].

However, finite scale theory does not deal with the velocity field at small length scales, but with an average velocity. In the finite scale picture, resolved kinetic energy turns into unresolved kinetic energy, for example, eddies smaller than the observer that continue the cascade to viscous length scales. Indeed, the closure theorem itself assumes the smoothness of the velocity field at small scales as a starting point of the renormalization. However, the similarity of results for the form of ε suggests that the two viewpoints are not necessarily contradictory. I will come back to these different points of view in Section 3.6. I note that there are similar results for shock theory, namely that the energy dissipated in a shock and associated entropy production are both proportional to Δu^3, and independent of molecular viscosity [4]. In other words, the rate of energy dissipation at small scales in high Reynolds number flows are *controlled* by the large scales of the flow, which is an expression of the idea of enslavement [18].

In [46], it is demonstrated that ILES simulations of incompressible decaying turbulence satisfy the 4/5ths law. Although there is no known analog of Kolmogorov's 4/5ths law for compressible fluids in Navier–Stokes theory [2], Eq. (3.32) represents the analog for compressible finite scale theory. In the case of shocks, it is well established that in numerical simulations of the Euler equations, the shock speed and jump conditions across the shock do not depend on the magnitude of the artificial viscosity, so long as it is sufficiently large to regularize the shock and sufficiently small to ensure numerical stability. Here another caveat must be added, that the shock is steady in time. Since the effect of artificial viscosity is to spread the shock over a fixed number of grid points, steadiness implies a uniform grid that in turn implies a fixed "observer." Simulation of shock propagation on a nonuniform grid was studied in [47], illustrating the importance of this caveat.

3.5   Numerical Considerations

Finite scale theory began in [45] with my attempt to develop a theoretical basis for the numerical technique now termed implicit large eddy simulation (ILES) [27]. ILES is a methodology that greatly simplifies the numerical simulation of turbulent fluid flows. In the years prior to 2000, a few authors had documented the ability of a particular class of numerical methods to simulate turbulent flow without need for any explicit subgrid scale model (artificial viscosity). These, the nonoscillatory finite volume (NFV) algorithms, have been successfully applied in diverse simulations of geophysical problems, aerodynamics, astrophysics, combustion, hydrodynamic instabilities, and so on. Despite this success, and even though the underlying NFV schemes are widely used in computational fluid dynamics, ILES is only slowly being accepted into the mainstream of the turbulence modeling community. Many of the theoretical ideas of the previous section have analogs in numerical methods. Here I will review some aspects of numerical methods relevant to finite scale theory and in particular to ILES. A recounting of my own involvement in ILES will follow in Section 3.5.2.

3.5.1   Nonoscillatory Finite Volumes

NFVs represent an amalgam of some of the most basic ideas of numerical approximation. Historically, this development begins with the work of von Neumann and Richtmyer [76] dating to the earliest days of computer simulation. Physical shock profiles are typically much too narrow to be accurately represented on computational grids and it is found that direct simulation of Navier–Stokes equations in flows with shocks shows large and unphysical oscillations in the postshock region. However, it is well known that the shock speed and the Hugoniot relations across a physical shock are independent of viscosity. The essential idea of [76] is the introduction of an "artificial viscosity" that smears the shock over several computational grid points and allows the proper energy dissipation and entropy increase. Artificial viscosity is the numerical analog of inviscid dissipation.

The form of this artificial viscosity appears magically in [76], although it is motivated to some degree in earlier internal Los Alamos reports by Peierls [57] and Richtmyer [62]. It has qualitatively the same form as the finite scale term in the momentum flux in Eq. (3.17), particularly in its dependence on Δx^2. This same form, with appropriate extensions for multidimensional application [79], remains in common use today where it is employed only in regions of fluid compression. A generic name for artificial viscosity methods is shock capturing. In the Russian literature, these are termed homogenized schemes [73].

An important, but largely forgotten addition to the theory of artificial viscosity was published by Bill Noh [56]. Noh studied a troublesome phenomenon in the numerical simulation of shocks termed wall heating, designed a still popular test problem to illustrate the problem, and recognized that wall heating could be mitigated by an "artificial heat conduction." However, it turned out that his particular form of the artificial heat conduction had the undesirable side effect of dissipating contact discontinuities and so far no more effective formulation has been found.

The extension of artificial viscosity to turbulent fluids first occurred in the field of weather prediction, sometime in the early 1950s. Joseph Smagorinsky realized that a method was needed to account for atmospheric turbulence that occurred on scales smaller than the computational grid, but still played an important role in the atmospheric energy cycle. The Smagorinsky turbulence model [68], the first large eddy simulation (LES) model, is a tensor extension of the von Neumann–Richtmyer artificial viscosity. This relation is not accidental, for earlier Smagorinsky had worked with Jule Charney and John von Neumann and was familiar with artificial viscosity concepts. Smagorinsky recounts the relation of his turbulence model to artificial viscosity in [69].

The origin of finite volume methods [40] is probably due to Peter Lax in another early Los Alamos report [35]. In any case, these ideas were in widespread use at both Los Alamos and Livermore Laboratories by the late 1950s. In the classic Volume 3 of Methods in Computational Physics (1964), almost all the authors are using finite volume approximations. The individual articles by Bill Noh [56] and Mark Wilkins [79] contain explicit descriptions.

The basic elements of finite volume methods are the use of integral forms of the PDEs and the application of Green's theorem to convert divergence terms to surface integrals. For two computational cells that share a common surface (edge in 2D), the flux into one cell is exactly the flux out of the other. This idea of detailed balance allows exact conservation of mass, momentum, and energy to the level of roundoff error. By contrast, finite difference methods only conserve to the level of truncation error.

The introduction of nonoscillatory methods revolutionized computational fluid dynamics. This part of the story begins with an important theorem proved by Sergei Godunov in his 1954 PhD thesis [24, 77]. The theorem states:

Linear numerical schemes for solving partial differential equations having the property of not generating new extrema (monotonicity preserving schemes), can be at most first-order accurate.

Preservation of monotonicity and high-order (or at least second-order) accuracy are both desirable properties. However, first-order schemes are too diffusive for most practical applications and the eighteen or so years after Godunov, numerical methods focused on accuracy at the expense of allowing unphysical oscillations. It was Jay Boris who first realized how to overcome the barrier of Godunov's theorem – *to give up linearity*. Here, linearity means that the numerical approximations are the same for every grid point. In the flux corrected transport (FCT) algorithm [7], Boris and David Book describe a method that mixed first- and second-order methods in such a way as to preserve monotonicity. The FCT algorithm by construction uses flow dependent coefficients to approximate the transport terms and so is not linear in the sense mentioned earlier. It is also not quite second-order accurate by standard measures, but can be made very high order in smooth regions of the flow. It is well known [42] that shock capturing methods can only be of first order near shocks.

The strategy for avoiding Godunov's barrier spread quickly in the computational fluid dynamics (CFD) community. Bram van Leer [75] developed a methodology based on Godunov's method (another important contribution by Godunov). Soon there was an alphabet soup of methods and today nonoscillatory algorithms are the mainstream for simulating high Reynolds number fluid dynamics. Some important properties shared by all NFV schemes are nonlinear stability (under appropriate time step constraints); exact conservation of mass, momentum, and energy; and absence of nonphysical oscillations. Further, NFV schemes are parameter free and relatively easy to implement.

It appears that most if not all NFV methods are suitable for ILES. A review of several methods and examples of applications is included in our book [27]. Of the many NFV methods available, I was particularly attracted to the MPDATA schemes of Piotr Smolarkiewicz [70]. I thought they had a good physical basis, were easy to modify and augment, and in particular were tractable for mathematical analysis. It was part of our long standing collaboration that Smolarkiewicz and I discovered ILES in the context of MPDATA calculations of the atmospheric boundary layer [50]. The basic formulation of MPDATA is not monotonicity preserving, but has the weaker constraint of positivity preservation. It is fully second order and can be easily made higher-order accurate. It is also, in the vernacular, spatially unsplit, meaning that it is more accurate in simulations where the computational grid does not reflect the symmetry of the flow. MPDATA can be made optionally montonicity preserving, either by coupling to FCT [70] or through the use of gauge transformations [48].

Many of the papers cited earlier have literally thousands of citations. Here are two papers that are undeservedly less well known, but which I found important. Tony Hirt noted as early as 1969 [30] that subgrid scale for turbulence models are of the same order of magnitude as the truncation errors of second-order accurate methods. The importance of this message was apparently missed by the turbulence modeling community until Sandip Ghosal's rediscovery in 1996 [22]. Predictably, the mainstream community reacted by striving to develop more accurate hydro models.

Here, again, Jay Boris was the first to make the crucial connection, that numerical truncation errors could in fact serve as an LES model. At an influential meeting of theoretical and numerical turbulence modelers convened by John Lumley in 1989, Boris states: "I do not believe it is practical to separate the formulation of the LES problem from the numerical method used for its solution." By this time, Boris had already discovered the ability of his FCT models to provide implicit models of turbulence.

But what properties would a numerical method require to act as an implicit turbulence model? The lack of an underlying theory for ILES was a barrier to the general acceptance of implicit turbulence methods by the turbulence modeling community. An essential clue was provided in another largely ignored paper by Marshall Merriam in 1987. In this paper [54] Merriam showed the connection of the numerical property of monotonicity to the second law of thermodynamics.

The origins of implicit turbulence modeling are not in doubt at all. Some time before 1989, Jay Boris had discovered this seemingly magical property of FCT algorithms.

He termed it MILES, for monotonically integrated large eddy simulation. Unlike his introduction of nonoscillatory difference methods, MILES met active resistance from the established turbulence modeling community, which nevertheless did not keep him from advocating and publishing his methodology. Jay has written detailed accounts of his MILES research included as chapter 1 of our book [27], and also a more general retrospective of FCT in [6]. In the next section, I will give my own account of how Piotr Smolarkiewicz and I discovered the ILES property of MPDATA.

"The most exciting phrase to hear in science, the one that heralds new discoveries, is not 'Eureka!' but 'That's funny...'" Isaac Asimov

3.5.2 An Early History and Rationalization of ILES

In 1998, neither Piotr nor I was aware of MILES and the discovery of the ILES property of our MPDATA code came as a complete surprise to us. For several years, Piotr Smolarkiewicz and I had been developing a numerical model to demonstrate that NFV methods like MPDATA could improve the quality of numerical simulations of atmospheric phenomena. In this period in the late 1980s, centered-in-time methods were preferred in the atmospheric modeling community. Although the meteorological community had been responsible for introducing many of the important ideas in numerical methods, they had not yet appreciated the advantages of nonoscillatory approximations as described in the previous subsection. As it turned out, Piotr and I had also not recognized all of those advantages.

By 1998, Piotr and I had demonstrated the ability of MPDATA to simulate large scale atmospheric processes and had turned our attention to smaller scale phenomena. The atmospheric boundary layer is a turbulent boundary layer that forms due to the daytime heating of the ground by the sun. A typical thickness of the layer is 1,000 m. This phenomena is particularly attractive for numerical study as there is both a large collection of experimental data and many other numerical studies with which to compare it.

We anticipated that a subgrid scale turbulence model was required for these studies and adopted the model of Piotr's colleague, Zbigniew Sorbjan. Sorbjan's model was an elaboration of Smagorinsky's original model with several adjustable parameters. We implemented this model and were pleased to get almost immediate agreement with both data and previous results. But as we modified the model parameters to optimize our agreement, we made a crucial mistake. In one run, we accidentally turned the model off, and yet still produced a result in excellent agreement with the data. We scoured the code over and over, and finally convinced ourselves: MPDATA was acting as an implicit and adaptive turbulence model.

We didn't publish right away, but I started to give talks on our results and soon learned two lessons. First, I generated the same negative response from the turbulence community as Boris had encountered. But on the positive side, I learned about MILES. Bill Rider has an encyclopedic knowledge of the CFD literature, and after attending one of my talks introduced me to MILES. Further, Bill also identified two other colleagues who had been using the MILES idea, albeit more circumspectly.

Paul Woodward, one of the originators of an early NFV method named PPM for piecewise parabolic method [12], and colleagues applied ILES to astrophysical problems. This is a regime of highly compressible flow with extremely high Reynolds numbers. In [59], the authors presented a 512^3 (highly resolved for this time period) simulation of decaying supersonic flow, demonstrating that their ILES simulations could reproduce analytical properties of turbulence, including an inertial range with Kolmogorov's $-5/3$ slope. At about the same time, David Youngs and colleagues [41] were applying van Leer methods to modeling the growth of turbulent regions and the mixing resulting from fluid instabilities, including Rayleigh-Taylor and Kelvin-Helmholtz. These calculations contain adjacent regions of very high and very low Reynolds number, illustrating another very useful feature of ILES simulation – that the same fluid solver can be used for smooth and for turbulent flows. Further, the growth in wavelength and amplitude of the instabilities demonstrates the ability to upscale the energy, for example, stochastic backscatter.

Gaining confidence from these previous publications, Piotr and I published our paper [50], writing in the abstract: "Other researchers have reported similar success, simulating turbulent flows in a variety of regimes while using only nonoscillatory advection schemes to model subgrid effects. At this point there is no theory justifying this success." But the fact that four different NFV methods, FCT, PPM, van Leer, and MPDATA, all worked as implicit turbulence models was an essential clue. That subgrid scale turbulence was the same order of magnitude as second-order truncation terms [30] suggested the next step of doing a modified equation analysis for each of the four NFV schemes and looking for the common term. However, when I finished the analysis of MPDATA first, the answer was already clear. The modified equation of MPDATA contained the von Neumann-Richtmyer artificial viscosity.

In 2001 I was invited to a special session on VLES (very large eddy simulation) in an ECCOMAS conference in Swansea, United Kingdom. Bill Rider and I used this opportunity to develop a rationale for ILES that was based on showing the similarity of evolution equations for finite volumes of fluid and the modified equations of NFV methods. This paper [45] combined renormalization techniques with inductive logic to derive the finite scale equations for Burgers's equation; the same technique was generalized in subsequent papers for incompressible [46] and compressible [43] Navier–Stokes. Renormalization methods were familiar to me from my graduate school studies of quantum field theory. The interpretation of the results in terms of an observer comes from my fascination with Heisenberg and uncertainty principles. The closure theorem and my concept of form invariance were original, and I hope will be subject to more mathematical scrutiny in the future.

My thesis advisor, Frank Harlow, once cautioned me that in presenting new ideas, one should as much as possible find common ground with the audience. The field of LES, that is, using *explicit* subgrid scale models, has been very successful and there are many such models derived, published, and employed in the literature [63]. It occurred to me that the finite scale momentum flux term in Eq. (3.19) should resemble at least one of the standard models, and indeed that is the case. The particular model is termed the tensor diffusivity model, first derived by Leonard [81]. However, it was found in computational experiments that this model was not sufficiently dissipative. When supplemented by a Smagorinsky type viscosity the modified model, now appropriately termed the hybrid model, has proved very successful in the turbulence modeling community [81]. The connection of ILES to the more conventional explicit subgrid scale models has aided in the further acceptance of ILES.

3.6   A Finite Scale Formulation at Molecular Length Scales

"It seems to me that right and wrong are fuzzy concepts."

Isaac Asimov

In Section 3.2 I raised some questions about Chapman–Enskog (CE) theory in the high Reynolds number limit. The Navier–Stokes equations are understood to be a first-order correction in Knudsen number to the equilibrium Euler equations. However, the Knudsen number does not appear either in the viscous terms or in the heat conduction. More precisely, the macroscopic scale length, denoted here as Δx, that appears in the denominator of the Knudsen number does not appear in these terms nor more generally in the Navier–Stokes equations at all.

On the other hand, Navier–Stokes equations were formulated empirically more than seventy years before Chapman and Enskog. The forms of Newtonian viscosity and Fourier heat conduction do not depend on CE, but are more fundamentally based in experiment. Recent interest in microfluidics has exposed some inadequacies in the Navier–Stokes description when it is applied to problems beyond its regime of validity [31], for example, when the physical scales of interest approach the molecular mean free path. For engineering applications on small scales, new hybrid models that combine Navier–Stokes with kinetic models are being developed [55, 61] that bypass the question of transport coefficients by directly calculating the transport flux. Still, such models are complex and computationally expensive; it is of interest to explore whether simpler modifications of Navier–Stokes theory can be effective in the regime of small length scales.

A particular problem that is of interest in the compressible flow community is the calculation of the width and profile shape of strong shocks. This is a problem where Navier–Stokes theory is known to fail, both with respect to experimental results [1] and on theoretical [74] grounds. In contrast, more fundamental calculations beginning with Boltzmann's equation show very excellent agreement with the data [1]. This indicates that the problem lies in CE theory and so is fertile ground to explore.

The lack of scale separation in turbulence is what distinguishes the applicability of finite scale theory from the inapplicability of eddy viscosity. A similar issue will arise in considering hydrodynamic models at length scales of the order of the mean free path. In the rest of this section I will present an alternate formulation of the finite scale theory that is more appropriate for small scales. At these scales, individual realizations of the flow will be noisy and I will want to deal with the averages of the flow that will be smoother. This transition implies that I must discard the Taylor series methods for characterizing the velocity fluctuations and employ a stochastic approach that ties more directly to kinetic theory by using the velocity probability distributions. In addition, this approach will enable an answer to how the pressure depends on finite system size. I will begin by contrasting two well-known procedures in the field of turbulence modeling, mixing length theory and Reynolds decomposition. The former corresponds to a viscosity (diffusive) theory that cannot be expected to be valid at small length scales. The latter represents a more general approach to calculating transport coefficients for a second-order hydrodynamics.

3.6.1 Eddy Diffusivity in Turbulence Models

The original ideas of eddy diffusivity models date back to giants such as Osborne Reynolds, Joseph Boussinesq, and Ludwig Prandtl. The original references to that early work can be found in standard textbooks on turbulence, for example, [21, 58, 72].

In Reynolds decomposition, the fluid velocity at a point is written as the sum of two parts, a mean velocity and a fluctuating velocity.

$$u(x,t) = \overline{u}(x,t) + u'(x,t). \tag{3.34}$$

The mean velocity $\overline{u}$ is defined as an ensemble average, but is more practically realized as either a spatial average or a temporal average depending on the situation at hand. Inserting this decomposition into the NS momentum equation and applying the averaging operator leads directly to an evolution equation for the mean velocity. In this equation, a new contribution to the pressure appears, the Reynolds stress tensor

$$\tau_{ij} \equiv -\overline{\rho u_i' u_j'}. \tag{3.35}$$

In order to close the mean momentum equation, it is necessary to express τ_{ij} in terms of the mean velocity and other mean quantities. Boussinesq suggested the idea now known as eddy viscosity, namely that

$$\tau_{ij} \equiv -\overline{\rho u_i' u_j'} \approx K\rho \left(\frac{\partial \overline{u}_i}{\partial x_j} + \frac{\partial \overline{u}_j}{\partial x_i} \right), \tag{3.36}$$

where the eddy diffusivity K is the product of a typical length scale and a typical velocity. Later, Prandtl developed a mixing length theory that provides a more physical picture and quantitative estimate of the eddy diffusivity.

These eddy diffusivity models, also termed gradient transport models, have been successfully used in many practical turbulence calculations. One example widely used in atmospheric calculations is the Smagorinsky model [68]. However, deficiencies in that model have been identified in *a priori* tests [11] where model predictions of the Reynolds stress itself were compared with a highly refined direct numerical simulation (DNS). Several papers generally critical of the use of gradient transport for turbulence calculations have been written. For example, Corrsin [13] and Schmitt [65] each note that the gradient transport theory is only justifiable when the characteristic macroscopic scales (Δx) are much bigger than the transporting mechanism, a situation that is almost never justified in turbulence simulations. Corrsin writes [13]:

it has long been realized that a gradient transport model requires (among other things) that the characteristic scale of the transporting mechanism (mean free path in gas kinetics: Lagrangian velocity integral time scale multiplied by root-mean-square velocity, in turbulence) must be small compared with the distance over which the mean gradient of the transported property changes appreciably.

Corrsin's comment is interesting for its transposition: if the CE derivation of the viscosity coefficient depends somehow on a gradient transport assumption, then it will not be justifiable on the scale of the molecular mean free path. This appears to be the case in the first-order (in Knudsen number) CE approximation; the particular solution of the first-order integral equation resulting from the CE expansion is *assumed* to depend linearly on the velocity and temperature gradients; see, for example, pages 90–91 in [28] or pages 85–86 in [35].

3.6.2 An Alternate Perspective

A single measurement of a shock profile will be *noisy*. A representative noisy profile may be generated numerically by solving the Landau–Lifshitz modifications of Navier–Stokes in which fluctuating terms in the momentum and energy equations are added [35]. However, the average of an ensemble of such simulations will yield a *smooth* profile; see, for example, figure 10 in [80]. Similarly, an ensemble of measurements will also yield a *smooth* profile; see, for example, figure 1 in [1]. The noisy profile is the "sufficiently irregular" situation that Eyink describes [15], cf. section 3.4.2. However, the smooth profile resulting from ensemble averaging is not the Navier–Stokes solution. This failure is well known and is illustrated (among many places) in figure 10 of [64] where the NS predicted shock width is seen to be about half the measured shock width.[6] This figure also shows the much better agreement that solutions directly based on the Boltzmann equation achieve. Similar agreement is seen in figure 4 of Alsmeyer [1].

The goal in this section is to develop the equations that describe the ensemble averaged smooth solution. The philosophy of finite scale theory is well suited for developing such coarse grained equations; however, the process described in Section 3.3.3 must be modified as the underlying velocity field here will not be smooth enough to justify convergent Taylor series expansions. Instead, I will employ the more general formalism of the Reynolds decomposition described in the previous section.

Start by considering a point x and an interval of (at the moment unspecified) width $\mathcal{L}$ centered on x. Divide the interval into $M \gg 1$ equal subintervals. Inside each sub-interval $j; j = 1, M$ there is a velocity distribution characterized by a PDF $f_j(x_j, c, t)$

---

[6] Note that it is inverse shock width that is plotted in this figure.

where x_j is the midpoint of subinterval j and the width of each subinterval is $\delta x = L/M$. As defined in Eq. (3.2), in each subinterval j there is the density

$$\rho_j = m \int f_j(x_j, c, t)\, dc, \tag{3.37}$$

and as in Eq. (3.3) the momentum density

$$\rho_j u_j = m \int c f_j(x_j, c, t)\, dc. \tag{3.38}$$

Then in analogy to Eq. (3.14), define

$$\widehat{\rho}(x) \equiv \frac{1}{M} \sum_{j=1,M} \rho_j. \tag{3.39}$$

Also, recalling the discussion in Section 3.3.2 where Favre averaging is shown to be the more useful averaging method for compressible flows (i.e., it does not introduce higher-order correlations into the continuity equation), let us define the average velocity in analogy with Eq. (3.15):

$$\overline{\rho}\widetilde{u} \equiv \frac{1}{M} \sum_{j=1,M} \rho_j u_j. \tag{3.40}$$

Then the fluctuating u_j' in the subinterval j is defined by

$$u_j' \equiv \widetilde{u} - u_j. \tag{3.41}$$

Now the discussion in Section 3.2.1 shows that the balance laws, Eqs. (3.7)–(3.9), are valid for these subinterval averages at every point; this validity results purely from the averaging over the velocity phase space. For example, the momentum Equation (3.8) in the subinterval j is written:

$$\frac{\partial \rho_j u_j}{\partial t} + \frac{\partial \rho_j u_j^2}{\partial x} = -\frac{\partial p_j}{\partial x}, \tag{3.42}$$

where the pressure from Eq. (3.10) is

$$p_j = m \int C^2 f_j(x_j, c, t)\, dc). \tag{3.43}$$

Finally, let us average the subinterval equations over the total interval L in coordinate space to derive equations for average velocity. This is most easily accomplished by substituting for $\widetilde{u}_j$ using Eq. (3.41) and then summing over j. The result for the momentum equation is:

$$\frac{\partial \widehat{\rho}\widetilde{u}}{\partial t} + \frac{\partial \widehat{\rho}\widetilde{u}^2}{\partial x} = -\frac{\partial}{\partial x}\left(p - \widehat{\rho}\widetilde{u'u'}\right). \tag{3.44}$$

The new term that appears on the right hand side, $\tau_{ij}^m = \left(\overline{\rho\, \widetilde{u'u'}} \right)$, is a *molecular Reynolds stress*.

There are several points to discuss about Eq. (3.44). Most significantly,

- There is a close analogy with the finite scale momentum flux (3.17), where the term $\frac{1}{3}\left(\frac{\Delta x}{2}\right)^2 \widehat{\rho}\, \widetilde{u}_x^2$ appears. In the case of smooth flow, $u' \approx \frac{\partial u}{\partial x}\delta x$. However, Eq. (3.44) is not restricted to smooth flow.

- If the distribution function f is a function of the size of the system, then the pressure in Eq. (3.44) is also a function of system size.

- It will be necessary to show that the averaging process does not depend on how the interval is subdivided – that is, on the value of M.

- Inserting the Reynolds decomposition into the energy equation will lead to a similar molecular heat transport term, analogous to that in the finite scale energy equation (3.20).

- One essential difference between the molecular Reynolds stress and its diffusive approximation is that the former introduces fluctuations into the flow while the latter is always dissipative; cf. the discussion in Section 3.4.1.

- A quantitative model for the molecular Reynolds stress on length scales of the mean free path depends ultimately on the form of the PDF $f_j(x_j, c, t)$. This is the subject of the next section.

Many items in this list point out the similarity of the development of finite scale equations at macro- and microscales. There is one important difference to emphasize as well, that being in the definition of the averaging process. In Section 3.3.3, the averaging length Δx was associated with the idea of the observer and could be specified arbitrarily. However, the constraint that the ensemble averaged result is a *smooth flow* places a lower limit on the averaging length. Qualitatively, as suggested by Corrsin, that minimum length scale should be of the order of the molecular mean free path.

3.7 Coarsening the Velocity Distribution

The critical step in constructing a hydrodynamic model is developing an approximate solution of the Boltzmann equation for the velocity distribution function. The CE theory uses a perturbation approach based on Knudsen number that I have argued is not correct for high Reynolds number fluid flows. Here I propose an alternate strategy based on form invariance. The basic idea is illustrated in Figure 3.1. There at the top of the figure are two 1D cells of equal "volumes" δx. The coarsening process consists of merging these two cells into one larger cell of volume $2\delta x$. Form invariance requires that the *functional form* of the velocity distribution is preserved. For example, if each of the two small cells in Figure 3.1 is Gaussian, then the homogenized cell should be Gaussian as well.[7] In the following, I will show that form invariance requires a

---

[7] In the stochastic calculus, form invariance corresponds to the idea of stable distributions. A stable distribution is infinitely divisible. See [33], 50–51.

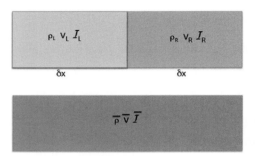

Figure 3.1 The coarsening process consists of merging two cells with possibly different values of state variables density, velocity and internal energy into a single larger cell. Values of the state variables in the larger cell are constrained, but not completely determined by conservation of mass, momentum, and energy.

perturbation of the Gaussian form that accounts for unresolved kinetic energy and so includes an embedded length scale.

3.7.1   The Coarsening Process

In the case of local thermodynamic equilibrium (LTE), the velocity distribution function is the Gaussian

$$f^{eq} = \rho \sqrt{\frac{1}{2\pi \mathcal{I}}} \exp - \frac{(v - \bar{v})^2}{2\mathcal{I}}. \tag{3.45}$$

The thermodynamic state of a cell is described by three hydrodynamic variables, for example, density, bulk velocity, and internal energy. For the two smaller cells in Figure 3.1, these are denoted $(\rho L, vL, \mathcal{I}_L)$ and $(\rho R, vR, \mathcal{I}_R)$, where the subscripts L and R indicate left and right, respectively. Then the corresponding hydrodynamic variables of the coarsened cell, $(\widehat{\rho}, \widehat{v}, \widehat{\mathcal{I}})$, are completely determined by conservation of mass, momentum, and energy. In particular,

$$\widehat{\rho} = \frac{1}{2}(\rho L + \rho R)$$

$$\widehat{\rho v} = \frac{1}{2}(\rho L vL + \rho R vR) \tag{3.46}$$

$$\widehat{\rho}\left(\widehat{\mathcal{I}} + \frac{1}{2}\widehat{v}^2\right) = \frac{1}{2}\left(\rho_L\left[\mathcal{I}_L + \frac{1}{2}v_L^2\right]\right) + \frac{1}{2}\left(\rho_R\left[\mathcal{I}_R + \frac{1}{2}v_R^2\right]\right).$$

Note, however, that information has been lost in this mapping. To have a unique inverse mapping, three additional coarsened variables are needed:

$$\delta\rho = \rho_R - \rho_L$$
$$\delta v = v_R - v_L \tag{3.47}$$
$$\delta\mathcal{I} = \mathcal{I}_R - \mathcal{I}_L.$$

Now Eqs. (3.46) and (3.47) can be inverted:

$$\rho_L = \widehat{\rho} - \frac{1}{2}\delta\rho \quad ; \quad \rho_R = \widehat{\rho} + \frac{1}{2}\delta\rho$$

$$v_L = \widehat{v} - \frac{1}{2}\delta v\left(1 + \frac{\delta\rho}{2\widehat{\rho}}\right) \quad ; \quad v_R = \widehat{v} + \frac{1}{2}\delta v\left(1 - \frac{\delta\rho}{2\widehat{\rho}}\right)$$

$$\mathcal{I}_L = \widehat{\mathcal{I}} - \frac{1}{2}\delta\mathcal{I}\left(1 + \frac{\delta\rho}{2\widehat{\rho}}\right) - \frac{1}{4}\delta v^2 \tag{3.48}$$

$$\mathcal{I}_R = \widehat{\mathcal{I}} + \frac{1}{2}\delta\mathcal{I}\left(1 - \frac{\delta\rho}{2\widehat{\rho}}\right) - \frac{1}{4}\delta v^2.$$

In this section, I will assume a "near equilibrium" situation where the gradients are small in the sense where

$$\delta\rho << \widehat{\rho} \quad ; \quad \delta v << \sqrt{\widehat{\mathcal{I}}} \quad ; \quad \delta\mathcal{I} << \widehat{\mathcal{I}}. \tag{3.49}$$

Identifying these cells with the subintervals of the previous section, these conditions can be assured by choosing M large enough. (Note that due to the random nature of the forces due to molecular collisions, the velocity may be continuous but is nowhere differentiable.) I will assume all variables in Eq. (3.47) are $\mathcal{O}(\delta x)$, or more briefly $\mathcal{O}(\delta)$.

3.7.2   Noninvariance of LTE

Here I test the form invariance of the LTE. Specifically, is the assumption that one has a Gaussian distribution in each of the two smaller cells of Figure 3.1 consistent at $\mathcal{O}(\delta^2)$ with the assumption of a Gaussian distribution in the coarsened cell? Note that inserting Eq. (3.12) into Eq. (3.5) leads to a simple relation between temperature and internal energy for the case of the equilibrium distribution in 1D,

$$\mathcal{I} = \frac{k\mathcal{T}}{m}.$$

Then in mathematical terms, form invariance of LTE implies the hypothesis

$$\delta x\left[\rho_L f_L^{eq}(v) + \rho_R f_R^{eq}(v)\right] \overset{?}{=} 2\delta x \overline{\rho}\,\overline{f}^{eq}(v)$$

or in 1D

$$\frac{\rho_L}{\sqrt{2\pi\mathcal{I}_L}}\exp-\frac{(v-v_L)^2}{2\mathcal{I}_L} + \frac{\rho_R}{\sqrt{2\pi\mathcal{I}_R}}\exp-\frac{(v-v_R)^2}{2\mathcal{I}_R} \overset{?}{=} \frac{2\overline{\rho}}{\sqrt{2\pi\overline{\mathcal{I}}}}\exp-\frac{(v-\overline{v})^2}{2\overline{\mathcal{I}}}. \tag{3.50}$$

In principle, a strategy for testing Eq. (3.50) is to eliminate the small cell variables, that is, those with subscript L or R, in terms of coarse cell variables, that is, those with hats or with δ by using the relations of Eq. (3.48). Verifying form invariance would require that the LHS and RHS be equal to $\mathcal{O}(\delta^2)$. However, for the purposes of constructing a hydrodynamic theory, it is sufficient to show that LHS and RHS of Eq. (3.50) produce the same low-order moments.

In the case of the equilibrium distribution, it is readily seen that this condition cannot be met. One can manipulate Eqs. (3.46), (3.47), and (3.48) to show:

$$\widehat{\mathcal{I}} = \frac{1}{2}(\mathcal{I}_L + \mathcal{I}_R) + \frac{1}{4}\delta v^2. \tag{3.51}$$

That is, the coarsened internal energy depends on δv^2. In the coarsening process, some resolved kinetic energy has been converted to unresolved kinetic energy. This is in contradiction with the macroscopic result of Eq. (3.25), which shows that the total internal energy does not include the unresolved kinetic energy. The internal energy in Eq. (3.51) contains the unresolved kinetic energy because there is no other place to put it. That is, the three macrosopic variables of the equilibrium PDF are not sufficient to uniquely specify the thermodynamic state of the finite size volume to $\mathcal{O}(\delta^2)$.

3.7.3   Invariance near Equilibrium

Here, I interpret *near equilibrium* to mean a perturbation of the Gaussian of Eq. (3.50), in which the Gaussian is recovered as small macroscopic gradients vanish. Following the ideas of Grad [25], I will focus on the Hermite polynomials. This is a convenient, but not a unique, representation. Noting the left–right symmetry of Eqs. (3.48), it is only necessary to consider even orders of the Hermite series. I will assume that, to $\mathcal{O}(\delta x^2)$, the form of the perturbed velocity distribution is:

$$f^G = \rho\sqrt{\frac{\pi}{2a}}\left[\frac{1 + \beta(v - \widehat{v})^2}{1 + \dfrac{\beta}{2a}}\right] \cdot \exp - a(v - \widehat{v})^2. \tag{3.52}$$

Here, in accordance with near equilibrium, $\beta \sim \mathcal{O}(\delta x^2)$.

Note that the parameter a is no longer simply related to the internal energy. Inserting f^G into Eq. (3.5) and recalling the distinction in Section 3.4.2 between $\mathcal{U}$ and $\mathcal{I}$ leads to

$$\mathcal{U} = \frac{1}{2a}\left(\frac{1 + 3\beta/2a}{1 + \beta/2a}\right) \approx \frac{1}{2a}(1 + \beta/a). \tag{3.53}$$

Clearly, Eqs. (3.52) and (3.53) recover an equilibrium form as $\beta/a \to 0$.

Next let us reconsider the consequences of conservation when the PDF is f^G. Conservation of mass and momentum still imply the first two relations of Eq. (3.46). Also, the equation of total energy conservation, Eq. (3.6), also holds for any form of the PDF. However, at this point it is convenient to separate the internal energy from the unresolved kinetic energy. In conformance with Eq. (3.25), I will define the part of the internal energy that is independent of δx^2 as $\mathcal{I}$. Then, from Eq. (3.53),

$$\mathcal{I} = 1/2a \quad ; \quad \mathcal{S} = \beta/2a^2 = 2\beta\mathcal{I}^2. \tag{3.54}$$

Now the conservation of energy implies, to $\mathcal{O}(\delta x^2)$,

$$\widehat{\rho}\left(\widehat{\mathcal{I}} + 2\widehat{\beta}\widetilde{\mathcal{I}}^2 + \frac{1}{2}\widetilde{v}^2\right) = \frac{1}{2}\left(\rho_L\left[\mathcal{I}_L + 2\beta_L\mathcal{I}_L^2 + \frac{1}{2}v_L^2\right]\right) + \frac{1}{2}\left(\rho_R\left[\mathcal{I}_R + 2\beta_R\mathcal{I}_R^2 + \frac{1}{2}v_R^2\right]\right).$$

$$(3.55)$$

To ensure that internal energy $\mathcal{I}$ remains independent of unresolved kinetic energy, I will break this into two separate relations:

$$\widehat{\rho}\widehat{\mathcal{I}} = \frac{1}{2}(\rho_L\mathcal{I}_L + \rho_R\mathcal{I}_R);$$

$$(3.56)$$

$$\widehat{\rho}\left(2\widehat{\beta}\widetilde{\mathcal{I}}^2 + \frac{1}{2}\widetilde{v}^2\right) = \widehat{\rho}\left(2\widehat{\beta}\widetilde{\mathcal{I}}^2\right) + \frac{1}{4}\rho_L v_L^2 + \frac{1}{4}\rho_R v_R^2.$$

Note that in the second relation, I have ignored $\delta\mathcal{I}$ and $\delta\rho$ in the coefficients of all the β terms; their presence would only lead to terms higher than second order. Also, I have set $\beta_L = \beta_R = \beta$ since $\delta\beta = \beta_R - \beta_L$ is higher than second order. The results for the inverse relations assuming f^G:

$$\rho_L = \widehat{\rho} - \frac{1}{2}\delta\rho \quad ; \quad \rho_R = \widehat{\rho} + \frac{1}{2}\delta\rho$$

$$v_L = \widehat{v} - \frac{1}{2}\delta v\left(1 + \frac{\delta\rho}{\widehat{\rho}}\right) \quad ; \quad v_R = \widehat{v} + \frac{1}{2}\delta v\left(1 - \frac{\delta\rho}{\widehat{\rho}}\right)$$

$$\mathcal{I}_L = \widehat{\mathcal{I}} - \frac{1}{2}\delta\mathcal{I}\left(1 + \frac{\delta\rho}{\widehat{\rho}}\right)$$

$$(3.57)$$

$$\mathcal{I}_R = \widehat{\mathcal{I}} + \frac{1}{2}\delta\mathcal{I}\left(1 - \frac{\delta\rho}{\widehat{\rho}}\right)$$

$$\widehat{\beta}\widetilde{\mathcal{I}}^2 = \widehat{\beta}\widehat{\mathcal{I}}^2 - \frac{\delta v^2}{8}.$$

It remains to specify the new parameter β and to verify it is $\mathcal{O}(\delta x^2)$.

3.7.4   Unresolved Kinetic Energy and Pressure

Let us next consider the source of the δ fluctuations. First, there are the thermal fluctuations, which are the continuum representation of the irregularity of the collision process. These fluctuations are stochastic variables. Second, there are the *sure* gradients of density, velocity, and internal energy. When these gradients are of $\mathcal{O}(\Delta)$, their effects have been considered in Section 3.3, whereas here I consider the effects of strong gradients at $\mathcal{O}(\delta)$.

In the particular problem of simulating shock structure, the thermal fluctuations of velocity as measured by $\sqrt{\mathcal{I}}$ and the velocity jump across the shock are of the same magnitude and so are likely coupled. Here, to simplify the development I will assume that the thermal fluctuations are well described (in the sense of the smooth averaged solutions that are sought) by the standard diffusional viscosity. This will allow a more straightforward treatment of the effects of the large gradients in widening the shock profile and providing a hydrodynamic model that is closer to the data. This decoupling is not meant to be an assumption for the long term.

If the δ terms are associated only with the sure variables of the gradients, then the β coefficient can be readily determined, either by solving the recursion relation of Eq. (3.57) or by requiring conservation of total kinetic energy for all resolutions. However, a simpler and more satisfying derivation follows from showing that the macroscopic form of the unresolved kinetic energy, Eq. (3.26), satisfies the recursion relation derived from the kinetic description. Note from Figure 3.1 that $\widetilde{\beta}$ corresponds to a cell of width $2\delta x$ and β to a cell of width δx. Then $\delta v = u_x \delta x$. Recalling Eq. (3.53), assume

$$\mathcal{S}(2\delta x) \approx \frac{1}{6}\left(\frac{2\delta x}{2}\right)^2 \widehat{\rho}\,\widetilde{u}_x^2 = 2\widetilde{\beta}\mathcal{I}^2$$

$$\mathcal{S}(\delta x) \approx \frac{1}{6}\left(\frac{\delta x}{2}\right)^2 \widehat{\rho}\,\widetilde{u}_x^2 = 2\beta\mathcal{I}^2. \tag{3.58}$$

Then subtracting the second equation from the first reproduces the recursion relation and verifies that $\beta \sim \mathcal{O}(\delta^2)$.

The calculations of this section tie nicely to the macroscopic results, but are more constructive as they also help elucidate the role of unresolved kinetic energy in the constitutive law for pressure. The unresolved kinetic energy per unit volume due to gradients, small or large, simply contributes directly to the pressure. It should be emphasized that it is necessary to distinguish the two terms $\mathcal{U}$ and $\mathcal{I}$. I note that in CFD codes, it is $\mathcal{U}$ that is calculated and so no changes need be made in those codes for the equation of state. However, when temperature dependent reaction processes are coupled to the hydrodynamics, temperature must be derived (via the specific heat) from $\mathcal{I}$, not $\mathcal{U}$.

The calculation of the molecular Reynolds stress that appears in Eq. (3.44) is straightforward, and in fact exactly mimics the macroscopic calculation when the length scale Δx is replaced by $\mathcal{L}$. Since this term only incorporates the effect of sure gradients, the fluctuating velocities can be calculated from those gradients leading to

$$\tau_{ij} = \frac{1}{3}\widehat{\rho}\,\widetilde{u_i'u_j'} = \frac{1}{3}\left(\frac{\mathcal{L}}{2}\right)^2 u_x^2. \tag{3.59}$$

Thus, the remarkable result of this careful treatment of the kinetics underlying a hydrodynamic theory is that the form of the finite scale equations, with a linear and quadratic viscosity, applies as well at small scales. To clarify, this result depends on the simplifying assumption that the effects of sure gradients and random thermal fluctuations can be separated. Also, a more quantitative estimate of the length scale $\mathcal{L}$ is required. For consistency, this estimate should be based in continuum theory, for example on correlation lengths, rather than on discrete dynamics.

3.8   Finite Scale Shock Width

One of the reasons the similar form of finite scale theory at small scales is remarkable is the existence of theory and computational capability for these equations. Most CFD codes for compressible flow include both linear and quadratic artificial viscosity and so are easily adapted to numerical studies at small scales. Two recent theoretical papers describe the self-similar solution for a piston driven finite scale shock. I will use the results of one of these papers to estimate the dependence of shock width on Mach number. The results of this section are qualitative, meant only to show that finite scale theory produces wider shocks more in line with measurements than does Navier–Stokes theory.

In [52] Diane Vaughan and I analyzed the problem of a piston pushing into a stationary gamma law gas with velocity u_p. Those solutions included the effects of viscosity, but neglected the (important) effects of heat conduction. The following solution for the velocity gradient is given as eq. (23) in [52]:

$$\frac{du}{dy} = \frac{\eta}{2A} - \frac{\sqrt{\eta^2 + 2(\gamma + 1)Au(u_p - u)}}{2A} \tag{3.60}$$

where η is the kinematic viscosity, $y = x - vt$ is the similarity coordinate, and $A = \mathcal{L}^2/12$ is the coefficient of quadratic viscosity. The velocity v is the shock speed and $\mathcal{L} = \phi\ell$ is a to-be-determined multiple of the mean free path length ℓ. The denominator of 12 in A represents only the advective contributions to the quadratic viscosity.

I will define the shock width as:

$$W = u_p / \left(\frac{du}{dy}\right)_{1/2}. \tag{3.61}$$

In words, the width is the jump in velocity across the shock divided by its slope at the midpoint. This is typical although other definitions have been used. When $u = \frac{1}{2}u_p$,

$$\left(\frac{du}{dy}\right)_{1/2} = \frac{\eta}{2A} - \frac{\sqrt{\eta^2 + \frac{1}{2}(\gamma + 1)Au_p^2}}{2A}.$$

Also, a simple estimate of viscosity is

$$\eta \approx \frac{1}{3}v_{rms}\ell; \tag{3.62}$$

see, for example, [60], 475. Here the coefficient 1/3 should not be taken too literally. Alsmeyer [1] invokes a similar equation, but with the coefficient $\approx .606$, nearly twice as large.

The rms velocity is the sound speed at the center of the shock and is proportional to the square root of the internal energy: $v_{rms} = \sqrt{\gamma(\gamma - 1)\mathcal{I}}$. The internal energy at the midpoint of the shock can be found from eq. (18) of [52] but this is an implicit relation and complicated. I will write more generically

$$v_{rms} \sim \mathcal{F}(M)u_p$$

where $\mathcal{F}$ is a function only of Mach number and the parameter γ. $\mathcal{F}$ is monotonically decreasing and tends to a constant of order unity in the limit of large Mach number.

Putting it all together,

$$W = \frac{4Au_p}{\gamma + 1} \left[\eta + \sqrt{\eta^2 + \frac{\gamma + 1}{2} Au_p^2} \right]$$

$$= \frac{\ell}{2} \left(\mathcal{F} + \sqrt{\mathcal{F}^2 + \phi^2} \right)$$

(3.63)

where γ has been set $= 5/3$. When $\phi = 0$, the width is the Navier–Stokes result. Since $\mathcal{F} \geq 1$, the Navier–Stokes width is always at least one mean free path. But real shocks are wider than one mean free path. This may be accounted for by the choice of the averaging length and underscores the importance of developing a theory for this quantity.

Two caveats should be mentioned about the finite scale solution in Eq. (3.60). First, the effects of temperature are not represented in this solution. Heat conduction is not included in the equations and the viscosity η is assumed independent of temperature. However, the temperature dependence of viscosity is included in the estimate of Eq. (3.62). Thus the analysis in this section may suffer from some inconsistency.

Second, the linear viscosity in the finite scale equations is assumed proportional to density. This choice is consistent with the form of linear artificial viscosity as used in CFD codes. However, physical viscosity is essentially independent of density, a result first recognized by Maxwell. In a recent publication [35], Jordan and Kieffer derive the finite scale solution for the more physical form of linear viscosity. This seeming simple change in the governing equations turns out to significantly complicate the analysis. One interesting result is the emergence of a critical value of the averaging length scale with respect to the mean free path. From dimensional analysis, the physical viscosity must involve some reference density that implies a length scale originating in the collision physics. Thus, this critical length scale may be related to the minimal averaging length that I termed L.

In unpublished results, Jordan (personal communication) investigated the difference in shock width between the two solutions using the two different linear viscosity formulations. In general, the widths differed by an order of 10 to 15% with the differences getting smaller as Mach number increases.

One final remark on Eq. (3.63); the length scale ℓ is not an absolute length scale, but depends on density and (weakly) on temperature. One cannot assess the dependence of the width of the shock in absolute terms from this equation.

3.9   Conclusions

"It is wrong to think that the task of physics is to find out how nature is. Physics concerns what we can say about nature."

Niels Bohr

The finite scale equations are an attempt to make mathematically more precise the role of the observer in the classical physics of fluid flow. My characterization of the observer is a length (or time) scale that is independent of fluid and of the flow, but which enters the evolution equations as a measure of the uncertainty of a theoretical or computational prediction. In short, the observer quantifies the "we" in the section opening quote by Bohr.

The ideas of finite scale theory began with my attempt to understand and analyze the numerical technique of ILES in which the truncation errors of particular numerical algorithms serve as an implicit turbulence model for high Reynolds number simulations. Bohr's quote also renders moot the philosophical question of whether "nature" is continuous or discrete. What we can say about nature is the result of discrete measurements. One might then say that it is the discrete equations that are *truth* and the PDEs that are the approximation. However, PDEs are easier to analyze and also they are unique. As described in Section 3.5, many different NFV methods can be used for ILES. Each gives a slightly different result, but all are qualitatively the same and quantitatively indistinguishable. The finite scale equations as a model of the discrete equations synthesizes this lack of ambiguity.

This chapter consists both of a review of previous results and of new material. In Section 3.3, I reviewed the finite scale theory for a compressible Navier–Stokes fluid and discussed their differences with respect to classical Navier–Stokes theory, both in the interpretation of the conserved variables and in the form of the flux terms. There I also gave a brief description of how the finite scale equations are derived. The derivation follows simply from a closure theorem proved in several previous references [45, 46, 27] that is, I believe, mathematically rigorous. In Section 3.4, I discussed some of the properties of the solutions of the finite scale equations. These properties encompass ideas such as inviscid dissipation and fluctuations that have been discussed previously, but that are not easily discernible in the Navier–Stokes equations.

In Section 3.5, I attempted to collect and briefly describe the most relevant ideas from numerical methods that together led first to the methodology of ILES and then to the finite scale equations. This section is in no way intended to be a complete textbook on numerical methods. Some of these ideas, for example, shock capturing and finite volume approximation, were implemented in algorithms at the very beginning of numerical simulation on computers. Some are more recent. In my opinion, the development of nonoscillatory methods in the early 1970s not only revolutionized the capability to simulate high Reynolds number fluid flows, but also opened a new perspective for theoretical treatment of nonlinearity in those flows.

There is one aspect of the original derivation of the finite scale equations that has troubled me, namely that the process begins with the Navier–Stokes equations. In earlier work, I suggested that there was no new physics in the finite scale equations versus Navier–Stokes, just a change of variables [44]. However, this assertion is not consistent with the principle of form invariance. Sections 3.2, 3.6, and 3.7 are meant to rectify that concern. Beginning with Boltzmann's equation, I redefined the averaging process in Section 3.6 in terms of the velocity distribution function. In Section 3.7, I used the principle of form invariance to derive a *near equilibrium* velocity distribution that explicitly includes the finite size of the system – I would term this a "discrete thermodynamics." The gratifying result of this section is that the equations of motion for smooth (ensemble–averaged) flow at molecular length scales have the same form as the finite scale equations at macroscopic scales, though the choice of the averaging length now requires some additional modeling.

There is a broad theme woven in this paper, namely the close connection between physics and successful numerical algorithms. In the world of CFD, this relationship is termed mimetic differencing. Many ideas, such as finite volume approximation discussed in Section 3.5.1, compatible differencing [49, 67], and discrete calculus [53], have been developed for the sake of better spatial approximations. The work by Merriam [54] mentioned in Section 3.5.1 shows how the second law of thermodynamics is built into nonoscillatory approximations, an example of mimetic differencing in the time domain. These and other ideas have been readily accepted and have moved smoothly from physics and mathematics to numerical methods. On the other hand, ILES is an example of an idea moving in the opposite direction, from numerics to physics, and shows how awkward that transition can be.

There is a second, understated theme in this paper that associates the concept of the observer with uncertainty. It is my experience that numerical predictions are often much more accurate than is estimated by standard procedures such as truncation analysis or by statistical methods such as forward propagation of errors. ILES and shock capturing techniques provide two compelling examples. The physical properties of inviscid dissipation and enslavement of the small scales of motion by the resolved scales express the idea that in many situations the details don't matter. Which situations and which details depend on the question being asked.

3.10   Dedication

It is with both pleasure and sadness I dedicate this chapter to my friend and colleague Howard Brenner. Howard's extraordinary career spanned more than sixty years, filled with accomplishment and creative ideas to the very end. Howard died in February 2014, but his legacy is still being written. "Creativity takes courage" (Henri Matisse).

3.11   Acknowledgments

Many people have taken the time to discuss these new ideas with me, especially Tim Clark, Dave Higdon, Daniel Israel, David Nicholaeff, Jon Reisner, and Ray Ristorcelli. Their insights have been valuable, although one should not necessarily assume their acceptance or agreement with my point of view. I am grateful to Pedro Jordan, Jason Reese, and Francisco Uribe for continuing discussions of the intriguing questions involving strong shock structure. This work was performed under the auspices of the U.S. Department of Energy's NNSA by the Los Alamos National Laboratory operated by Los Alamos National Security, LLC under contract number DE-AC52-06NA25396.

References

[1] H. Alsmeyer, 1976, "Density profiles in argon and nitrogen shock waves measured by the absorption of an electron beam," *J. Fluid Mech.* 74, 497–513.

[2] H. Aluie, 2013, "Scale decomposition in compressible turbulence," *Physica D* 247, 54–65.

[3] A. Aspden, N. Nikiforakis, S. Dalziel, and J. B. Bell, 2008, "Analysis of implicit LES methods," *Comm. App. Math. Comp. Sci.* 3, 103–126.

[4] H. Bethe, 1942 [1998], "On the theory of shock waves for an arbitrary equation of state," Technical Report NDRC-B-237, office of Scientfic Research and Development. Reprinted in *Classic Papers in Shock Compression Science*, Springer-Verlag.

[5] J.P. Boris, 1990, "On large eddy simulation using subgrid turbulence models," in *Whither Turbulence? Turbulence at the Crossroads*, ed. by J.L. Lumley, Springer-Verlag, 344–353.

[6] J.P. Boris, 2013, "Flux-corrected transport looks at forty," *Computers & Fluids* 84, 113–126.

[7] J.P. Boris, D.L. Book, 1973: "Flux-corrected transport: 1. SHASTA, A fluid transport algorithm that works," *J. Comp. Phys.* 11, 38–69.

[8] H. Brenner, 2005, "Kinematics of volume transport," *Physica A* 349, 11–59.

[9] C. Cercignani, 2000, *Rarefied Gas Dynamics*, Cambridge Texts in Applied Mathematics.

[10] S. Chapman and T.G. Cowling, 1991, *The Mathematical Theory of Non-uniform Gases: An Account of the Kinetic theory of Viscosity, Thermal Conduction and Diffusion in Gases*, Cambridge University Press.

[11] R.A. Clark, J.H. Ferziger, W.C. Reynolds, 1979, "Evaluation of subgrid scale models using an accurately simulated turbulent flow," *J. Fluid Mech.* 91, 1–16.

[12] P. Colella and P.R. Woodward, 1984, "The piecewise parabolic method (PPM) for gas-dynamical simulations," *J. Comput. Phys.* 54, 174–201.

[13] S. Corrsin, 1974, "Limitations of gradient transport models in random walks and turbulence," *Adv. Geophys.* 18 A, 25–60.

[14] J.W. Deardorff, 1972, "Theoretical expression for the countergradient vertical heat flux," *J. Geophysical Res.* 77, 5900–5904.

[15] G.L. Eyink, 1994, "Energy dissipation without viscosity in ideal hydrodynamics," *Physica D* 78, 222–240.

[16] G.L. Eyink, 2003, "Local 4/5-law and energy dissipation anomaly in turbulence," *Nonlinearity* 16, 137–145.

[17] A. Favre, 1983, "Turbulence: Spacetime statistical properties and behavior in supersonic flows," *Phys. Fluids* 26, 2851–2863.

[18] C. Foias, G.R. Sell, R. Temam, 1988, "Inertial manifolds for nonlinear evolutionary equations," *J. Differential Eqs.* 73, 309–353.

[19] C. Foias, D.D. Holm, E.S. Titi, 2001, "The Navier–Stokes–alpha model of fluid turbulence," *Physica D* 152, 505–519.

[20] J.S Frederiksen and A.G. Davies, 1997, "Eddy viscosity and stocastic backscatter parameterizations on the sphere for atmospheric circulation models," *J. Atmos. Sci.* 54, 2475–2492.

[21] U. Frisch, 1998, *Turbulence*, Cambridge University Press.

[22] S. Ghosal, 1996, "An analysis of numerical errors in large-eddy simulations of turbulence," *J. Comput. Phys.* 125, 187–206.

[23] J. Gleick, 2008, *Chaos: Making a New Science*, Penguin Books.

[24] S.K. Godunov and K. Sergei, 1954, "Different Methods for Shock Waves," PhD Dissertation, Moscow State University.

[25] H. Grad, 1958, "Principles of the kinetic theory of gases," in *Handbuch der Physik Vol. XII*, ed. by S. Flugge, Springer, 205–294.

[26] C.J. Greenshields and J.M. Reese, 2007, "The structure of shock waves as a test of Brenner's modifications to the Navier-Stokes equations," *J. Fluid Mech.* 580, 407–429.

[27] F.F. Grinstein, L.G. Margolin, and W.J. Rider, 2007, *Implicit Large Eddy Simulation: Computing Turbulent Fluid Dynamics*, Cambridge University Press.

[28] S. Harris, 1971, *An Introduction to the Theory of the Boltzmann Equation*, Holt, Rinehart and Winston,.

[29] C.W. Hirt, 1968, "Heuristic stability theory for finite difference equations," *J. Comput. Phys.* 2, 339–355.

[30] C.W. Hirt, 1969, "Computer studies of time-dependent turbulent flows," *Phys. Fluids Supplement II*, 219–227.

[31] C-M. Ho and Y-C. Tai, 1998, "Micro-electro-mechanical systems (MEMs) and fluid flows," *Ann. Rev. Fluid Mech.* 30, 579–612.

[32] D.D. Holm, J.E. Marsden, and T.S. Ratiu, 1998, "Euler–Poincaré models of ideal fluids with nonlinear dispersion," *Phys. Rev. Lett.* 80, 4173–4176.

[33] K. Itô, 1957, *Essentials of Stochastic Processes*, American Mathematical Society.

[34] D.A. Jones, L.G. Margolin, and A.C. Poje, 2002, "Accuracy and nonoscillatory properties of enslaved difference schemes," *J. Comput. Phys.* 181, 705–728.

[35] P.M. Jordan and R.S. Kieffer, 2015, "A note on finite-scale Navier–Stokes theory: the case of constant viscosity, strictly adiabatic flow," *Phys. Letters A* 379, 124–130.

[36] G.M. Kremer, 2010, *An Introduction to the Boltzmann Equation and Transport Processes in Gases*, Springer.

[37] L.D. Landau and E.M. Lifshitz, 1959, *Fluid Mechanics*, Course of Theoretical Physics Vol. 6, Addison-Wesley, chapter 17.

[38] P.D. Lax, 1952, "On discontinuous initial value problems for nonlinear equations and finite differences," Los Alamos Scientific Laboratory report LAMS-1332.

[39] C.E. Leith, 1990, "Stochastic backscatter in a subgridscale model," *Phys. Fluids A* 2, 297–299.

[40] R.J. LeVeque, 2002, *Finite Volume Methods for Hyperbolic Problems*, Cambridge University Press.

[41] P.F. Linden, J.M. Redondo, and D.L. Youngs, 1994, "Molecular mixing in Rayleigh-Taylor instability," *J. Fluid Mech.* 265, 97–124.

[42] A. Majda and S. Osher, 1977, "Propagation of error into regions of smoothness for accurate difference approximations to hyperbolic equations," *Comm. Pure Appl. Math.* 30, 671–705.

[43] L.G. Margolin, 2009, "Finite-scale equations for compressible fluid flow," *Philosophical Transactions of the Royal Society A* 367, 2861–2871.

[44] L.G. Margolin, 2014, "Finite scale theory: The role of the observer in classical fluid flow," *Mech. Res. Comm.* 57, 10–17.

[45] L.G. Margolin and W.J. Rider, 2002, "A rationale for implicit turbulence modeling," *Int. J. Num. Meth. Fluids* 39, 821–841.

[46] L.G. Margolin, W.J. Rider, and F.F. Grinstein, 2006, "Modeling turbulent flow with implicit LES," *J. Turbulence* 7, 1–27.

[47] L.G. Margolin, H.M. Ruppel, and R. B. Demuth, 1985, "Gradient Scaling for Nonuniform Meshes," *Proc. 4th International Conference on Numerical Methods in Laminar and Turbulent Flow*, The University of Wales, Swansea, 1477.

[48] L.G. Margolin and M. Shashkov, 2005, "MPDATA: Gauge transformation, limiters and monotonicity," *Int. J. Num. Meth. Fluids* 50, 1193–1206.

[49] L.G. Margolin and M. Shashkov, 2007, "Finite volume methods and the equations of finite scale: a mimetic approach," *Int. J. Num. Methods Fluids* 56, 991–1002.

[50] L.G. Margolin, P.K. Smolarkiewicz, and Z., Sorbjan, 1999, "Large eddy simulations of convective boundary layers using nonoscillatory differencing," *Physica D* 133, 390–397.

[51] L.G. Margolin, P.K. Smolarkiewicz, and A.A. Wyszogradzki, 2006, "Dissipation in implicit turbulence models: A computational study," *J. Appl. Mech.* 73, 469–473.

[52] L.G. Margolin and D.E. Vaughan, 2012, "Traveling wave solutions for finite scale equations," *Mech. Res. Comm.* 45, 64–69.

[53] C. Mattiussi, 1997, "An analysis of finite volume, finite element and finite differences methods using some concepts from algebraid topology," 133, 289–309.

[54] M.L. Merriam, 1987, "Smoothing and the Second Law," *Comp. Meth. Appl. Mech. Eng.* 64, 177–193.

[55] X.B. Nie, S.Y. Chen, W.N. E, and M.O. Robbins, 2004, "A continuum and molecular dynamics hybrid method for micro- and nano-fluid flow," *J. Fluid Mech.* 500, 55–64.

[56] W.F. Noh, 1964, "CEL: a time-dependent two-dimensional coupled Eulerian-Lagrangian code," in *Methods in Computational Physics*, ed. B.J. Alder. Academic Press, 117–180.

[57] R.E. Peierls, 1945, "Theory of von Neumann's method of treating shocks," Los Alamos Scientific Laboratory report LA-332.

[58] S. B. Pope, 2000, *Turbulent Flows*, Cambridge University Press, 124–128.

[59] D.H. Porter, A. Pouquet, and P.R. Woodward, 1994, "Kolmogorov-like spectra in decaying three-dimensional supersonic flows," *Phys. Fluids* 6, 2133–2142.

[60] F. Reif, 1965, *Fundamentals of Statistical and Thermal Physics*, McGraw-Hill, 475–476.

[61] W. Ren and Weinan E, 2005, "Heterogeneous multiscale method for the modeling of complex fluids and micro–fluidics," *J. Comput. Phys.* 204, 1–26.

[62] R.D. Richtmyer, 1948, "Proposed numerical method for calculation of shocks," Los Alamos Scientific Laboratory report LA-671.

[63] P. Sagaut, 2006, *Large Eddy Simulation for Incompressible Flows*, Springer Verlag.

[64] B. Schmidt, 1969, "Electron beam density measurements in shock waves in argon," *J. Fluid Mech.* 39, 361–373.

[65] F.G. Schmitt, 2007, "About Boussinesq's turbulent viscosity hypothesis: Historical remarks and a direct evaluation of its validity," *C.R. Mecanique* 335, 617–627.

[66] X. Shan, X-F. Yuan, and H. Chen, 2006, "Kinetic theory representation of hydrodynamics: A way beyond the Navier-Stokes equation," *J. Fluid Mech.* 550, 413–441.

[67] M. Shashkov, 1996, *Conservative Finite Diference Methods on General Grids*, CRC Press.

[68] J. Smagorinsky, 1963, " General circulation experiments with the primitive equations I. The basic experiment," *Mon. Wea. Rev.* 91, 99–164.

[69] J. Smagorinsky, 1983, "The beginnings of numerical weather prediction and general circulation modeling: early recollections," *Adv. Geophys.* 25, 3–37.

[70] P.K. Smolarkiewicz and L.G. Margolin, 1998, "MPDATA: A finite-difference solver for geophysical flows," *J. Comput. Phys.* 140, 459–480.

[71] R.B. Stull, 1991, "Review of non-local mixing in turbulent atmospheres: transilient turbulence theory," *Boundary-Layer Meteorology* 62, 21–96.

[72] H. Tennekes and J.L. Lumley, 1972, *A First Course in Turbulence*, MIT Press.

[73] A.N. Tikhonov and A.A. Samarskii, 1962, "Homogeneous difference schemes," *USSR Computational Mathematics and Mathematical Physics* 1, 5–67.

[74] F.J. Uribe, 2011, "The shock wave problem revisited: The Navier–Stokes equations and Brenner's two velocity hydrodynamics," in *Coping with Complexity: Model Reduction and Data Analysis*, ed. by A.N. Gorban and D. Roose, Springer–Verlag, 207–233.

[75] B. van Leer, 1979, "Towards the ultimate conservative difference scheme 5: 2nd-order sequel to Godunov's method," *J. Comput. Phys.* 32, 101–136.

[76] J. von Neumann and R.D. Richtmyer, 1950, "A method for the numerical calculation of hydrodynamic shocks," *J. Appl. Phys.* 21, 232–237.

[77] P. Wesseling, 2001, *Principles of Computational Fluid Dynamics*, Springer-Verlag.

[78] M.L. Wilkins, 1964, "Calculation of Elastic-Plastic Flow," in *Methods in Computational Physics*, ed. B.J. Alder, Academic Press, 211–263.

[79] M.L. Wilkins, 1980, "Use of artificial viscosity in multidimensional fluid dynamic calculations," *J. Comput. Phys.* 36, 281–303.

[80] S.A. Williams, J.B. Bell, and A.L. Garcia, 2008, "Algorithm refinement for fluctuating hydrodynamics," *Multiscale Model. Simul.* 6, 1256–1280.

[81] G. Winckelmans, A. Wray, O. Vasilyev, and O. Jeanmart, 2001, "Explicit filtering large edddy simulation using the tensor diffusivity model supplemented by a dynamic Smagorinsky term," *Phys. Fluids* 13, 1385–1403.

[82] Y. Zhou, F.F. Grinstein, A.J. Wachtor, and B.M. Haines, 2014, "Estimating the effective Reynolds number in implicit large eddy simulation," *Phys. Rev. E* 89, 013303.

4 Material Conservation of Passive Scalar Mixing in Finite Scale Navier Stokes Fluid Turbulence

J. Raymond Ristorcelli

4.1 Introduction

The finite scale Navier–Stokes (FSNS) equations are the equations of a second-order fluid and represent the equations of motion of a Navier–Stokes (NS) fluid when observed at finite length and time scales. This chapter is concerned with the scalar variance transfers in the stirring and mixing by an incompressible turbulent flows governed by the FSNS equations. If the FSNS equations are interpreted as a coarse grained grid dependent form of NS, then the FSNS fluid can be related to diverse LES [1] and ILES [2] strategies used in the numerical simulation of turbulent flows.

Our purpose is to understand the statistics of the stirring and mixing that arise when filtering or averaging procedures are applied to NS. To this end we introduce a new application of the well-known 1895 idea of Osborne Reynolds [3], and derive the moment equations of the FSNS fluid. In particular, the Reynolds averaging procedure is used to illustrate the exchanges in positive definite variance type quantities in passive scalar mixing. We focus in this paper on the evolution of the moments of a materially conserved (*modulo diffusion*) passive scalar field. This serves to make clear – with a minimum of mathematical complexity – the properties of mixing in FSNS turbulence. We show that the moment equations of FSNS passive scalar mixing have the same conservation property as NS.

To broaden the scope of applicability of the moment analysis we also analyze an isotropic eddy viscosity model (EVM) and contrast its flow physics with that of NS and FSNS turbulence. Although the shortcomings of EVM are well known [4], our analysis provides a new perspective of the essential differences and their consequences between FSNS/ILES and EVM. We show that the moment equations of EVM passive scalar mixing do not have the same conservation property as NS.

Our moment analysis addresses several well-known coarse grained strategies. Combinations of the EVM and FSNS results allow one to assess the "nonlinear" LES model [6], the mixed model [5, 13, 1], and the rational approximation model [7, 1]. In addition, terms in the FSNS have the same form as some truncation terms in certain finite volume (FV) codes and thus the FSNS analysis addresses the ILES strategies [8] in which it is proposed that the leading numerical truncation terms accurately account for unresolved fluid physics.

4.2   The FSNS Fluid

Generalizing the spatial averaging results of [8] to three-dimensional tensor coordinates and adding temporal averaging, the instantaneous finite scale equations for the transport of a passive scalar and momentum by an incompressible fluid ($u_{j,j} = 0$) to $\mathcal{O}(L^4,,T^4)$ are

$$\frac{\partial}{\partial t}\,\overline{c} + \overline{u}_j\overline{c}_{,j} = D\overline{c}_{,jj} - L^2\left[\overline{u}_{j,k}\overline{c}_{,k}\right]_{,j} - T^2\left[\overline{u}_{j,t}\overline{c}_{,t}\right]_{,j} \tag{4.1}$$

$$\frac{\partial}{\partial t}\,\overline{u}_i + \overline{u}_j\overline{u}_{i,j} = -\overline{p}_{,i} - L^2\left[\overline{u}_{j,k}\overline{u}_{i,k}\right]_{,j} - T^2\left[\overline{u}_{j,t}\overline{u}_{i,t}\right]_{,j} + v\overline{u}_{i,kk} \tag{4.2}$$

$$-\nabla^2\overline{p} = \overline{u}_{j,i}\overline{u}_{i,j} + L^2\overline{u}_{j,ik}\overline{u}_{i,jk} + T^2\overline{u}_{j,ti}\overline{u}_{i,tj}\,. \tag{4.3}$$

In these equations, commas indicate differentiation, $u_{j,k} \equiv \frac{\partial u_j}{\partial x_k}$, and the overbars indicate spatial and temporal averaging in a fixed (Eulerian) coordinate system over length scale L and time scale T.

Equations (4.1) and (4.2) describe the evolution of the FSNS fluid governed by the equations of a second-order fluid with material parameters L and T. We emphasize that L and T should be considered as user defined "material" parameters reflecting the scale at which the velocity and concentration fields are described. We emphasize that L and T are properties not of the flow, but rather of an "observer" of the flow using finite instruments or employing a computational grid in which case one might identify $L = \Delta x$ and time step $T = \Delta t$.

The FSNS equations represent the same physics as NS; that is, they are derived from NS and they reduce to NS in the limits that L and T become small. The derivations of Equations (4.1) and (4.2) are a direct consequence of a mathematically well-defined process of iterative application of Taylor series and Taylor series inversions to the NS equations. Each successive iterative application of the Taylor series approximation is applied to a new set of equations, which are smooth on the length scale of the flow described by the set of equations of the previous iteration. The FSNS equations are form invariant with respect to these succesive Taylor series approximations and so L and T can become large with respect to the first application of the Taylor series expansion to the original unsmoothed NS equations, which are only valid at the Kolmogorov scales of NS. For more details, see [9].

4.2.1   FSNS and ILES Models

The FSNS equations with the appropriate choice of L and T have been used to justify the ILES strategy [8]. The new terms in Equations (4.1) and (4.2), with $L^2 = const._x(\Delta x)^2$ and $T^2 = const._t(\Delta t)^2$, with constants determined by numerical scheme specifics, represent truncation terms in the modified equation analysis of [10] of typical nonoscillatory finite volume (NFV) methods used for ILES [2]. MEA is a technique for generating the equations whose continuous solution closely approximates the discrete solution of the numerical algorithm underlying the simulation model. MEA was used in the early formal comparisons in [11] between ILES and traditional LES, to show that a class of NFV algorithms with certain leading order truncation terms provides appropriate built-in (implicit) subgrid models of a generalized anisotropic eddy viscosity tensor. Volume integrals in the FV representation naturally link with the discrete spatial filtering operation in LES, and their use recasts leading order truncation terms in divergence form.

4.2.2   FSNS and LES Models

If we view the FSNS equations as the LES equations for a fluid whose microstructure is NS and for which filtering produces the instantaneous subgrid terms, then the residual scale flux implied is

$$q_j = \overline{u_j c} - \overline{u}_j \overline{c} = L^2 \overline{u}_{j,k}\, \overline{c}_{,k} + T^2 \overline{u}_{j,t}\, \overline{c}_{,t} = L^2 \left[\overline{s}_{jk}\, \overline{c}_{,k} + \overline{w}_{jk}\, \overline{c}_{,k} \right] + T^2 \overline{u}_{j,t}\, \overline{c}_{,t}. \quad (4.4)$$

We have used $s_{ij} = \frac{1}{2} \left[u_{i,j} + u_{j,i} \right]$ and $w_{ij} = \frac{1}{2} \left[u_{i,j} - u_{j,i} \right]$. The residual scale momentum flux is

$$\tau_{ij} = \overline{u_i u_j} - \overline{u}_i \overline{u}_j = L^2 \overline{u}_{j,k}\, \overline{u}_{i,k} + T^2 \overline{u}_{j,t}\, \overline{u}_{i,t} \quad (4.5)$$

$$= L^2 \left[\overline{s}_{ij}^2 - \overline{w}_{ij}^2 + \overline{w}_{jk} \overline{s}_{ki} + \overline{w}_{ik} \overline{s}_{kj} \right] + T^2 \overline{u}_{j,t}\, \overline{u}_{i,t}. \quad (4.6)$$

Note that there is no term linear in s_{ij} as is occurs in an EVM; we demonstrate the consequences of this in the following. In [12] a multiscale expansion procedure yields the same tensor bases to reach the same conclusion.

4.2.3   Nonlinear LES Model

In Equations (4.1) and (4.2) we can relate the averaging scales L and T to the computational grid size Δx and computational time step Δt. If $T = 0$ then the "nonlinear" or "gradient" model of Clark [6] is obtained; see also [13]. The Clark model results from the Taylor series expansion of NS and is, in principle, limited (by derivation if not application) to L being on the order of the Kolmogorov microscales. This suggests that (4.4) and (4.5) can be viewed as realizability constraints: the residual stress model must have (asymptotically) the earlier form as the spatial and temporal filter widths approach the Kolmogorov microscales. EVMs do not satisfy such a limit. Margolin, Rider, and Grinstein [9] show by induction that (4.1) and (4.2) are form invariant under the averaging operation and hold for arbitrary L and T to order L^4 and T^4.

4.3   Some Preliminaries

Background information necessary for our comparison of the ensemble mixing and transport processes and spectral fluxes of FSNS and NS is now presented. While the

FSNS equations are valid for all Reynolds numbers (Re) [9] our analysis below is for high Re flow in which there is a useful decorrelation between large and small scales of the flow as occurs in turbulent flows with an inertial subrange.

4.3.1   The Reynolds Decomposition

The passive scalar equation is nonlinear and the standard Reynolds decomposition is used to study the ensemble statistics.

$$\bar{u}_i = U_i + u_i, \quad \bar{c} = C + c, \tag{4.7}$$

$$\langle \bar{u}_i \rangle = U_i, \langle \bar{c} \rangle = C. \tag{4.8}$$

The angle brackets represent an ensemble average. Here u_i and c represent fluctuations of the FSNS fluid velocity and concentration $\bar{u}_i, \bar{c}$ about the ensemble means, U_i and C.

4.3.2   Taylor's Hypothesis

In the statistical analysis that follows, we replace temporal derivatives with spatial derivatives using Taylor's hypothesis

$$\bar{c}_{,t} = -\bar{u}_q \bar{c}_{,q} \quad \bar{u}_{j,t} = -\bar{u}_k \bar{u}_{j,k} + \mathrm{O}(\lambda/\ell). \tag{4.9}$$

The Taylor hypothesis is applied solely for the statistical analysis as much more is known about spatial statistics that can be applied to understand the fluid physics of FSNS. More progress can be made – statistically – working with the spatial rather than the temporal terms.

For the scalar equation, Taylor's hypothesis for modest and high Schmidt number in many flows is an excellent approximation. For the momentum equation the neglect of the pressure terms in Taylor's hypothesis is not a valid approximation *on an instantaneous basis*; it is, however, applicable for statistics with the fluctuating pressure gradient in homogeneous flows. A demonstration of this for homogenous flows is given in [14]. In computations, for example by [17, 18], the temporal terms are kept.

For the ensemble analysis below we begin with the following equations, derived from (1) and (2) using the Taylor hypothesis:

$$\frac{\partial}{\partial t} \bar{c} + \bar{u}_j \bar{c}_{,j} = D\bar{c}_{jj} - L^2 \left[\bar{u}_{j,k} \bar{c}_{,k} \right]_{,j} + T^2 \left[\bar{u}_k \bar{u}_{j,k} \bar{u}_q \bar{c}_{,q} \right]_{,j} \tag{4.10}$$

$$\frac{\partial}{\partial t} \bar{u}_i + \bar{u}_j \bar{u}_{i,j} = -\bar{p}_{,i} - L^2 \left[\bar{u}_{j,k} \bar{u}_{i,k} \right]_{,j} + T^2 \left[\bar{u}_k \bar{u}_{j,k} \bar{u}_q \bar{u}_{i,q} \right]_{,j} + \nu \bar{u}_{i,kk}. \tag{4.11}$$

Everything that follows is a direct mathematical consequence of (4.10) and (4.11) employing well-accepted assumptions about statistics and scalings in the inertial range. It is assumed in our statistical analysis[1] that L and T are in the higher wavenumber portions of the inertial subrange.

---

[1] The derivation of FSNS in [9] is more general and makes no assumption about flow Re or the existence of an inertial range.

4.3.3   Energy Transfer Rates in FSNS

As is standard practice, the production terms are understood as "spectral fluxes" and thus the reduction of the variances by cascade processes can be written as

$$\tau_{ij}\overline{s}_{ij} = L^2\left[\overline{s}_{kk}^3 - \overline{s}_{ij}\,\overline{w}_{ij}^2\right] + T^2\left[u_p u_q u_{j,k}\,u_{j,p}\,u_{k,q}\right] \tag{4.12}$$

$$q_k\overline{c}_{,k} = L^2\overline{c}_{,k}\,\overline{s}_{kj}\overline{c}_{,j} + T^2\left[u_k u_p u_{j,k}\,c_{,p}\,c_{,j}\right]. \tag{4.13}$$

It is easier to understand these terms by taking the ensemble mean. Doing so and assuming that, at the level of a leading order result, the fifth-order cumulant vanishes, as per Lumley [15], we find

$$\left\langle\tau_{ij}\overline{s}_{ij}\right\rangle = L^2\left\langle\overline{\omega}_k\,\overline{s}_{kq}\,\overline{\omega}_q\right\rangle + T^2\left\langle u_p u_q\right\rangle\left\langle u_{j,k}u_{j,p}\,u_{k,q}\right\rangle \tag{4.14}$$

$$\left\langle q_k\overline{c}_{,k}\right\rangle = L^2\left\langle\overline{c}_{,k}\,\overline{s}_{kj}\,\overline{c}_{,j}\right\rangle + T^2\left\langle u_k u_p\right\rangle\left\langle u_{j,k}\,\overline{c}_{,p}\,\overline{c}_{,j}\right\rangle. \tag{4.15}$$

We use the relation that $\left\langle u_{j,k}\,u_{j,p}\,u_{k,q}\right\rangle = \left\langle\overline{s}_{kk}^3\right\rangle - \left\langle\overline{s}_{ij}\,\overline{w}_{ij}^2\right\rangle = \left\langle\overline{\omega}_i\,\overline{s}_{ij}\,\overline{\omega}_j\right\rangle$, which presumes small scale isotropy. In an isotropic turbulence, the term $\left\langle\overline{u}_{j,k}\overline{u}_{j,q}\overline{s}_{kq}\right\rangle = \left\langle\omega_i\,s_{ij}\,\omega_j\right\rangle$ represents the enstrophy generation rate. Using Lumley's anisotropy tensor [16],

$$b_{ij} = \frac{\left\langle\overline{u}_i\overline{u}_j\right\rangle}{2k} - \frac{1}{3}\delta_{ij}, \tag{4.16}$$

one can write for the spectral fluxes

$$\left\langle\tau_{ij}\overline{s}_{ij}\right\rangle = \left(L^2 + \frac{2}{3}kT^2\right)\left\langle\overline{\omega}_k\overline{s}_{kq}\overline{\omega}_q\right\rangle + 2kT^2 b_{pq}\left\langle\overline{u}_{j,k}\,\overline{u}_{j,p}\,\overline{u}_{k,q}\right\rangle \tag{4.17}$$

$$\left\langle q_k\overline{c}_{,k}\right\rangle = \left(L^2 + \frac{2}{3}kT^2\right)\left\langle\overline{c}_{,k}\,\overline{s}_{kj}\overline{c}_{,j}\right\rangle + 2kT^2 b_{pk}\left\langle\overline{u}_{j,k}\,\overline{c}_{,p}\,\overline{c}_{,j}\right\rangle. \tag{4.18}$$

The first terms in $\left\langle\tau_{ij}\overline{s}_{ij}\right\rangle$ and $\left\langle q_k\overline{c}_{,k}\right\rangle$ represent the well-known production of enstrophy and mixing rate – root mean square (*RMS*) scalar gradient [3, 25]. Physically they represent the net stirring by the fluctuating strain in a NS fluid.

We note that the temporal terms in (4.17) and (4.18) increase the spectral flux or "dissipation" by a factor of $\frac{2}{3}kT^2$. The scale similarity and gradient models have been found to lack enough dissipation. The mixed model was proposed on phenomenological grounds to add dissipation to scale similarity and gradient models, [5, 13, 22, 19]. The second set of covariances suggests that stirring in a turbulence whose large scales are anisotropic has additional geometrical effects that contribute to the scalar cascade that is a direct result of the temporal filter. These temporal terms vanish for isotropic flow, consistent with the anisotropy tensor vanishing.

The above results are mathematically rigorous given the conventional approximations: (1) a zero fifth-order cumulant, (2) homogeneity, and (3) a sufficiently high *Re* such that the correlation between quantities at different ends of the spectrum can be neglected.

4.4   Passive Scalar Mixing in a Turbulent NS Fluid

In this section we discuss the Navies Stokes fluid in order to provide the background and show how, in the following section, the FSNS fluid captures the same processes and conserves the same invariants with slightly different partitioning. The material in this section is primarily a review. For more detail on statistical approaches to turbulence, see [3, 13].

4.4.1   Variance of the NS Scalar Field

The variance of the mean and the fluctuating scalar NS fields in our homogeneous model problem [24] is

$$\frac{D}{Dt} C^2 \overset{NS}{=} 2\langle u_j c \rangle C_{,j} - 2D C_{,j} C_{,j} \tag{4.19}$$

$$\frac{D}{Dt} \langle c^2 \rangle \overset{NS}{=} -2\langle u_j c \rangle C_{,j} - 2D \langle c_{,j} c_{,j} \rangle. \tag{4.20}$$

The appearance of the production term, $2\langle u_j c \rangle C_{,j}$, with opposite signs in both equations reflects the reversible exchange between C^2 and $\langle c^2 \rangle$ [3]. In many equilibrium flows of the self-similar simple shear type, the production term is negative and converts C^2 into $\langle c^2 \rangle$, which in turn is molecularly dissipated by $-D \langle c_{,j} c_{,j} \rangle$. Symbolically,

$$C^2 \leftrightarrow 2\langle u_j c \rangle C_{,j} \leftrightarrow \langle c^2 \rangle. \tag{4.21}$$

This is a well-known second moment equation result [3] and a consequence of the fact that the underlying quantity is $c^* = C + c$ and any $f(c^*)$ and moments thereof are materially conserved. The Reynolds decomposition and averaging takes the original c^* field and partitions it into the two fields, C^2 and $\langle c^2 \rangle$, with a reversible exchange between them that reflects the underlying the material conservation of the sum

$$\frac{D}{Dt} \left[C^2 + \langle c^2 \rangle \right] \overset{NS}{=} 0. \tag{4.22}$$

By materially conserved we mean the equations with $D = 0$.

The Reynolds averaging of NS has partitioned the variance field into two variances:

$$\langle c^{*2} \rangle \overset{NS}{=} C^2 + \langle c^2 \rangle. \tag{4.23}$$

The energy exchanges between the two components of the total variance are accomplished as in the following ways:

$$C^2 \leftrightarrow 2\langle u_j c \rangle C_{,j} \leftrightarrow \langle c^2 \rangle. \tag{4.24}$$

The reduction of the fluctuating variance is accomplished by the molecular mixing $D \langle c_{,j} c_{,j} \rangle$. In stationary turbulence the production term is typically single signed and serves to increase the fluctuating variance.

4.4.2 Scalar Gradient Variance of a NS Scalar

The equation for the mixing rate $\langle c_{,j} c_{,j} \rangle$ in statistically homogeneous NS turbulence is:

$$\frac{\partial}{\partial t} \langle c_{,k} c_{,k} \rangle |_{NS} \stackrel{NS}{=} -\langle c_{,j} s_{kj} c_{,k} \rangle - \langle u_{k,j} c_{,j} \rangle C_{,k} - 2D \langle c_{,kk} c_{,jj} \rangle. \tag{4.25}$$

The first term on the right hand side is the production of the mixing rate – the net stirring of the scalar field by the fluctuating strain. The object $\langle c_{,j} s_{kj} c_{,k} \rangle$ is well known as the analog to the enstrophy production term $\langle \omega_i s_{ij} \omega_j \rangle$ of the statistically isotropic hydrodynamic problem; [3, 25] and references therein. For our purposes we call the production of the enstrophy and the production of the mixing rate the *stirring*. For stationary turbulent flows the enstrophy (or dissipation) rate and the mixing rate track what is fed to it from larger scales by an effectively inviscid nondiffusive cascade [3, 21, 13]. Thus one writes

$$\varepsilon \stackrel{NS}{=} \nu \langle \omega^2 \rangle \qquad \varepsilon_c \stackrel{NS}{=} D \langle c_{,j} c_{,j} \rangle \tag{4.26}$$

where ε and ε_c represent the cascade rates from the large scale, which is the rate limiting step that sets the dissipation rates $\nu \langle \omega^2 \rangle, D \langle c_{,j} c_{,j} \rangle$.

4.5 Passive Scalar Mixing in a Turbulent FSNS Fluid

In an analog to the earlier analysis of the NS variance we analyze the variance equations that result from the ensemble averaging of the FSNS equations.

4.5.1 Variance of the FSNS Scalar Field

The following assumptions are made: (1) isotropy of the large scales $\langle u_p u_q \rangle$, (2) small scale isotropy of $\langle u_{j,q} c_{,q} \rangle$, (3) constant mean scalar gradient, $C_{,ij} = 0$, (4) no mean velocity gradient. The equation for the variance of the scalar FSNS field is

$$\frac{D}{Dt} C^2 \stackrel{FSNS}{=} 2 \langle u_j c \rangle C_{,j} - 2D C_{,j} C_{,j} \tag{4.27}$$

$$\frac{D}{Dt} \langle c^2 \rangle \stackrel{FSNS}{=} -2 \langle u_j c \rangle C_{,j} - 2D \langle c_{,j} c_{,j} \rangle + 2 \left(L^2 + \frac{2}{3} kT^2 \right) \langle c_{,j} s_{jk} c_{,k} \rangle. \tag{4.28}$$

See the Appendix for details. The appearance of the production term, $2 \langle u_j c \rangle C_{,j}$, with opposite signs in both equations reflects the reversible exchange between $C^2 |_{FSNS}$ and $\langle c^2 \rangle |_{FSNS}$ as was seen for the NS fluid.

In a NS turbulence the mixing term, $D \langle c_{,j} c_{,j} \rangle$, is solely responsible for the reduction of the variance. In a FSNS turbulence the mixing is comprised of two terms $D \langle c_{,j} c_{,j} \rangle$ and the stirring term $\left(L^2 + \frac{2}{3} kT^2 \right) \langle c_{,j} s_{ij} c_{,i} \rangle$ reduces the variance. The finite scaling procedure has produced a stirring term $\langle c_{,j} s_{ij} c_{,i} \rangle$ in the $\langle c^2 \rangle$ equation compensating for the reduction of the gradients (responsible for molecular dissipation) by the averaging process. As we have seen, the stirring term $\langle c_{,j} \, s_{ij} \, c_{,i} \rangle$ is the nonlinear production term in the small scale gradient $\langle c_{,j} c_{,j} \rangle$ equation of a NS fluid turbulence.

A few observations are pertinent:

- In analog to (4.26) the mean cascade rate is now written

$$\varepsilon_c \overset{FSNS}{=} 2D\langle c_{,j} \, c_{,j} \rangle - 2\left(L^2 + \frac{2}{3} kT^2\right)\langle c_{,j} \, s_{jk} \, c_{,k} \rangle. \tag{4.29}$$

Instantaneously the term $c_{,k} \, s_{jk} \, c_{,j}$ is positive or negative and allows either an up or down scale transfer. The extraction of energy from the large scales is increased by a factor $\frac{2}{3} kT^2$ over the nonlinear model. Interpreting L and T as grid parameters and applying a CFL argument shows the spatial and temporal terms to be of the same order.

- To show that ε_c is positive and performs the same function as a dissipation the stirring term can be written in terms of the skewness coefficient; see [25] and references therein.

$$S_c = \frac{\langle c_{,j} \, s_{jk} \, c_{,k} \rangle}{\langle s^2 \rangle^{1/2} \langle c_{,k} \, c_{,k} \rangle}. \tag{4.30}$$

For a developed inertial range in a stationary turbulence, $S_c \sim -\frac{1}{2}$ and

$$\varepsilon_c \overset{FSNS}{=} \left[2D + 2\langle s^2 \rangle^{1/2}\left(L^2 + \frac{2}{3} kT^2\right)|S_c|\right]\langle c_{,j} c_{,j} \rangle > 0 \tag{4.31}$$

for an equilibrium homogeneous turbulent flow. ILES simulations of passive scalar mixing by [17] have verified that S_c exhibits a stationary value of $S_c \approx -\frac{1}{2}$ in line with the NS value. For additional material, see [18].

- Based on the notions of an inertial subrange, we hypothesize that

$$\varepsilon_c \overset{FSNS}{=} 2D\langle c_{,j} c_{,j} \rangle - 2\left(L^2 + \frac{2}{3} kT^2\right)\langle c_{,j} \, s_{jk} \, c_{,k} \rangle \neq f(L, T), \tag{4.32}$$

that is, it is independent of L^2 and T^2 for L and T in the inertial range. The equation here shows that the net stirring has moved from the $\langle c_{,j} c_{,j} \rangle$ equation to the "large" scale $\langle c^2 \rangle$ equation in which stirring reduces the variance in a way consistent with the notions of a stationary spectral cascade, $\varepsilon_c \neq f(L, T)$ [3, 21, 13].

These findings apply, with appropriate choices of L and T, to the various LES and ILES models consistent with the FSNS form.

4.5.2   Scalar Gradient Variance of a FSNS Fluid

The FSNS mixing rate $\langle c_{,k} c_{,k} \rangle$ equation, which is the FSNS analog to (4.25), is now derived. The equation requires considerable algebraic manipulation; see Ristorcelli et al. [14] for an overview. The variance of the scalar gradient in stationary homogeneous isotropic turbulence obeys

$$
\begin{aligned}
\frac{D}{Dt}\langle c_{,k}\, c_{,k}\rangle \overset{FSNS}{=} &-2\langle c_{,j}\, s_{kj}\, c_{,k}\rangle - \langle u_{k,j}\, c_{,j}\rangle C_{,k} + 2\left(L^2 + \frac{2}{3}kT^2\right)\langle c_{,jp}\, s_{jk}\, c_{,kp}\rangle \\
&-T^2\left[\langle c_{,q}\, c_{,i}(\nabla u)_{iq}^3\rangle + \langle u_{p,k}\, c_{,pk}(c_{ik} - c_{,qq}\,\delta_{ik})\rangle\right] \\
&-T^2 C_{,i}\langle c_{,q}(\nabla u)_{iq}^3\rangle - 2D\langle c_{,kk}\, c_{,jj}\rangle.
\end{aligned}
$$

$$(4.33)$$

Inspecting the equation we see that the FSNS $\langle c_{,k}\, c_{,k}\rangle$ is produced by the usual NS stirring mechanism, $\langle c_{,j}\, s_{kj}\, c_{,k}\rangle$, and a new stirring term, $2\left(L^2 + \frac{2}{3}kT^2\right)$, $\langle c_{,pj}\, s_{jk}\, c_{,pk}\rangle$, built on the scalar Hessian appears.

The fact that $2\left(L^2 + \frac{2}{3}kT^2\right)$, $\langle c_{,j}\, s_{jk}\, c_{,k}\rangle$ appears as a production term in the $\langle c^2\rangle|_{FSNS}$ and with opposite sign in the $\left(L^2 + \frac{2}{3}kT^2\right)\langle c_{,k}\, c_{,k}\rangle|_{FSNS}$ equation has some important consequences. In addition to the exchange between the C^2 and the $\langle c^2\rangle$ fields in NS and FNSN turbulence we see an additional exchange between the $\langle c^2\rangle|_{FSNS}$ variance and $\left(L^2 + \frac{2}{3}kT^2\right)\langle c_{,k}\, c_{,k}\rangle|_{FSNS}$ variance field at a rate proportional to the NS stirring term $2\left(L^2 + \frac{2}{3}kT^2\right)\langle c_{,i}\, s_{ij}\, c_{,j}\rangle$.

As a consequence, the FSNS equations have the material conservation invariant in analog to (4.22) NS:

$$
\frac{D}{Dt}\left[C^2 + \langle c^2\rangle + \left(L^2 + \frac{2}{3}kT^2\right)\langle c_{,k}\, c_{,k}\rangle\right] \overset{FSNS}{=} 0
$$

$$(4.34)$$

to leading order in $\left(L^2 + \frac{2}{3}kT^2\right)$.

4.5.3   What Does This All Mean?

The FSNS procedure has partitioned the variance field into three variances:

$$
\langle c^{*2}\rangle \overset{FSNS}{=} C^2 + \langle c^2\rangle + \left(L^2 + \frac{2}{3}kT^2\right)\langle c_{,j} c_{,j}\rangle.
$$

$$(4.35)$$

The energy exchanges between the three components of the total variance are accomplished in the following ways

$$
C^2 \leftrightarrow 2\langle u_j c\rangle C_{,j} \leftrightarrow \langle c^2\rangle.
$$

$$(4.36)$$

$$
\langle c^2\rangle \leftrightarrow \left(L^2 + \frac{2}{3}kT^2\right)\langle c_{,i} s_{ij} c_{,j}\rangle \leftrightarrow \left(L^2 + \frac{2}{3}kT^2\right)\langle c_{,j}\, c_{,j}\rangle.
$$

$$(4.37)$$

The reduction of the variance of the scalar fluctuations, $\langle c^2\rangle = \langle c^2\rangle|_{FSNS}|$, is accomplished by the molecular mixing observed at scales L and T $D\langle c_{,j}\, c_{,j}\rangle|_{FSNS}$ and the stirring $\left(L^2 + \frac{2}{3}kT^2\right)\langle c_{,i}\, s_{ij}\, c_{,j}\rangle|_{FSNS}$ observed at scales L and T.

Based on these results it appears that we can make the following observations about FSNS mixing in high Reynolds number, homogeneous, stationary mixing:

- The FSNS equations satisfy the underlying material conservation of c^* to leading order.

- The material field is partitioned into three separate positive definite variances, C^2, $\langle c^2 \rangle$, and $\left(L^2 + \frac{2}{3} kT^2 \right) \langle c_{,j} c_{,j} \rangle$, with reversible exchanges between them. The new exchange term, $\langle c_{,i} s_{ij} c_{,j} \rangle$, can be negative or positive. For isotropic stationary turbulence it is negative (in the mean).
- The FSNS equations extract resolved scale variance at a NS stirring rate $\langle c_{,i} s_{ij} c_{,j} \rangle$ and convert it into the unresolved scales in the form $\left(L^2 + \frac{2}{3} kT^2 \right) \langle c_{,j} c_{,j} \rangle$, which is then dissipated.
- In homogeneous turbulence (for which $C^2|_{NS} = C^2|_{FSNS}$) the materially conserved property in both NS and FSNS leads to the consequence

$$\langle c^2 \rangle NS = \langle c^2 \rangle \big|_{FSNS} + \left| \left(L^2 + \frac{2}{3} kT^2 \right) \langle c_{,j} c_{,j} \rangle \right|_{FSNS}, \qquad (4.38)$$

which serves to obtain the variance of the physical flow from the resolved and residual FSNS computation.

- The materially conserved variance in both NS and FSNS leads to the consequence

$$2D \langle c_{,j} c_{,j} \rangle_{NS} = 2D \langle c_{,j} c_{,j} \rangle_{FSNS} - 2 \left(L^2 + \frac{2}{3} kT^2 \right) \langle c_{,j} s_{jk} c_{,k} \rangle_{FSNS} \qquad (4.39)$$

for a stationary turbulence with an inertial range.

4.6   Passive Scalar Mixing in a Turbulent SMG Fluid

In an analog to the earlier analysis of the FSNS variance we study the variance equations of an isotropic EVM. We shall use the popular Smagorinksy (SMG) model [26] and investigate passive scalar mixing in high Reynolds number Smagorinsky fluid turbulence.

4.6.1   Variance of the SMG Scalar Field

We apply similar Reynolds averaging procedures to the isotropic EVM of the SMG fluid model:

$$\frac{\partial}{\partial t} \overline{c} + \overline{u}_j \overline{c}_{,j} \stackrel{SMG}{=} D \overline{c}_{jj} + L^2 \left[\left(\overline{s}^2 \right)^{1/2} \overline{c}_{,j} \right]_{,j}. \qquad (4.40)$$

In the earlier equations $L^2 = C_s(\Delta, x)^2$ and is used symbolically in the spirit of illustrating comparisons between the two methods. We denote the tensor contraction $\overline{s}^2 = \overline{s}_{pq} \overline{s}_{pq}$ and for convenience define $\sigma = \left(\overline{s}^2 \right)$. The Reynolds decomposition indicates

$$\sigma = \overline{\sigma} + \sigma' \quad \overline{\sigma} = \left(S^2 + \langle s^2 \rangle \right)^{1/2} \quad \sigma' = \left(S^2 + 2 S_{pq} s_{pq} + s^2 \right)^{1/2} - \overline{\sigma}. \qquad (4.41)$$

The Reynolds averaged mean equations for the scalar in a SMG fluid is then

$$\frac{\partial}{\partial t} C + \left[U_j C + \langle u_j c \rangle \right]_{,j} \overset{SMG}{=} DC_{,jj} + L^2 \left[\bar{\sigma} C_{,j} + \langle \sigma' c_{,j} \rangle \right]_{,j}. \tag{4.42}$$

The mean variance in our homogenous constant mean scalar gradient model problem is

$$\frac{D}{Dt} C^2 \overset{SMG}{=} 2\langle u_j c \rangle C_{,j} - 2DC_{,j} C_{,j}. \tag{4.43}$$

Following the usual procedures, the fluctuating variance equation for our homogenous model problem is written

$$\frac{D}{Dt} \langle c^2 \rangle \overset{SMG}{=} -2\langle u_j c \rangle C_{,j} - 2L^2 \langle \sigma' c_{,j} \rangle C_{,j} - 2\left(D + L^2 \bar{\sigma} \right) \langle c_{,j} c_{,j} \rangle - 2L^2 \langle \sigma' c_{,j} c_{,j} \rangle. \tag{4.44}$$

This is the SMG analog to (4.28). Using $\sigma' = s^2 - \langle s^2 \rangle$ and that $\bar{\sigma} C_{,j} = -\langle \sigma' c_{,j} \rangle$ in homogeneous turbulence the equation is written

$$\frac{D}{Dt} \langle c^2 \rangle \overset{SMG}{=} -2\langle u_j c \rangle C_{,j} - 2D\langle c_{,j} c_{,j} \rangle - 2L^2 \left\langle \left(s^2 \right)^{\frac{1}{2}} c_{,j} \right\rangle C_{,j} - 2L^2 \left\langle \left(s^2 \right)^{\frac{1}{2}} c_{,j} c_{,j} \right\rangle. \tag{4.45}$$

This difference, common to all isotropic EVMs, has several important consequences affecting issues of material conservation and the capability to reconstitute the NS scalar variance.

4.7 Turbulent Mixing in NS, FSNS, and SMG Fluids

The variances of the various fields are now compared. For a stationary forced isotropic turbulence mixing a scalar field with a prescribed mean scalar gradient, NS, FSNS, and SMG mean variance equations are the same:

$$\frac{D}{Dt} C^2 = 2\langle u_j c \rangle C_{,j} - 2DC_{,j} C_{,j}. \tag{4.46}$$

We address the inhomogeneous form in another venue. The variance equations are

$$\frac{D}{Dt} \langle c^2 \rangle \overset{NS}{=} -2\langle u_j c \rangle C_{,j} - 2D\langle c_{,j} c_{,j} \rangle, \tag{4.47}$$

$$\frac{D}{Dt} \langle c^2 \rangle \overset{FSNS}{=} -2\langle u_j c \rangle C_{,j} - 2D\langle c_{,j} c_{,j} \rangle + 2\left(L^2 + \frac{2}{3} kT^2 \right)_{,} \langle c_{,j} s_{jk} c_{,k} \rangle, \tag{4.48}$$

$$\frac{D}{Dt} \langle c^2 \rangle \overset{SMG}{=} -2\langle u_j c \rangle C_{,j} - 2D\langle c_{,j} c_{,j} \rangle - 2L^2 \left\langle \left(s^2 \right)^{\frac{1}{2}} c_{,j} \right\rangle C_{,j} - 2L^2 \left\langle \left(s^2 \right)^{\frac{1}{2}} c_{,j} c_{,j} \right\rangle. \tag{4.49}$$

A few comments are useful.

- The three dissipation rates: the NS, FSNS, and SMG dissipation rates are

$$\varepsilon_c \stackrel{NS}{=} 2D\langle c_{,j}\, c_{,j}\rangle, \tag{4.50}$$

$$\varepsilon_c \stackrel{FSNS}{=} 2D\langle c_{,j}\, c_{,j}\rangle + 2\left(L^2 + \frac{2}{3}kT^2\right)\langle c_{,j}\, s_{jk}\, c_{,k}\rangle, \tag{4.51}$$

$$\varepsilon_c \stackrel{SMG}{=} 2D\langle c_{,j}\, c_{,j}\rangle + 2L^2\left\langle \left(s^2\right)^{1/2}c_{,j}\, c_{,j}\right\rangle. \tag{4.52}$$

The SMG model for the stirring term in the variance equation does not account for the orientation of the strain with concentration gradients and the cascade/dissipation rate is not related to the stirring of the material field. In the FSNS fluid the contribution to the dissipation of $\langle c^2\rangle$ is by the triple moment NS stirring term.

- There is no material conservation principle for mixing in SMG turbulence. Because there is no stirring term in the SMG scalar variance equation one cannot write the equivalent of

$$\frac{D}{Dt}\left[C^2 + \langle c^2\rangle + \left(L^2 + \frac{2}{3}kT^2\right)\langle c_{,k}\, c_{,k}\rangle\right] \stackrel{FSNS}{=} 0 + h.o.t. \tag{4.53}$$

for SMG fluid turbulence. The unresolved material fluctuations represented as they are by the isotropic form $\left[(s^2)^{\frac{1}{2}}c_{,j}\right]_{,j}$ for SMG as opposed to $\left[u_{j,k}\, c_{,k}\right]_{,j}$ for FSNS obviates the possibility of material conservation for all scales of the motion. At the level of principle this is of concern; at the level of practice the lack of material invariance has not produced problems that have been seen (or produces problems that have not been looked for).

- The mean production term the $\langle c^2\rangle$ equation of SMG turbulence. If the term $\left\langle (s^2)^{\frac{1}{2}}c_{,j}\right\rangle \neq 0$ then there is an additional small scale production term $\left\langle (s^2)^{\frac{1}{2}}c_{,j}\right\rangle C_{,j}$ in the $\langle c^2\rangle|SMG$ equation dependent on the mean scale gradient in a way that is not consistent with the notion of that the computational cut off is in the far inertial range.

4.8   Summary and Conclusions

In this chapter we have demonstrated the use of an ensemble moment equation analysis to obtain an objective understanding of coarse grained models used for the simulation of high Re NS turbulence.

The FSNS fluid exhibits many features seen in NS stirring. The spectral flux, $\langle c_{,j}\, s_{ij}\, c_{,i}\rangle$, in the FSNS $\langle c^2\rangle$ equation has the same form as the production of $\langle c_{,j}c_{,j}\rangle$, rigorously demonstrating a conservation property for the FSNS fluid that partitions the scalar variance field, $\langle c^2\rangle$, into three different variances, C^2, $\langle c^2\rangle$, and $\left(L^2 + \frac{2}{3}kT^2\right)\langle c_{,j}\, c_{,j}\rangle$. The reversible exchange between the last two terms, the resolved and unresolved variance, is determined by

$$\langle c^2 \rangle \leftrightarrow \left(L^2 + \frac{2}{3} kT^2 \right) \langle c_{,i} \, s_{ij} \, c_{,j} \rangle \leftrightarrow \left(L^2 + \frac{2}{3} kT^2 \right) \langle c_{,j} \, c_{,j} \rangle. \tag{4.54}$$

The FSNS second moment equations show that the stirring and cascade are moved from the small scale $\langle c_{,j} c_{,j} \rangle$ equation to the "larger" resolved scale $\langle c^2 \rangle$ equation in a way that is consistent with the notion of a materially conservative inertial subrange governed by NS.

Comparison of the FSNS and SMG second moment results has produced some new results regarding the shortcomings of SMG and strengths of FSNS. The new finding is that NS second order moments are reconstituted from the triple partition of the resolved FSNS moments

$$\langle c^2 \rangle_{NS} = \left[\langle c^2 \rangle + \left(L^2 + \frac{2}{3} kT^2 \right) \langle c_{,j} \, c_{,j} \rangle \right]_{FSNS}, \tag{4.55}$$

where $\left(L^2 + \frac{2}{3} kT^2 \right) \langle c_{,j} c_{,j} \rangle_{FSNS}$ represents the spatial and temporal portions of unresolved variance and no explicit subgrid model is required. Due to the materially conservative nature of the FSNS one has an exact estimate for the unresolved variance and thus the scalar variance of the orginal NS problem. The reconstitution results hold for LES and ILES types that fall in the FSNS class.

4.9   Acknowledgments

The author acknowledges the Laboratory Directed Research and Development (LDRD) program at Los Alamos for support of this research through exploratory project 20100441ER. This work was performed under the auspices of the U.S. Department of Energy's NNSA by the Los Alamos National Laboratory operated by Los Alamos National Security, LLC under contract number DE-AC52-06NA25396.

Appendix: Reynolds Averaging for FSNS Fluid Turbulence

We begin with Equation (4.10) and show how the first moment equation is derived:

$$\frac{\partial}{\partial t}\overline{c} \; + \; \overline{u}_j\,\overline{c}_{,j} \overset{FSNS}{=} D\overline{c}_{,jj} - L^2\left[\overline{u}_{j,k}\,\overline{c}_{,k}\right]_{,j} + T^2\left[\overline{u}_k\,\overline{u}_{j,k}\,\overline{u}_q\,\overline{c}_{,q}\right]_{,j}. \tag{4.56}$$

The Reynolds decomposition for concentration is $\overline{c} = C + c$, $\overline{u}_j = U_j + u_j$, where $C \equiv \langle\overline{c}\rangle$, $U_j = \langle\overline{u}_j\rangle$. Applying the Reynolds decomposition to (56) produces the spatial terms given in the first moment equation in the text. The temporal terms require explanation as similar procedures are applied to the second moment equations and the manipulations, while in principle the same, are in practice far more complicated. The following is therefore an example of how the final form of the equation is arrived at and demonstrates the simplifications involved.

$$\begin{aligned}
\langle\overline{u}_k\,\overline{u}_{j,k}\,\overline{u}_q\,\overline{c}_{,q}\rangle = {} & [U_pU_jU_{k,p} + \langle u_pu_j\rangle U_{k,p} + U_p\langle u_ju_{k,p}\rangle + \langle u_pu_{k,p}\rangle U_j + \langle u_pu_ju_{k,p}\rangle]C_{,j} \\
& + U_pU_{k,p}\langle u_jc_{,j}\rangle + U_jU_{k,p}\langle u_pc_{,j}\rangle + U_pU_j\langle u_{k,p}c_{,j}\rangle \\
& + U_p\langle u_ju_{k,p}\,c_{,j}\rangle + U_j\langle u_pu_{k,p}\,c_{,j}\rangle + U_{k,p}\langle u_pu_jc_{,j}\rangle \\
& + \langle u_pu_ju_{k,p}\,c_{,j}\rangle.
\end{aligned} \tag{4.57}$$

Now the higher order terms are neglected with the following arguments. Line 1: the third, fourth, and fifth terms are neglected as they are covariances between quantities that come from very different scales of motion that will be negligible in high Reynolds number turbulence. Using continuity, the fourth term is neglected on the grounds of homogeneity – our proof of concept problem is homogeneous turbulence. Line 2: the first and second terms on Line 2 are neglected as these are covariance between quantities from opposite ends of the spectrum. Line 3: all terms are negligible for the same reason. One then obtains

$$\langle\overline{u}_k\overline{u}_{j,k}\,\overline{u}_q\overline{c}_{,q}\rangle = [U_pU_jU_{k,p} + \langle u_pu_j\rangle U_{k,p}]C_{,j} + U_pU_j\langle u_{k,p}\,c_{,j}\rangle + \langle u_pu_ju_{k,p}\,c_{,j}\rangle. \tag{4.58}$$

For the last term one assumes that the fourth cumulant vanishes and

$$\langle u_pu_ju_{k,p}\,c_{,j}\rangle = \langle u_pu_j\rangle\langle u_{k,p}\,c_{,p}\rangle + \langle u_{p,k}\,u_k\rangle\langle u_jc_{,j}\rangle + \langle u_p\,c_{,j}\rangle\langle u_ju_{k,p}\rangle. \tag{4.59}$$

This approximation does not presume that the probability distribution function (PDF) is Gaussian; it uses the Gaussian functional relation between the second and fourth moments. Again assuming the correlation between the energy containing scales and the dissipative scale motions is small, we neglect the second and third term to obtain

$$\langle u_p u_j u_{k,p} \, c_{,j} \rangle = \langle u_p u_j \rangle \langle u_{k,p} \, c_{,p} \rangle \tag{4.60}$$

and thus

$$\langle \bar{u}_k \bar{u}_{j,k} \, \bar{u}_q \bar{c}_{,q} \rangle = U_p U_j [U_{k,p} \, C_{,j} + \langle u_{k,p} \, c_{,j} \rangle] + \langle u_p u_j \rangle [U_{k,p} \, C_{,j} + \langle u_{k,p} \, c_{,p} \rangle]. \tag{4.61}$$

The mean the FSNS concentration equation becomes

$$\frac{\partial}{\partial t} C + [U_j C + \langle u_j c \rangle]_{,j} \stackrel{FSNS}{=} DC_{,jj} - L^2 [U_{j,k} \, C_{,k} + \langle \bar{u}_{j,k} \, c_{,k} \rangle]_{,j}$$
$$- T^2 [U_p U_j (U_{k,p} \, C_{,j} + \langle u_{k,p} \, c_{,j} \rangle)]_{,k}$$
$$- T^2 [\langle u_p u_j \rangle (U_{k,p} \, C_{,j} + \langle u_{k,p} \, c_{,p} \rangle)]_{,k}. \tag{4.62}$$

The mean spatial term is

$$L^2 [\bar{u}_{j,k} \, \bar{c}_{,k}]_{,j} = L^2 [U_{j,k} \, C_{,k} + \langle \bar{u}_{j,k} \, c_{,k} \rangle]_{,j} = L^2 [S_{jk} C_{,jk} + \langle s_{jk} c_{,jk} \rangle] \tag{4.63}$$

using continuity and symmetry of $C_{,jk}$ and $\bar{c}_{,jk}$. One expects $\langle u_{j,k} c_{,k} \rangle = 0$, which is true for isotropic high Reynolds number turbulence in which the mean scalar gradient is small compared with the *RMS* fluctuating gradient and we will neglect $\langle s_{jk} c_{,jk} \rangle$ in the following.

If (1) the mean flow is of the unidirectional boundary layer type, (2) gradients in the turbulence quantities are of second order, and (3) the turbulence is weakly anisotropic, one obtains

$$\frac{\partial}{\partial t} C + [U_j C + \langle u_j c \rangle]_{,j} \stackrel{FSNS}{=} DC_{,jj} - \left(L^2 + \frac{2}{3} kT^2 \right) [U_{j,k} \, C_{,k} + \langle u_{j,k} c_{,k} \rangle]_{,j} \tag{4.64}$$

$$\stackrel{FSNS}{=} DC_{,jj} - \left(L^2 + \frac{2}{3} kT^2 \right) [S_{jk} C_{,kj} + \langle s_{jk} c_{,kj} \rangle]. \tag{4.65}$$

In the homogeneous limit the $\left(L^2 + \frac{2}{3} kT^2 \right)$ terms vanish. We have assumed the turbulence is to leading order isotropic and thus the rotation of the scalar Hessian does not contribute: $\langle u_p u_j \rangle \langle u_{k,p} c_{,jk} \rangle = \langle u_p u_j \rangle [\langle s_{kp} c_{,jk} \rangle + \langle w_{kp} c_{,jk} \rangle] = \frac{2}{3} k$ $[\langle s_{kp} c_{,pk} \rangle + \langle w_{kp} c_{,pk} \rangle] = \frac{2}{3} k \langle s_{kp} c_{,pk} \rangle$ due to the antisymmetry of the rotation tensor.

The mean variance equation is found by multiplying the earlier primitive form of the C equation by C.

$$\frac{\partial}{\partial t} C^2 + [U_j C^2 + \langle u_j c \rangle C]_{,j} \stackrel{FSNS}{=} -2\langle u_j c \rangle C_{,j} - 2DCC_{,jj} - \left(L^2 + \frac{2}{3} kT^2 \right) S_{jk} C_{,kj} C.$$

In the primitive equation we have used $\langle u_{j,k} c_{,k} \rangle = 0$, which is true for isotropic high Reynolds number turbulence in which the mean scalar gradient is small compared with the *RMS* fluctuating gradient. In DNS this quantity is not isotropic as there is some imprint of the direction mean scalar gradient on the scalar fluctuating gradients. One then obtains

$$\frac{\partial}{\partial t} C^2 + [U_j C^2 + \langle u_j c \rangle]_{,j} \overset{FSNS}{=} -2\langle u_j c \rangle C_{,j} - 2DCC_{,jj} - \left(L^2 + \frac{2}{3} kT^2\right) S_{jk} C_{,kj} C.$$

For homogeneous turbulence with constant mean scale gradient – our model problem – one obtains

$$\frac{\partial}{\partial t} C^2 \overset{FSNS}{=} -2\langle u_j c \rangle C_{,j} - 2DC_{,j} C_{,j} - \left(L^2 + \frac{2}{3} kT^2\right) S_{jk} C_{,j} C_{,k}.$$

The last term vanishes in the absence of deformation. This completes our example of the sorts of statistical manipulations we use to obtain equations in the text.

A.1   Averaging in FSNS

Finite scale theory arose in the context of a popular methodology employed in computational fluid dynamics, the so-called finite volume methods [23]. In these methods, the dependent variables represent averages in space and time over the computational cells. The question naturally arises: if every point of a volume obeys NS equations, what equations govern the averages?

It is clear from the nonlinearity of advection that these averages do not obey NS; the product of the averages is not the average of the products. The derivation of the finite scale equations is described in Reference 8. The following theorem allows one to proceed directly from the underlying PDEs to their finite scale analogs.

Theorem: Let $A(x, y, z, t)$ and $B(x, y, z, t)$ be continuous variable fields, which might be scalars or components of vector fields, and define the average variable $\overline{A}$ in a cubic domain

$$\overline{A}(x, t) \equiv \frac{1}{\Delta t \Delta x^3} \int_{t-\frac{1}{2}\Delta t}^{t+\frac{1}{2}\Delta t} dt' \int_{x-\frac{1}{2}\Delta x}^{x+\frac{1}{2}\Delta x} dx' \int_{y-\frac{1}{2}\Delta x}^{y+\frac{1}{2}\Delta x} dy' \int_{z-\frac{1}{2}\Delta x}^{z+\frac{1}{2}\Delta x} dz' A(x', y', z', t') \qquad (4.66)$$

and similarly for $\overline{B}$. Then

$$\overline{AB} = \overline{A}\,\overline{B} + \left(\frac{\Delta x^2}{12}\right) [\overline{A}_x \overline{B}_x + \overline{A}_y \overline{B}_y + \overline{A}_z \overline{B}_z] + \left(\frac{\Delta t^2}{12}\right) \overline{A}_t \overline{B}_t + \mathcal{O}(L^4, T^4). \qquad (4.67)$$

Here, we denote partial derivatives with subscripts, for example,

$$\overline{A}_x = \frac{\partial \overline{A}}{\partial x}.$$

Note that the spatial averaging is over a cube of volume L^3. More general volumes could be assumed and a metric tensor would be involved. The proof of this theorem is based on an unusual synthesis of renormalization and induction. The underlying idea is that the averaging is a smoothing process; although A may not be smooth on scales $\Delta x, \Delta t$, $\overline{A}$ is smooth enough to expand in a convergent Taylor series on those scales. Details are described in References 8 and 9.

References

[1] Sagaut, P. (2006). *Large Eddy Simulation for incompressible flows*. Springer Verlag.

[2] Grinstein, F.F, L.G. Margolin, W.J. Rider (2010). *Implicit Large Eddy Simulation*, 2nd printing. Cambridge University Press.

[3] Tennekes, H. and J.L. Lumley (1972). *A First Course in Turbulence*. MIT Press.

[4] Liu, S., C. Meneveau, and J. Katz (1994). "On the properties of similarity subgrid-scale models as deduced from measurements in a turbulent jet." *J. Fluid Mech.* 275:83–119.

[5] Bardina, J., J.H. Ferziger, and W.C. Reynolds (1980). "Improved turbulence models for large eddy simulation." AIAA paper 80-1357.

[6] Clark, R.A., J. H. Fertziger, and W. C. Reynolds (1979). "Evaluation of subgrid models using accurately simulated turbulent flow." *J. Fluid Mech.* 91:1–16.

[7] Galdi, G.P., W.J. Layton (2000). "Approximating the larger eddies in fluid flows II: a model for space filtered flow." *Math. Models and Meth. in Appld. Sciences* 10(3):343–350.

[8] Margolin, L.G. and W.J. Rider (2002). "A rationale for implicit turbulence modeling." *Int. J. Num. Methods Fluids* 39:821–841.

[9] Margolin, L.G., W.J. Rider, and F.F. Grinstein (2006). "Modeling turbulent flow with implicit LES." *J. Turbulence* 7:1–27.

[10] Hirt, C.W. (1968). "Heuristic stability theory for finite difference equations." *J. Comput. Phys.* 2, 339355.

[11] Fureby C. and F.F. Grinstein (1999). "Monotonically integrated large eddy simulation of free shear flows." *AIAA Journal* 37:544.

[12] Eyink, G. (2006). "Multi-scale gradient expansion of the turbulent stress tensor." *J. Fluid Mech.* 549:159–190.

[13] Pope, S.B. (2000). *Turbulent Flows*. Cambridge University Press.

[14] Ristorcelli, J.R., L.G. Margolin, and F.F. Grinstein (2011). "Moment analysis of Finite Scale Navier Stokes for coarse grained computations of turbulent mixing." Los Alamos National Laboratory report: LA-UR-11-03387.

[15] Lumley, J.L. (1970). *Stochastic Tools in Turbulence*. Academic Press.

[16] Lumley, J.L. (1978). "Computational modeling of turbulent flows." *Adv. Appl. Mech.* 18:123–176.

[17] Wachtor, A.J., F.F. Grinstein, C.R. DeVore, J.R. Ristorcelli, and L.G. Margolin (2013). "Implicit large-eddy simulation of passive scalar mixing in statistically stationary isotropic turbulence." *Physics of Fluids* 25(2), 025101.

[18] Grinstein, F.F., A.A. Gowardhan, J.R. Ristorcelli, A.J. Wachtor, and A.J. Gowardhan (2012). "On coarse-grained simulations of turbulent material mixing." *Physica Scripta* 86(5).

[19] Bensow, R.E. and C. Fureby (2008). "On the justification and extension of mixed models in LES." *J Turbulence*, 8(54):1–17.

[20] Boris, J.P. (1990). "On large eddy simulation using subgrid turbulence models." In *Whither Turbulence? Turbulence at the Crossroads*, ed. by J.L. Lumley, Springer, 344.

[21] Frisch, U. (1995). *Turbulence*. Cambridge University Press.

[22] Layton, W.J. (2000). "Approximating the larger eddies in fluid motion V: Kinetic energy balance of scale similarity models." *Math. Comp. Modelling* 31:1–7.

[23] Leveque, R.J. (2002). *Finite Volume Methods for Hyperbolic Problems*. Cambridge University Press.

[24] Overholt, M.R. and S. B. Pope (1996). "Direct numerical simulation of a passive scalar with imposed mean gradient in isotropic turbulence." *Physics of Fluids* 8, 3128.

[25] Ristorcelli J.R. (2006). "Passive scalar mixing: analytic study of time scale ratio, variance and mix rate." *Phys. Fluids* A 3:1269–1277.

[26] Smagorinsky J. (1963). "General circulation experiments with primitive equations: I. Basic equations." *Mon. Weather Rev* 91:99–264.

Part II

Challenges

5 Subgrid and Supergrid Modeling

Fernando F. Grinstein

5.1 Introduction

Availability of effective predictive tools is a crucial aspect in the turbulent flow applications. Laboratory studies typically demonstrate the end outcome of complex nonlinear three-dimensional (3D) physical processes with many unexplained details and mechanisms. Flow experiments based on numerical simulations carried out with precise control of initial and boundary conditions (BCs) are ideally suited to provide insights into the underlying dynamics of laboratory observations. A crucial aspect in this collaborative context is that of adequately characterizing the laboratory and numerical experiments, so that potential sources of discrepancies can be clearly evaluated and controlled.

Capturing the dynamics of all relevant scales of motion, based on the numerical solution of the Navier–Stokes equations, constitutes direct numerical simulation (DNS), which is prohibitively expensive in the foreseeable future for practical flows of interest at moderate-to-high Reynolds number (*Re*). The microscopic simulation case must be made allowing to consider isolating small constrained regions of the flow on which DNS can be focused in practice. On the other end of computer simulation possibilities, the Reynolds-averaged Navier–Stokes (RANS) approach is employed for turbulent flows of industrial complexity – with averaging carried out over time, homogeneous directions, or across an ensemble of equivalent flows.

Coarse grained simulation (CGS) – based on computing the macroscale large energy containing portions of the unsteady turbulent motion and using subgrid scale (SGS) closure models for the rest – has become the effective intermediate approach between DNS and RANS, capable of simulating flow features that cannot be handled with RANS, such as significant flow unsteadiness and strong vortex–acoustic couplings. CGS includes classical large eddy simulation (LES) [1] focusing on explicit use of SGS models, implicit LES (ILES) [2] relying on SGS modeling and filtering provided by physics capturing numerical algorithms, and more general LES combining explicit and implicit SGS modeling. CGS is based on the view that physically meaningful scales of turbulence can be split into two groups:

- resolved geometry, initial conditions, BCs, and regime specific scales – the so-called energy containing scales;

- unresolved conditions, and eddies, for which presumably more-universal dynamics can be represented with SGS models; SGS modeling is primarily empirical activity within pragmatic practice.

Practical CGS readiness for complex full-scale flows of interest is expected far in the future, for example, 2045 for simulating flow over a whole aircraft wing [3]. Hybrid RANS/CGS strategies, effectively blending computed and modeled dissipation by using explicit SGS models generated from RANS turbulence models, are the current industrial standard to drastically cut computational costs for such simulations [4, 5].

High *Re* turbulent flow complexity requires achieving accurate and dependable large scale predictions of highly nonlinear processes with underresolved computer simulation models. Laboratory observations demonstrate the outcome of complex nonlinear 3D physical processes with many unexplained details and mechanisms. Carefully controlled computational experiments based on the numerical solution of the conservation equations for mass, momentum, and energy provide insights into the underlying flow dynamics.

Relevant computational fluid dynamics issues to be addressed relate to the modeling of the unresolved flow conditions at the subgrid SGS level – within a computational cell – and at the supergrid (SPG) scale – at initialization and beyond computational boundaries. SGS and SPG information must be prescribed for closure of the equations solved numerically. SGS models appear explicitly or implicitly as additional source terms in the modified flow equations solved by the numerical solutions being calculated, while SPG models provide the necessary set of initial and BCs that must be prescribed to ensure unique, well-posed solutions. From this perspective, it is clear that the simulation process is inherently determined by the SGS and SPG information prescription process. Likewise, observables in laboratory experiments are always characterized by the finite scales of the instrumental resolution of measuring/visualizing devices and by SPG aspects. It is thus important to recognize the inherently intrusive nature of observations based on computational or laboratory experiments [6].[1] Ultimately, well-characterized and resolved (laboratory and computational) experiments, frameworks for verification, validation, and uncertainty quantification, as well as dedicated metrics for the specific problems at hand, are needed to establish predictability.

The possible transient and long term effects of the particular initial conditions (ICs) of computational and laboratory experiments need also be addressed. In what follows, an overview of SGS and SPG issues is presented, and relevant discretization aspects are noted. The discussion then focuses on fundamental turbulent IC characterization and modeling difficulties arising when attempting to integrate LES and laboratory experiments in complex flow problems of current practical interest.

---

[1] The quote from Werner Heisenberg in Chapter 3 is also a very appropriate one here: "What we observe is not nature itself, but nature exposed to our method of questioning."

5.2   Subgrid Scale

In LES we are constrained to simulate a flow with the smallest characteristic resolved scale determined by a resolution cutoff wavelength determined by a spatial filtering process. Formal analysis of LES can be based on modified equation analysis (MEA) [7–10], a technique for generating the approximate equations for the computed solutions – actual solutions of the numerical algorithm underlying the simulation model. To simplify the discussion, we focus on a conceptually simple case, that of incompressible flow with scalar mixing – which we regard as example of other physics to be simulated coupled with the flow. The modified equations – satisfied by the computed numerical solutions – are the following:

$$\partial \overline{u}_i / \partial t + \overline{u}_j \overline{u}_{i,j} = -\overline{p}_{,i} + v\overline{u}_{i,kk} - (\nabla \cdot \boldsymbol{\sigma}_\mathbf{u} + \tau_\mathbf{u})_i$$
$$\partial \overline{c} / \partial t + \overline{u}_j \overline{c}_{,j} = \kappa \overline{c}_{,jj} - \nabla \cdot \boldsymbol{\sigma}_c + \tau_c,$$

where the bar denotes a filtering operation, $\mathbf{u}$ is the (*solenoidal*) velocity field, c is a conserved material scalar concentration, p is the pressure, and v and κ denote momentum and material diffusivity, respectively. Furthermore, $u_{i,j} \equiv \partial u_i / \partial x_j$, sum over repeated ($i$, j, or, k) indices is assumed, $\tau_\mathbf{u}$ and τ_c address effects of discretization and commutation between differentiation and filtering. To ensure closure of the equations in filtered unknowns, an equation of state and models for $\boldsymbol{\sigma}_\mathbf{u} = \overline{\mathbf{u} \times \mathbf{u}} - \overline{\mathbf{u}} \times \overline{\mathbf{u}}$ and $\boldsymbol{\sigma}_\mathbf{u} = \overline{c\mathbf{u}} - \overline{c}\,\overline{\mathbf{u}}$ have to be provided, and depending on v based Re and Schmidt number $Sc = v/\kappa$, different SGS models for the scalar and the momenta may be needed in general. A spatial filtering operation, $\overline{f} = \int G(\mathbf{x}, \mathbf{x}') f(\mathbf{x}') d\mathbf{x}'$, characterized by a prescribed kernel, $G(\mathbf{x}, \mathbf{x}')$, is traditionally involved in LES. More detailed MEA discussion (including compressibility issues) can be found elsewhere [11, 12].

A crucial practical computational aspect is the need to distinctly separate the effects of spatial filtering and SGS reconstruction models from their unavoidable implicit counterparts due to discretization. Indeed, it has been noted that in typical LES strategies, $\tau_\mathbf{u}$ and τ_c, due to discretization and filtering, have contributions directly comparable with those of the explicit models [8]. Seeking to address the seemingly insurmountable issues posed to LES by underresolution, the possibility of using the SGS modeling and filtering provided implicitly by the numerics has been considered as an option [13, 14] and generally denoted numerical LES [15]. However, arbitrary numerics will not work for LES: *good or bad SGS physics can be built into the simulation model depending on the choice and particular implementation of the numerics.*

The monotone integrated LES (MILES) approach [13, 14] incorporates the effects of the SGS physics on the resolved scales through functional reconstruction of the convective fluxes using locally monotonic finite volume (FV) schemes. The more broadly defined ILES [2] proposes using high resolution nonoscillatory FV (NFV) algorithms [16] to solve the unfiltered Euler or NS equations. Popular physics capturing methods have been used in ILES, such as flux corrected transport (FCT), the piecewise parabolic method, Godunov, hybrid, and total variation diminishing algorithms. By focusing on inertially dominated flow dynamics and regularization of underresolved flow, ILES follows on the precedent of using NFV methods for shock capturing – requiring weak solutions and satisfaction of an entropy condition. Other strategies closely related to ILES use vorticity confinement [17], approximate deconvolution [18, 19], combined compact differencing and filtering [20], or finite difference WENO methods [21].

MEA provides a framework to reverse engineer desirable physics into the numerics design. MEA was used in the early formal comparisons [9] between MILES and traditional LES, to show that a class of NFV algorithms with dissipative leading order terms provides appropriate built-in (implicit) SGS models of a mixed tensorial (generalized) eddy viscosity type. The analysis examined specific implementation aspects of the numerics, for example, specific temporal integration schemes and synchronization issues, and can address how prescribed anisotropies introduced by nonuniform adaptive gridding – affecting *any* LES – contribute to the implicit SGS stress tensor in the modified equations [9, 11]. Key features in the early analysis were the MEA framework and the use of the FV formulation. Volume integrals in the FV representation naturally link with the discrete spatial filtering operation in LES – the top hat filtering associated with the kernel $G(\mathbf{x}, \mathbf{x}') \equiv 1$, and their use readily allows recasting leading order truncation terms in divergence form. MEA of the implicit SGS models shows that FV discretizations are to be preferred over finite differences for ILES [12].

Depending on grid resolution and flow regimes involved, additional explicit SGS modeling ($\sigma \neq 0$) may be needed to address SGS driven flow physics – for example, near walls, and when simulating backscatter, mixing, and chemical reaction. A major research focus is on evaluating the extent to which the particular SGS physical effects can be implicitly modeled as the turbulent velocity fluctuations, recognizing when additional explicit models and/or numerical treatments are needed, and when so, addressing how to ensure that explicit and implicit SGS models effectively act in collaborative mixed fashion.

5.2.1 Finite Scale Navier Stokes

FVs and MEA have been also crucial ingredients in the analysis connecting ILES with the solution of finite scale NS (FSNS) equations for laboratory observables [10, 22]. To $O(L^4, T^4)$ the FSNS equations have the form (see Chapters 3 and 4)

$$\partial \overline{u}_i / \partial t + \overline{u}_j \overline{u}_{i,j} = -\overline{p}_{,i} + v \overline{u}_{i,kk} - L^2 \left[\overline{u}_{j,k} \overline{u}_{i,k} \right]_{,j} - T^2 \left[\overline{u}_{j,t} \overline{u}_{i,t} \right]_{,j},$$
$$\partial \overline{c} / \partial t + \overline{u}_j \overline{c}_{,j} = \kappa \overline{c}_{,jj} - L^2 \left[\overline{u}_{j,k} \overline{c}_{,k} \right]_{,j} - T^2 \left[\overline{u}_{j,t} \overline{c}_{,t} \right]_{,j},$$
$$-\nabla^2 \overline{p} = \overline{u}_{j,i} \overline{u}_{i,j} + L^2 \overline{u}_{j,ik} \overline{u}_{i,jk} + T^2 \overline{u}_{j,ti} \overline{u}_{i,tj},$$

where L and T denote the FSNS characteristic length and temporal scales, respectively, and the bar is used here to denote averaging (top hat filtering) carried out over *space* and *time*:

$$\overline{f} = \frac{1}{\Delta t \Delta x^3} \int\limits_{t-\Delta t/2}^{t+\Delta t/2} dt' \int\limits_{x-\Delta x/2}^{x+\Delta x/2} dx' \int\limits_{y-\Delta y/2}^{y+\Delta y/2} dy' \int\limits_{z-\Delta z/2}^{z+\Delta z/2} dz' f(x',y',z',t').$$

The FSNS framework reframes the computational questions of turbulence. In the continuum, the usual questions relate to the velocity at a point, a concept limited computationally by FVs, mesh spacing Δx, and timestep Δt, and there are associated convergence issues as well as issues of consistent definition of FV averaged velocities. The question FSNS proposes to answer is different (Chapter 4): what is the velocity in a FV (L^3) over a finite time T? There are no convergence issues in the FSNS framework, as the true and desired solution is not the NS solution but rather an average over space and time (L and T) of its solution, and the definition of averaged velocity will not change once $\Delta x < L$ and $\Delta t < T$ – that is, there is no need for SGS modeling once the latter resolution limit can be achieved.

Motivation for FSNS relates to directly addressing the nature of laboratory observables: measurement devices always involve finite space/time scales and cannot compute arbitrarily small scales of high Re turbulence. By design, FSNS actually computes what is observable for a measurement device characterized by length and time scales L and T (e.g., thermocouple or hotwire width L, instrumentation inertia T). Additional motivation for such approaches (beyond addressing the noted inherent limitations of the measuring process) come from practical applications with inherently discrete (noncontinuum) requirements: for example, the needed contaminant dispersal prediction for urban consequences management is dosages (contaminant in volume L^3 over time T).

The demonstrated connections between ILES and the solution of FSNS equations [10, 22] provide a rationale for ILES [23]: ILES works because it solves the equations that most accurately represent the dynamics of FVs of fluid, that is, dynamics governing the behavior of measurable physical quantities on the computational cells.

5.2.2 Coarse Grained Convergence

Since nature controls the flow physics independently of SGS and SPG constraints in the laboratory or numerical experimental process, a legitimate question to ask relates to whether instances exist for some sort of convergence of the larger scale observed flow features and dynamics, that is, of scales much larger than characteristic numerical or instrumental resolution cutoff scales, but presumably small enough, spatially and temporally, that we can assume that they have not been heavily affected by SPG features.

Convergence issues versus resolution in turbulent flow simulations are typically problem dependent and difficult to address in general. Grid independence can be achieved in principle in a properly designed DNS framework. Otherwise, inherent to any CGS approach is the fact that the smallest characteristic resolved turbulence scale is determined by the resolution cutoff wavelength prescribed by an explicit or implicit spatial filtering process. For high Re turbulent flow the conventional wisdom is expecting convergence of mean and second-order moments, and self-similar behavior of higher order turbulence measures is achieved once the cutoff wavenumber lies within the inertial range of the turbulent kinetic energy spectra.

Ideally, well-resolved CGS will involve explicit spatial filtering operations with grid independent filter lengths [24]. However, resolution requirements for such CGS are typically not achievable for problems involving complex geometries and moderate-to-high Re – for which underresolved conditions are unavoidable. In this context, the effective cutoff is most likely determined implicitly by discretization, although the earlier practical convergence ideas are still regarded as relevant. The concept of (resolution dependent) turbulent Re, Re_{eff}, has been extensively invoked in the past [25–27] and in Chapter 1, and formal framework for analysis and definition of Re_{eff} were proposed [28], as in Chapter 2. A reasonable expectation is that Re_{eff} can be evaluated and compared with a critical value such as the mixing transition threshold [29], $Re_{\lambda} = Re_{mix} \sim 100$; developed turbulence metrics become useful once resolution is fine enough to ensure that $Re_{eff} > Re_{mix}$ [27–28]; see Chapter 1. These issues must be directly projected into the process of establishing suitable procedures and metrics for CGS validation.

5.2.3 Transition to Turbulence

Transition to turbulence is traditionally viewed in terms of rapid increase in the energy and the enstrophy production by mode coupling of a spectrum of smaller length scale motions, which can lead to an inertial subrange in the turbulent kinetic energy spectrum for sufficiently high Re [30, 31] – for example, above the cited mixing transition threshold, $Re_{\lambda}=100$ [29].

Transition has been examined in the CGS context for the Taylor–Green vortex (TGV). The TGV is a well-defined flow that has been used as prototype for vortex stretching, instability, and production of small scale eddies to examine the dynamics of transition to turbulence based on DNS [32, 33]. The TVG case has also been used to demonstrate how convective numerical diffusion effects of certain algorithms can be effectively used by themselves to emulate the dominant SGS physics of transition to turbulence for high (but finite) Re flows [26, 34].

CGS strategies have been verified in the TGV context, for a wide range of classical LES and ILES [26]. The studies demonstrated that CGS can robustly capture DNS predicted kinetic energy dissipation rates and transition time – with slight effects of varying (explicit or implicit) SGS model specifics. ILES involved monotonic numerics over uniform grids, including Eulerian (characteristics based Godunov, varying accuracy FCT with or without directional splitting) and van Leer based Lagrange remap algorithms – using Euler or NS equations. ILES results are exemplified in Figure 5.1(a), where effects of varying implicit SGS model specifics are apparent.[2]

Results of the (low subsonic) TGV studies indicated that some sort of Re independent regime is asymptotically attained with ILES with increasing grid resolution

---

[2] Following MEA of LES [11], performance is sensitive to *combined* explicit *and* implicit SGS model specifics [26].

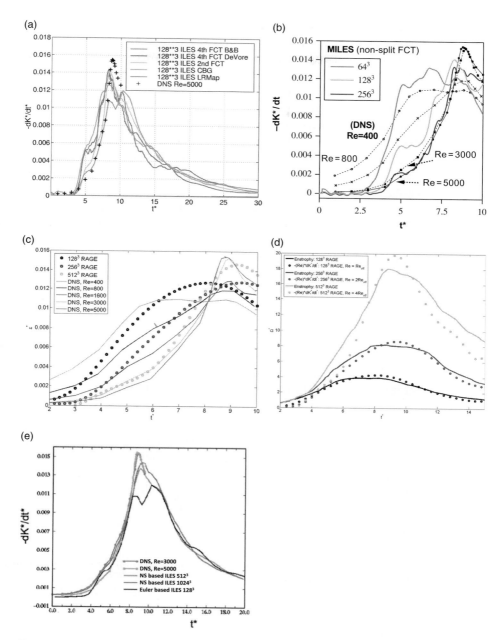

Figure 5.1 TGV kinetic energy dissipation with ILES [26, 34] compared with DNS [32, 33]; Re based on the integral scale. (a) Eulerian third-order characteristics based Godunov, second-/fourth-order FCT codes versus third-order van Leer based Lagrange remap code on uniform 128^3 grid; ILES Euler based, except for NS based split second-order FCT. (b) Fourth-order nonsplit FCT versus resolution. (c) and (d), RAGE. (e) NS and Euler based ILES using the third-order van Leer Lagrange remap code. A black and white version of this figure will appear in some formats. For the color version, of 5.1 (c) and (e) please refer to the plate section.

(Fig. 5.1(b), (c)). Lower observed transition time at dissipation peaks (as well as lower and wider peaks) were predicted by the coarser grid ILES – a trend consistently exhibited by the DNS results as Re is lowered. Moreover, we also found a consistent correlation between nondimensional profiles of mean kinetic energy dissipation rates $-dK^*/dt^*$ and mean enstrophy $\Omega^* = \frac{1}{2}\langle\omega^2\rangle$ [26, 34], where $K = \frac{1}{2}\langle V^2\rangle$. The simple scaling demonstrated in Figure 5.1(d) fits the 128^3 grid results with a selected $Re = Re_{eff} \sim 350$ and then uses $Re = 2Re_{eff}$ and $Re = 4Re_{eff}$ for the corresponding 256^3 and 512^3 data, respectively. Peak Ω^* values increase with grid resolution, and the correlation between Ω^* and $-dK^*/dt^*$ [26] reflects on the unsteady dissipation being resolved with ILES – that is, $Re\ (dK^*/dt^*) = -2\Omega^*$ for an (incompressible) NS fluid with Re is effectively satisfied.

Figure 5.1(e) from [26] compares Euler and NS based ILES with the DNS results [32, 33] near what appears to be a suggested Re independent regime. The NS based ILES for 1024^3 is very close to being fully resolved DNS and extends the reference DNS results beyond the dissipation peak. Good agreement between the fairly coarse 128^3 Euler-based ILES results and the high Re DNS (except near the peak where there is $\sim$27% maximum difference) demonstrate that unresolved viscous effects by ILES are relatively important only in the neighborhood of the dissipation peak – further supporting the view that cascade mechanisms in the high Re limit are driven by Re independent dissipation [35].

5.3   Supergrid Scale: Initial and Boundary Conditions

Although SGS issues have motivated intense research, less attention has been devoted to the equally relevant SPG scale BC modeling aspects, the importance of which is often overlooked. Because SPG choices select flow solutions, emulating particular flow realizations demands precise characterization of initial and asymptotic conditions, as well as conditions at solid and other relevant boundaries – see [36] for a recent survey. The flow characterization issue is a particularly challenging one when laboratory realizations are involved because the available SPG info is typically incomplete.

The impact of SPG specifics in driving the flow dynamics has been recognized in laboratory experiments [37–41], and is clearly emphasized in [40]: "Unlike the theoretician, the experimentalist already knows the solution, for it is the flow he has realized. His objective is to find which equations and which boundary and ICs his solution corresponds to, and then to compare them and his results to those dealt with by the theoretician."

In studying flows developing in space and time the simulated solution must be initialized and only a finite spatial portion of the flow can be investigated. We must ensure that the presence of artificial open boundaries adequately bounds the computational domain without polluting the solution in a significant way [42–44]. Physical issues involved in the BC analysis involve issues of numerical consistency and difficulties in mathematically prescribing actual physical models. To ensure that specifically desired flow realizations are simulated, the boundary condition models must be capable of:

- prescribing effective turbulent ICs,
- emulating eventual feedback effects from presumed virtual flow events outside of the finite sized computational domains at open inflow, outflow, and cross-stream boundaries [44],
- enforcing appropriate flow dynamics and energy transfer near walls [45–49],
- minimizing spurious numerical reflections at all computational boundaries through use of suitable discretized representations.

Because of discretization derivatives can only be approximated at the boundaries. Additional numerical BC (NBC) need to be specified to ensure closure of the discretized system of equations. NBC are distinct from the discretized representations of the physical BCs (PBC) required to uniquely define a solution of the continuum fluid dynamical problem traditionally used as reference. The goal is to ensure that the expected behavior of the latter solution outside the computational domain be properly and consistently imposed on the solution inside. The consistency require- ment demands that, in the continuum limit, the NBC be compatible with the flow equations and PBC, in such a way that they do not generate new BCs that overspecify the fluid dynamical problem. For hyperbolic equations, a relatively simple framework for BC implementation focuses on the terms of the flow equations containing deriva- tives with respect to the (local) direction perpendicular to the boundary [50]. Despite its limitations – being one-dimensional and based on characteristic analysis [43] – this strategy offers a systematic approach [42] to the problem of imposing PBC and NBC in practical simulations.

Traditionally, the loss of memory assumption has been made in turbulence research, that is, IC effects eventually wash out as the turbulence develops. However, a growing body of fundamental research indicates that only very special turbulent flows are truly (universally) self-similar. Far field (late time) sensitivity to ICs has been extensively reported in recent years [39, 51–53]. Robustness of CGS results is an important unsettled issue in this context, and particularly so when nonnegligible backscatter effects become significant – see also Chapter 6. If the IC information contained in the filtered out smaller and SGS spatial scales can significantly alter the evolution of the larger scales of motion and practical integral measures, then the utility of any CGS for their prediction as currently posed is dubious and not rationally or scientifically justifiable.

The selected case studies that follow illustrate crucial IC characterization and model- ing difficulties encountered when attempting to integrate LES and laboratory experi- ments in complex flow problems of interest.

5.3.1   Characterizing Inflow

To illustrate typical PBC prescription requirements, we focus on the open BC problem; [42] presents a discussion of typical requirements for other cases. The number of inflow/ outflow open BCs required to ensure that the flow solution of interest is well posed and uniquely determined within a given finite domain is well known for both Euler and NS equations [54, 55] – and are listed for reference in Table 5.1.

For the sake of discussion, consider the 3D Euler equations, and the problem of specifying the necessary open BCs in the x direction, which we assume to be the streamwise direction. At the inflow boundary, the four inflow PBCs can be chosen to specify the freestream velocity components, plus one additional prescribed quantity – for example, mass density, temperature, *or* pressure. The additional viscous inflow BC condition needed for the case of the NS equations is expected to have small effects on inflow characterization [42].

Different inflow PBC choices providing closure are not equivalent, that is, they do not necessarily lead to the same solution. This is a very important aspect to keep in mind when comparing computational *or* laboratory experiments with presumably very similar but not necessarily identical BCs. Nonreactive General Electric aircraft engines (GEAE), LM-6000 swirl combustor simulations [56, 57] are used in what follows to exemplify this issue. At the subsonic combustor inlet, four primary flow variables can be prescribed through Dirichlet conditions, and at least one other physical quantity must be allowed to float. The available information from the GEAE LM-6000 laboratory studies consisted of the mean (time averaged) profiles of the velocity components at a selected transverse inlet plane, where turbulent velocity fluctuations were only reported to be low. Standard temperature and pressure conditions were expected, but there were no indications from the laboratory data on whether choosing any particular fourth flow variable to specify at the inflow – other than the velocity components – was to be preferred.

Inflow turbulence was either emulated with broad band random fluctuations [56] or neglected altogether [57] in the LM-6000 simulations. In [56] the authors chose to prescribe inlet velocity components and temperature (Suresh Menon, private communication). Two other inlet BC approaches were tested in [57] based on: (1) prescribing the inflow velocity components and mass density and allowing pressure and temperature to float through a characteristic analysis based condition [42], and (2) allowing the inlet radial velocity to float (suggested by Zvi Rusak, private communication). Comparison of mean centerline velocity results in Figure 5.2(a), from nonreactive LES in [56] and [57] clearly demonstrate that the near inlet combustor flow can be quite sensitive to the choice of inlet floating condition; on the other hand, Figure 5.2(b) indicates that LES predictions become fairly robust once adequately similar inflow BCs were involved in the simulations.

5.3.2   Turbulent Inflow

As noted, the sensitivity of turbulent flows to particular IC choices is now well recognized [52]. Far field portions of turbulent flows remember their particular near field features, and the mechanism by which the transition from ICs to particular associated asymptotic flow occurs involves unsteady large scale coherent structure dynamics – which can be captured by LES but not by single point closure turbulence modeling – typical in RANS strategies. As a very particular consequence, starting with

Table 5.1 Number of PBC Required for Well-Posed 3D Subsonic Flow

BC type	Euler	NS
Inflow	4	5
Outflow	1	4

(a) (b)

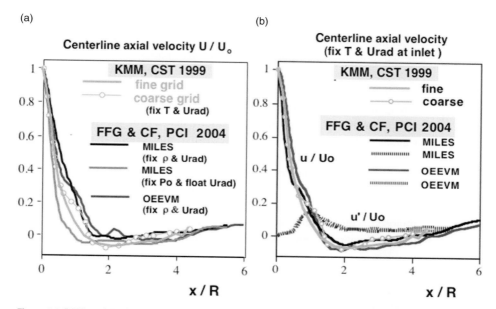

Figure 5.2 LES and MILES of a swirl combustor flow. Sensitivity of the simulated centerline axial velocity within a combustor to actual choices of steady inflow BCs; streamwise variable is scaled with inlet diameter R. Results from [57] were obtained with two different codes based on MILES and one-equation-eddy-viscosity model LES, respectively, and compared with previous (dynamic Smagorinsky) LES results from [56]. A black and white version of this figure will appear in some formats. For the color version, please refer to the plate section.

the typical availability of single point statistical data, there is no unique way to reconstruct a 3D unsteady velocity field with turbulent eddies to define realistic inflow BCs; such data are typically insufficient to parameterize turbulent inflow BCs for LES of inhomogeneous flows [58]. Approaches to modeling turbulent inflow are extensively surveyed in [1]. Because the flow is more or less driven by inflow conditions, prescribed realistic turbulent fluctuations must be able to achieve some sort of equilibrium with imposed mean flow and other BCs. The inherent inability to carry out this inflow BC reconstruction properly in the simulations has led to using a transition inflow region where imposed flow conditions evolve into realistic turbulent velocity fluctuations after allowing for feedback effects to occur as the simulation progresses. The challenge is how to minimize the length of this developmental region, since its presence adds to overall computational cost through additionally required grid points and data processing involved. At a more fundamental level, however, using an artificial transitional inflow region may not be adequate to emulate the actual turbulent inflow conditions involved – a serious issue if ICs are not forgotten. Typical difficulties with characterizing and modeling turbulent inflow conditions in recent practical simulations are exemplified in what follows.

The upwind atmospheric boundary layer characterization directly affects the inflow condition prescription required in urban scenario simulations. Wind fluctuation specifics are major factors in determining urban contaminant transport. The important length scales (tens of meters to kilometers) and time scales (seconds to minutes) in wind gusts can in principle be resolved. However, the flow data from actual field trials or wind tunnel experiments are typically inadequate and/or insufficient to fully characterize the boundary layer conditions required in the urban flow simulation model. In the recent simulations of flow and dispersal over an urban model (cube arrangement) in wind tunnels [59], the available datasets from the laboratory experiments consisted of high quality, spatially dense (but not time resolved) single point statistical data. The inflow velocity in the simulations modeled mean profiles and superimposed fluctuations. An iterative process was used, in which a phenomenological unsteady wind model was calibrated to provide a fit to the experimentally observed *rms* values at locations upstream of the urban model. Agreement was achieved by adjusting amplitude, spatial wavelength, and temporal frequency of the imposed wind fluctuations in an unsteady wind model. This approach provided a practical approximation to the turbulent inflow boundary condition specification problem consistent with the available laboratory data. Particularly relevant insights follow from this work, when comparing predicted and measured volume fractions of a tracer scalar in the first few urban model canyons using various different inflow condition models – shown in Figure 5.3. Prescribing some reasonable inflow turbulence – as opposed to prescribing steady inflow – was found to be critical. On the other hand, the fluid dynamics within the cube arrangement, that is, beyond the first canyon, becomes somewhat insulated from flow events in the boundaries, that is, it is less dependent on the precise details of the modeled inflow turbulence and largely driven by the urban geometry specifics within the urban arrangement.

As noted earlier, using an artificial transitional inflow region to emulate turbulent inflow physics may not be an option This is the case in the context of simulation of complex combustor flows with multiswirl inlets [41, 57, 60, 61], where inlet length controls the coupling between swirl motion and sudden expansion, and thus cannot be arbitrarily varied to accommodate an inflow BC implementation. Promising pseudodeterministic approaches propose generating an interface of inflow BCs for LES based on the use of two point statistics and combined linear stochastic estimation and proper orthogonal decomposition techniques [58]. A linear stochastic estimation approach was used to postprocess the multiswirl inlet velocity data [62]. It enabled the reconstruction of the complete coherent flow fields at the multiswirler outlet from the knowledge of only a few near field acoustic signals, thus providing a promising, effective tool for laboratory data reduction to formulate turbulent inflow BCs for practical LES combustor studies. However, hybrid approaches attempting to represent interface physics with effective BCs should somehow also include the effects of

(a)

(b)

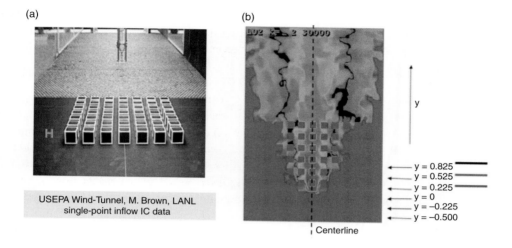

USEPA Wind-Tunnel, M. Brown, LANL
single-point inflow IC data

y = 0.825
y = 0.525
y = 0.225
y = 0
y = −0.225
y = −0.500

Centerline

(c)

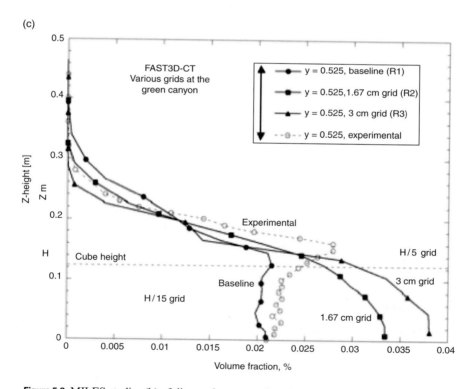

Figure 5.3 MILES studies (b) of dispersal over an urban (cube arrangement) wind-tunnel model (a), from [59]. Colors correspond to predicted and measured dispersal of tracer volume fractions in first few urban canyons indicated on (b); effects of grid resolution (c) and initial conditions (d) are shown.

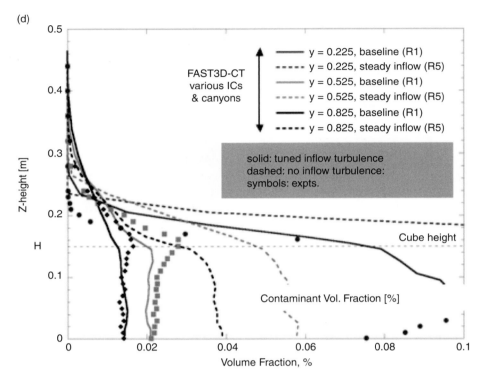

Figure 5.3 (*cont.*)

flow couplings across the BC plane on the development of acoustic or combustion instabilities. Depending on the importance of the latter couplings, and on the specific questions that the simulations are meant to address, carrying out the more expensive larger system LES – including the simulation of the complex inlet geometry flow – may be unavoidable [61].

5.3.3   Initial Material Interface Characterization

In many applications of interest SGS and SPG issues cannot be dealt with independently. This is typically the case when studying near wall flows or material mixing driven by underresolved multiscale turbulent velocity fields, where multiscale resolution issues become compounded with the inherent sensitivity of turbulent flows to ICs. An example is turbulent mixing generated by shock waves via Richtmyer–Meshkov instabilities (RMIs) [63], where vorticity is introduced at material interfaces by the impulsive loading of a shock wave or energy deposition (see, e.g., Chapters 8 and 9). A critical feature of this impulsive driving is that the turbulence decays as dissipation removes kinetic energy from the system. RMI add the complexity of shock waves and other compressible flow effects to the basic physics associated with mixing; compressibility further affects the basic nature of material mixing when mass density and material mixing fluctuation effects are not negligible. Because RMI are shock driven, resolution requirements make DNS unpractical even on the largest supercomputers. State of the art simulations use hybrid methods [64] – switching between shock capturing schemes and conventional LES depending on local flow conditions, or ILES – combining shock and turbulence emulation capabilities [25, 34, 65].

The accurate and reliable simulation of material interfaces is an essential aspect in simulating the material mixing fluid dynamics. Interfaces between fluids can be miscible or immiscible, with the character changing during the evolution of a system. This causes numerical approximations to the physics to be extremely challenging because the basic features of the approximations need to be inherently dynamic, that is, adapt to the evolution of the materials. For example, in many applications of interest (e.g., inertial confinement fusion capsules) an interface may begin as sharp and immiscible, but evolves into a state where it mixes at an atomic level with neighboring material due to the effects of temperature and difussion, or as it becomes a plasma. There exist a number of outstanding challenges in the material interface modeling area arising from the relative weakness of mathematical foundations associated with the techniques available for solutions [66]. This is contrasted by shock wave theory, where convergence guarantees are based upon conservation and entropy production [67, 68]; for example, guidance given by the Lax-Wendroff theorem [67] for captured shock waves (conservation leads to weak solutions, entropy leads to uniqueness) is not yet available for the mixing problem. Moreover, among the most crucial issue associated with interfaces is the sensitivity of most problems to the ICs: relatively small variations in the initial state of the interface can result in quite significant changes to even the integral character of a mixing layer at late times [34, 69, 70]. The inherent difficulties with the open problem of predictability of material stirring and mixing by underresolved multiscale turbulent velocity fields are now compounded with the inherent sensitivity of turbulent flows to ICs [52].

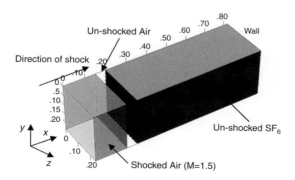

Figure 5.4 Typical planar shock-tube computational domain; lengths in m.

A typical shock tube configuration used in the investigations involves gases of significantly different density – for example, low (air) and high (SF6) in Figure 5.4 from [34] – see also Chapter 8, presumed ICs at the material interface initially separating the gases, periodicity in the transverse (y, z) directions, and allowing eventual shock reflection off an end wall to emulate reshock. The shock propagates in the (x) direction through the contact discontinuity, and the detailed later evolution of the shocked mixing layer is investigated. The surface displacement of the material interface in the RMI experiments [71] has been modeled using well-defined modes often combined with random perturbation components. In the early simulation work [25] the shocked planar RMI phase was examined in detail using local S+L deformations,

$$\begin{aligned} dx(y, z) &= S + L \\ &= A_1 |\sin(k_S y)\sin(k_S z)| + A_2 \sin(k_L y)\sin(k_L z), \end{aligned}$$

combining a short (S) egg crate mode – chosen to represent the result of pushing the membrane through the wire mesh, and superimposed distortion of the wire mesh on a longer characteristic shocktube scale L – chosen to be the transverse periodicity dimension of the computational domain. In later papers the reshocked mixing layer was also simulated, and the IC modeling strategies included S+ϕ deformation [34, 64], where ϕ is a spatially random distribution intended to break the presumed characteristic interface symmetries; simulations were performed for a variety of random perturbations superimposed [34] *or not* superimposed [72] to the S mode – see Chapter 8.

The prescribed range of modes in ϕ, typically requires that wavelengths be resolved by the computational grid (e.g., $\lambda_{min} \geq 8$ computational cells, $\lambda_{max} \leq L$), including $\lambda_o = \pi/k_S$, the characteristic length scale of the egg crate mode. Typically, A_1, A_2, and $\lambda_o = L/12 \sim 2$ cm are chosen identical or very similar [25, 34, 64] and for the sake of computational resolution efficiency, the characteristic initial interface (A_1) was chosen thicker than that in the laboratory experiments (~ 0.1 cm); moreover, the perturbing amplitude A_2 is chosen as a fraction, say 10%, of A_1, and only comparisons with laboratory mixing layer growth rates were attempted.

Mixing analysis can be carried out in terms of suitable variables and ensemble averaging,

$$\begin{aligned} \langle f \rangle(x) &= \frac{1}{A}\int f(x, y, z)dydz, \quad A = \int dydz, \\ Y_{SF_6} &= \rho_{SF_6}/\rho, \\ \theta &= 4Y_{SF_6}[1 - Y_{SF_6}], \\ \psi(x) &= \langle Y_{SF_6} \rangle, \\ M(x) &= 4\psi(x)[1 - \psi(x)], \end{aligned}$$

where $<\,>$ denotes ensemble (cross-stream plane) averaging, ρ_{SF6} is the SF_6 mass density, Y_{SF6} is the SF_6 mass fraction, and θ and $M(x)$ are local and averaged mixedness measures.

A key issue first addressed in [34] is that of effects of underresolving the $|\sin(k_o y)\sin(k_o z)|$ function defining the egg crate mode S. In [25] the characteristic transverse length of the computational domain, $10 \times \lambda_o$, was resolved with 512, 1024, or 2048 computational cells in each direction – that is, at least fifty cells per characteristic scale λ_o. Figure 5.5(a) from [25] shows distinct effects of underresolution on the predicted mixedness measures when using the coarsest resolution, ~50 cells per λ_o; differences are fairly apparent in particular, for nondimensional time < 5 corresponding to physical time $<$~4 ms. The results in Figure 5.5(a) correspond to purely shocked growth of RMI (no reshock) and associated with having only the egg crate mode as IC.

In [25] evidence was shown of both laminar and transitional behavior on first shock for a given Atwood number A and shock Mach number Ma depending on grid resolution – and associated effective Re [27, 28] (see Chapters 1 and 2). A comparison of renderings of vertical slices of the simulated flow at a selected late time for various resolutions (Fig.5.5(b)) indicates qualitatively different flow behavior at higher resolutions (1024^3 and $2048^2 \times 1920$) compared with lower resolution (384^3). The latter case is characterized by well-defined structures that become sharper with resolution; the higher resolution simulations depict abundant finer scale "turbulent" features consistent with suggested inertial ranges in the characteristic spectra – which are longer for higher resolution (see Fig. 5.5(c), also from [25]). This resembles the mixing transition with increasing Re suggested in [29], with high Re associated here with the numerical algorithm residual dissipation.

In [64] the transverse domain length $12 \times \lambda_o$ was resolved with 128–256 cells (~ 10–20 cells per λ_o); in [21], a single grid was used for which $10 \times \lambda_o$ was resolved with 257 cells (~ 13 cells per λ_o), whereas $12 \times \lambda o$ was resolved with 10–20 cells per λ_o in the present work. It could be argued that high order methods used in [21] and [64] – fifth- and ninth-order WENO, respectively – compared using second-order Godunov methods in [25] and [34] would presumably require using fewer cells per wavelength to achieve comparable resolution of the egg crate term $| \sin(k_o y) \sin(k_o z)|$; however, as already noted, all shock capturing methods degrade to first-order in the vicinity of shocks, and in particular, at the very important stage at which shocks interact with the material interface.

(a)

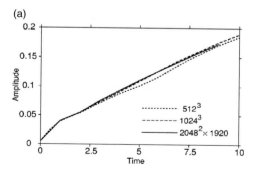

Figure 5.5 Computed shocked driven mixing versus grid resolution. The coarsest case (512^3) used more than ~50 cells per characteristic scale λ_o from [25].

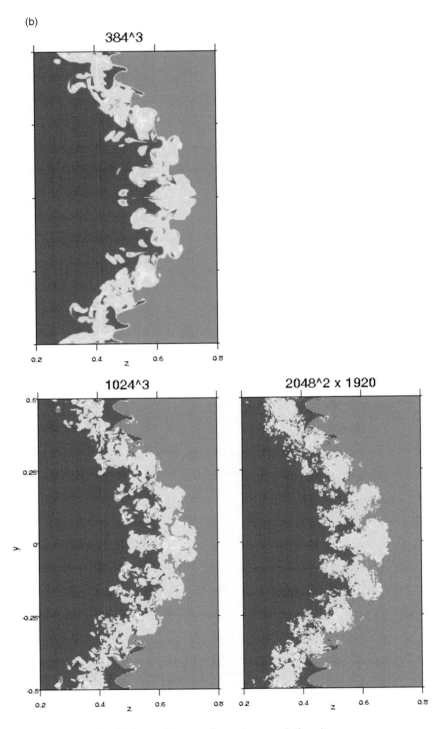

Figure 5.5 (*cont.*, distributions of entropy for various resolutions.)

(c)

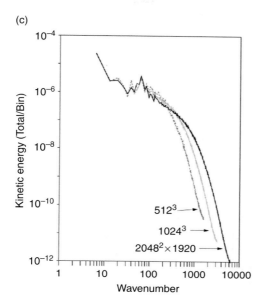

Figure 5.5 (*cont.,* power spectra for various resolutions.)

In [64] the IC perturbation ϕ was generated once for all grids and there is only indirect suggestion of consistency of mixedness results for one particular (Ma=1.5) case with 0.2 cm and 0.1 cm resolutions. Based on private communications (Dale Pullin) – see [34] – it was further clarified that, compared with [25], the term $|\sin(k_o y)\sin(k_o z)|$ was similarly underresolved in [21] and [64]. Because of the grid dependent spatial filtering and IC definition, the potential sensitivity of the reported results to grid resolution is expected – for example, by noting that smaller scales have faster growth according to Richtmyer's formula [73].

On Figure 5.6(a) spectral characteristics are shown for the egg crate function $|\sin(k_o y)\sin(k_o z)|$ as resolved on a 0.1 cm grid, 240×240 cell transverse stencil, depicting a major peak associated with the characteristic egg crate wavelength λ_o and higher wavenumber peaks at its harmonics. On Figure 5.6(b) a similar analysis is shown for a coarsened egg crate version based on keeping only a few leading terms of the 2D fast Fourier expansion, while the arrows indicate the relevant cutoffs for the 0.1 cm, 0.05 cm, and 0.025 cm grids. We denote these as baseline egg crate (BEC) and baseline coarsened egg crate (BCEC), respectively.

For the 2D analysis that follows (Fig. 5.7(a) and (b)), we use the appropriate analogue of the IC deformation defined earlier as

$$dx(y)^{2D} = S^{2D} + \phi^{2D} = A_1|\sin(k_o y)| + A_2\sum_{n,m} a_{n,m}\sin(k_n y + \varphi_n)$$

Figure 5.7(a) shows mixedness results of 2D simulations carried out with the BEC IC (no perturbation), on the 0.1 cm, 0.05 cm, and 0.025 cm resolutions, which clearly

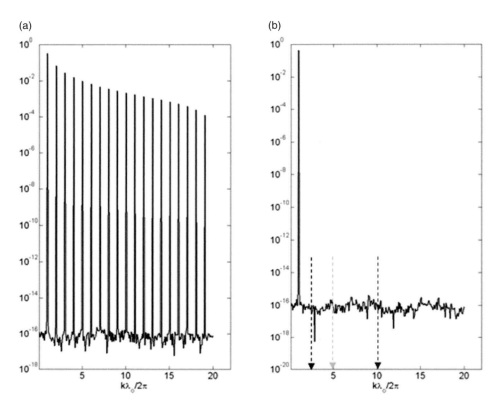

Figure 5.6 Spectral characteristics of the egg crate function $|\sin(k_o y)\sin(k_o z)|$; (a) resolved for the 0.1 cm grid, 240×240 transverse stencil (left); (b) similar analysis of a coarsened egg crate mode definition is shown on the right, where the arrows indicate the cutoffs for the 0.1 cm, 0.05 cm, and 0.025 cm grids.

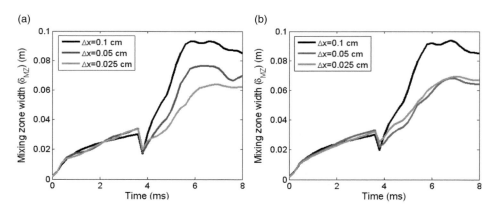

Figure 5.7 (a) 2D computed shocked mixing layer width with non-perturbed BEC egg crate mode vs. grid resolution. (b) 2D computed shocked mixing layer width with non-perturbed BCEC egg crate mode vs. grid resolution.

depict nonconverged results after reshock. Corresponding results shown on Figure 5.7 (b) for the BCEC IC suggest convergence for the two finest resolutions once twenty or more cells per λ_o are involved in the IC prescription.

The egg-crate functional prescription was similarly under-resolved historically [21, 34, 64]. In hindsight, our earlier discussion and analysis indicates that ICs should be defined in robust grid independent fashion (e.g., as the BCEC) to ensure well-posed flow realizations. On the other hand, it is clear that actual predicted mixing layer thickness growth thus depends on IC resolution determining the effective IC associated with the computed solutions (i.e., effective egg crate definition involved); moreover, the fact that solutions associated with different resolutions are associated with correspondingly different values of some characteristic (initial) effective Re – for example, Chapter 1 – should be reiterated in the present context.

Growth rates of the shocked mixing layer depend on magnitude and sign of A, and are consistently much higher after reshock [74]; transition after reshock is typically observed [34, 64, 75] and extreme sensitivity to detailed conditions at reshock time has been reported [34, 72, 77, 78]. Before reshock, transition has been [25, 75] or not been [25, 34, 64] observed; transition occurrence depends on particular combination of IC specifics such as A [74], Ma [79, 80], and Re [25]; as noted, IC resolution issues are also likely relevant in the transition context. The work in [79, 80] reports very notable RMI growth sensitivity to Ma, particularly with regards to the transition process [80]: as little as a 15% variation in Ma can make the difference for a shocked interface to undergo or not undergo transition.

Our own low η_o first shocked results – see also Chapter 8 – are consistent (in not showing transition before reshock) not only with our own previously reported ($A>0$, low-η_o) ILES results *before reshock* [34] but also with those of in [64] using a classical LES strategy, different numerics, and similar resolution for the same planar RMI problem, and (as in [64]) predict mixing width growth rates in agreement with the laboratory experiments [71]. The fact that both cited simulations [34, 64] show *nontransition* before reshock suggests an effective (resolution dependent) Re below the mixing transition Re threshold [29]. For the sake of completeness of the IC prescription process – that is, ensuring that uniquely defined physical realizations are simulated – it appears that initial Re values based on ICs before first shock should also be prescribed. Initial Re was not explicitly specified in the reported work [21, 25, 34, 64, 77]. Although less sensitivity to IC may result for higher initial effective Re – as turbulent (more universal) IC are involved [29, 31], in the absence of a reason to choose particular initial Re values – dictated by the laboratory experiments being simulated, both classes of shock-driven flows (initially laminar and transitional) constitute physically legitimate flow realizations.

5.4   Concluding Remarks

Accurate predictions with quantifiable uncertainty are essential to many practical turbulent flow applications exhibiting extreme geometrical complexity and broad ranges of length and time scales. Underresolved computer studies are unavoidable in such applications, and LES becomes the effective simulation strategy mostly by necessity rather than by choice. We have noted the inherently intrusive nature of both computed and laboratory experiments due to their potential sensitivity to SGS and SPG aspects, which often appear intertwined. Crucial challenges in this context are then related to identifying and modeling the unresolved SGS and SPG features to fully characterize the flow realizations to be simulated and assessing the associated uncertainties of the LES predictions.

SGS issues have motivated intense research in the last four decades. A crucial practical computational issue is the need to distinctly separate the effects of explicit filtering and SGS reconstruction models from the unavoidable implicit ones due to discretization. ILES addresses the seemingly insurmountable issues posed to LES by underresolution, by focusing on the use of SGS modeling and filtering provided implicitly by physics capturing numerics. ILES has been successfully applied to broad range of free and wall bounded flows, ranging from canonical benchmark flows to extremely complex flows at the frontiers of current flow simulation capabilities. The performance of representative LES and ILES approaches is found to be equally good in emulating physical statistics of turbulent velocity fluctuations, and there is no discriminating characteristic favoring one or another. Looking toward practical complex flows and regimes, however, the ability of ILES to offer a simpler computational environment should be clearly emphasized.

Depending on grid resolution and flow regimes involved, additional explicit SGS modeling may be needed with ILES to address SGS driven flow physics – for example, near walls, and when simulating backscatter, material mixing, or chemical reaction. A major research focus is on evaluating the extent to which the particular SGS physical effects can be implicitly modeled as the turbulent velocity fluctuations, recognizing when additional explicit models and/or numerical treatments are needed, and when so, addressing how to ensure that the mixed explicit and implicit SGS models effectively act in collaborative rather than interfering fashion. An important challenge in this context is that of improving MEA, the mathematical and physical framework for its analysis and development, further understanding the connections between implicit SGS model and numerical scheme, and addressing how to build SGS physics into it.

Less attention has been historically devoted to SPG modeling aspects. A special focus of our discussion here has been on turbulent initial and inflow condition issues. We illustrated crucial characterization and modeling difficulties encountered when integrating LES and laboratory experiments in complex inhomogeneous flow problems of interest. The laboratory data is typically insufficiently characterized; different IC and BC choices consistent with the available information are not equivalent and can lead to significantly different flow solutions. One difficult issue is that of identifying complete datasets required to close the ICs and BCs in the simulations, the other involves appropriate laboratory data acquisition and reduction strategies to characterize the relevant upwind flow physics.

SGS and SPG modeling issues are unavoidably intertwined as modeling difficulties due to underresolution become compounded with the inherent sensitivity of turbulent flows to choice of ICs and BCs. Idealized fundamental problems, such as the decay of turbulence simulated in a box domain with mathematically well-defined periodic BCs, may appear to avoid confronting questions of initial and BCs. However, periodic box simulations actually involve all the difficult SGS and SPG issues. As noted, it is now well recognized that turbulent flows remember their ICs. Moreover, it has also been established that for sufficiently long simulation times, the integral scale of turbulence will eventually saturate since the larger simulated scales cannot have unaffected growth beyond the box size, and will eventually distort the characteristic power law of the turbulence decay [26, 81].

5.5   Acknowledgements

Los Alamos National Laboratory is operated by the Los Alamos National Security, LLC for the U.S. Department of Energy NNSA under Contract No. DE-AC52-06NA25396.

References

1. Sagaut P. 2006, *Large Eddy Simulation for Incompressible Flows*, 3rd ed., Springer.
2. Grinstein, F.F, Margolin, L.G., and Rider, W.J., editors, 2010, *Implicit Large Eddy Simulation: Computing Turbulent Fluid Dynamics*, 2nd printing, Cambridge University Press.
3. Spalart, P.R. 2000, "Strategies for turbulence modeling and simulations," *Heat and Fluid Flow*, 21, 252–263.
4. Spalart, P.R. 2009, "Detached Eddy Simulation," *Annu. Rev. Fluid Mech.* 41, 181–202.
5. Leschziner, M., Li, N., and Tessicini, F. 2009, "Simulating flow separation from continuous surfaces: Routes to overcoming the Reynolds number barrier," *Phil. Trans. R. Soc. A* 367, 2885–2903.
6. Grinstein, F.F. 2009, "On integrating large eddy simulation and laboratory turbulent flow experiments," *Phil. Trans. R. Soc. A*, 67, 2931–2945.
7. Hirt, C.W. 1969, "Computer studies of time-dependent turbulent flows," *Phys. Fluids* II, 219–227.
8. Ghosal, S. 1996, "An analysis of numerical errors in large-eddy simulations of turbulence," *J. Comp Phys.*, 125, 187–206.
9. Fureby, C. and Grinstein, F.F., 1999, "Monotonically integrated large eddy simulation of free shear flows," *AIAA Journal*, 37, 544–56.
10. Margolin, L.G. and Rider, W.J. 2002, "A rationale for implicit turbulence modeling," *Int. J. Numer. Methods Fluids*, 39, 821–841.
11. Grinstein, F.F. and Fureby, C. 2007, "On flux-limiting-based implicit large eddy simulation," *J. Fluids Eng.* 129, 1483–1492.
12. Margolin, L.G. and Rider, W.J. 2010, "Numerical regularization: The numerical analysis of implicit subgrid models," ch. 5 in *Implicit Large Eddy Simulation: Computing Turbulent Fluid Dynamics*, ed. by F.F. Grinstein, L.G. Margolin, and W.J. Rider, 2nd printing, Cambridge University Press.
13. Boris J.P. 1990, "On Large Eddy Simulation Using Subgrid Turbulence Models," in *Whither Turbulence? Turbulence At the Crossroads*, ed. by J.L. Lumley, Springer, 344.

14. Boris, J.P., Grinstein, F.F., Oran, E.S., and Kolbe, R.J. 1992, "New insights into large eddy simulation," *Fluid Dyn. Res.* 10, 199–228.
15. Pope, S.B. 2004, "Ten questions concerning the large eddy simulation of turbulent flows," *New J. Phys*, 6, 35.
16. Harten, A. 1983, "High resolution schemes for hyperbolic conservation laws," *J. Comput. Phys.* 49, 357–393.
17. Fan, M., Wenren, Y., Dietz W., Xiao, M., and Steinhoff, J. 2002, "Computing blunt body flows on coarse grids using vorticity confinement," *Journal Fluids Engineering* 124, 876–885.
18. Domaradzki, J.A. and Adams, N.A. 2002, "Direct modeling of subgrid scales of turbulence in large eddy simulations," *Journal of Turbulence*, 3, 024.
19. Stolz, S., Adams, N.A., and Kleiser, L. 2001, "The approximate deconvolution model for LES of compressible flows and its application to shock-boundary-layer interaction," *Phys. Fluids*, 13(2985).
20. Visbal, M. and Rizzetta, D. 2002, "Large-eddy simulation on curvilinear grids using compact differencing and filtering schemes," *Journal of Fluids Engineering*, 124, 836–847.
21. Schilling, O. and Latini, M. 2010, "High-order WENO simulations of three-dimensional reshocked Richtmyer–Meshkov instability to late times: Dynamics, dependence on initial conditions, and comparisons to experimental data," *Acta Mathematica Scientia*, 30B(2), 595–620.
22. Margolin L.G., 2009 "Finite-scale equations for compressible fluid flow," *Phil. Trans. R. Soc. A*, 367, 2861–2871.
23. Grinstein, F.F., Margolin, L.G., and Rider, W.J. 2010, "A rationale for implicit LES," ch. 2 in *Implicit Large Eddy Simulation: Computing Turbulent Fluid Dynamics*, ed. by F.F Grinstein, L.G. Margolin, and W.J. Rider, 2nd printing, Cambridge University Press.
24. Bose, S.T., Moin, P., and You, D. 2010, "Grid-independent large-eddy simulation using explicit filtering," *Phys. Fluids*, 22, 1–11.
25. Cohen, R.H., Dannevik, W.P., Dimits, A.M., Eliason, D.E., Mirin, A.A., Zhou, Y., Porter, D.H., and Woodward, P.R. 2002, "Three-dimensional simulation of a Ritchmyer–Meshkov instability with a two-scale initial perturbation," *Physics of Fluids*, 14, 3692–3709.
26. Drikakis, D., Fureby, C., Grinstein, F.F., and Youngs, D. 2007, "Simulation of transition and turbulence decay in the Taylor–Green vortex," *Journal of Turbulence*, 8, 020.
27. Wachtor, A.J., Grinstein, F.F., Devore, C.R., Ristorcelli, J.R., and Margolin, L.G. 2013, "Implicit large-eddy simulations of passive scalar mixing in statistically stationary isotropic turbulence," *Physics of Fluids*, 25, 025101.
28. Zhou, Y., Grinstein, F.F., Wachtor, A.J., and Haines, B.M., 2014, "Estimating the effective Reynolds number in implicit large eddy simulation," *Physical Review E*, 89, 013303.
29. Dimotakis, P.E. 2000, "The mixing transition in turbulent flows," *J. Fluid Mech.*, 409, 69–98.
30. Tennekes, H. and Lumley, J.L. 1972, *A First Course in Turbulence*, MIT Press.
31. Zhou, Y. 2007, "Unification and extension of the similarity scaling criteria and mixing transition for studying astrophysics using high energy density laboratory experiments or numerical simulations," *Phys. of Plasmas*, 14, 082701.
32. Brachet, M.E., Meiron, D.I., Orszag, S.A., Nickel, B.G., Morg, R.H., and Frisch, U.J. 1983, "Small-scale structure of the Taylor-Green vortex," *J. Fluid. Mech.* 130, 411.
33. Brachet, M.E. 1991, "Direct numerical simulation of three-dimensional turbulence in the Taylor–Green vortex," *Fluid Dyn. Res.* 8(1).

34. Grinstein, F.F., Gowardhan, A.A., and Wachtor, A.J. 2011, "Simulations of Richtmyer–Meshkov instabilities in planar shock-tube experiments," *Physics of Fluids* 23, 034106.

35. Kaneda, Y., Ishihara, T., Yokokawa, M., Itakura, K., and Uno, A. 2003, "Energy dissipation rate and energy spectrum in high resolution direct numerical simulations of turbulence in a periodic box," *Phys. Fluids*, 15, L21.

36. Grinstein, F.F., editor, 2004, "Boundary conditions for large eddy simulation, special section," *AIAA Journal*, 42, 437–492.

37. Hussain, A.K.M.F. and Zedan, M.F. 1978, "Effects of the initial condition on the axisymmetric free shear layer: effect of the initial fluctuation level," *Phys. Fluids*, 21, 1475–1481.

38. Gutmark, E. and Ho, C.M. 1983, "Preferred modes and the spreading rates of jets," *Phys. Fluids*, 26, 2932.

39. Wygnanski, I., Champagne, F., and Marasli, B., 1986, "On the large-scale structures in two-dimensional, small-deficit, turbulent wakes," *Journal Of Fluid Mechanics* 168, 31–71.

40. George, W.K. 1990, "Governing equations, experiments, and the experimentalist," *Exp. Thermal Fluid Sc.* 3, 557–566.

41. Li, G. and Gutmark, E.J. 2006, "Experimental study of boundary condition effects on non-reacting and reacting flow in a multi-swirl gas turbine combustor," *AIAA. J.* 44, 444.

42. Poinsot, T.J. and Lele, S.K. 1992, "Boundary conditions for direct simulations of compressible viscous flows," *J. Comp. Phys.* 101, 104.

43. Colonius, T., Lele, S.K., and Moin P. 1993, "Boundary conditions for direct computations of aerodynamic sound," *AIAA J.*, 31, 1574.

44. Grinstein, F.F. 1994, "Open boundary conditions in the simulation of subsonic turbulent shear flows," *J. Comp. Phys.* 115, 43–55.

45. Spalart, P.R., Jou, W.H., Strelets, M., and Allmaras, S.R. 1997, "Comments on the Feasibility of LES for Wings, and on Hybrid RANS/LES Approach," in *Advances in DNS/LES, First AFOSR International Conference in DNS/LES*, Greyden Press.

46. Kong, H., Choi, H., and Lee, J.S. 2000, "Direct numerical simulation of turbulent thermal boundary layers," *Phys. Fluids*, 12, 2555.

47. Fureby, C., Alin, N., Wikström, N., Menon, S., Persson, L., and Svanstedt, N. 2004, "On large eddy simulations of high Re-number wall bounded flows," *AIAA. J.* 42, 457.

48. Sagaut, P., Garnier, E., Tromeur, E., Larchevêque, L., and Labourasse, E. 2004, "Turbulent inflow conditions for large-eddy simulation of compressible wall-bounded flows," *AIAA J.* 42, 469.

49. Sidwell, T., Richards, G., Casleton, K., Straub, D., Maloney, D., Strakey, P., Ferguson, D., Beer, S., and Woodruff, S. 2006, "Optically accessible pressurized research combustor for computational fluid dynamics model validation," *AIAA. J*, 44, 434.

50. Thompson, K. 1990, "Time dependent boundary conditions for hyperbolic systems, II," *J. Comp. Phys.* 89, 439.

51. Slessor, M.D., Bond, C.L., and Dimotakis, P.E. 1998, "Turbulent shear-layer mixing at high Reynolds numbers: effects of inflow conditions," *J. Fluid Mech.*, 376, 115–138.

52. George, W.K. and Davidson, L., 2004, "Role of initial conditions in establishing asymptotic flow behavior," *AIAA Journal* 42, 438–446.

53. Ramaprabhu, P., Dimonte, G., and Andrews, M.J. 2005, "A numerical study of the influence of initial perturbations on the turbulent Rayleigh–Taylor instability," *J. Fluid Mech.* 536, 285-319-23.

54. Strikwerda, J.C. 1977, "Initial value boundary value problems for incompletely parabolic systems," *Comm. Pure Appl. Math.* 30, 797.

55. Oliger, J. and Sundstrom, A. 1978, "Theoretical and practical aspects of some initial boundary value problems in fluid dynamics," *SIAM J. Appl. Math.* 35, 419.

56. Kim W.-W., Menon, S., and Mongia, H.C. 1999, "Large-eddy simulation of a gas turbine combustor flow," *Combust. Sci. Tech.* 143, 25–62.

57. Grinstein, F.F. and Fureby, C. 2004, "LES studies of the flow in a swirl gas combustor," *Proc. of the Comb. Inst.* 30, 1791.

58. Druault, P., Lardeau, S., Bonnet, J.P., Coiffet, F., Delville, J., Lamballais, E., Largeau, J.F., and Perret, L. 2004, "Generation of three-dimensional turbulent inlet conditions for large-eddy simulation," *AIAA Journal*, 42, 447–456.

59. Patnaik, G., Boris, J.P., Young, T.R., and Grinstein, F.F. 2007, "Large scale urban contaminant transport simulations with MILES," *J. Fluids Eng.* 129, 1524–1532.

60. Grinstein, F.F., Young, T.R., Gutmark, E.J., Li, G., Hsiao, G., and Mongia, H.C. 2002, "Flow dynamics in a swirl combustor," *J. of Turbulence* 3, 1468.

61. Fureby, C., Grinstein, F.F., Li, G., and Gutmark. E.J. 2007, "An experimental and computational study of a multi-swirl gas turbine combustor," *Proc. of the Combustion Institute*, 31, 3107–3114.

62. Verfaillie, S., Gutmark, E.J., Bonnet, J.P., and Grinstein, F.F. 2006, "Linear stochastic estimation of a swirling jet," *AIAA J.*, 44, 457–468.

63. Brouillete, M. 2002, "The Ritchmyer–Meshkov instability," *Annu. Rev. Fluid Mech.*, 34, 445–468.

64. Hill, D.J., Pantano, C., and Pullin, D.I. 2006, "Large-eddy simulation and multiscale modelling of a Ritchmyer–Meshkov instability with reshock," *J. Fluid Mech.* 557, 29–61.

65. Youngs, D.L. 2007, Rayleigh–Taylor and Richtmyer–Meshkov mixing," ch. 13 in *Implicit Large Eddy Simulation: Computing Turbulent Fluid Dynamics*, ed. by F.F. Grinstein, L.G. Margolin, and W.J. Rider, 2nd printing, Cambridge University Press, 2010.

66. Rider, W. and Kothe, D. 1998, "Reconstructing volume tracking," *J. Comp. Phys.* 141, 112–152.

67. Lax, P. and Wendroff, B. 1960, "Systems of conservation laws," *Comm. Pure Appl. Math.* 13, 217–237.

68. Leveque, R.J. 1990, *Finite Volume Methods for Hyperbolic Problems*, Cambridge University Press.

69. Drikakis, D.,Grinstein, F.F., and Youngs, D. 2005, "On the computation of instabilities and symmetry-breaking in fluid mechanics," *Progress in Aerospace Sciences* 41(8), 609–641.

70. Dimonte, G. 1999 "Nonlinear evolution of the Rayleigh–Taylor and Richtmyer–Meshkov instabilities," *Phys. Plasma* 6(5), 2009–2015.

71. Vetter, M. and Sturtevant, B. 1995, "Experiments on the Richtmyer–Meshkov instability of an air/ SF6 interface," *Shock Waves* 4, 247–252.

72. Gowardhan, A.A., Ristorcelli, J.R. and Grinstein, F.F. 2011, "The bipolar behavior of the Richtmyer–Meshkov instability," *Physics of Fluids* 23 (Letters), 071701.

73. Richtmyer, R.D. 1960, "Taylor instability in shock acceleration of compressible fluids," *Commun. Pure Appl. Math.* 13, 297.

74. Lombardini, M., Hill, D.J., Pullin, D.I., and Meiron, D.I. 2011, "Atwood ratio dependence of Richtmyer–Meshkov flows under reshock conditions using large-eddy simulations," *J. Fluid Mech.*, 670, 439–480.

75. Thornber, B., Drikakis, D., Youngs, D.L., and Williams, R.J.R. 2012, "Physics of the single-shocked and reshocked Richtmyer–Meshkov instability," *J. of Turbulence* 13(10), 1–17.

76. Orlicz, S.G.C., Prestridge, K.P., and Balakumar, B.J. 2012, "Experimental study of initial condition dependence on Richtmyer-Meshkov instability in the presence of reshock," *Physics of Fluids* 24, 034103.

77. Gowardhan, A.A. and Grinstein, F.F. 2011, "Numerical simulation of Richtmyer–Meshkov instabilities in shocked gas curtains," *Journal of Turbulence* 12(43), 1–24.

78. Ristorcelli, J.R., Gowardhan, A.A., and Grinstein, F.F. 2013, "Two classes of Richtmyer–Meshkov instabilities: A detailed statistical look," *Phys. Fluids* 25, 044106.

79. Leinov, E.,Malamud, G., Elbaz, Y., Levin, I.A., Ben-dor, G., Shvarts, D., and Sadot, O. 2009, "Experimental and numerical investigation of the Richtmyer–Meshkov instability under re-shock conditions," *J. Fluid Mech.* 626, 449–475.

80. Orlicz, G.C., Balakumar, B.J., Tomkins, C.D., and Prestridge, K.P. 2009 "A Mach number study of the Richtmyer–Meshkov instability in a varicose, heavy-gas curtain," *Phys. Fluids* 21, 064102.

81. Wang, H.L. and George, W.K. 2002, "The integral scale in homogeneous isotropic turbulence," *J. Fluid Mech.* 459, 429–443.

6   Cloud Modeling
An Example of Why Small Scale Details Matter for Accurate Prediction

Jon Reisner

6.1   Introduction

One of the basic tenets of weather forecasting is that predictions can be made only over a finite length of time. In particular, errors limiting predictive capability of numerical models are associated with the following: initial and boundary conditions, numerical formulation, and physical parameterizations. The impact of these errors on model predictability can become even more significant when simulating highly nonlinear weather systems such as a hurricane interacting with a mid-latitude system; the errors can be rapidly amplified, leading to a forecast that is completely out of phase with the subsequent event. For example, several days before the actual event hurricane Sandy was forecasted by the the United States Global Forecast System (GFS) model to harmlessly go out to sea, whereas the European Medium Range Forecast model (ECMWF) had the hurricane curving back to the coast. Of importance to the findings discussed later in this chapter is that subsequent analysis has suggested that the fields of latent heat release associated with hurricane Sandy needed to be adequately resolved to enable a forecast model to simulate the correct westward track [15]. But though the ECMWF was able to get the right track, in general, predicting rapid changes in hurricane intensity has been one of the few areas in numerical weather prediction that have not seen significant improvements over the past fifty years [4]. Further, in contrast to the ongoing theme in this book that small scale turbulent processes are enslaved by larger scale processes, for clouds, small scale processes such as evaporation can have a significant impact on the larger scale system.

When water vapor changes phase from gas to liquid via attachment onto aerosol – that can be the size of individual gas molecules – this very small scale and fast condensational process releases energy. Combined with the reverse process, evaporation, the integral impact of this latent heat release is what drives weather systems such as hurricanes. Note that because of particle size dependence in models representing condensation (or evaporation), during evaporation small droplets quickly disappear leaving larger droplets that fall rapidly, forming small scale filaments at cloud boundaries and/or within regions of downward motions (e.g., virga). In this chapter, an attempt to resolve these filaments will be made and their impact on dynamical fields via a subgrid parameterization will not be examined.

Unfortunately, when simulating clouds numerical models can suffer from large errors, especially at boundaries. Specifically, advection of clouds in a Eulerian

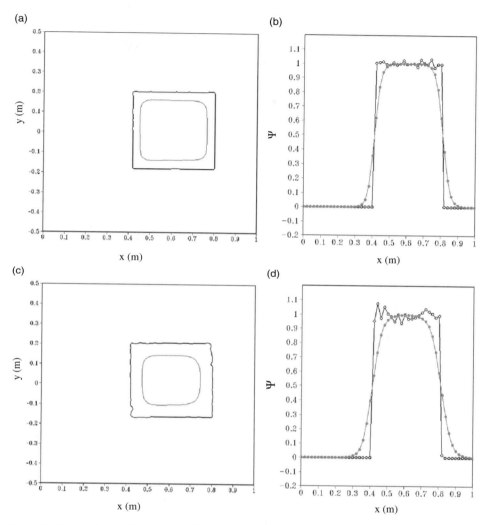

Figure 6.1 Comparison of Lagrangian (black lines) and Eulerian (gray lines) approaches for advection of a rotating square of field Ψ with a fifth-order WENO scheme being used in the Eulerian formulation. The contour plot ($\Psi = 0.9$) and cross-section (through y=0) in panels (a) and (c) are after the first revolution, whereas panels (b) and (d) are after ten revolutions.

framework can lead to dissipation and/or distortion of a simulated cloud deck. Furthermore, when this advection error feeds back into latent heat release, large oscillations can develop near cloud boundaries – the so-called advection condensation problem. For example, Figure 6.1 demonstrates how an initially square feature becomes diffused upon rotation in a Eulerian framework, even when using a fifth-order weighted essentially nonoscillatory (WENO) advection routine [27, 13]. In contrast, Figure 6.1 shows that a Lagrangian particle based method does not induce significant diffusion of the square; however, small oscillations associated with sampling errors are now found within the square.

Though advective errors may be the largest source of error in a multidimensional calculation, even in a one-dimensional setting the feedback of advective errors into parameterizations for latent heat release can be significant. One of the first attempts by the author to reduce the impact of this advection–condensation error was application of the volume of fluid (VOF) approach to represent the subgrid movement of cloud interfaces [16]. Note that the VOF approach is the most common approach within numerical codes used by the Department of Energy for simulating the evolution of interfaces between solids and gases and tends to be more efficient than particle based solution procedures [24]. Figure 6.2 shows a comparison between a Lagrangian approach (exact), a traditional forward-in-time Eulerian approach used in models such as the Weather Research and Forecasting (WRF) model [28] and the VOF approach for the advection of an one-dimensional cloud. As evident in Figure 6.2(b), the result of Eulerian advective errors feeding back into evaporation results in large oscillations at the cloud interfaces, whereas the VOF approach, by tracking and subsequently geometrically limiting advection fluxes near cloud interfaces, produces a solution with much smaller oscillations and no spurious evaporation.

In this chapter, one of the emphases will be to demonstrate that the numerical errors shown in Figure 6.1 and Figure 6.2(b) play a role in predictability and, depending on the circumstance, it may be difficult to quantify how much of the latent heat is produced by either numerical errors or actual physical mechanisms. Another important source of error, like what was shown in hurricane Sandy, is the need to employ high resolution to resolve the latent heat release. However, unlike Sandy, in which condensation is the primary driver, for the examples shown in this chapter the importance of evaporation for helping to maintain systems such as stratus decks or hurricanes will also be illustrated. Further, it will be demonstrated that the spatial scales associated with evaporation can be difficult to resolve using traditional cloud modeling approaches and this is one of the primary reasons for why cloud resolving models cannot get the right result for the right reason, for example, both intensity and structure of a hurricane are properly represented within a simulation.

While stratus decks and hurricanes have different time scales with regard to latent heat release and advection, the impact of numerical errors can be large for either phenomenon. For example, stratus decks contain weak vertical motions but are relatively long lived over vast regions of the Earth. The decks are the principle drivers of Earth's climate [30] and require a delicate balance between evaporation of cloud droplets at cloud edges and the ocean surface, large scale subsidence (leading to inversions in temperature and water vapor), radiation, and cloud droplet formation at cloudtop for their continual formation.

In fact, one of the first attempts by the author to solve numerical issues and/or the advection–condensation problem was the application of the VOF approach to an observed stratus deck. In [14], the VOF method was used to track and differentiate potential temperature, water vapor, and cloud water within the free atmosphere and the cloud. Figure 6.3 shows results from a VOF simulation that can be contrasted against results from a more traditional cloud modeling approach shown in Figure 6.4. As is evident in the comparison of the two figures, when the VOF approach is applied to a

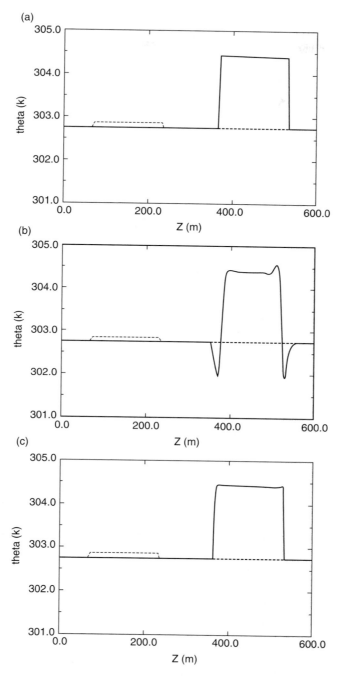

Figure 6.2 Potential temperature fields at the initial (dashed lines) and final states (solid lines) for the idealized advection–condensation problem: (a) exact solution; (b) the non-VOF solution produced by a typical numerical method used in atmospheric science; and (c) the VOF solution. From fig. 3 of [16]. Note the large amount of spurious evaporation in panel (b) produced by advection errors feeding back into the evaporation model.

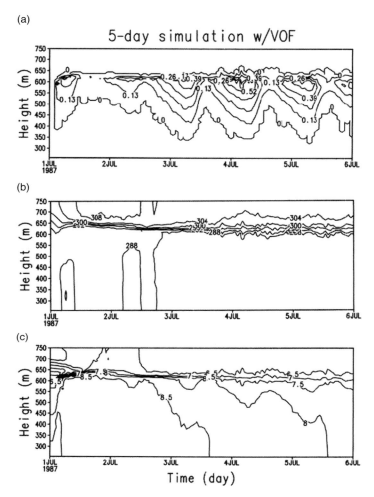

Figure 6.3 The time evolution of domain-averaged (a) cloud water from 0 to 0.65 g kg^{-1} with a contour interval of 0.13 g kg^{-1}; (b) potential temperature from 288 to 308 K with an interval of 4 K; and (c) water vapor from 3.5 to 9 g kg^{-1} with an interval of 0.5 g kg^{-1} versus height for a 5 day simulation with VOF. From fig. 2 of [14].

stratus field, the deck no longer dissipates via spurious evaporation associated with the advection–condensation problem and is maintained over a 5 day time period. Further, the stratus deck now is able to respond to changes in short wave heating with clear changes, in reasonable accord with observations (see Fig. 6.5), of the cloudtop and base.

While stratus clouds play a key role in the Earth's climate, hurricanes by their ability to transport large quantities of water vapor toward the poles also play a role in Earth's large scale energy balance. Additionally, hurricanes are one of the most destructive natural phenomena on Earth and can lead to large losses in life and property. Unfortunately, the impact of hurricanes on either short predictive time periods or longer climate time scales is difficult to predict. For example, debate exists whether more or fewer

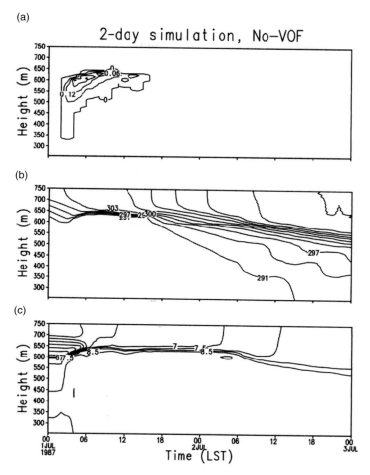

Figure 6.4 The time evolution of domain-averaged (a) cloud water from 0 to 0.24 g kg^{-1} with a contour interval of 0.06 g kg^{-1}; (b) potential temperature from 291 to 321 K with an interval of 3 K; and (c) water vapor from 3.5 to 9 g kg^{-1} with an interval of 0.5 g kg^{-1} vs height for a 2 day simulation without VOF. From fig. 6 of [14].

hurricanes will occur in a future climate, with one of the keys to settling this debate involving the integrated impact of wind shear on the number of hurricanes that may occur in a future climate [5].

Because of the difficulty in predicting systems containing clouds, the utilization of one modeling approach may not be sufficient to quantify uncertainty. Hence, two disparate approaches to model clouds will be presented in this chapter, a smooth Eulerian cloud model (ECM) approach and a stochastic Lagrangian cloud model (LCM) methodology, based on using particles in cell (PIC) to represent the various aerosol and cloud particles found in the atmosphere. As previously mentioned, latent heat release can induce large errors in numerical models with these errors being enhanced further by conditional switches, for example, if cloud water is above a prescribed value then rain is formed. Hence, over time a numerical model has been

(a)

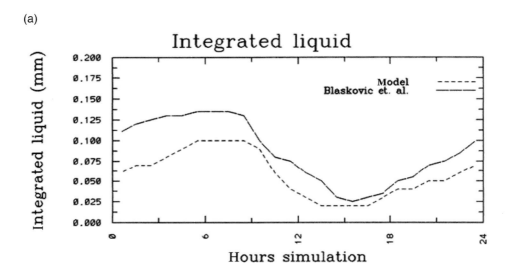

(b)

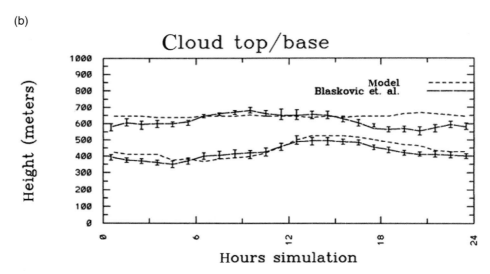

Figure 6.5 (a) Modeled domain averaged, vertically integrated cloud liquid water; and (b) cloud base height and cloud top height, compared with the observations by Blaskovic et al. 1991. From fig. 5 of [14].

developed [22] that removes switches and can enable nonlinear convergence of the entire model via sophisticated nonlinear solution procedures, such as Jacobian free Newton Krylov [23], leading to a significant reduction in the growth of numerical errors that can have the potential to degrade a given forecast.

In contrast to the Eulerian approach, a stochastic LCM using so-scaled superparticles (each particle can potentially represent billions of cloud particles) has also been developed that uses sampling within various cloud processes such as condensation,

(a)   (b)

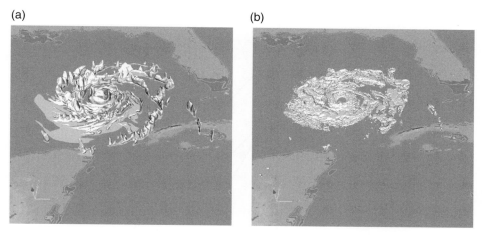

Figure 6.6 Isosurfaces of cloud water (0.0005 g kg^{-1}) from hurricane Rita on September 21, 2007, at 1800 UTC; (a) Eulerian cloud model simulation and (b) Lagrangian cloud model simulation. A black and white version of this figure will appear in some formats. For the color version, please refer to the plate section.

collision, and movement of particles [2, 1]. Note that because of its ability to resolve the motion and phase changes of individual particles in a subgrid manner, the LCM has the potential to be more accurate than either the smooth ECM or an ECM employing a VOF solution procedure. Specifically, as will later be shown, simulating evaporation at cloud edges and its feedback into particle size is very difficult to reproduce in an Eulerian setting using a reasonable mesh size.

Neither cloud modeling approach is perfect (truncation versus sampling errors as demonstrated in Fig. 6.1), but the old adage "you get what you pay for" does appear to occasionally hold true. For example, Figure 6.6 shows simulations of hurricane Rita (setup described in [6]) using either modeling approach with the LCM simulation being about 10 times more expensive than the Eulerian simulations; however, this simulation does reproduce the observed minimum sea level pressure (mSLP) versus time (see Fig. 6.7). The lower mSLP can be attributed to the following numerical and physical model errors: the LCM produces lower amounts of numerical diffusion associated with advection; cloud processes including condensational heating are better represented in the model; and evaporation is also reasonably resolved, as opposed to being possibly misrepresented via advection–condensation errors in the ECM simulations. Note that Figure 6.7 shows the application of an evaporative limiter [22] that attempts to minimize the spurious evaporative cooling shown in Figure 6.2(b) and any impact on hurricane intensity. But, even though the mSLP is lower in the simulation using the evaporative limiter, the simulation still does not have the resolution to resolve the release and subsequent transport of latent heat that is being produced in a subgrid manner by the LCM, resulting in slightly higher mSLP. Likewise, hurricane Rita was a relatively symmetric hurricane with wind shear not having significant impact at its maximum intensity, but for sheared hurricanes for which numerical errors may be significantly larger the differences in mSLP from a LCM and an ECM simulation may be considerable.

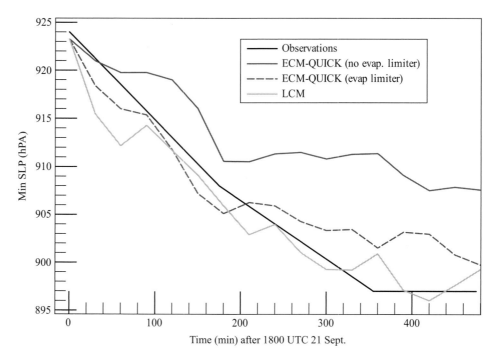

Figure 6.7 Minimum sea level pressure versus time for hurricane Rita with the light gray line from a Lagrangian cloud model (LCM) simulation; the dark gray lines from Eulerian cloud model (ECM) simulations employing no evaporative limiter (solid gray line) or an evaporative limiter (dashed gray line); and the observations (black line).

Because of the complexity of a hurricane, the impact of the differences between the two model formulations will be first shown in a somewhat simpler framework representative of a stratus deck. And a key aspect for both the subsequent stratus deck and hurricane simulations will be that results are compared against some of the best data sets ever obtained for the respective phenomena. These data sets will help illustrate why it is so difficult to accurately simulate small scale features associated with clouds, especially when numerical errors associated with advection distort these features.

The remainder of the chapter is organized as follows. In the next section examples from both modeling approaches will be given for a stratus deck observed during DYCOMS-II (Second Dynamics and Chemistry of Marine Stratocumulus field study) [29], a field program designed from the onset to provide data for cloud model validation. Next, ensemble results from an ECM will be presented for hurricane Guillermo for which high resolution dual doppler radar has been processed into a variety of different fields at 10 half-hour increments [20]. In particular, a unique aspect of this data processing was the development of fields of latent heat that can be used to examine issues related to the advection–condensation problem [9]. Based on the findings from the ensemble, results from both ECM and LCM high resolution implicit large eddy simulations of hurricane Guillermo will be subquently shown. And finally, a few concluding remarks will be given regarding the difficulty in predicting systems for which significant amounts of small scale energy can feed back onto the larger scales.

6.2   Stratus Simulations

6.2.1   ECM Simulations

This section is intended to demonstrate the impact of numerical errors that are present in a majority of ECMs currently used in atmospheric science and why an ECM has difficulty in resolving evaporation of small cloud droplets near cloud edges, a prime driver for the long term maintenance of a stratus deck. Both the ECM and LCM models to be presented in this section were developed at Los Alamos and employ several numerical techniques, such as PIC, that were also developed by prior researchers at the lab [11, 12].

In [22], a smooth ECM utilized in simulations of stratus clouds was presented. This model was composed of momentum, energy, and mass conservation equations for the gas (dry air and water vapor) along with mass and number conservation equations for cloud water. A simple bulk microphysical model (specified size distribution) that converted water vapor into cloud water or vice versa within a time scale that could be resolved by the model was also utilized in the simulations; as well as a simple functional relationship for long wave cooling. The primary developments that enabled the entire cloud model to converge within a nonlinear solution procedure were as follows: the replacement of switches found in the flux corrected advective scheme with a simple artificial viscosity scheme [18]; all conditional switches typically found in the bulk microphysical model were replaced by hyperbolic tangent relationships; and an evaporative limiter was utilized that restricts evaporation in regions where artificial viscosity is active (impact on hurricane Rita was already shown in Fig. 6.7). Another unique feature of the discrete version of the model was that numerous higher order time stepping and/or spatial differencing schemes could be employed in a simulation. In fact, as demonstrated in the paper, the use of higher order temporal differencing reduced overall temporal error (see fig. 7 in [22]) by a factor of 100 with respect to more traditional approaches, for example, split explicit formulation used in the community based WRF model [32].

Besides demonstrating that a relatively complex model could indeed be made differentiable, another important point of the paper was that the simple artificial viscosity approach together with the evaporative limiter could reproduce an observed stratus deck. Indeed Figures 6.8 and 6.9 show that the smooth ECM with these two options active produces a stratus deck that agrees reasonably well with the DYCOM-II observations and enables the lower part of the stratus deck to be maintained. In contrast, the previous figures also reveal that more traditional cloud modeling approaches produced a stratus deck that slowly dissipated with time via the

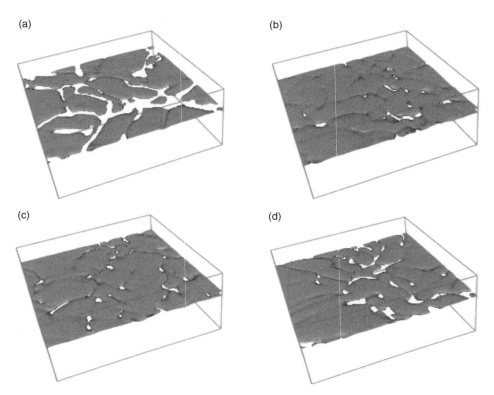

Figure 6.8 Isosurfaces of cloud water (10^{-4} kg kg^{-1}) at the end of the DYCOMS-II simulations from (a) an ECM simulation without the evaporative limiter active; (b) an ECM simulation with the limiter active; and (c)/(d) ECM simulations employing more traditional numerical approaches. From fig. 1 of [22].

advection–condensation problem. Note that in a paper by [29] that compared several traditional cloud models against the DYCOMS-II observations, all the cloud models appear to suffer from the advection–condensation problem with their simulated stratus decks also slowing thinning during the course of the simulations (fig. 3 of [29]).

Unlike the highly idealized advection of a square cloud shown in Figure 6.2 in which all evaporation is spurious, for real clouds evaporation does occur at cloud edges and helps drive the overall circulation. Recognizing this physics, the evaporative limiter was designed such to limit evaporation with respect to the magnitude of the wind normal to a cloud boundary. Unfortunately, even when this construction of the evaporative limiter is utilized, the lower panels of Figure 6.9 show a significant reduction in vertical motion statistics with regard to the other simulations and observations (see fig. 5 in [29]). But, though the traditional simulations are producing significant vertical motions, some of these motions are driven by spurious evaporation produced via the advection–condensation problem and not real evaporation.

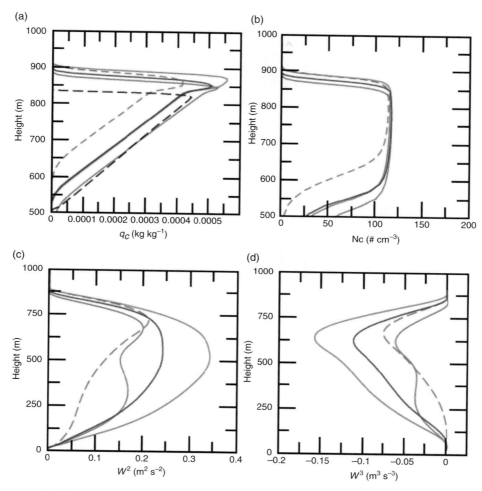

Figure 6.9 Mean vertical profiles averaged over the last hour of (a) cloud water; (b) cloud droplet number; (c) variance of vertical motions; and (d) third moment of vertical motions from DYCOMS-II simulations from ECM simulations with or without evaporative limiter active (red or dashed red lines) and from ECM simulations employing more traditional approaches (green or blue lines). From fig. 2 of [22]. A black and white version of this figure will appear in some formats. For the color version, please refer to the plate section.

6.2.2   LCM Simulations

Per Figure 6.1, the ability of the Lagrangian particle method to accurately represent the movement of a distinct feature on a Eulerian mesh is one of its strengths and when this ability is combined with its subgrid treatment of reaction physics it produces a numerical method that is very difficult to replicate, unless very high spatial resolution is employed in an ECM simulation. One of the first applications of a LCM in atmospheric science that clearly demonstrated its abilities was in simulating the stratus deck observed during DYCOMS-II. In these stratus deck simulations a LCM was used to

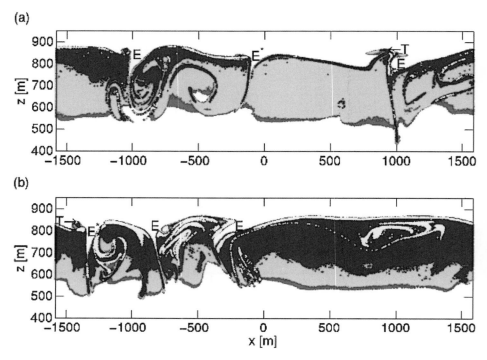

Figure 6.10 Spatial locations of cloud droplets over a portion of the vertical domain from LCM simulation using either (a) 1040 aerosol or (b) 260 initial aerosol. The size of the droplets is indicated by the following: $r < 3$ (intermediate-gray thin-layer on bottom), $3 \leq r < 9$ (darkest and next-lighter gray central thick regions), and $r > 9$ (lightest gray thin-layer on top). The locations marked by "E" denote regions, where entrainment events are taking place with T denoting the two areas blown up in the next figure. From fig. 5 of [2].

represent the movement of aerosol and any subsequent condensation of water vapor onto these particles. Within the LCM super particles were coupled to a Navier–Stokes equation set via interphase coupling terms that enabled a two way interaction of momentum, energy, and mass to occur between the gas and particle phases. A stochastic sampling procedure coupled to a turbulence kinetic energy model was used to produce subgrid fluctuations of wind and water vapor that were used in equations representing particle movement and growth.

Figure 6.10 shows two different realizations of the DYCOMS-II stratus deck when different background aerosol numbers are used with the resulting changes in aerosol number producing rather stark differences in the size of cloud droplets. However, though the sizes are different the spatially and temporally averaged (last hour of the simulations) cloud water mass in either simulation is very close to the observed field (not shown). Figure 6.11 shows expanded views near entrainment events of the simulations shown in the previous figure; the events are the result of evaporation induced cooling. What is evident in both figures is that evaporation induces the depletion of smaller droplets but not the larger droplets, for example, time scale for evaporation is inversely proportional to the size of the droplet. The resulting larger

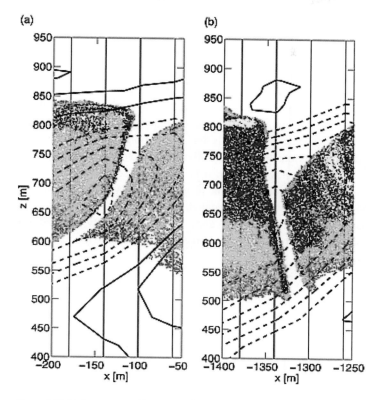

Figure 6.11 Enlargement of two entrainment events occurring within simulations using initially (a) 1040 aerosol or (b) 260 aerosol with colors indicating cloud droplet size as defined in the previous figure and the thin vertical lines denoting the grid cell boundaries in the x direction. Thick black lines represent contours of vertical velocity with solid/dashed contours signaling areas of positive/negative motions for a contour interval of 0.2 m s^{-1}. From fig. 7 of [2].

droplets are now present in very narrow filaments that are significantly smaller than the grid resolution used by the Navier–Stokes solver.

These large droplets fall rapidly and induce enhanced downward motion in the gas via multiphase drag terms along with a secondary maximum in droplet size at the cloud bottom that agree with observations taken during DYCOMS-II (see figs. 3 and 4 of [2]). Hence, though computationally more demanding than the Eulerian model, the Lagrangian PIC model can "somewhat" capture physics that would require a much finer Eulerian mesh to resolve. And by enhancing downward motions, these small scale filaments help feed energy back into the larger scale circulation that maintains the stratus deck. Note that vertical motion statistics produced by the LCM simulations are in line with those observed during DYCOMS-II (compare fig. 3 of [2] with fig. 5 of [29]); suggesting the LCM approach, unlike traditional ECM approaches, is able to produce dynamical and cloud fields that agree with observations.

6.3   Hurricane Guillermo

In this section, a series of simulations will be presented that demonstrate why hurricanes, and especially hurricanes encountering wind shear such as hurricane Guillermo, are so difficult to predict. In particular, results from a large 120 member ensemble using an ECM will first be presented in which key parameters related to the intensity and structure are estimated and shown to produce a hurricane whose structure, but not intensity, agrees reasonably well with the observations. Next, to reduce the impact of numerical diffusion associated with the advection of cloud features in a circular motion around the eye (error shown in Fig. 6.1) and produce a more intense storm whose structure still agrees with observations, results from very high resolution implicit large eddy ECM simulations will be presented. While results from these high resolution simulations produce simulated hurricanes that are indeed closer to observations, the simulations still suffer from the inability to accurately resolve the impact of evaporation.

6.3.1   Observations

Hurricane Guillermo was an Eastern Pacific hurricane that occurred primarily during early August 1997. Guillermo was a sheared hurricane with the large scale deep layer shear being near 7.5 m s^{-1} during its rapid intensification phase that lasted from 0600 UTC on August 2 to 1200 UTC on August 3 (see Fig. 6.12). Because wind shear typically counteracts other environmental factors that lead to intensification, hurricane models have difficulty forecasting the balance between these various factors, occasionally leading to significant errors in predicted intensity. Further, the

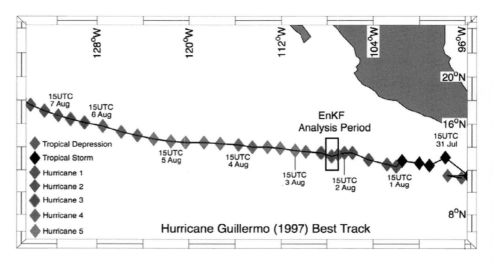

Figure 6.12 Best track for hurricane Guillermo (1997). Hurricane intensity is color-coded based on the Saffir-Simpson scale with legend shown on the bottom left of the figure. Data courtesy of the Tropical Prediction Center (TPC), NOAA. The EnKF analysis period is denoted by the rectangle. From fig. 1 of [8].

occasional presence of an open eyewall in sheared hurricanes promotes the evaporation of cloud water associated with horizontal advection and can lead to significant advection–condensation errors.

One of the primary reasons for examining hurricane Guillermo was that during the rapid intensification phase the Tropical Experiment field program was ongoing and was able to deploy two aircraft during a 6 hour time period. The dual Doppler radar data obtained by the aircraft have been extensively processed and analyzed by [20] with the paper illustrating the development of strong eyewall convection on the downshear-left quadrant of Guillermo. This strong convection appeared to be at least partially responsible for enabling Guillermo to intensify in the presence of shear.

In addition to enabling the dynamics of a sheared hurricane to be understood from an observational standpoint, this unique radar data set – horizontal winds and derived latent heat fields (see Fig. 6.13) have also been processed [20, 9] – is also of sufficient quality to quantify various errors during the simulation of the hurricane, such as those associated with numerical errors. In particular, due to the presence of an incomplete eyewall and a wind field rapidly advecting cloud quantities across the edges of a possible open eyewall, the impact of numerical diffusion and evaporation (real or spurious) should be significant. For example, Figure 6.13 shows integrated latent heat fields and reveals small pockets of intense evaporative cooling occurring immediately next to broader areas of condensational heating. Hence Guillermo is an ideal candidate to assess whether an ECM and/or a LCM can accurately simulate both fields of condensational heating and the small pockets of evaporation while producing an intensity that agrees with the observations.

6.3.2 EnKF Description

In [8] a study using an ECM within an ensemble Kalman filter (EnKF) system was undertaken to examine whether the large and unique data set obtained during the hurricane could be used to optimally determine key parameters that would enable a model to reproduce both the intensification rate and latent heat structure of the hurricane. In particular, surface moisture and friction, the prescribed turbulent length scale, and the environmental wind shear were varied over a 120 member ensemble. Note that latent heat release is a direct function of surface moisture availability and note how much of this moisture is transported from the ocean's surface. Transport is both resolved by the conservation equations and modeled via a subgrid turbulence model in which a turbulent length scale needs to be specified. Further, the eyewall size and shape is a function of the magnitude of surface friction as well as the background wind shear impacting the vortex.

Given the unique data set provided by the careful analysis of [20] the following questions that were addressed for the first time in [8] are as follows: how much data are required for parameter estimation; how different the four parameter estimates are when using latent heat versus wind data; how the parameter estimates vary as a function of time; and whether the optimally estimated parameters reduce model error. A key aspect to addressing these questions is that the Kalman filter used to estimate the parameters

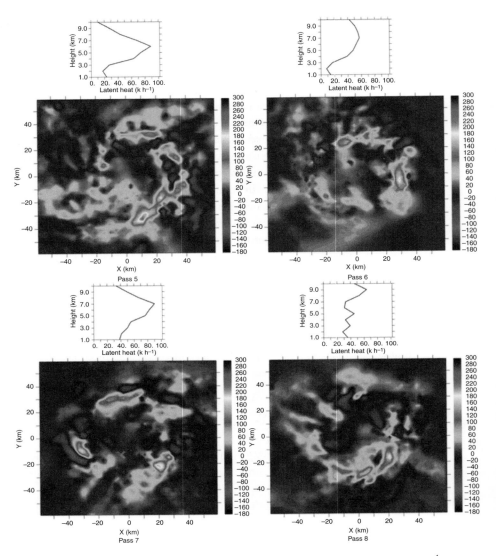

Figure 6.13 Horizontal views (averaged over all heights) of the latent heating rate (K h^{-1}) of condensation/evaporation retrieved from airborne Doppler radar observations in hurricane Guillermo (1997) at four select times of the dual-Doppler radar data, that is, pass 5 corresponds to 2117 UTC from fig. 6 of Reasor et al. (2009). Note that grid points without latent heating were assigned zero values after the vertical averaging. The vertical profile of the azimuthal mean latent heating rage at RWM (30 km) is shown above each contour plot. The first level of data is at 1 km due to ocean surface contamination. From fig. 2 of [8].

was formulated in a matrix free manner and hence could incorporate nearly 200,000 data points during the optimization [7]. Additionally, the EnKF procedure was formulated to take advantage of a large parallel computing environment; all 120 members ran concurrently using upwards of 27,000 processors and the Kalman filter was also able to run over multiple processors.

The base equation set used in the ECM ensemble was similar to what was used in the ECM stratus cloud simulations, except that additional conservation equations and source terms were added for new microphysical variables, for example, rain, cloud ice, snow, and graupel (small frozen rain droplets), using a cloud model that closely follows [31] and [21]. The computational domain employed $300 \times 300 \times 86$ grid points over $500 \times 500 \times 22$ km with horizontal and vertical stretching to enable highest resolution near the hurricane eyewall (about 1 km horizontal resolution and slightly lower than the 3 km resolution used in operational hurricane models (http://storm.aoml.noaa.gov/hwrfxpro jects/?projectName=BASIN), and at the ocean's surface (about 20 m). One hundred and twenty members were concurrently run in parallel with parameters for each member being generated using a Latin hypercube sampling technique (see Fig. 6.14). Initial conditions for a given ensemble member were obtained by using both fields of latent heat and winds obtained from the radar data in a simple nudging procedure for the first hour of the simulation. Once the nudging was completed, the vortex slowly began to organize and at the end of 6 hours the initialization period was over, with a simulation running for an additional 5 hours for comparison against the radar data.

Though the numerical model was similar to what was used in the stratus simulations, the setup employed for the EnKF simulations was designed to mimic what is used in operational hurricane models, for example, flux limiting [33] in place of the simple artificial viscosity scheme for ensuring cloud quantities stay positive and evaporative limiter is not active, to help quantify whether the data assimilation procedure employed

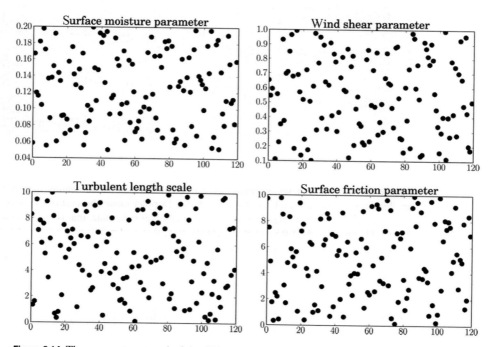

Figure 6.14 The parameter spread of the 120 ensemble members obtained by utilization of the Latin hypercube sampling strategy within the limits shown in the figure. From fig. 3 of [8].

in [8] could help in the future improve predictability of an operational model. While a few simulations using the smooth ECM variant were undertaken, the results were like the stratus results in that fields of evaporation were barely resolved and the evaporative limiter even further reduced the magnitude of this field to the point of producing a significantly stronger hurricane.

6.3.3   EnKF Results

Figure 6.15 shows minimum sea level pressure (mSLP) versus time for the 120 members and the observed pressure represented by red dots. What is obvious is that there exists quite a few members that can reproduce the observed pressure, but what is not readily apparent is what member among this reduced set also reproduces the best overall structure. Based on comparison against the wind and latent heat data sets, it was deemed that ensemble member 44 had the best overall structure and intensity during the 5 hour radar comparison time period. For example, Figure 6.16 shows fields of averaged in height vertical motions from both the ensemble average and from member 44 with the figure illustrating that the fine scale intense vertical motions found in member 44 become washed out during the ensemble averaging; however, the ensemble averaged

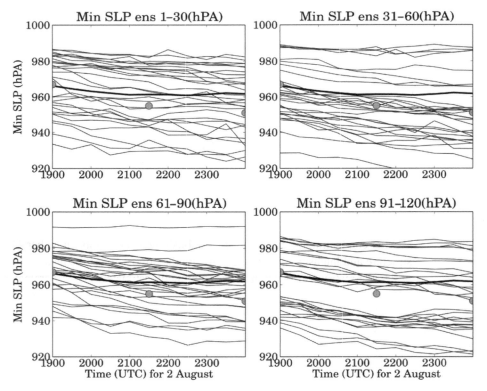

Figure 6.15 Minimum sea level pressure versus simulation time for each ensemble member (gray lines), ensemble average (thick black line), and observations (gray dots). From fig. 4 of [8].

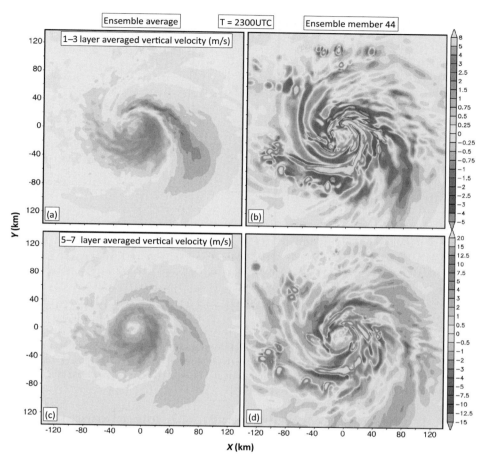

Figure 6.16 Ensemble average vertical motion fields at 2300 UTC (11 hours into the simulations) averaged between 1–3 km (a) or 5–7 km (c). Corresponding layer averaged vertical motion fields from ensemble member 44 between 1–3 km (b) or 5–7 km (d). From fig. 5 of [8].

structure in terms of having strongest motions in the southeast quadrant and also its intensity does bear some resemblance to the observations.

After conducting the ensemble and establishing that ensemble member 44 produced a structure and intensity that best agreed with the observations, the next step in [8] was to utilize the matrix free Kalman filter to optimally estimate the four parameters and then conduct new simulations based on these parameter estimates. The hope is that the new estimates will produce a structure and intensity that is an even better comparison against the observations than what was produced by ensemble member 44. Further, note that while member 44 produced the lowest error of the ensemble, the diameter of the eye was smaller than observations and hence suggests that the eyewall structure can be improved via the Kalman filter parameter estimation procedure.

Once it was established that a large amount of data were needed to produce parameter estimates with low variance across the 120 members and that the associated estimates

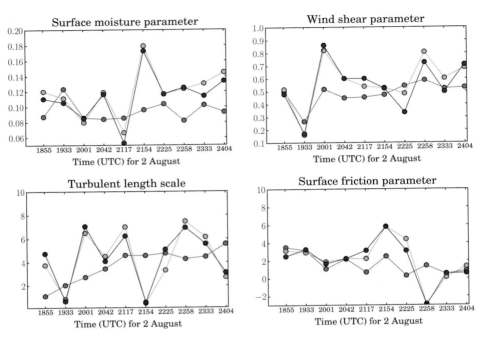

Figure 6.17 Time distribution of EnKF parameter estimates using latent heat data only (blue line), horizontal winds (red line), and or both (green line). From fig. 10 of [8]. A black and white version of this figure will appear in some formats. For the color version, please refer to the plate section.

have some temporal varability (see Fig. 6.17), temporal averaged estimates using latent heat, wind, and a combination of the two data sources along with their variance were computed and compared against the parameter values from member 44 (see Fig. 6.18). On comparison it was evident that three of the four parameter values used in member 44 were near the edges of the temporally averaged estimates, whereas one parameter, surface moisture, was far outside the estimate. As previously discussed, surface moisture availability is key to the intensification of a simulated hurricane with the additional moisture possibly being needed via condensational heating to offset any spurious evaporation produced within member 44. Another point evident in Figure 6.18 is that the differences between the three estimates of the four parameters are relatively small, but, as will be shown next, these differences lead to significant variations in simulated intensity and structure.

For example, when the three temporally averaged parameter estimates are used in new standalone simulations, the simulations that employed estimates via latent heat data produced an error in wind speed that decreased markedly during the simulations, whereas the error for all three stand alone simulations with regard to observed latent heat fields increased during the simulations, but was still lower than the error produced by member 44 (see Fig. 6.19). Note that the only differences between the standalone simulations and member 44 was the parameters; no additional state information from the Kalman filter such as the horizontal winds was employed in the standalone simulations. When comparing minimum surface pressure versus time from the simulations

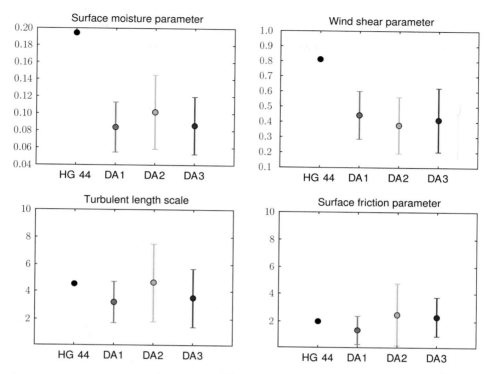

Figure 6.18 Analysis parameters averaged over time for using latent heat data only (blue dot), horizontal winds (red dot), or both (green dot), and ensemble member 44 (black dot). The vertical lines from the dots indicate the temporal variance of the parameter estimates from the various data sources. From fig. 11 of [8].

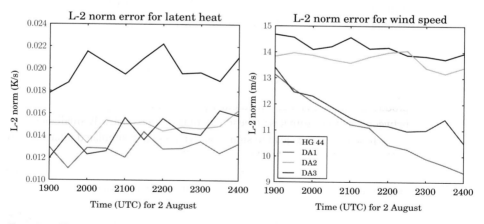

Figure 6.19 Error estimates as a function of time computed using Eq. 20 of [22] and observational data as the reference for ensemble member 44 and three simulations using parameters estimate from latent heat (blue lines), horizontal winds (red lines), or both (red lines). From fig. 16 of [8]. A black and white version of this figure will appear in some formats. For the color version, please refer to the plate section.

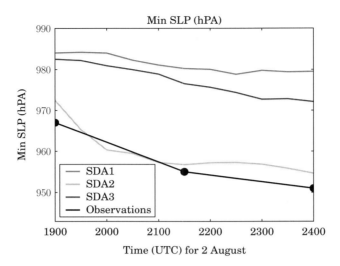

Figure 6.20 Minimum sea level pressure versus time for the same three simulations shown in Fig. 6.19 along with observations from hurricane Guillermo. From fig. 17 of [8].

against the observations (see Fig. 6.20), only the member that used wind data produced a pressure that agreed with the observations with the two other stand alone simulations that used latent heat producing weaker hurricanes. Hence, given that the two simulations that lead to a reduction in error in wind speed did not produce the right intensity, it shows the difficulty – even when using a horizontal resolution that is finer than that employed in an operational setting – of getting the right result in terms of both structure and intensity for a sheared hurricane. Likewise, these results illustrate how small changes in key parameters can make significant impacts in simulated intensity and structure, suggesting why operational hurricane models face long odds when trying to predict changes in either metric for sheared hurricanes.

6.4   Implicit Large Eddy Simulations of Hurricane Guillermo

6.4.1   ECM Implicit Large Eddy Simulations

Given the complexity of the system, there could be a number of reasons that the model has difficulty with regard to producing the right structure and intensity. One of the reasons may be the impact of diffusion associated with both the advection operator and the turbulence model. To reduce the impact of diffusion, implicit large eddy simulations (ILES) were undertaken in which the source of diffusion associated with the explicit turbulence model was no longer active, for example, dissipation is now only associated with the advective operator. Thus, transport of water vapor from the ocean surface was only a process of the advective operator and the primary parameter now determining the source of this transport was the one found in the parameterization for the surface moisture flux. Note that parameter values for surface moisture flux, surface friction, and wind shear use the temporal averaged estimates of latent heat data obtained in the EnKF procedure shown in Figure 6.18.

Hence the numerical model used in the ILES simulations is identical to the previous ensemble, except no turbulence model is active. Four ILES simulations were conducted with each simulation employing increasingly higher resolution within the domain encompassed by the radar data. Specifically, horizontal grid resolutions of 640, 320, 160, and 80 m (order magnitude less than an operational model) were used in the four simulations with 300 × 300 × 86 grid points being used in the coarsest simulation up to 2400 × 2400 × 86 grid points in the finest simulation. Vertical resolution was the same in each of the four simulations and each simulation does employ horizontal stretching to capture the larger circulation of Guillermo present away from the radar domain. Of note, for the higher resolution runs the impact of diffusion produced by the Eulerian advective operator should be considerably smaller than what was produced by an individual member of the previous ensemble. Simulations were again run for a total length of 11 hours with 6 hours required to spin up the vortex and 5 hours to match the radar observations.

Compared with the observed field shown in Figure 6.21, Figure 6.22 shows horizontal fields of latent heat at 3 km obtained at the beginning of the forecast cycle (6 h into the simulations). What is obvious is that at the coarser resolutions the model is struggling to even produce a hurricane, while at the higher resolutions the latent heat fields are well developed and composed of small convective pockets away from the more uniform eyewall convection. Note that the radar data is being sampled at an effective resolution of 2 km and hence may be averaging out the smaller pockets of latent heat produced by the ILES simulation using the highest spatial resolution. For example, Figure 6.23 shows fields of latent heat from the 80 m simulation averaged

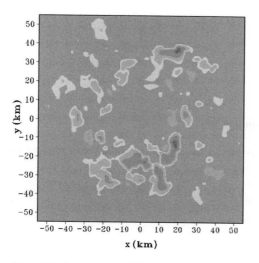

Figure 6.21 Latent heat at 3 km observed during hurricane Guillermo with evaporation contour of grey being −0.025 K s^{-1}, whereas condensational heating is shown in white (0.005 K s^{-1}), light grey (0.01 K s^{-1}), dark grey (0.025 K s−1), and black (0.05 K s^{-1}).

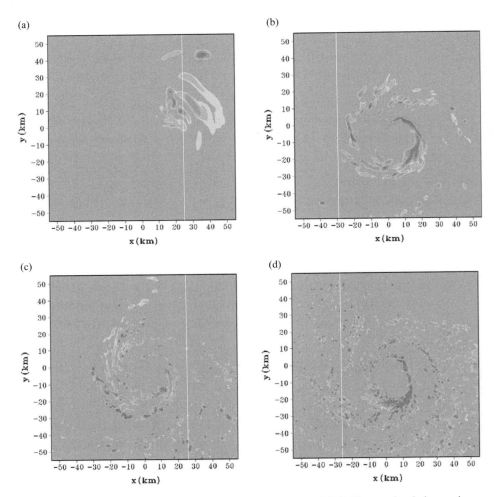

Figure 6.22 Same contour intervals as Fig. 6.21 except from ILES Guillermo simulations using 640 m (a), 320 m (b), 160 m (c), and 80 m (d) horizontal resolutions.

to the radar resolution and the points that are apparent from this figure are as follows: eye size is similar to observed size, the strongest heat release in both the model and observations is in the southeast quadrant, and the model does not produce significant amounts of evaporative cooling.

Table 6.1 shows mSLP at the time the fields of latent heat were computed as well as ratios of simulated latent heat and evaporation to the observed quantities (integrated over the radar domain). As was shown in the previous figures, Table 6.1 demonstrates that with increased resolution the model does better with respect to observations; however, even at the highest resolution the model is still unable to resolve the magnitude of the evaporative cooling. Though errors in radar processing algorithms may have led to an overestimation of evaporative cooling, the model in all probability is underestimating the magnitude of this cooling – eventually producing a hurricane at the

Table 6.1 Statistics of Hurricane Guillermo ILES Simulations with Obs./Sim. Denoting Either Observations or Simulations and Conds./Evap. Denoting Either Sondensation or Svaporation

Resolution	mSLP	Ratio of Sim./ Obs. Conds.	Ratio of Sim./ Obs. Evap.
640 m	998		
320 m	972	0.52	0.30
160 m	959	0.56	0.44
80 m	950	1.10	0.54

(a)

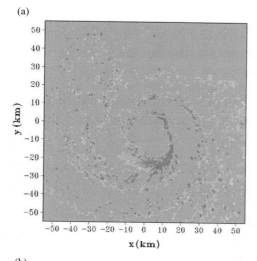

(b)

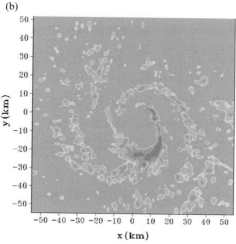

Figure 6.23 Same contour intervals as Fig. 6.21 except from an ILES Guillermo simulation using 80 m horizontal resolution (a) and from the same simulation but averaged to the resolution as the radar data (b) shown in Fig. 6.21.

highest resolution that is stronger than the observed system. And, once again, it is not entirely obvious how much of this evaporation is even physically based; for example, some may be associated with the advection–condensation problem shown in Figure 6.2b and not due to the evaporation of cloud water in downdrafts.

6.4.2   LCM Implicit Large Eddy Simulation

As was shown in the DYCOMS-II stratus simulations, the LCM was able to reasonably distinguish real evaporation occurring in narrow filaments at cloud edges, as opposed to the ECM simulations that either suffer from evaporation due to the advection operator and/ or need very high spatial resolution to resolve these filaments. Likewise, as is shown in Figure 6.1, diffusion associated with the advector operator is larger in an ECM and will also lead to distortion/dissipation of cloud fields. Hence, as was the case for the stratus deck, it is expected that a LCM approach should be able to more accurately simulate the various microphysical fields, such as cloud water, associated with hurricane Guillermo.

But the utilization of the LCM at the highest resolution ILES run shown in the previous section is currently impractical and for demonstration purposes a coarser resolution grid using 600×600 horizontal grid points (320 m resolution within the eyewall) will be utilized for the current LCM simulation. But, at this resolution, the ability to resolve evaporation even within the LCM framework is limited and the primary difference between the previous ECM simulation and the LCM simulation should be the diffusion associated with the advective operator. Note that the LCM resolves the spectra or particle number versus size for a variety of cloud species, for example, cloud water, rain, cloud ice, snow, and graupel, and in some grid cells over 1,000 particles are being used to resolve these distributions. The complexity and cost required to undertake the collision operations between these various particles is extremely time consuming and is one of the reasons for using a coarser resolution. But, unlike in an ECM that would need 1,000 three-dimensional arrays over the entire domain to conduct the calculation, the localization of the LCM makes the current calculations tractable. Further, since the current version of the LCM has not been described in the open literature, a brief description will be given next.

6.4.2.1   Model Description

The Lagrangian cloud model is composed of predictive equations for particle location, velocity, cloud droplet radius, and collision coalescence with these equations being expressed as

$$\frac{dx_{i'}}{dt} = v_{i'} + v'_{i'} \tag{6.1}$$

$$\frac{dv_{i'}}{dt} = \frac{1}{\tau_p} \left(u_{i'}^* - v_{i'} \right) - g\delta_{i,3} \tag{6.2}$$

$$\frac{dr}{dt} = \frac{GS^* F}{r + a} \tag{6.3}$$

$$\frac{dn_{new}}{dt} = K(r_{id}, r_{jd})n_{id}n_{jd} \tag{6.4}$$

where $x_{i'}$ is the particle location in space ($i'=1,3$); $v_{i'}$ is the particle velocity; $v_{i'}'$ is a velocity fluctuation computed by randomly sampling from a Gaussian distribution with mean value zero and standard deviation $\sigma_{v_{i'}} = TKE$ (taken to be zeros in the current ILES simulation); $u_{i'}^*$ is the gas velocity at the particle location; and $\tau_p = \frac{2\rho_p r^2}{9\mu}(1 + Re^{.687})$ [3] is the velocity relaxation time with ρ_p the density of a given particle, r the droplet radius, μ dynamic viscosity of the gas, and $Re = \frac{2\rho_g r|v_{i'} - u_{i'}^*|}{\mu}$ the Reynolds number (ρ_g is the gas density).

The ice phase in the LCM is represented as ice, snow, and frozen liquid particles and the liquid phase is denoted by either cloud or rain droplets ($r > 1.e^{-04}$ m). The primary impact regarding the binning of the various particles into these chosen types is within the collision operator with little distinctions being made regarding particle types within other parts of the LCM, that is, the velocity relaxation time is at present independent of particle type. Likewise, no distinction within the ice or snow phases is currently made regarding the formation of various ice particle habits, that is, plates or dendrites, with this option possibly being added in future LCM simulations that are, unlike the present LCM simulations addressing the impact of numerical errors, designed to understand how various detailed microphysical processes impact hurricane dynamics.

Unlike in [2], the activation of aerosol in the LCM hurricane simulations is not via a complex condensation model. Instead, a specified number of Lagrangian aerosol particles, for ice and liquid activation, are initially introduced (40 in the current simulations) into each grid cell with each particle becoming instantly activated when the supersaturation at a particle location becomes positive and the grid cell averaged cloud or ice number concentration is less than a prescribed value. Once activated, the diffusional growth of a particle is given by Eq. 6.3; S^* is the supersaturation at a particle location; G is the condensational time scale, for example, Eq. (A12) of [22]; F is the ventilation factor [25, 26] for the various particle types; and a is a constant typically taken to be a micron.

As illustrated in [2], the accuracy of the LCM is a function of the actual number of droplets being represented by each particle (sampling error) with each particle roughly being equivalent to a single bin used in detailed Eulerian cloud simulations. Note that, because of precipitation and/or movement of particles, the number of actual particles representing, for example, aerosol (required for condensation) in a grid cell can trend toward zero or very large values; however, to avoid these extremes the number of aerosol particles is kept either to a minimum number (20 in the current simulations) by introducing aerosol particles randomly in space within a grid cell or a maximum number (240 in the current simulations) by "rebinning" two particles of approximately the same size into one particle.

Though a relatively complex melting option [19] can be activated during a LCM simulation, for the simulations to be presented frozen particles were assumed to instantaneously melt. Freezing of liquid particles is also instantaneous, once a liquid particle of a given radius falls below its "freezing" temperature. The primary formation of ice is due to freezing of cloud droplets; however, ice can be formed via activation [17]. For efficiency, the collision–coalescence procedure in the particle model uses a discrete Monte Carlo like approach in which two random particles (n_{id} and n_{jd}) are selected until a certain percentage (20%) of the total number of allowable collisions is reached. Depending on the properties of the two random particles and the magnitude of the collision kernal, ($K(r_{id}, r_{jd})$, [10]), new particles, n_{new}, of various types are created with any contributions to the gas energy equation also being calculated, e.g, due to the freezing of cloud droplets when colliding with snow particles at temperatures below freezing. Once created, if the number of new particles in combination with the number of existing particles of a given type exceeds a preset value (240 in the present simulations), then the new particles are combined with existing particles that are closest in size with the newly generated particles.

6.4.2.2   Results

As a cost saving measure, the LCM simulation to be presented in the section was initialized using data from the ECM ILES 320 m calculation obtained at 4 hours. The LCM simulation was then run for an additional two hours. Figure 6.24 shows fields of latent heat from both the LCM and ECM simulations at 5 and 6 hours, revealing that the two disparate cloud modeling approaches produce similar fields of latent heat at the two time periods. But, in somewhat of a contrast to the fields of latent heat, Table 6.2 shows that the LCM approach produces a much stronger vortex at hour 6 than the ECM approach, while producing roughly similar integrated fields of latent heat.

The weaker hurricane produced by the ECM simulation may be the result of numerical diffusion associated with the advection of the cloud water field. For example, Figure 6.25 shows isosurfaces of cloud water from both ILES simulations and reveals that LCM ILES simulation produced a larger shield of cloud water, especially to the southeast of the simulated hurricane. Likewise, the cloud water field from the ECM simulation appears elongated in the eyewall and not composed of smaller and stronger convective type cells that are evident in the LCM simulation. Even though the LCM simulation produces a stronger system that is closer to the observed pressure, the simulation, per Table 6.2, still does not produce a latent heat field that even in an integral sense agrees with the observations. But, as shown in the previous section, higher spatial resolution – even when using an LCM framework – may be needed to resolve the latent heat release observed by the radar, especially the fields of evaporation.

Table 6.2 Statistics of Hurricane Guillermo ILES 600 m Simulations from Either LCM or ECM Calculations with Obs./Sim. Denoting Either Observations or Simulations and Conds./Evap. Denoting Either Condensation or Evaporation.

Time	mSLP	Ratio of Sim./ Obs. Conds.	Ratio of Sim./ Obs. Evap.
hour5-LCM	972	0.44	0.37
hour6-LCM	960	0.53	0.34
hour5-ECM	978	0.52	0.44
hour6-ECM	972	0.52	0.30

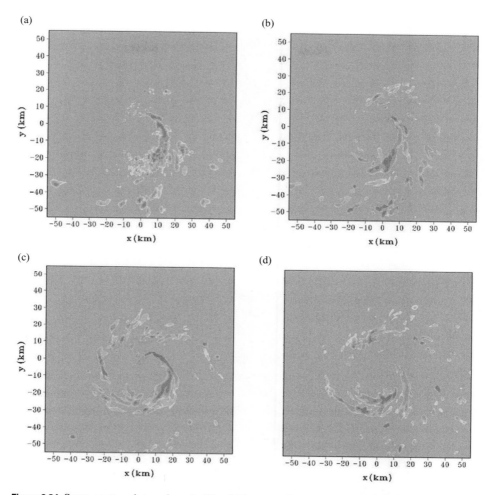

Figure 6.24 Same contour intervals as in Fig. 6.21 except from an ILES LCM Guillermo simulation at hour 5 (a) or hour 6 (c) and from an ILES ECM Guillermo simulation at hour 5 (b) or hour 6 (d).

6.5   Concluding Remarks

Though establishing initial and/or boundary conditions is extremely important for accurate forecasts of cloud systems such as hurricanes, taken as a whole this chapter illustrates that numerical errors and/or lack of resolution can lead to predictions that do not agree with observations and can produce errors as significant as those associated with initial conditions. As discussed at the start of this chapter, the track of hurricane Sandy was not accurately forecasted by the GFS model using its standard operational resolution, with higher resolution being needed to resolve latent heat release and its subsequent impact on track. Fortunately, for hurricane Sandy, unlike what was shown in the stratus and hurricane simulations of this chapter, resolving evaporative cooling was not

(a)

(b)

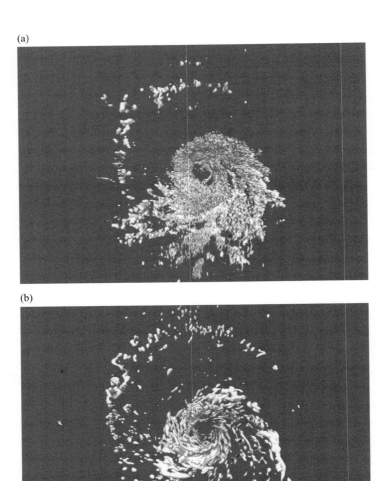

Figure 6.25 Isosurface of cloud water
(10^{-4} kg kg^{-1}) from an ILES LCM Guillermo simulation (a) and an ILES Guillermo ECM
simulation (b).

as important as resolving condensational heating. And under circumstances when evaporation does play a significant role in the overall energy balance, Eulerian cloud models
may still have difficulty resolving evaporation at the edges of clouds, e.g., as shown in
the ILES of hurricane Guillermo using 80 m resolution. Further, it is not entirely obvious
how much of the evaporation in a Eulerian framework using a traditional approach is real
or the product of the advection–condensation problem shown in Figure 6.2.

Hence, a solution to this problem is the use of a LCM approach employing particles
that can reasonably resolve evaporation occurring at cloud edges and not distort via

advection fields of cloud water. But, as shown in the ILES LCM simulation of hurricane Guillermo, while the approach can better represent the motion of cloud water and its impact on hurricane intensity, the resolution employed in the simulation, unlike the stratus simulations, is still not sufficient to resolve fields of latent heat release. Thus, an outstanding question is whether the impact of this energy release needs to be accurately represented at coarser model resolutions via some type of subgrid model or that under-resolving these fields can still lead to a predictive model solution. It is also interesting to note that both the PIC and VOF numerical approaches were developed in the 1950s and 1960s and have yet to become established within atmospheric science. Based on the findings shown in this chapter, it appears further investigation using either numerical approach to model clouds is warranted and could lead to significant improvements in overall model predictability.

6.6   Acknowledgments

This chapter summarizes some of work undertaken at Los Alamos in cloud modeling over the past twenty years. This work involved numerous researchers at LANL and elsewhere; as well as significant computational resources at Los Alamos and others such as Oakridge's Jaguar computing cluster. A partial list of some of the researchers and their current affiliations are as follows: Len Margolin (LANL), Piotr Smolarkiewicz (ECMWF), Jim Kao (retired), Chris Jeffery (LANL), Humberto Godinez (LANL), Alex Fierro (NOAA), Mirek Andrejczuk (Oxford), Andii Wyszogrodski (UCAR), Chick Keller (retired), Will (Scott) Smith (LANL), Manvendra Dubey (LANL), and Stephen Guimond (NASA).

References

[1] M. Andrejczuk, W. Grabowski, J. Reisner, and A. Gadian, "Cloud-aerosol interactions for boundary-layer stratocumulus in the Lagrangian cloud model," *J. Geophys. Res*, 115, 2010.

[2] M. Andrejczuk, J. Reisner, B. Henson, M. Dubey, and C. Jeffery, "The potential impacts of pollution on a non-drizzling stratus deck: Does aerosol number matter more than type?," *J. Geophys. Res*, 113, 2008.

[3] C. Crowe, M. Sommerfeld, and Y. Tsuji, *Multiphase Flows with Droplets and Particles*, CRC Press, New York, 1998.

[4] M. DeMaria, C. Sampson, J. Knaff, and K. Musgrave, "Is tropical cyclone intensity guidance improving?," *Bull. Amer. Meteor. Soc.*, 95:387–398, 2014.

[5] K. Emanual, "Global warming effects on U.S. hurricane damage," *Weather, Climate, and Society*, 3:261–268, 2011.

[6] A. Fierro and J. Reisner, "High resolution simulation of the electrification and lightning of hurricane Rita during the period of rapid intensification," *J. Atmos. Sci.*, 137:477–494, 2010.

[7] H. Godinez and D. Moulton, "An efficient matrix-free implementation of the ensemble kalman filter," *Comput. Geosci.*, 2012.

[8] H. Godinez, J. Reisner, A. Fierro, S. Guimond, and J. Kao, "Optimally estimating key model parameters of rapidly intensifying hurricane Guillermo (1997) using the ensemble kalman filter," *J. Atmos. Sci.*, 2848–2868, 2012.

[9] S. Guimond, M. Bourassa, and P. Reasor, "A latent heat retrieval and its effects on the intensity and structure of hurricane Guillermo (1997). part 1: The alogithm and observations," *J. Atmos. Sci.*, 68:1549–1567, 2011.

[10] W. Hall, "A detailed microphysical model within a two-dimensional dynamic framework: Model description and preliminary results," *J. Atmos. Sci.*, 37:2486–2507, 1980.

[11] F. Harlow, "Particle-in-cell method for two-dimensional hydrodynamic problems," *LA-02082-MS*, 1956.

[12] F. Harlow, "The particle-in-cell computing method for fluid dynamics," *Meth. Comput. Physics*, 3:319–342, 1964.

[13] C. Hu and C.-W. Shu, "Weighted essentially non-oscillatory schemes on triangle meshes," *J. Comput. Physics*, 150:97–127, 1999.

[14] C. Kao, Y. Hang, J. Reisner, and W. Smith, "Test of the volume-of-fluid method on marine boundary layer clouds," *Mon. Wea. Rev.*, 128:1960–1970, 1999.

[15] L. Magnusson, J.-R. Bidlot, S. Lang, A. Thorpe, N. Wedi, and M. Yamaguchi, "Evaluation of medium-range forecasts for hurricane Sandy," *Mon. Wea. Rev.*, 142:1962–1981, 2014.

[16] L. Margolin, J. Reisner, and P. Smolarkiewicz, "Application of the volume-of-fluid method to the advection-condensation problem," *Mon. Wea. Rev.*, 125:2265–2273, 1997.

[17] M. Meyers, P. DeMott, and W. Cotton, "New primary ice-nucleation parameterizations in an explicit cloud model," *J. Appl. Meteor.*, 31:708–721, 1992.

[18] J. V. Neumann and R. Richtmyer, "A method for the numerical calculation of hydrodynamic shocks," *J. of Appl. Physics*, 21:232–237, 1950.

[19] V. Phillips, A. Pokrovsky, and A. Khain, "The influence of time-dependent melting on the dynamics and precipitation production in maritime and continental storm clouds," *J. Atmos. Sci.*, 64:338–359, 2007.

[20] P. Reasor, M. Eastin, and J. Gamache, "Rapidly intensifying hurricane Guillermo (1997). Part 1: Low-wavenumber structure and evolution," *Mon. Wea. Rev.*, 137:603–631, 2009.

[21] J. Reisner, R. Bruintjes, and R. Rasmussen, "An examination on the utility of forecasting supercooled liquid water in a mesoscale model," *Q. J. R. Meteorol. Soc.*, 124:1071–1107, 1998.

[22] J. Reisner and J. Jeffery, "A smooth cloud model," *Mon. Wea. Rev.*, 137:1825–1843, 2009.

[23] J. Reisner, A. Wyszogrodzki, V. Mousseau, and D. Knoll, "An efficient physics-based preconditioner for the fully implicit solution of small-scale thermally driven atmospheric flows," *J. Comput. Phys.*, 189:30–44, 2003.

[24] W. Rider and B. Kothe, "Reconstructing volume tracking," *J. Comput. Physics*, 141:112–152, 1998.

[25] S. Rutledge and P. Hobbs, "The mesoscale and microscale structure and organization of clouds and precipitation in midlatitude cyclones. VIII: A model fo the seeder-feeder process in warm frontal rainbands," *J. Atmos. Sci.*, 40:1185–1206, 1983.

[26] S. Rutledge and P. Hobbs, "The mesoscale and microscale structure and organization of clouds and precipitation in midlatitude cyclones. XI: A diagnostic modeling study of precipitation development in narrow cold-frontal rainbands," *J. Atmos. Sci.*, 41:2949–2972, 1984.

[27] C.-W. Shu and S. Osher, "Efficient implementation of essentially non-oscillatory shock-capturing schemes, II" *J. Comput. Physics*, 83:97–127, 1999.

[28] W. Skamarock and J. Klemp, "A time-split non-hydrostatic atmospheric model for weather research and forecasting applications," *J. Comput. Physics*, 227:3465–3485, 2008.

[29] B. Stevens, C.-H. Moeng, A. Ackerman, C. Bretherton, A. Chlond, S. Roode, J. Edwards, J.-C. Golaz, H. Jiang, M. Khairoutdinov, M. Kirkpatrick, D. Lewellen, A. Lock, F. Muller, D. Stevens, E. Whelan, and P. Zhu, "Evaluation of large-eddy simulations via observations of nocturnal marine stratocumulus," *Mon. Wea. Rev.*, 133:1443–1462, 2005.

[30] B. Stevens and S. Schwartz, "Observing and modeling earth's energy flows," *Surveys in Geophysics*, 41:779–816, 2012.

[31] G. Thompson, R. Rasmussen, and K. Manning, "Explicit forecasts of winter precipitation using an improved bulk microphyiscs scheme. part ii: Implementation of a new snow parameterization," *Mon. Wea. Rev.*, 136:5095–5115, 2008.

[32] L. Wicker and W. Skamarock, "Time-splitting methods for elastic models using forward time schemes," *Mon. Wea. Rev.*, 130:2088–2097, 2002.

[33] S. Zalesak, "Fully multidimensional flux-corrected transport algorithm for fluids," *J. Comput. Phys.*, 31:335–362, 1979.

7 Verification, Validation, and Uncertainty Quantification for Coarse Grained Simulation

William J. Rider, James R. Kamm, and V. Gregory Weirs

7.1 Overview

Verification, validation, and uncertainty quantification (VVUQ) is a complex and sometimes controversial set of procedures for determining and documenting the overall quality of a simulation activity. In this chapter we focus on determining a model's sufficiency by comparing it with experimental data (validation), while incorporating the effects of numerical error (solution verification) and parameter uncertainty (uncertainty quantification). While there is broad consensus on the general framework and its value, the definitions of what properly constitutes VVUQ are subject to debate and the field continues to evolve. Aspects of the overall process we describe have been developed in numerous documents, for example, [1, 2, 5, 9, 10, 14–20, 31–33, 38, 39]. Despite all that has been written about VVUQ, there remain diverse, often imperfect, and potentially conflicting definitions for each element of the process.

From the practitioner's point of view, though, this state of affairs is not a barrier to using the processes and techniques in his or her own work. The context of the validation activity (the specific model/simulation code compared with a specific set of experimental data for a specific purpose) dominates the choices of the VVUQ practitioner. Within this context, the definitions of specific processes are included and justified as a part of the documentation of the validation assessment.

This chapter contains a general workflow for VVUQ that is based on the concepts by previous investigators mentioned earlier. The intention of our presentation is not to be dogmatic regarding a particular VVUQ approach (including the one espoused), but rather to inform scientists of the issues and to suggest approaches by which to enhance the overall VVUQ content of their work.

7.2 The Scientific Simulation Context

Scientific simulations are run for a variety of reasons, including scientific discovery (e.g., in astrophysics) and in a capacity to inform decision making (e.g., in many engineering, national security, and manufacturing fields). In this chapter, we focus on scientific simulation in the latter role. In a pure scientific context, the decision that must be made is whether the simulation is trustworthy for discovery, or the degree of trustworthiness. The goal of scientific simulation is to provide answers to particular questions; these answers must be inferred from numerical values output by the simulations of the specified problems. The numerical values are rarely precise and, as such, contain intrinsic uncertainty, which should be regarded as essential to quantify as part of the simulation activity. We echo the terminology and sentiment of Trucano et al. [39], who focus on "confident prediction" and whose observation captures the essence of the use of scientific simulations:

> In computational science, of course, to some degree confidence is correlated with belief in the quantitative numerical, mathematical and physical accuracy of a calculated prediction. Intuitively, confident prediction then implies having some level of confidence or belief in the quantitative accuracy of the prediction. This further implies a willingness to use the prediction in some meaningful way, for example in a decision process.

Many modeling efforts necessarily involve complex, multiphysics simulations, the results of which are unavoidably *approximate* solutions. This immediately raises important questions. How good are the approximations? Is the approximate solution good enough (in the context of the decision)? Or, considering the quality of the approximate solution, can a decision be comfortably made? The first question is difficult to address because the exact solution is not known (if it were, scientific simulation would not be needed). Differences between the simulation results and reality arise from many sources. VVUQ are different forms of assessment intended to quantify various aspects of these differences; see [1, 2, 5, 9, 10, 14–20, 31–33, 38, 39]. These assessments address the first question and help inform the decision maker regarding the second question.

The accuracy of the approximate solutions (and their corresponding uncertainties) for a complex, multiphysics problem of interest is directly related to the fidelity of the solutions for simpler "component" problems, with additional errors arising from coupling and the increased complexity. For simpler foundational problems, different (usually better and inherently less error prone) sources of information can be used to provide the estimation of uncertainty without the complications of the coupled models. These circumstances lead to a hierarchical view of validation problems, with the problem of interest at the top of a notional pyramid and the simplest problems at the base. Scenarios at the base of this conceptual pyramid can be solved (simulated or examined experimentally) with demonstrably greater confidence and more certainty, as these simpler problems are, by definition, less complex and better understood.

In this context, "simpler" means, for example, fewer physical phenomena, less complex submodels, limited physical scale, or reduced geometric complexity. Uncertainty and empiricism naturally increase as the top of the pyramid is approached. At the highest level the problem is, by definition, complicated and not well understood. This hierarchy is a useful paradigm by which to identify where knowledge (e.g., of a particular physics model) is insufficient and where empiricism is a placeholder for this missing knowledge. The purpose of hierarchical validation is not to eliminate empiricism, but to restrict it to the parts of the hierarchy where it truly represents a lack of knowledge. This concept is reflected in the general validation hierarchy, shown in Figure 7.1.

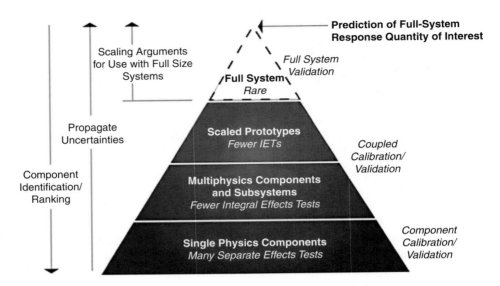

Figure 7.1 The overall validation hierarchy as an example of how validation could be structured. From the CASL project centered at ORNL & INL [36].

In consideration of this hierarchical paradigm, validation can be understood as an intrinsically multiscale process. The following conceptual approach can be followed to address the overall validation process:

1. Establish a multiscale hierarchy of physics and phenomena to be validated.
2. Determine simulations and experiments that can be conducted to examine the hierarchy.
3. Execute the validation process on each level of the hierarchy.
4. Assess the overall validation.

A goal of this framework is to identify, estimate, and ultimately reduce uncertainties at each conceptual level of the pyramid. The practical purpose of this hierarchical approach is not to eliminate uncertainty, but to identify where uncertainty cannot be reduced. VVUQ can be used to help identify where, in the hierarchical modeling and simulation decomposition, empiricism and uncertainty are the greatest: these are the elements where improvements can have the largest impact. This process is necessarily imperfect and, at times, ambiguous; nevertheless, it can be used to focus efforts, for example, by identifying where additional model calibration may be particularly useful. It is important to note that the intermediate portions of this pyramid present the most difficult validation activities and, so, are absent from the process. Part of the challenge of VVUQ is to fill in the missing portions of the hierarchy and provide a more complete picture of the nature and sources of uncertainty.

Regarding this observation, we caution that calibration is a potentially dangerous activity when carried out in an ad hoc manner, for example, as a process by which to determine modeling coefficients where no (or limited) first principles guidance is available. Calibration can account for inappropriate effects and, so, can mask what is in fact uncertainty and lack of knowledge. Consequently, calibration experiments are appropriate and well suited at the lowest level in a modeling hierarchy. Calibration at the higher levels of the hierarchy may be less rigorous if not applied in a principled fashion. Thus, calibration is an acknowledged necessity, but it is increasingly problematic as the modeling moves up the problem hierarchy.

It is important to acknowledge that empirical/calibrated models may also be appropriate when they replace better grounded models, but do not increase the overall uncertainty of the problem of interest. This endeavor should be conducted with exceptional care because of its seeming violation of scientific principle and potential for harm when the conditions fall outside the range for the empirical calibration. The validation of the subproblem may demonstrate the sufficiency of the empirical model for replacing the better grounded model; however, this is different from calibrating an empirical model to experimental data because a better grounded model does not exist. In this case, the empirically calibrated model actually results in a lower uncertainty, but the risks intrinsic in this approach must be kept clearly in mind.

In the following, we present general workflows for applying validation and uncertainty quantification. These workflows become more specific by the context in which they are applied. The validation of the top level problem in the pyramid will be quite different than that at the lowest level, because the context is different. Likewise, uncertainty quantification will be context dependent: it can be applied as a part of validation at any level of the hierarchy but also across levels.

7.3   VVUQ Workflow

At the highest level, the workflow for conducting validation at each level of a hierarchy can be described concisely in a short series of steps, each of which will be described in more detail in its own workflow. The conceptual validation workflow we advocate is described at this level in Figure 7.2. The steps are the following:

1.   Begin the validation process.
2.   The simulation path of the process.
   a.   The conduct of the simulations defines the baseline simulation model.
   b.   The determination of the simulation uncertainty defines the sources and magnitude of uncertainty in the simulation model.
3.   The experimental path of the process.
   a.   The conduct of the experiment provides data to compare the simulation results to reality.
   b.   The determination of the experimental uncertainty provides the quality of the data used for simulation comparison.
4.   The validation assessment provides the overall context for the simulation and validation outcomes in relation to their intended application.

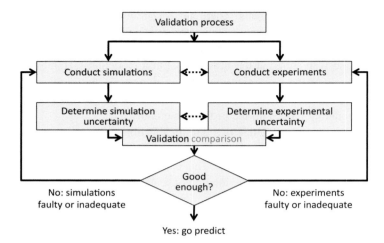

Figure 7.2 The overall validation process as described in this chapter, encompassing both simulations and experiments, the associated uncertainty quantification analyses, and the ultimate validation assessment.

5.   The determination of the sufficiency of the validation status based on the defined requirements for the application of the simulation: if "yes," then proceed to prediction; if "no," then address the shortcomings in the process, i.e., refine simulations and/or experiments.

One complication to acknowledge at the outset of the process is the difference between concurrent and legacy experiments. Concurrent experiments are conducted in collaboration with the simulations, so that the two processes experience direct feedback from each other. In particular, dedicated validation experiments[1][1, 14] are those concurrent experiments that are specifically designed to validate code calculations; they are executed and analyzed in concert with the corresponding simulations. [14, 15, 20]. The value of dedicated validation experiments cannot be overstated, as the corresponding data are particularly valuable for evaluating the simulation capability of a given code. Moreover, validation experiments are designed to be simulated with relatively less difficulty than traditional experiments, which are often designed for different purposes. Unfortunately, it can be challenging to identify resources for dedicated validation experiment campaigns, as these efforts are sometimes perceived to be less valuable than, say, scientific discovery or design experiments (i.e., traditional experimental science). Legacy data, on the other hand, come from experiments in the past; for the purposes of this discussion, legacy data will be more generally characterized as being from experiments that are not for the express purpose of validation or are not influenced by the simulation activity. Since legacy data may not have been generated in

---

[1] As pointed out by Oberkampf [1, 14], "validation experiments" comprise a unique class that does not fall into the usual categories of scientific discovery experiments, model improvement experiments, or performance/acceptance tests.

coordination with modern simulation tools and current VVUQ perspectives, such data may be of more limited value, as they often lack sufficient documentation to be used in support of a defensible simulation–validation effort. Last, we note that the use of experimental data to conduct calibration of simulation models will have a nearly identical structure to that for validation, but the purpose of the activity is distinctly different in character.

Each of the steps in this overall process can be expanded into its own process, which we describe and depict in the following. This process can be applied at any level of the notional hierarchy presented earlier in the chapter.

7.3.1   Begin the Validation Process

Validation is almost always conducted by a heterogeneous group of specialists, such as the end application, physics phenomena relevant to the application, numerical methods, experimental techniques, and uncertainty quantification (see Fig. 7.2). Additional stakeholders include project managers, model (simulation code) customers, and others who have a vested interest in the information provided by the model. The diversity of stakeholders is an essential part of validation because expertise in different areas is required. But as a consequence, it is critical to spend time defining the objectives and the scope of the effort so that people with different roles and with different technical backgrounds can fully contribute. Rather than a detailed workflow for this section, a list is presented identifying a number of activities and communication tools that can focus the discussion and identify issues that need to be considered, as well as guide the technical work:

1. Identify stakeholders and establish a consensus on the purpose and objectives of the validation activity.
2. Define the framework used for the assessment (e.g., code scaling applicability and uncertainty (CSAU) [3], predictive capability maturity model (PCMM) [18], PCMM++ [7, 18], quantified margins and uncertainties (QMU) [6, 21]).
3. Conduct a phenomena identification and ranking table (PIRT) [Tru02] analysis of the simulation (this could/should be conducted as part of the conduct of simulations or experiment). If this has already been conducted, reassess the previous PIRT.
4. Develop an initial plan for each element of the simulation and experimental work.
5. Define an initial set of system response quantities (SRQs, also called quantities of interest or QoIs), which will be extracted from both simulations and experiments, and quantitative metrics for comparing them.

The context for the validation activity should emerge from these activities and tools: What physical phenomena is it most important for the model to capture? How much do typical simulations cost, and how many can be run in the time available? What experiments can be conducted and what can the diagnostics measure? Approximately how many input parameters will be treated as uncertain, which are thought to be the most important, and are they correlated? It is helpful to start initial discussions well in advance of the main effort for several reasons. The main reason is that it takes significant time to get all stakeholders on the same page. As these discussions evolve, final code verification, model calibration, experimental facility and diagnostic calibration, and other preparatory work can be tackled with an improving sense of the specific goals of the validation effort.

As indicated in the second item, there are various frameworks available that provide guidelines for validation assessments. CSAU [3] is the framework accepted by the Nuclear Regulatory Committee for validation by dimensionally scaled experiments. PCMM [18] identifies four qualitative levels of modeling and simulation capability maturity that gauge (1) model representation and geometric fidelity, (2) physics and material model fidelity, (3) code verification, (4) solution verification, (5) model validation, and (6) uncertainty quantification and sensitivity analysis. The enhanced predictive capability maturity model (PCMM++) [18] is PCMM with additional evaluations of software modularity and extensibility (an evaluation of the ease with which software modifications can be made) and readiness of the software for HPC platforms (e.g., how easily the software is adapted to different machines). QMU [6, 21] is a framework used at the Defense Program National Laboratories to support the certification of the nuclear stockpile. The emphasis in QMU is the incorporation of uncertainties into a decision framework. More recently the PCMM framework has been critically examined and extended within the context of advanced engineering simulations [7].

PIRT analysis mentioned in the third item is an approach to define the importance of phenomena and their relative importance to the situation at hand. The PIRT should be guided by the relative impact of the identified phenomena on the system response quantities of interest. Progress toward implementation and UQ of identified phenomena is also identified in the PIRT. The PIRT, because it ranks the relative impact of phenomena and the status of the modeling, can help guide additional model development, VV, and UQ.

The fourth and fifth items are specified in detail in later stages of the workflow, but initial coordination between experimentalists and modelers can eliminate options that are not possible from both approaches.

7.3.2　Conduct Simulations

1. The starting point is a simulation code that has been sufficiently software quality assured (SQAed) and the code verified. (Fig. 7.3.)
2. The problem of interest (the experiment) is defined, including models, submodels, and initial and boundary conditions, as well as any supporting data.
3. The simulation codes' input is defined, incorporating models, data, initial and boundary conditions, and mesh/geometric description; this input should undergo a verification/SQA process.
4. The system response characteristics to be analyzed in the simulation should be defined, as should a means of comparison.
5. The simulation should be run on the computer.

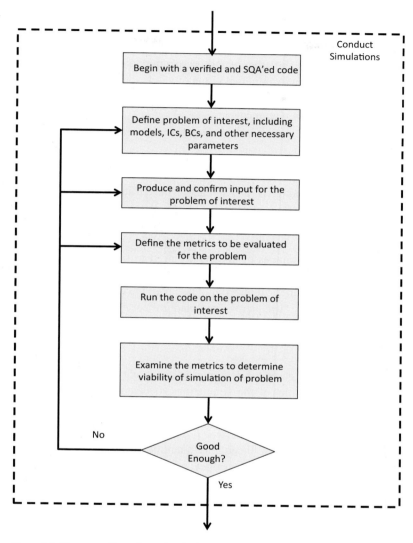

Figure 7.3 The overall conduct simulations process as described in this chapter is shown. This process clearly identifies three possible aspects that must be investigated if the quality of the simulation does not meet the modeler's expectations.

6.   The metrics are used to determine the viability and quality of the simulation for the problem of interest.

7.   The simulations results are determined to either be or not be sufficient to move to the determination of uncertainty.

It is important that the simulations of the problem of interest be made as faithful as possible to the corresponding experiment. This issue can pose particular difficulties for validation and calibration of complex phenomena. This concept is characterized by

Trucano et al. [39] as the "alignment" between a simulation and an intended application. They categorize the code input parameters into two (possibly overlapping) sets, p_A (input parameters that specify alignment with the intended model application) and p_N (numerical parameters and other quantities required to execute a calculation and control its numerical accuracy). A principal virtue of validation experiments is that they may be designed and executed in coordination with the simulation tool to increase the experiment–simulation alignment.

7.3.3   Determination of Simulation Uncertainty

1.   The starting point is a simulation of sufficient quality or importance to examine simulation uncertainty. Care should be taken if the simulation's numerical cost is too high to reasonably explore nonnumerical uncertainties. (Fig. 7.4.)
2.   The types and sources of uncertainty need to be determined. Sources of uncertainty include intrinsic variability, model form, model parameters, simulation choices, and numerical uncertainties. These uncertainties can be broadly categorized as aleatory ("irreducible" or "random") or epistemic ("reducible" or "lack of knowledge"); see, for example, [27, 39] and references therein.
3.   The metrics and methodology for examining for uncertainty determination should be established.
4.   The simulation cases for determining uncertainty should be conducted.
5.   Process the results of the simulations to determine the uncertainties from the defined sources.
6.   Examine the sufficiency of the uncertainty estimation and their magnitude.

The literature on simulation uncertainty quantification is vast (see the earlier citations and their references) and the challenges associated with it are well recognized in the VVUQ community. It is important to acknowledge that addressing simulation uncertainty increases the technical difficulty of simulation activity (as well as the difficulty of subsequent simulation–experiment comparisons). The intrinsic statistical or probabilistic nature of uncertainty information is the unavoidable source of this issue.

7.3.4   Conduct Experiments

1.   Begin the experiment at a well-characterized facility with defined quality controls. (Fig. 7.5.)
2.   Define the experimental setup including the physics of interest and initial and boundary conditions.
3.   Identify the experimental parameters that can be explicitly controlled as part of the experimental operation.
4.   Define the diagnostics used in the experiment including data capture, postprocessing, and related information.

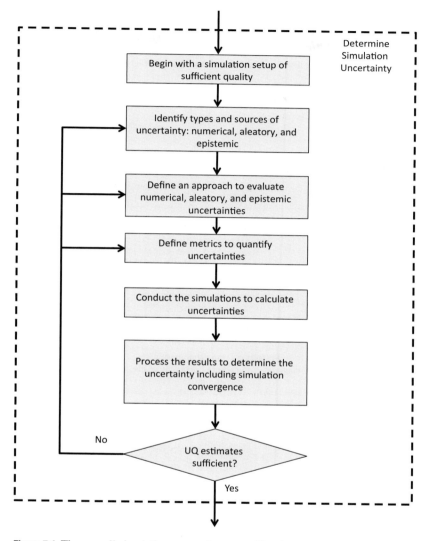

Figure 7.4 The overall simulation uncertainty quantification (UQ) process as described in this chapter is displayed. This process clearly identifies three possible aspects that must be investigated if the quality of the simulation UQ estimates does not meet the modeler's expectations.

5. Conduct an instance of the experiment.
   a. Conduct replicate experiments (if possible) to determine the repeatability of the results, and/or assess the aletory uncertainty in the experiment.
   b. Where experiments are single time and unique, the physical phenomena must be carefully scrutinized.
6. Postprocess the experimental data and assess the quality of the experimental data providing uncertainty estimates.

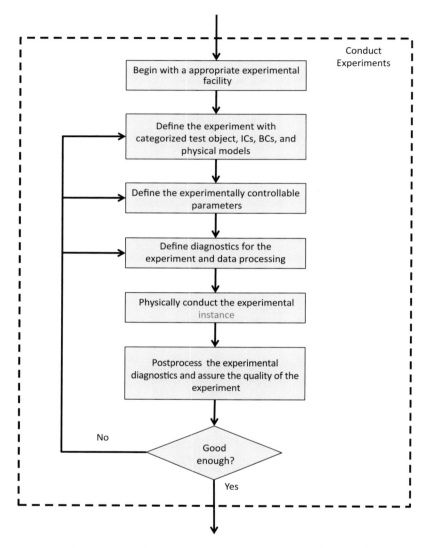

Figure 7.5 The overall conduct experiments process as described in this chapter is shown here. This process clearly identifies three possible aspects that must be investigated if the quality of the experimental data does not meet the experimentalist's expectations.

Ideally, experimental data used for validation are generated by experiments designed expressly for validation as opposed to scientific discovery (although experiments serving both purposes are inherently desirable). Such experiments are focused on generating high quality data, i.e., with high accuracy, with well-quantified errors and uncertainties, and of a highly repeatable nature. Experiments and simulations should be carried out in concert, with dynamic feedback between the two. This requires close coordination and frequent constructive discussion between computational scientists/ engineers and experimentalists in the roles of collaborators. Of course, the larger the scale and greater the complexity of the experiment, the more difficult this objective is to achieve. Nonetheless, the greater the common understanding of techniques, issues, and practice in both experiments and simulations among all parties involved, the higher the quality of the VVUQ results.

7.3.5   Determination of Experimental Uncertainty

1.   Begin with well-defined experimental results. (Fig. 7.6.)
2.   Identify the various sources of error in the experiment (measurement, data reduction, etc.) as well as their type (aleatory or epistemic).

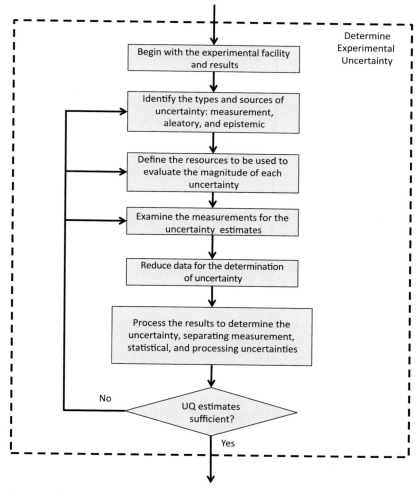

Figure 7.6 The overall experimental uncertainty quantification process as described in this chapter is displayed here. This process clearly identifies three possible aspects that must be investigated if the quality of the experimental UQ estimates does not meet the experimentalist's expectations.

3. Define resources for determining the magnitude of uncertainty.
4. Examine the experimental measurements for each uncertainty.
5. Reduce the data associated with the determination of uncertainty.
6. Postprocess the uncertainty to separate the different effects, and produce an overall experimental uncertainty budget.
7. Determine the sufficiency of the uncertainty estimates.

The complete definition of experimental uncertainties and any detailed discussion of their determination is beyond the scope of this chapter. A few broadly based observations regarding the nature of experimental uncertainty are, however, in order. Generally speaking, experimental data have errors in their measurement, processing, inference, and statistics. Each of these errors has a fundamentally different character and has an analog in the computational domain. Measurement error is the most obvious, being the relative inability to produce an exact measurement of the real condition present in the experiment. Processing errors are introduced when raw signals are processed (often, through several steps) into values used in subsequent analysis or inference. Inference is often used to produce a "measurement" from the original or processed data. For example, particle image velocimetry (PIV) is based on an inference from the actual measurement. This inference is usually software and algorithm based, and as such is subject to many of the same sources of error as the computations being validated. Finally, the statistical error is the variability of a measurement over time, or with repeated conduct of the experiment. This type of error is strongly associated with aleatory uncertainty.

7.3.6   Validation Comparison

1. Define the framework used for the assessment (e.g., CSAU [3], PCMM [18], PCMM++ [7, 18], QMU [6, 21]). (Fig. 7.7.)
2. Conduct a PIRT [Tru02] analysis of the simulation (this could/should be conducted as part of the conduct of simulations or experiment). If this has already been conducted, reassess the previous PIRT.
3. Examine each element of the simulation and experimental work in the context of the assessment framework.
4. Evaluate the validation in terms of the requirements associated with the intended application.

The assessment framework and PIRT analysis should be reviewed as they set the context for the validation comparison and validation assessment. A complexity inherent in any of these assessment frameworks is that the comparison between simulations and experiment cannot be expressed simply as difference of the two quantities, but instead involves comparison of the probability distributions associated with simulations and experiments. The mapping of metrics used in simulations to those used in experiments is problematic. Simulation science and its associated mathematics provide error measures in the form of error norms (a functional measure of a mathematical function), and experiments measure quantities made available through instruments. Fortunately, the

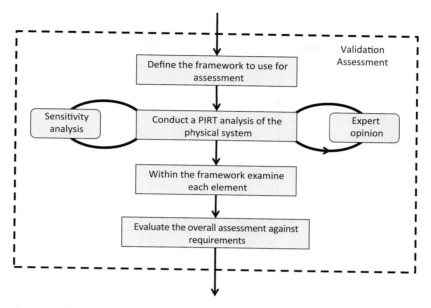

Figure 7.7 The validation assessment process as described in this chapter. This process clearly defines a course of action for assessment, including a well-defined framework for assessment and a PIRT analysis of the physical system.

error norms typically involve some sort of integration in time or space, which mimics the processes involved in instrumental science. As a result there is a rough correspondence between experimental and simulation measures. Nonetheless, the correspondence is approximate and notional, leading to yet another source of uncertainty in the simulation of physical circumstances. The end result of these differences leads to the failure of mathematical theory to provide assurances to simulation quality. This is most acutely felt in the uncertainty estimation due to numerical approximation (calculation verification), but influences all uncertainties arising through numerical approximation.

The determination of metrics requires the close interaction of computational and experimental scientists. The coordination strives to provide a balance between what is possible with experimental measurement science and metrics that are accurate and well behaved mathematically. As experiments reach the upper echelon of the validation hierarchy this becomes increasingly difficult, as measurements become more and more related to the underlying application specific measures of success (see Section 7.4.2).

7.3.7   Validation Assessment

The validation comparison provides quantitative information to the stakeholders and decision makers on the differences between the model and the experimental data (see Fig. 7.7). The interpretation of those differences is a distinct and subjective step. One question is if the level of the differences is small enough to predict subsequent experiments with confidence.

A second, much more difficult question is if the level of the differences is small enough to predict the application scenario with confidence. The experiments provide information in one area of parameter space, and the application scenario is often in a different area, although usually close in some sense. Experiments may be carefully designed to provide some justification for extrapolating to the application regime, such as the scaling arguments in the CSAU framework, or by choosing the experimental sample points to minimize extrapolation errors. Such approaches are based on the most important differences between the experimental parameter space and the application space, but are usually chosen by expert opinion and cannot capture all the dimensions in which the experimental and application spaces differ.

7.4   Discussion of the Workflow in Practice

The idealized workflow guidelines we have outlined are both resource intensive and time consuming. This observation should not be surprising. The intent of VVUQ analysis is to provide value that is commensurate with the effort put into the process. This value is found through producing a defensible level of confidence in the analysis, which then informs any subsequent decision making. Any such decision reached via modeling should be made with a quantitative definition of the associated uncertainties.

The level (and expense) of assessment should be appropriate to the risks and consequences of the decision. This chapter is not about how to make decisions, but it is about how to provide input to the decision making process by encouraging that well-defined uncertainties be associated with modeling.

It is the authors' observation that the quantity and quality of verification, validation, calibration, sensitivity analysis, data assimilation, and uncertainty quantification in scientific simulation has historically been extremely variable. It is undeniable, however, that VVUQ is growing in both recognition and importance. It is our impression that, in practice, scientific simulation studies:

- Often confuse several of these activities with each other;
- Regularly include one of these;
- Sometimes include two of these;
- Virtually never include all of these.

The objective of this chapter is not for all analysts to become experts in all of these (i.e., to do a perfect job in all of them), but to strive to include and become competent in each of them and apply the concepts appropriately, mindful of the following generalizations.

- VVUQ is usually at its best in the lowest level of the hierarchy of Figure 7.1 (i.e., for simple, single physics problems).
- VVUQ often breaks down in the middle of the hierarchy, where theory and experiments are sparse.
- Aspects of VVUQ activity are often absent from the application to the highest level.

7.4.1   What Is Validation?

We will endeavor to define the context for VVUQ in the following sections, providing selected excepts from key papers and discussion to clarify matters on terms that remain difficult to separate from each other conceptually. In a nutshell, compare simulation and experimental data to assess the adequacy of the simulation model. The following quotations are from Trucano et al.'s article, "Calibration, validation, and sensitivity analysis: what's what" [39].

7.4.1.1   Validation's Purpose

First, we address the purpose of validation, which is described well as follows [39]:

Validation is the process of quantifying the physical fidelity and credibility of a code for particular predictive applications through the comparison with defined sets of physical benchmarks, consisting of experimental data.

This offers a concise definition along with the previously described definition of benchmark. Next, we move to a purpose for validation:

The purpose of validation is to quantify our confidence in the predictive capability of a code for a given application through comparison of calculations with a set of experimental data.

Additionally,

Validation deals with the question of whether the implemented equations in a code are correct for an intended application of the code.

We note that the intended application will have an increasing influence on the nature of experiments used in validation as one moves higher in the validation hierarchy. We define VVUQ and its impact on simulation based decision making by addressing the impact of evidence and an analogy to the legal system: "V&V is a process of evidence accumulation similar to the formation of legal cases." The workflow should ideally have a logical and linear path of activities with a well-defined sequence,

In an ideal setting, validation should not be considered until verification analyses have been satisfactorily addressed. In practice, however, for modern simulation codes used to model complex physical phenomena, full resolution of verification and validation questions is essentially impossible. Instead, verification and, more acutely, validation are ongoing processes that should be subject to continual improvements.

7.4.2   Metrics

Oberkampf and Trucano [15] present an analysis of useful general characteristics of a validation comparison (called a metric in that paper) and a specific example. Roy and Oberkampf [27] provide a concrete example validation metric use. A different formulation of validation comparisons is found in Zhang and Mahadevan [40] and Mahadevan and Rebba [12].

A validation metric ideally includes the following properties:

1. Has a physically meaningful interpretation, that is, one that is thought to be relevant to the SRQ of interest/feature of merit (FOM);
2. Is experimentally achievable;
3. Is sufficiently sensitive to discriminate meaningfully different results, but not so sensitive that a "meaningful" result cannot be obtained.

These are characteristics (or shortcomings) of a (community's) given (favorite) metrics; one needs to know how to weigh them when making a validation evaluation (even if this procedure is heuristic).

They are used to compare experiment with simulation in an application of interest for a purpose of interest. That is, the context and purpose of the comparison should be known at the outset. In the long term, the metrics evolve over time and their definition is iterative in nature based on feedback from the results found applying the VVUQ processes.

7.4.3   Role of Other Assessment Techniques in Validation

One obtains a "best estimate model" in validation, which does include uncertainties. (UQ within verification and validation.)

Verification aspect:

- Calculation verification is to be used to quantify numerical uncertainties;
- Calculation verification provides a means by which to defensibly estimate the contribution of the numerical algorithm to the uncertainty budget.

7.4.4   Hierarchical View

VVUQ is necessarily hierarchical in nature just as modeling. Multiscale modeling is an important emphasis in computational science, and by the same token validation is multiscale. For most integrated engineering applications many physical models are joined together. The issue of design of experiment, quality of experimental data, and comparison is simplest for a single physical process. As more physical processes are joined together, the entire process becomes more difficult, and uncertainty systematically increases. Ultimately, the specter of calibration becomes a necessity in an increasingly unprincipled manner moving toward the multiphysics end of the spectrum. This alone complicates the VVUQ process substantially.

7.4.5   Distinguishing Calibration and Data Assimilation from Validation and Uncertainty Quantification

There is some disagreement among experts as to the proper interaction of validation and calibration. Trucano et al. [39] maintain that validation must be undertaken before calibration: "calibration is logically dependent on the results of validation" for the reasons that "[v]alidation provides important information to calibration accounting for model-form uncertainty" and "[v]alidation provides information that is necessary to understand the ultimate limitations of calibration." Nelson et al. [13], however, see calibration and validation as being different aspects of the same overall "calibration/validation/prediction process," maintaining that "calibration is a task that is considered part of validation."

The process of calibration and validation can be brought together into a single self-consistent framework via data assimilation. The structure of the solution and data is considered as a whole providing an optimal calibration with the simultaneous assessment of the state of validation for the model.

To start, given a calibrated model, you need to assess it before applying it to prediction.

The following quotations are from Trucano et al.'s article [30] and are relevant to calibration and validation: "Calibration and validation are essentially different" and "[c]alibration and validation are distinct." I think the issue comes down to the fact that validation must be done with data *not* used in the putative calibration. The data used for calibration can contain useful information about aleatory uncertainty, but not epistemic uncertainty due to the fundamental nature of each category of uncertainty. The definition of "not" in this case could be problematic depending on the difficulty of achieving independent experimental data for a given circumstance. This goes to the ultimate validation providing faith in that the calibration is producing something akin to a good prediction to the system in question. The particular issue that is the most difficult to examine is the quality of the extrapolation of a simulation to situations where experimental data do not exist, yet the situation has application significance. For nuclear reactor operations this includes most severe accident scenarios.

Validation and calibration in computational science and engineering (CS&E) both depend on results of verification. We also claim that calibration is logically dependent on the results of validation, which is one way of emphasizing that calibration cannot be viewed as an adequate substitute for validation in many CS&E applications. . . .

Calibration is conditioned by the probability that validation has been successfully performed. . . .

Validation provides important information to calibration accounting for model form uncertainty. . . .

Validation provides information that is necessary to understand the ultimate limitations of calibration.

7.4.6   What Is Uncertainty Quantification and Its Purpose?

Establish defensible and credible bounds on a prediction. A prediction can be based on simulations, experiments, or their combination. The focus is to establish credible bounds on predictability of the model on an intended application in a regime of interest. We have little to add that has not already been stated at length by other authors. With this in mind we provide a series of relevant quotations from Trucano et al. [39].

The following quotations are from Trucano et al.'s article [39] regarding the definition and reasoning behind UQ:

Quantification of Uncertainty is driven by the identification, characterization, and quantification of the uncertainties that appear in the code predictions of "Best Estimate" calculations. The thrust of Best Estimate plus Uncertainty (BE+U) is that prediction is probabilistic precisely because of our inability to complete V&V in some definitive sense and because of uncertainties intrinsic to complex modeling activities.

The definition of prediction is central to the use: "Here, a *prediction* is a calculation that predicts a value a set of these values prior to or in lieu of their physical measurement." We echo the terminology and sentiment of Trucano et al., who focus on "confident prediction":

In computational science, of course, to some degree confidence is correlated with belief in the quantitative numerical, mathematical and physical accuracy of a calculated prediction. Intuitively, confident prediction then implies having some level of confidence or belief in the quantitative accuracy of the prediction. This further implies a willingness to use the prediction in some meaningful way, for example in a decision process.

A key word associated with UQ is "predictability."

UQ is about the model (but it must include experimental data and their uncertainties.) UQ does not (cannot) replace validation or verification; model is accepted as input to the UQ process.

What is the role of sensitivity analysis? This can be illuminated by quotations from Trucano et al.'s article [39]:

Sensitivity analysis underlies the determination of the importance, hence priorities, of code elements that must be subjected to V&V in particular studies. . . .

Parameter sensitivity is also important in guiding our studied reaction to model uncertainty. Parsimony, the reduction of the size of the parameter vector, is guided by sensitivity analysis and remains an important model selection principle. Sensitivity analysis is required for understanding the extent to which a model is complicated enough to be credible but not too complicated. . . .

First, sensitivity analysis directly contributes to the definition of planned validation activities that culminate in the definition application of validation benchmarks as defined above. This centers on the use of a Phenomenology identification and Ranking Table (PIRT) in defining the key requirements of planned validation tasks. . . .

Second, we stress that from the perspective of prediction, calculation of parametric uncertainties of calculations of benchmarks, either local or global, suggests the need to predict these sensitivities off the chosen benchmark sets.

7.5   Conclusions and Recommendations

In this chapter, we have described the concepts and flow of activities in VVUQ through a process by which uncertainties in modeling can be evaluated. Our context is broad and applies to the full spectrum of the modeling hierarchy. Moreover, the workflow we have described is general and is intended in the spirit of more of guidelines than as a "cookbook." In particular, each component activity can be conducted with a fair amount of flexibility to define uncertainties in a manner appropriate for the physical model and the available experimental data. The important aspect to be emphasized is that each element – VVUQ – must be part of an overall assessment of simulation quality.

While the suggested approach to VVUQ is moderately well codified and used in some quarters, practical details of VVUQ vary widely. Most validation cases encountered by modelers require significant application of domain specific knowledge and experience. Unless the analyst has chosen very simple problems, each particular modeling problem will likely present its own challenges that will require insight, innovation, and determination on the part of the analyst to resolve. Despite these obstacles, VVUQ is a necessary part of the "due diligence" of scientific modeling. The outcome of the VVUQ process is a quantitative assessment of uncertainty, which provides the decision making authority a degree of confidence to place in the modeling activity's contribution to the process.

References

[1] AIAA, *Guide for the Verification and Validation of Computational Fluid Dynamics Simulations*, American Institute of Aeronautics and Astronautics, Reston, VA, AIAA-G-077–1998 (1998).

[2] ASME, *V&V 10 - 2006 Guide for Verification and Validation in Computational Solid Mechanics*, American Society of Mechanical Engineers, New York (2006).

[3] Brock, J.S., Kamm, J.R., Rider, W.J., Brandon, S.T., Woodward, C., Knupp, P., and Trucano, T.G., *Verification Test Suite for Physics Simulation Codes*, Los Alamos National Laboratory report LA-UR-06–8421 (2006).

[4] Boyack, B.E., "Quantifying reactor safety margins Part 1: An overview of the code scaling, applicability, and uncertainty evaluation methodology," *Nucl. Engrng Design*, 119, 1–15 (1990).

[5] Hanson, K.M. and Hemez, F.M., "A framework for assessing confidence in computational predictions," *Exp. Tech.*, 25, 50–55 (2001).

[6] Helton, J., *Conceptual and Computational Basis for the Quantification of Margins and Uncertainty*, Sandia National Laboratories report SAND2009-3055 (2009).

[7] Hills, R., Witkowski, W., Rider, W., Trucano, T., and Urbina, A., "Development of a fourth generation predictive capability maturity model," Sandia National Laboratories report SAND2013-8051 (2013).

[8] Hemez, F.M., *Non-Linear Error Ansatz Models for Solution Verification in Computational Physics*, Los Alamos National Laboratory report LA-UR-05–8228 (2005).

[9] Kamm, J.R., Brock, J.S., Brandon, S.T., Cotrell, D.L., Johnson, B., Knupp, P., Rider, W.J., Trucano, T.G., and Weirs, V.G., *Enhanced Verification Test Suite for Physics Simulation Codes*, Los Alamos National Laboratory report LA-14379, Lawrence Livermore National Laboratory report LLNL-TR-411291, Sandia National Laboratories report SAND2008-7813 (2008).

[10] Knupp, P. and Salari, K., *Verification of Computer Codes in Computational Science and Engineering*, Chapman & Hall/CRC, Boca Raton, FL (2003).

[11] Knupp, P., Ober, C., and Bond, R., "Measuring progress in order-verification within software development projects," *Engrng. Comp.*, 23, 283–294 (2007).

[12] Mahadevan, S. and Rebba, R., "Validation of reliability computational models using Bayes networks," *Reliability Engineering & System Safety* 87(2), 223–232 (2005).

[13] Nelson, R., Unal, C., Stewart, J., and Williams, B., *Using Error and Uncertainty Quantification to Verify and Validation Modeling and Simulation*, Los Alamos National Laboratories report LA-UR-10-06125 (2010).

[14] Oberkampf, W.L., "What are validation experiments?," *Exp. Tech.*, 25, 35–40 (2001).

[15] Oberkampf, W.L. and Trucano, T.G., "Verification and validation in computational fluid dynamics," *Prog. Aerospace Sci.*, 38, 209–272 (2002).

[16] Oberkampf, W.L., Trucano, T.G., and Hirsch C., "Verification, validation, and predictive capability in computational engineering and physics," *Appl. Mech. Rev.*, 57, 345–384 (2004).

[17] Oberkampf, W.L. and Trucano, T.G., "Verification and validation benchmarks," *Nucl. Design Engrng*, 23, 716–743 (2007); also available as Sandia National Laboratories report SAND2007-0853 (2007).

[18] Oberkampf, W. L., Pilch, M., and Trucano, T. G., *Predictive Capability Maturity Model for Computational Modeling and Simulation*, Sandia National Laboratories report SAND 2007–5948 (2007).

[19] Oberkampf, W.L. and Roy, C.J., *Verification and Validation in Scientific Computing*, Cambridge University Press, New York (2010).

[20] Oberkamf, W.L. and Smith B., Assessment criteria for computational fluid dynamic validation benchmark experiments, 52nd Aerospace Sciences Meeting, *AIAA* 2014-0205, January 2014.

[21] Pilch, M., Trucano, T.G., and Helton, J.C., *Ideas Underlying Quantification of Margins and Uncertainties (QMU): A White Paper*, Sandia National Laboratories report SAND2006-5001 (2006).

[22] Rider, W.J., Kamm, J.R., and Weirs, V.G., *Code Verification Workflow in CASL*, Sandia National Laboratories report SAND2010-7060P (2010).

[23] Roache, P., "Code verification by the method of manufactured solutions," *J. Fluids Engrng*, 124, 4–10 (2002).

[24] Roache, P., "Building PDE Codes to be verifiable and validatable," *Comput. Sci. Engrng.*, 6, 30–38 (2004).

[25] Roache, P., *Fundamentals of Verification and Validation*, Hermosa Publishers, Albuquerque, NM (2009).

[26] Roy, C.J., "Review of code and solution verification procedures for computational simulation," *J. Comput. Phys.*, 205, 131–156 (2005).

[27] Roy, C.J. and Oberkampf, W.L., "A comprehensive framework for verification, validation, and uncertainty quantification in scientific computing," submitted to *Comp. Meth. Appl. Mech. Engrng.* (2010).

[28] Salari, K. and Knupp, P., *Code Verification by the Method of Manufactured Solutions*, Sandia National Laboratories report SAND2000-14444 (2000).

[29] Sargent, R.G., "Verification and validation of simulation models," in *Proceedings of the 1998 Winter Simulation Conference*, ed. by Medeiros, D.J., Watson, E.F., Carson, J.S., and Manivannan, M.S., 121–130 (1998).

[30] Sargent, R.G., "Some approaches and pardigms for verifying and validation simulation models," in *Proceedings of the 2001 Winter Simulation Conference*, ed. by Peters, B.A., Smith, J.S., Medeiros, D.J., and Rohrer, M.W., 106–114 (2001).

[31] Schwer, L.E., "An overview of the PTC 60 / V&V 10 guide for verification and validation in computational solid mechanics," Reprint by ASME, available at http://cstools.asme.org/csconnect/pdf/CommitteeFiles/24816.pdf (2006).

[32] Sornette, D., Davis, A.B., Ide, K., Vixie, K.R., Pisarenko, V., and Kamm, J. R., "Algorithm for model validation: Theory and applications," *Proc. Nat. Acad. Sci. USA* 104, 6562–6567 (2007).

[33] Stern, F., Wilson, R.V., Coleman, H.W., Paterson, E.G., "Comprehensive approach to verification and validation of CFD simulations—Part 1: Methodology and procedures," *J. Fluids Engrng*, 123, 793–802 (2001).

[34] Stern, F., Wilson, R.V., and Shao, J., "Quantitative V&V of CFD simulations and certification Of CFD codes," *Int. J. Num. Meth. Fluids*, 50, 1335–1355 (2005).

[35] Stewart, J.W., *Measures of Progress in Verification*, Sandia National Laboratories report SAND2005-4021P (2005).

[36] Szilard, R., Zhang H., Kothe D., and Turinsky P., *The Consortium for Advanced Simulation of Light Water Reactors*, Idaho National Laboratory (United States): Funding organization: DOE-NE (United States) (2011).

[37] Trucano, T.G., Pilch, M., and Oberkampf, W.L., *General Concepts for Experimental Validation of ASCI Code Applications*, Sandia National Laboratories report SAND2002-0341 (2002).

[38] Trucano, T.G., Pilch, M., and Oberkampf, W.L., *On the Role of Code Comparisons in Verification and Validation*, Sandia National Laboratories report SAND2003-2752 (2003).

[39] Trucano, T.G., Swiler, L.P., Igusa, T., Oberkampf, W.L., and Pilch, M., "Calibration, validation, and sensitivity analysis: What's what," *Reliab. Engrng. Syst. Safety*, 92, 1331–1357 (2006).

[40] Zhang, R. and Mahadevan S., "Bayesian methodology for reliability model acceptance," *Reliability Engineering & System Safety* 80(10, 95–103 (2003).

Part III

Complex Mixing Consequences

8 Shock Driven Turbulence

Fernando F. Grinstein, Akshay A. Gowardhan, and J. Raymond Ristorcelli

8.1   Introduction

The challenging problem of nonequilibrium, underresolved material mixing promoted by underresolved velocity and underresolved (or insufficiently characterized) initial conditions (ICs) in shock driven turbulent flows is the subject of what follows. In many areas of interest, such as inertial confinement fusion, understanding the collapse of the outer cores of supernovas, and supersonic combustion engines, vorticity is introduced at material interfaces by the impulsive loading of shock waves, and turbulence is generated via Richtmyer–Meshkov instabilities (RMI) [1]. RMI add the complexity of shock waves and other compressible effects to the basic physics associated with mixing; compressibility further affects the basic nature of material mixing when mass density differences and material mixing fluctuation effects are not negligible. Because RMI are shock driven, resolution requirements make DNS approaches prohibitively expensive even on the largest supercomputers.

Classical large eddy simulation (LES) [2] is particularly inadequate for flows driven by RMI because of the dissipative numerics needed for shock capturing; hybrid methods switching between shock capturing schemes and conventional LES depending on the local flow conditions [3] have been proposed; high order shock capturing (e.g., fifth-order weighted essentially nonoscillatory [WENO]) methods are typically chosen to ensure a smooth transition and matching of the inherently different simulation models. However, all shock capturing methods degrade to first-order in the vicinity of shocks because of the monotonicity requirements – and, in particular, at the very important initial stage at which shocks first interact with the material interface and generate the velocity field. Thus, severe resolution demands to address the competition between the (numerics provided) implicit and explicit subgrid models can be expected in the hybrid context. Alternatively, by combining shock and turbulence emulation capabilities based on a single (physics capturing numerics) model, implicit LES (ILES) [4] has the potential of providing a natural effective simulation strategy for RMI [5].

The accurate and reliable simulation of material interfaces is also an essential aspect in simulating the material mixing fluid dynamics – see Chapter 5. The crucial issue associated with initial interfaces is the sensitivity of most problems to the ICs: relatively small variations in the initial state of the interface can result in quite significant changes to even the integral character of a mixing layer at late times [5–7]. As we consider simulating RMI in the laboratory experiments, we must thus consider the effects of modeling insufficiently characterized initial contact discontinuity deformations.

In the recent simulations of RMI in shock tube experiments revisited in the following, the focus has been on IC effects on transition, and late time mixing and turbulence characteristics. The particular ILES strategy is based on using the Los Alamos National Laboratory (LANL) radiation adaptive grid Eulerian (RAGE) code [8]. RAGE solves the multimaterial compressible conservation equations for mass density, momenta, total energy, and partial mass densities, using a second-order Godunov scheme, adaptive mesh refinement (AMR), a variety of numerical options for gradient terms – limiters, and material interface treatments (not used here). The van Leer limiter option was chosen for the present simulations. The simulation model is nominally inviscid, effectively models high Reynolds number (Re) convection driven flow, with miscible (Sc~1) material interfaces. Simulations of planar [7–9] and gas curtain [10] shock tube experiments are examined, involving high (SF_6) and low density (air) gases, with sharp and diffusive initial material interfaces, respectively, and Atwood number $A = (\rho_{heavy} - \rho_{light})/(\rho_{heavy} + \rho_{light}) \sim 0.67$.

We deal here with inherently unsteady transitional and decaying turbulent flow. Turbulence metrics and analysis for this kind of flows are not established, and the state-of-the-art analysis largely relies on using suitable (unsteady) versions of diagnostics designed for the homogeneous isotropic regimes. We first focus on effects of IC resolved spectral content and initial interfacial morphology on transitional and late time turbulent mixing in the fundamental planar shock tube configuration. Next, we examine typical practical challenges encountered in predictability assessments in the more complex three-dimensional (3D) shocked gas curtain (GC) problem for which state-of-the-art laboratory data and diagnostics are available.

8.2   Shocked Planar Interface

Richtmyer's [11] impulse analysis shows that the initial growth rate of RMI is given by $\dot{a} = a_o \Delta U A^+ \kappa_o$, where a_o is the initial amplitude of the interface perturbation between the two fluids, ΔU is the change in the velocity of the mixing layer due to the shock, A^+ is the post-shock A, κ_o is the initial wavenumber of the interface, and $\dot{a}$ is growth rate of the instability. The analysis is based on single mode perturbation and is valid in the linear regime for very low initial wave number and amplitude ($\kappa_o a_o$) or low values of the material interface *rms* slope at early time. Beyond the work surveyed in [1], investigation of IC effects on RMI have been subject of many experimental [12–16], numerical [7, 16–22], and theoretical studies [21, 24].

We first focus here on the planar shock tube experiments of [12], involving presumed geometries of the membranes and the wire mesh initially separating the gases, and reshock off an end wall (Fig. 8.1). The mixing layer growth is affected by the initial interaction of the shock and material interface – with a direct distinct imprint of the initial contact discontinuity deformation specifics and significant further effects occurring after reshock. The contact discontinuity between air and SF_6 is modeled as a jump

Table 8.1 Planar Shock Tube Simulations I

Case		Δ_{min}(cm)	AMR	NX_{max}	NY_{max}	NZ_{max}	ϕ (cm)	$(\lambda_{min}, \lambda_{max}) / \lambda_o$
RUN1	3D	0.1	YES	1640	480	480	$A_2 = 0.025$	(0.2, 12)*
RUN2	3D	0.05	YES	1640	480	480	$A_2 = 0.025$	(0.2, 12)*
RUN3	3D	0.1	YES	820	240	240	$A_2 = 0.025$	(0.4, 4)
RUN4	3D	0.05	YES	1640	480	480	$A_2 = 0.025$	(0.4, 4)
RUN5	3D	0.1	YES	820	240	240	$A_2 = 0.025$	(4, 12)
RUN6	3D	0.1	YES	820	240	240	$A_2 = 0$	–

* Contributing perturbation modes are weighted with $k^4 \exp[-(k/k_o)^2]$.

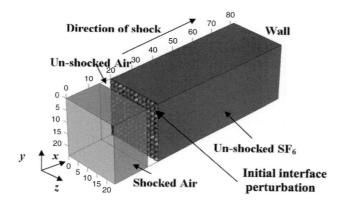

Figure 8.1 Typical planar shock-tube computational domain; lengths in m.

in density using ideal gases with $\gamma = 1.4$ and $\gamma = 1.076$, respectively, with constant pressure across the initial interface at rest. Typical 3D grids between 2.8–4.5×10^7 and 1.7–3.2×10^8 computational cells with smallest spacing $\Delta_{min} = 0.1$ cm and $\Delta_{min} = 0.05$ cm, respectively, were used in the simulations of the experiments in [12] (Table 8.1).

A shocked air region is created upstream satisfying the Rankine–Hugoniot relations for a Mach number $Ma = 1.5$ shock. The shock propagates in the x direction through the contact discontinuity and reflects at the end of the simulation box on the right, where purely reflecting boundary conditions are enforced. Periodic boundary conditions are imposed in the transverse (y, z) directions. By design, the location of the left boundary of the computational domain is chosen far away enough to ensure that eventual reflections there are unable to affect the mixing region during the reported times of interest. Up to three levels of AMR were used.

8.2.1 Initial Interface Modeling

The surface displacement of the material interface in RMI experiments [12] has been modeled using well-defined modes often combined with random perturbation components. In [18] the shocked planar RMI phase was examined in detail using S+L deformation, combining a short (S) egg crate mode, chosen to represent the result of pushing the membrane through the wire mesh, and superimposed distortion of the wire mesh on a longer characteristic shock tube scale L, chosen to be the transverse periodicity dimension of the computational domain. In later papers the reshocked mixing layer was also simulated, and the IC modeling strategies have included (1) S+L+ϕ deformation [17], and (2) S+ϕ deformation [3, 7], where ϕ is a spatially random distribution intended to break the presumed characteristic interface symmetries. Other studies have also used broadband ϕ multiscale perturbations with prescribed standard deviation and power spectra by themselves (e.g., [19–21]).

We examine here an S+ϕ strategy to model the initial interface conditions. We define the perturbation in terms of a prescribed range of modes $\lambda_{min} < \lambda < \lambda_{max}$, that is,

$$dx(y,z) = S + \phi$$
$$= A_1|\sin(k_o y)\sin(k_o z)| + \delta_o \sum_{n,m} \sin(k_n y + \varphi_n)\sin(k_m z + \chi_m),$$

where $dx(y,z)$ is the local interface deformation, $\lambda_o = \pi/\kappa_o = L/12 = 2$ cm is the characteristic length scale of the egg crate mode, $A_1 = 0.25$ cm, $\delta_0 = 0.025$ cm is the standard deviation of the perturbation, $\kappa_n = 2\pi n/L$, $\kappa_m = 2\pi m/L$, random coefficients $-1/2 < a_{m,n} < +1/2$, random phases φ_n, χ_m. The participating perturbation modes in the top hat band of wavelengths are constrained by

$$\lambda_{min} \leq L/\left[2\pi\sqrt{(n^2 + m^2)}\right] \leq \lambda_{max}$$

and by requiring that wavelengths be adequately resolved by the computational grid. Amplitudes and fundamental perturbation lengths (A_1, δ_0, and λ_o) were identical or very similar in [3, 7].

Simulations were performed for a variety of grid resolutions and perturbations superimposed to the S mode (Table 8.1, Fig. 8.2). The selected representative cases discussed here involve the following:

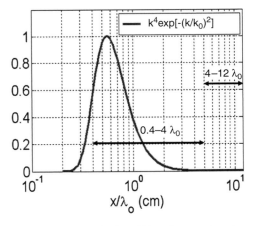

Figure 8.2 Schematic describing the wavelength content of the IC perturbations added to the fundamental component egg crate mode $k\lambda_o/\pi=1$ (Table 8.1).

- *short* perturbation: $(\lambda_{\min}, \lambda_{\max}) = (0.4\lambda_o, 4\lambda_o)$,
- *long* perturbation: $(\lambda_{\min}, \lambda_{\max}) = (4\lambda_o, 12\lambda_o)$,
- *weighted short* perturbation,
- *no* perturbation: $\delta_0 \equiv 0$.

The *weighted short perturbation* case emulates the ICs in [3] by requiring a top hat superposition of mode contributions in the band $(\lambda_{\min}, \lambda_{\max}) = (0.2\lambda_o, 12\lambda_o)$ weighted by $\sim k^4 \exp[-(k/k_o)^2]$, and with peak wavelength $\sim 0.55\lambda_o$ – as opposed to a range of randomly (equally weighted) superimposed short wavelengths in the *short perturbation* case.

8.2.2   Evolution of Mixing and Turbulence Characteristics

Analysis of the simulated mixing data is based on transverse plane averaged quantities,

$$\langle f \rangle (x) = \int f(x, y, z) dy dz / \int dy dz$$
$$Y_{SF_6} = \rho_{SF_6}/\rho, \quad \psi(x) = \langle Y_{SF_6} \rangle,$$
$$\theta = 4 Y_{SF_6}(1 - Y_{SF_6})$$
$$M(x) = 4\psi(x)[1 - \psi(x)],$$

where ρ is the mass density, ρ_{SF6} is the SF$_6$ partial mass density, Y_{SF_6} is the *SF$_6$* mass fraction, and $\theta, M(x)$, are local and transverse averaged mixedness measures, respectively. The instantaneous mixing region is defined by a slab of volume V about the center of the mixing layer constrained in the x direction by requiring $M(x) > 0.75$. Analysis of turbulence characteristics is based on data deviations around transverse planes within this *mixing slab* region,

$$\tilde{u} = \langle \rho u \rangle / \langle \rho \rangle, \quad u = \tilde{u} + u', \quad \omega = \nabla \times u, \quad 2K = \rho u_i' u_i'$$
$$\rho = \langle \rho \rangle + \rho', \quad R = \rho'^2,$$

where u is the velocity field, K is the local turbulent kinetic energy, R is the mass density variance, ω is the vorticity, and summation over repeated indices is assumed.

The instantaneous material mixing zone thickness δ_{MZ} used in [3, 7] is defined in terms of the transverse averaged mixedness $M(x)$, by $\delta_{MZ} = \delta = \int M(x) dx$, designed to yield $\delta_{MZ} = h$ for $\psi(x) = [1 + \tanh(2(x - x_c)/h)]/2$, where x_c defines the center of the mixing layer.

The diagram in Figure 8.3 shows the evolution and interaction of the Ma=1.5 shock and an air/SF$_6$ interface for a selected representative case, in good agreement with similar ones for the other cases considered here as well as in the previous cited work [3, 12]. The air/SF$_6$ interface is shocked at t = 0 ms, reshocked by the primary reflected shock at t ~ 3.5 ms, and then by the reflected rarefaction at t ~ 5ms. The material mixing layer is further affected at later times by reflected compression and weaker secondary reflected shock waves (see also the layouts in Fig. 8.4a). Flow visualizations of the vorticity magnitude shown in Fig. 8.4b at selected late times depict the persistence of large scale coherent structures in the long IC perturbation case.

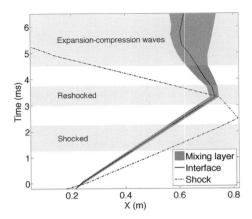

Figure 8.3 Shock interface diagram for a representative planar shock tube simulation; the dark and light blue traces indicate, center and approximate edges of the mixing layer, as specified in terms of $M(x)$ by maximum and 1% value locations, respectively.

Figure 8.5 compares the variation of the 3D mixing layer thickness δ_{MZ} as function of time for the different ICs considered at the baseline 0.1 cm resolution. Overall, the predicted mixing widths for all cases are very similar before reshock, when mixing width for the short (and weighted short) perturbation case is slightly greater, and a thicker mixing layer is associated with breaking of the egg crate mode by the perturbations. Late time mixing measures are higher for the long IC perturbation case; short and weighted short perturbation cases are fairly comparable, with the short case mixing measures being somewhat larger for late times – reflecting on higher longer wavelength IC content. Results for the nonperturbation case consistently appear as a limiting case of the long perturbation cases.

More detailed analysis is shown in Figure 8.6(a) and (b). Figure 8.6(a) exhibits the mixing zone width evolution before reshock as function of the varying IC spectral content. Smaller wavelength IC content leads to faster mixing zone growth. This is consistent with Richtmyer's analysis, the first impulsive model proposed for RMI growth of a single mode perturbation [11], $d\delta_{MZ}/dt \sim k$, where k is the characteristic interface deformation wavenumber. However, this model is only valid shortly after the material interface is shocked when $k\delta_{MZ} \ll 1$, and breaks down at later time (particularly near reshock) when $k\delta_{MZ} \gg 1$ due to growth of high wave number modes. During reshock, there is compression and vorticity baroclinicity deposited at the mixing layer (with opposite vorticity sign to that when first shocked). The simulated mixing width shows the opposite growth behavior as function of characteristic wavelength (Fig. 8.6(b)) after reshock. We first argued [7] that this reflects on higher mixing at reshock time for higher wavenumber IC being associated with smaller density gradients and consequently less baroclinic torque production. More recently, we demonstrated [25] that the results depicted in Figure 8.6 can be also explained in terms of the *bipolar behavior* of RMI [9] discussed further in the following.

Effects of IC resolution are discussed in detail in [7]; see also Chapter 5, where it was found that a key IC posing issue that needs to be addressed is that of adequately resolving the $|\sin(k_o y)\sin(k_o z)|$ function used to prescribe the egg crate mode S

(a)

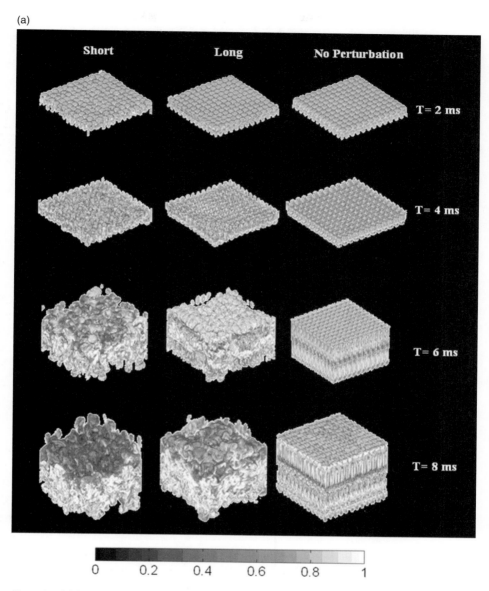

Figure 8.4 (a) Instantaneous isosurfaces of the local mixedness function θ for the short, long, and no perturbation cases at the 0.1 cm resolution. (b) Volume renderings of the vorticity magnitude at t = 8 ms, for the short (top) and long (bottom) perturbation cases at the 0.1 cm resolution. A black and white version of this figure will appear in some formats. For the color version, of Figure 8.4(b) please refer to the plate section.

(introduced in [18] and historically used since in this context). Actual predicted mixing layer thickness growth was shown to depend on the effective IC associated with the resolved egg crate function. Figure 8.7 shows spectra of K as function of time and resolution for the weighted short perturbation IC case. Relevant issues of grid resolution and convergence are further addressed in Figure 8.8. Our spectral and PDF analysis are

(b)

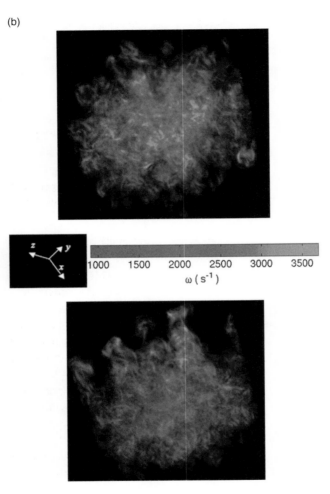

Figure 8.4 (*cont.*)

based on data in the mixing slab at each time. The reported spectra are the result of evaluating 2D spectra for each cross-stream plane within the slab, and then averaging over planes in the slab. In all cases, the peak spectral amplitude at $k\lambda_o/\pi = 1$ at t = 0 corresponds to the dominant egg crate mode (wavelength λ_o), lower amplitude regularly spaced peaks are associated to the egg crate harmonics, and (for reference) the IC (gray) trace corresponds to the spectra of the actual interface perturbation function at the baseline resolution 0.1 cm. Before reshock (t < 3.5 ms), the random interface perturbation distorts the S mode and leads to less turbulent kinetic energy growth. Figure 8.7 shows that the turbulent kinetic energy decreases after shocked time (t=0), and then increases at reshock, as energy is again deposited through baroclinic production of vorticity at the material interfaces; shorter wavelengths are populated as time evolves, and self-similar decay becomes increasingly apparent for later times.

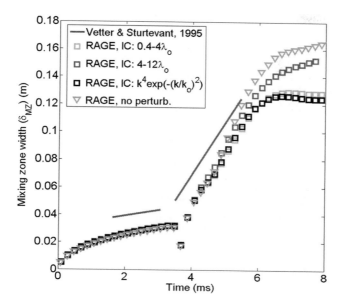

Figure 8.5 Variation of the 3D mixing layer thickness as a function of time vs. IC perturbation spectral content for the baseline 0.1 cm grid resolution vs. laboratory [12] results. A black and white version of this figure will appear in some formats. For the color version, please refer to the plate section.

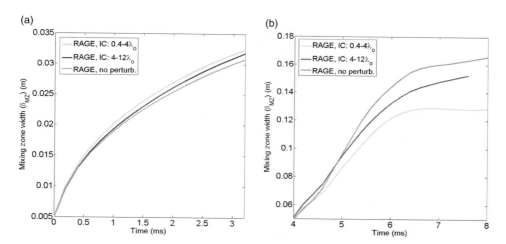

Figure 8.6 (a) 3D mixing zone width evolution before reshock for varying IC spectral content. (b) 3D mixing zone width evolution after reshock for varying IC spectral content.

Worm vortex dominated flow regimes are suggested for late times in the perturbed IC cases, with persisting larger scale structures apparent for the long perturbation case (e.g., Fig. 8.4(b)). The detailed late time analysis for the weighted short perturbation case shows that the presence of worms is accompanied by spectra consistent with Kolmogorov's $\sim k^{(-5/3)}$ power law (Fig. 8.7) for both the 0.05 cm and 0.1 cm resolutions, and by the noted good agreement of characteristic PDFs with those of

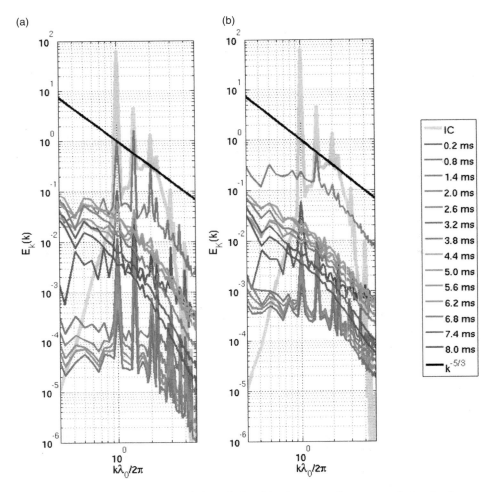

Figure 8.7 Time-dependent spectra of the turbulent kinetic energy K versus resolution for the weighted short case. (a) 0.1 cm grid; (b) 0.05 cm grid. A black and white version of this figure will appear in some formats. For the color version, please refer to the plate section.

(incompressible) isotropic turbulence DNS data (Fig. 8.8). As finer grids are used with ILES, increasingly smaller structures can be resolved and characteristic cross-sections of the smallest resolved vortical worm structures scale with the ILES cutoff [26, 27]. The observed resolution effects can be consistently characterized by a higher effective Re associated with the higher resolution – [30] (see also Chapters 1 and 2). As self-similar regimes are achieved for the late times, spatial velocity derivative PDFs exhibit expected trends when scaling values with the *rms* vorticity (as in the analysis of the DNS data): (1) the non-Gaussian tails, and (2) the effects of increasing grid resolution (i.e., increasing effective Re and resolved *rms* vorticity) emulating effects of increasing physical Re. We regard this as a clear demonstration that predictive underresolved simulations of the velocity in turbulent fluid flows are possible with (sufficiently resolved) ILES such as is used here – see also [26–28] and chapters in [4].

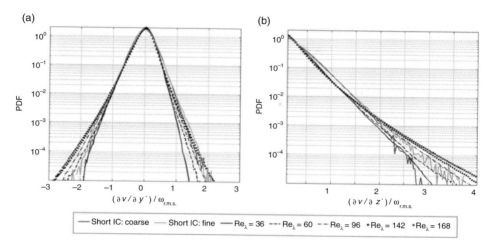

Figure 8.8 Effects of grid resolution on late time results for the short perturbation case (t = 8.0 ms). (a) PDFs of longitudinal $(\partial u_i/\partial x_i, i = 1, 2, or, 3)$ velocity derivatives; (b) PDF's of transverse velocity derivatives $(\partial u_i/\partial x_j, i,j = 1, 2, or, 3, i \neq j)$. The reference PDFs in (a) and (b) are based on DNS data of incompressible isotropic turbulence [23].

8.2.3   Bipolar Behavior

Our planar RMI studies [7] involving initially thin material interface showed mixing layer growth rates fairly insensitive to IC specifics before reshock, very sensitive after reshock (Fig. 8.5), and having very different *growth trends* as function of IC spectral content at shock and reshock time (Fig. 8.6) – pointing to the potentially significant role of morphological details in the thicker more complex mixing layers at reshock. These results motivated our systematic numerical study of the evolution of RMI [9] by varying characteristic *multiscale* spectral content and thickness of the initial interface between the two fluids, for fixed *Ma* and A. The focus was again on the planar shock tube configuration investigated earlier (Fig. 8.1), but now with IC prescribed only in terms of ϕ (no egg crate mode S) and no reshocks for the times considered.

The initial interfacial morphology is defined statistically by $\eta_o = \kappa_o \delta_o \sim \langle \nabla \chi \nabla \chi \rangle^{1/2}$, the initial *rms* slope, where $\chi(y, z) \equiv dx(y, z)$ is the previously defined local deviation of the initial material interface around the mean interface location at "x", $\kappa_o = 2\pi/\lambda_o$, λ_o is a representative wavelength of the multiscale perturbation in the initial interface, and $\delta_o = \delta(t = 0)$ denotes the initial interface thickness (Fig. 8.9(a)). Thus, a high value of η_o denotes a highly corrugated interface with high *rms* slope (Fig. 8.9(b), (c)).

The quantity κ_o is used in the study of homogeneous stochastic processes, where it is called the mean zero-crossing frequency [29]. In practice, the initial material interface value of κ_o is computed by checking for sign changes of the mass density fluctuation over lines within transverse planes and averaging their occurrences (Fig. 8.10); for $t > 0$, $\kappa = \kappa(t)$ is similarly evaluated within the mixing region. Mathematically, κ is associated with $\kappa^2 = \int q^2 E_R(q)dq/\int E_R(q)dq$, where $E_R(q)$ is the instantaneous R spectra in the mixing region, q is the wavenumber, and, thus, $\kappa(t)$ is related to the Taylor microscale; such spectra and connections are exemplified and discussed in the following.

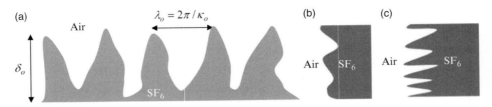

Figure 8.9 (a) Schematic of the initial interface; (b) low η_o; (c) high η_o.

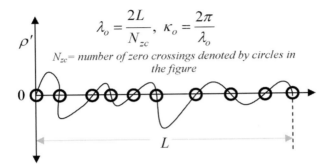

Figure 8.10 Zero-crossings of ρ'; L is the transverse dimension of the computational domain.

Various simulation based experiments were performed in terms of well-defined initial material interface perturbations. The material interface deformation is defined by:

$$\chi(y,z) = \Gamma \sum_{n,m} a_{m,n} \sin\left(\kappa_n y + \phi_n\right) \sin\left(\kappa_m z + \varphi_m\right), \quad \kappa_i = 2\pi i / L,$$

where we use random coefficients $-\frac{1}{2} < a_{m,n} < \frac{1}{2}$, Γ is used to prescribe δ_o, $\{\phi_n, \varphi_m\}$ are random phases, and the participating modes are constrained by the requirement

$$\lambda_{\min} \leq L / \left[2\pi \sqrt{(n^2 + m^2)} \right] \leq \lambda_{\max}.$$

For completeness, we note that for the single mode analysis [11] cited earlier, $\delta_o = a_o$. A variety of IC perturbations and grid resolutions (using up to two levels of AMR refinement) were considered. The baseline resolution involved a $820 \times 240 \times 240$ grid ($\Delta_{\min}$=0.1 cm); a much more finely resolved $1640 \times 480 \times 480$ grid ($\Delta_{\min}$=0.05 cm) was used for selected representative cases. The various cases are organized into two distinct categories having significantly different (low and high) initial *rms* slope (η_o) with prescribed spectral content but different interfacial thickness (see Table 8.2).

Instantaneous visualizations of Y_{SF_6} at the selected time, $t = 3000$ μs shown in Figure 8.11 suggest material (interpenetration) mixing increasing with η_o. Following [11], the passage of a shock through the material interface has the effect of having baroclinic vorticity $\sim \kappa_o \dot{a} = \Delta U A^+ \kappa_o \eta_o$ deposited as function of η_o. For *low* η_o, less baroclinic vorticity is generated. When the initial characteristic wavelength is greater than its characteristic amplitude, crests and troughs are well separated and the vortices are weaker and consequently do not interact strongly. In the *low* η_o regime, the modes

Table 8.2 Planar Shock Tube Simulations II

$(\lambda_{min}, \lambda_{max})$	$L\left(\frac{1}{24}, \frac{1}{6}\right)$	$L\left(\frac{1}{12}, \frac{1}{4}\right)$	$L\left(\frac{1}{6}, \frac{1}{3}\right)$	$L\left(\frac{1}{4}, \frac{1}{2}\right)$	$L\left(\frac{1}{24}, \frac{1}{6}\right)$	$L\left(\frac{1}{12}, \frac{1}{4}\right)$	$L\left(\frac{1}{6}, \frac{1}{3}\right)$	$L\left(\frac{1}{4}, \frac{1}{2}\right)$
δ_o (cm)	0.5 (low η_o)				5 (high η_o)			
κ_o (cm^{-1})	π	$\pi/2$	$\pi/4$	$\pi/6$	π	$\pi/2$	$\pi/4$	$\pi/6$

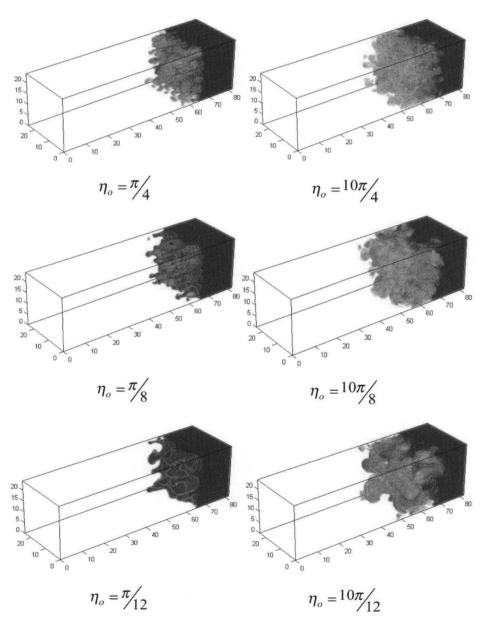

Figure 8.11 Effects of initial material interface conditions. Volume visualization of Y_{SF6} at time $t = 3000$ μs after the first shock.

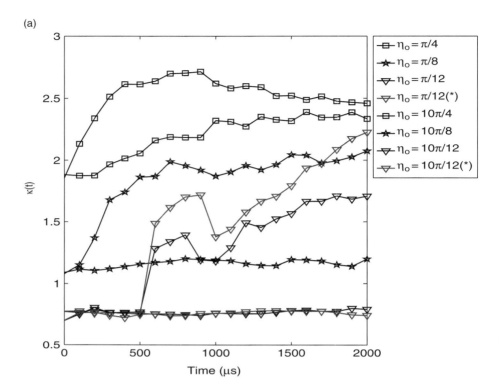

Figure 8.12 (a) Zero crossing frequency evolution; (b) representative R spectra; (*) finer resolution. A black and white version of this figure will appear in some formats. For the color version, please refer to the plate section.

mainly grow in the shock direction in a ballistic (noninteracting) fashion. For these (*low η_o*) flows, the zero crossing frequency κ is virtually unchanged with time as shown in Figure 8.12(a), indicating that no new modes are produced. For *high η_o*, more baroclinic vorticity is generated. The initial characteristic wavelength is less than its characteristic amplitude, thus crests and troughs of the perturbations are closer together and the vortices are stronger and create new modes through nonlinear processes. For these flows, there is an unmistakable jump in κ indicating rapid production of smaller scales. Consistent with this notion, Figure 8.12(a) shows that the higher the η_o, the sooner the jump occurs. The border-line case is $\lambda_o/2 \sim \delta$, *or $\eta_o \sim \pi$*.

Representative selected cases were also analyzed based on the higher resolution simulations. Integral measures such as the mixing width were found to be virtually insensitive to grid resolution; as the flow becomes nonlinear (high η_o), finer resolution results depict increased small scale production as indicated by higher κ (Fig. 8.12(a)) and longer self-similar ranges in the spectra of R (Fig. 8.12(b)) – which can be associated with higher effective Re [7, 28, 30].[1]

---

[1] The spectra is obtained by averaging (q-shelled) 2D $E_R(q)$ evaluated at cross-stream planes within the mixing slab region defined earlier [7].

(b)

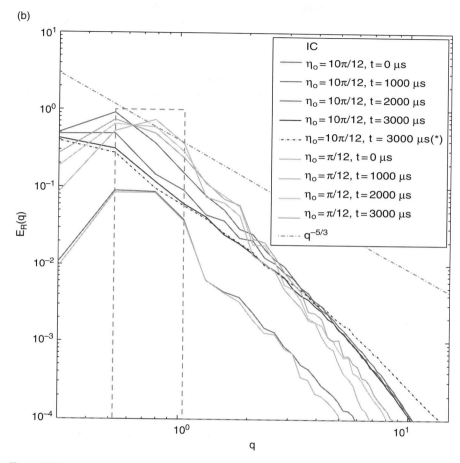

Figure 8.12 (*cont.*)

Inspection of the evolution of the mixing layer width $\delta(t)$(Fig. 8.13(a)) for relatively small η_o (for $\eta_o \ll \pi$) shows growth rates ordered in agreement with predictions of classical linear impulse theory (growth proportional to η_o) [11]. For the high η_o cases, higher η_o results in larger deposition of baroclinic vorticity and leads to thicker mixing layers initially. The energy produced by passage of the shock is mostly in the shock direction and leads to growth of the initial modes in that same direction and increased the mixing layer widths. However, for $\eta_o = \pi/2$, the theory is valid only for a very short time after the material interface is shocked. Soon thereafter, the growth rate drops and this is not consistent with linear theory. In contrast, for high η_o ($\eta_o \gg \pi$), the growth is found to be inversely proportional to η_o. For higher η_o there is a much larger deposition of baroclinic vorticity. With higher *rms* slope, the vortex centers are closer to each other and they interact, giving rise to the production of more smaller scale modes as is only possible through nonlinear processes. Here the energy of the flow, because of the

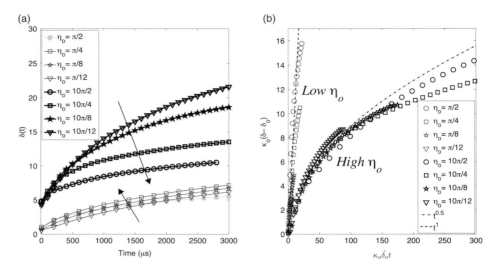

Figure 8.13 Mixedness measures as a function of time: (a) low η_o; (b) high η_o. The air/SF$_6$ interface is shocked at $t = 0$ μs. (b) Mixedness measures suitably shifted and scaled with IC parameters (time scaling first proposed in [13]).

nonlinear interaction, is more isotropically distributed. Interestingly, the increase in η_o does not result in increase in the mixing layer width but leads to production of more small scales and thus more dissipation.

As in the work of Jacobs and Sheeley [13], we found it useful to plot each layer thickness $\kappa_o(\delta - \delta_o)$ versus time scaled with $\kappa_o \dot{\delta}_o$ (Fig. 8.13(b)), using the corresponding computed early time shocked mixing layer growth rate $\dot{\delta}_o$. Such plotting tends to collapse the data into two distinctly different groups. Depending on the initial *rms* slope of the interface, RMI evolves into linear or nonlinear regimes, with different flow features, growth rates, turbulence statistics, and material mixing rates. We called this the bipolar RMI behavior.

For interfaces with low initial *rms* slope η_o, the mixing layer growth is ballistic with no mode coupling, the evolution of RMI is linear ($\sim t$), and linear scaling is followed [11]. Increasing η_o in the low η_o group increases the deposition of baroclinic vorticity on the initial material interface and leads to higher layer growth. In contrast, increasing η_o in the high η_o group also increases the deposition of baroclinic torque, but this leads to a reduced mixing width growth rate ($\sim t^{0.5}$) associated with the production of small scales by nonlinear mode coupling that are additionally dissipative. Early time data scatter apparent in Figure 8.13(b) reflects a historical (rather than fundamental) issue: our original scope did not focus on early time aspects and (in hindsight) frequent enough dumps in time to ensure uniformly accurate $\dot{\delta}_o$ evaluations for all cases were not available to us.

Transition to turbulence is traditionally viewed in terms of a rapid increase in the population of motions with smaller length scales, which can lead to an inertial subrange in the turbulent kinetic energy spectra (e.g., [31, 32]). The spectral bandwidth of fully

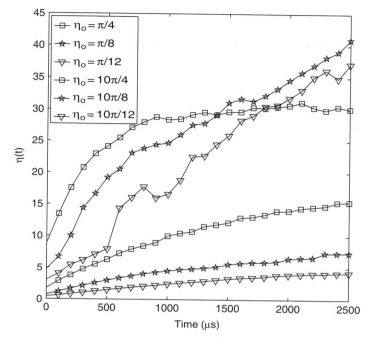

Figure 8.14 Evolution of the spectral bandwidth measure $\eta(t) = \kappa(t)\delta(t)$.

developed turbulence can be scaled by the turbulent Re [33], usually taken as a ratio of integral to Kolmogorov length scales. In our context, we use the thickness of the layer, $\delta(t)$ as a measure of the integral scale, and the mass density Taylor microscale $\lambda(t)$ – related to the spatial zero crossing frequency through $\lambda(t) \sim 1/\kappa(t)$ – as proxy for the small scales. We use $\eta(t) = \kappa(t)\delta(t)$ as measure suggestive of the spectral bandwidth. Figure 8.14 shows the evolution of $\eta(t)$ after the mixing layer is shocked. The fact that $\eta(t)$ increases with time for all cases suggests that the spectral bandwidth is increasing (which can also be understood as an increase in turbulent Re). The late time saturation of $\eta(t)$ for the highest η_o in Figure 8.14 indicates faster disappearance by dissipation of the smallest scales of the flow. Due to mode coupling there is a much larger (and faster) increase in spectral bandwidth for the high η_o group. Our observations suggest that sudden increase in $\eta(t)$ (and $\kappa(t)$) can be consistently used to indicate flow transition. Figure 8.15 exemplifies PDFs of Y_{SF_6} over the mixing region slab (at t=2500 μs). Similar to the spectral bandwidths (Fig. 8.14), the PDFs also increase monotonically with η_o, indicating that as we increase the initial *rms* slope we get more material (interpenetration) mixing for both linear (low η_o) and nonlinear (high η_o) regimes.

Vorticity production at shock time and eventual mode coupling thereafter will depend on the initial interfacial characteristics, as well as on the particular A, Ma, and (light/heavy or heavy/light) configuration considered. However, the initial *rms* slope of the material interface η_o appears to be a relevant parameter in determining whether the flow is in the linear ballistic regime, or in nonlinear mode coupling regime. In the linear

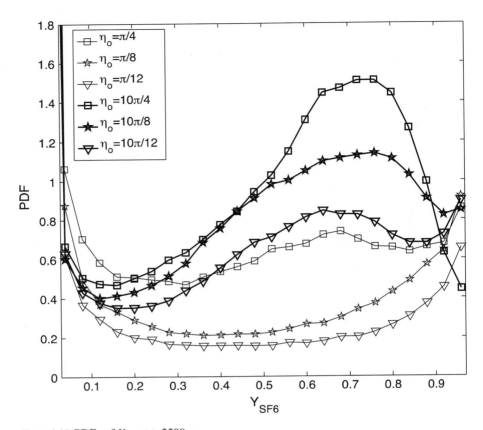

Figure 8.15 PDFs of Y_{SF_6} at t=2500 μs.

regime, the impulsive theory [11] predicts the mixing layer growth trends: as the initial *rms* slope increases the growth rate increases. Less mode coupling is seen, as inferred by the smaller spectral bandwidth, and the primary production of enstrophy is baroclinic. In the nonlinear regime, the mixing layer growth rate trends are the inverse of that predicted by [11]. There is significant mode coupling and our proxy for spectral bandwidth makes a sudden jump; this suggests that stretching becomes an important enstrophy production mechanism.

Some of our findings for the nonlinear regime are not consistent with heuristic notions one might have of statistically steady turbulence, and are likely a consequence of the nonequilibrium nature of RMI. The fact that slower growth rate of the mixing layer width is associated with more material (interpenetration) mixing demonstrates that mixing layer width growth rate, bulk *Re*, and material mixing are not causally connected. Our observations have a very straightforward physical explanation: higher initial material interface slopes lead to production of more smaller scales, which dissipate turbulent kinetic energy faster, reducing the growth rate.

An important practical consequence of our analysis [25] is that reshock effects on mixing and transition can be emulated at first shock if η_o is high enough. First shock

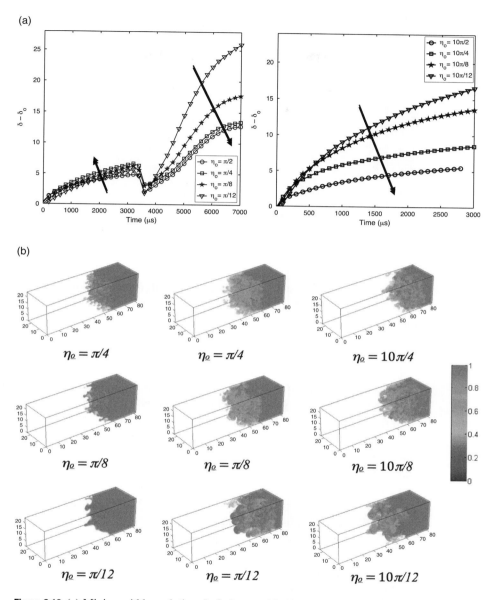

Figure 8.16 (a) Mixing width evolution. Left: low η_o (shocked and reshocked); right: high η_o (first shocked only). (b) Mixing visualizations. Left: low η_o (first shocked only); center: low η_o (shocked and reshocked); right: high η_o (first shocked only). Selected times are $t=3000$ μs after first shock, or $t=3000$ μs after reshock. (c) PDFs of Y_{SF_6}. Left: first shocked only (high η_o vs. low η_o); right: first shocked only (high η_o) vs. shocked and reshocked (low η_o); selected representative times as in Fig. 8.15. A black and white version of this figure will appear in some formats. For the color version, please refer to the plate section.

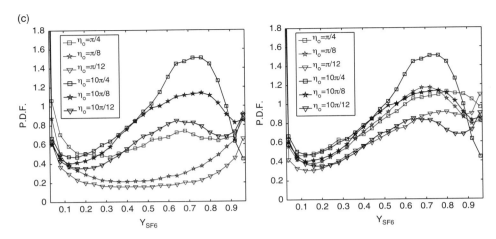

Figure 8.16 (*cont.*)

high η_o results were systematically compared with reshocked low η_o results in a shock tube configuration with a reflecting wall allowing for reshock at the material interface (at t ~ 4000 μs). Figure 8.16(a) shows integral mixedness growth trends for shocked high η_o consistent with those of reshocked low η_o. Mixing visualizations in shocked high η_o and shocked/reshocked low η_o cases are shown in Figure 8.16(b) in terms of volume distributions of mass fraction of SF_6 at selected times, t=3000 μs after the first shock (or t = 3000 μs after reshock). Figure 8.16(b) exhibits qualitative similar mixing features for both cases; for shocked low η_o there is distinctly less mixing as compared with the other two cases. Figure 8.16(c) shows the PDF of mass fraction of SF_6 at the same selected times. PDFs for the shocked high η_o and reshocked low η_o show similar features, whereas the PDF for shocked low η_o shows predominantly higher values indicating less mixing.

8.2.4 The Reshock Group of Instabilities

The reshock group of instabilities (RGI) occurs when the shock accelerated interface [9, 25] is not flat but highly corrugated and stochastic. This is the situation most characteristic of the reshock and the asymmetrical implosion, and exhibits an acute sensitivity to ICs seen in laboratory [34] and numerical [7, 10, 25] experiments.

The RGI is actually a combination of several instabilities, and is one about which the least is known: no statistical theories and no parameter space to describe the balance of fluid physics precludes a priori predictability. Despite the similarity in problem geometry, the RGI shares no dynamical or statistical features with the RMI. The RGI contributing instabilities are delineated in the enstrophy budget equation,

$$\frac{D}{Dt}\left\langle \omega^2 \right\rangle \sim \left\langle \omega_i s_{ij} \omega_j \right\rangle + \left\langle \omega_i \omega_j \right\rangle S_{ij} + \varepsilon_{ijk} \overline{\rho}^{-2} \left\langle \omega_i \rho_{,j} \right\rangle P_{,k} + \varepsilon_{ijk} \overline{\rho}^{-2} \left\langle \omega_j p_{,i} \right\rangle \overline{\rho}_{,k},$$

$$\text{I} \qquad\qquad \text{II} \qquad\qquad \text{III} \qquad\qquad\qquad \text{IV}$$

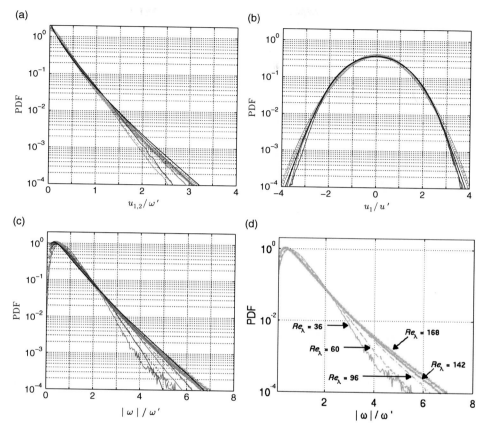

Figure 1.5 PDF velocity analysis shows trends with varying Re predicted with DNS as function of grid resolution. ILES – blue: 32^3; red: 64^3; green: 128^3; black: 256^3. DNS by Jimenez et al. [19] (gray) – solid line: $Re_\lambda = 36$; dashed line: $Re_\lambda = 60$; dash-dot line: $Re_\lambda = 96$; diamond: $Re_\lambda = 142$; X: $Re_\lambda=168$; $W' = W_{rms}$ and results are for the $Ma=0.27$ case.

(a) Transverse velocity derivatives; (b) velocity fluctuations; (c, d) vorticity magnitude; (e, f) longitudinal velocity derivatives. PDFs of longitudinal velocity derivatives are based on the full velocity field in (e), whereas the corresponding figure (f) is based only on its solenoidal part; the solenoidal velocity is extracted by inverse-Fourier transformation after a Helmholtz decomposition of the velocity field in Fourier space. (g, h) Vortex stretching scaled with the strain rate magnitude (based on full and solenoidal velocity, respectively). ILES – blue: 32^3; red: 64^3; green: 128^3; black: 256^3. DNS (gray) – solid line: $Re_\lambda = 36$; dashed line: $Re_\lambda = 60$; dash-dot line: $Re_\lambda = 96$; diamond: $Re_\lambda = 142$; X: $Re_\lambda = 168$.

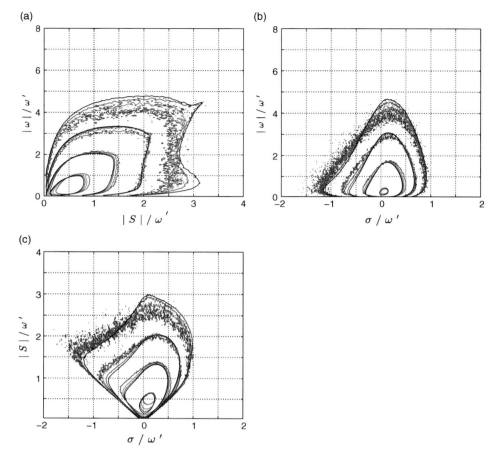

Figure 1.7 JPDFs of (a) vorticity magnitude and strain rate magnitude; (b) vorticity magnitude and vortex stretching; and (c) strain rate magnitude and vortex stretching. All axes are scaled with the *rms* vorticity ω'. Blue: 32^3; red: 64^3; green: 128^3; black: 256^3. Contour levels are plotted at 10^{-3}, 10^{-4}, 10^{-5}, and 10^{-6}.

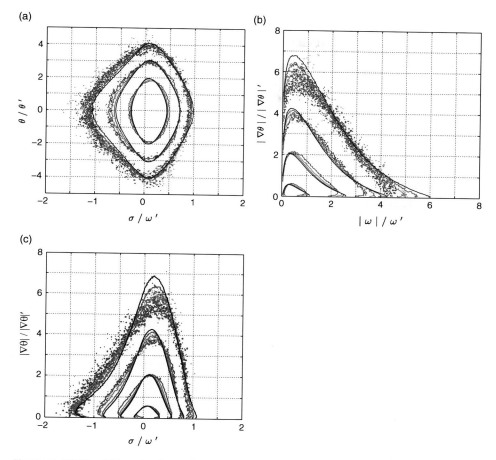

Figure 1.8 JPDFs of (a) scalar fluctuation scaled by the *rms* of the scalar fluctuation and vortex stretching scaled by the *rms* vorticity; (b) magnitude of the scalar gradient scaled with the *rms* of the scalar gradient and vorticity scaled with the *rms* vorticity; and (c) magnitude of the scalar gradient scaled with the *rms* of the scalar gradient and vortex stretching scaled with the *rms* vorticity. Blue: 32^3; red: 64^3; green: 128^3; black: 256^3. Contour levels are plotted at 10^{-3}, 10^{-4}, 10^{-5}, and 10^{-6}.

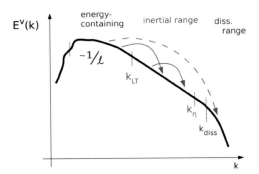

Figure 2.8 Sketch of a kinetic energy spectrum indicating the energy containing, inertial, and dissipation ranges and their wavenumber boundaries. The idea behind the minimum state is that the inertial range should be long enough that direct interactions between modes in the energy containing and dissipation ranges are energetically weak, indicated by the dashed (green) arrow. Some strong interactions are indicated via the solid (green) arrows. From [40].

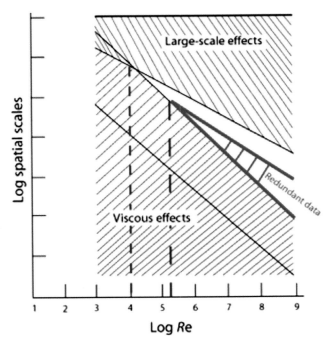

Figure 2.9 An illustration of how an arbitrarily high Re flow can be scaled down to a manageable one (shown in green), but still captures the physics of the large scale. Based on a figure in Dimotakis with added marks in color. From [40].

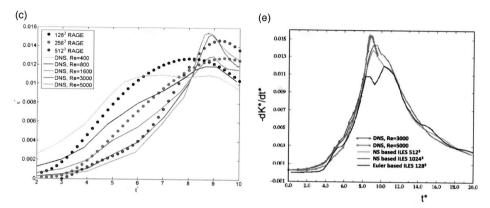

Figure 5.1 TGV kinetic energy dissipation with ILES [26, 34] compared with DNS [32, 33]; Re based on the integral scale. (a) Eulerian third-order characteristics based Godunov, second-/fourth-order FCT codes versus third-order van Leer based Lagrange remap code on uniform 128^3 grid; ILES Euler based, except for NS based split second-order FCT. (b) Fourth-order nonsplit FCT versus resolution. (c) and (d), RAGE. (e) NS and Euler based ILES using the third-order van Leer Lagrange remap code.

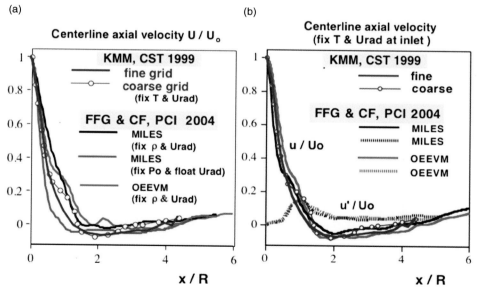

Figure 5.2 LES and MILES of a swirl combustor flow. Sensitivity of the simulated centerline axial velocity within a combustor to actual choices of steady inflow BCs; streamwise variable is scaled with inlet diameter R. Results from [57] were obtained with two different codes based on MILES and one-equation-eddy-viscosity model LES, respectively, and compared with previous (dynamic Smagorinsky) LES results from [56].

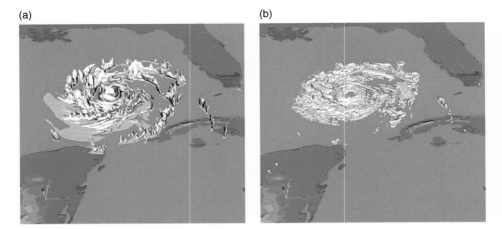

Figure 6.6 Isosurfaces of cloud water (0.0005 g kg^{-1}) from hurricane Rita on September 21, 2007, at 1800 UTC; (a) Eulerian cloud model simulation; (b) Lagrangian cloud model simulation.

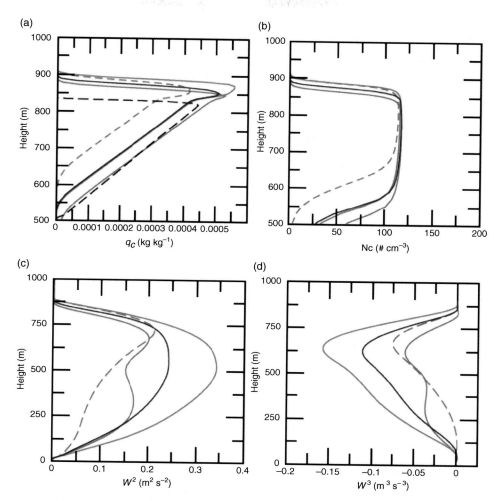

Figure 6.9 Mean vertical profiles averaged over the last hour of (a) cloud water; (b) cloud droplet number; (c) variance of vertical motions; and (d) third moment of vertical motions from DYCOMS-II simulations from ECM simulations with or without evaporative limiter active (red or dashed red lines) and from ECM simulations employing more traditional approaches (green or blue lines). From fig. 2 of [22].

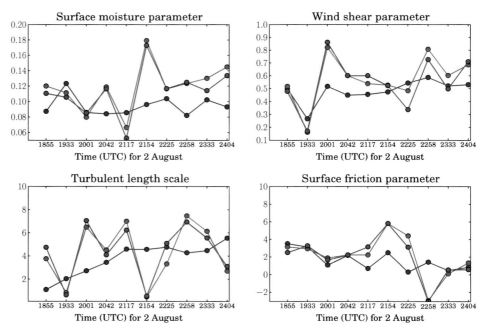

Figure 6.17 Time distribution of EnKF parameter estimates using latent heat data only (blue line), horizontal winds (red line), and or both (green line). From fig. 10 of [8].

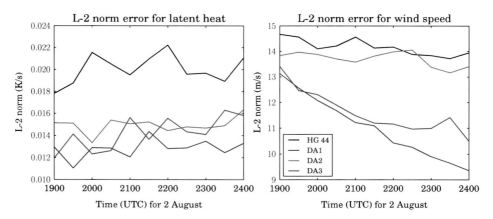

Figure 6.19 Error estimates as a function of time computed using Eq. 20 of [22] and observational data as the reference for ensemble member 44 and three simulations using parameters estimate from latent heat (blue lines), horizontal winds (red lines), or both (red lines). From fig. 16 of [8].

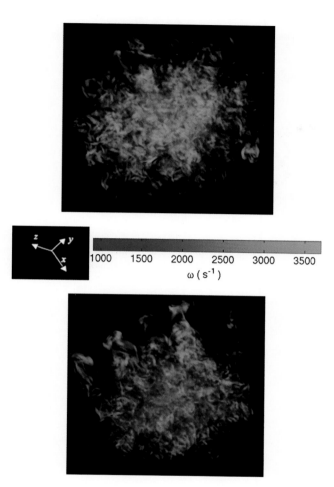

Figure 8.4 (b) Volume renderings of the vorticity magnitude at t = 8 ms, for the short (top) and long (bottom) perturbation cases at the 0.1 cm resolution.

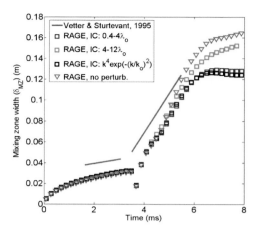

Figure 8.5 Variation of the 3D mixing layer thickness as a function of time vs. IC perturbation spectral content for the baseline 0.1 cm grid resolution vs. laboratory [12] results.

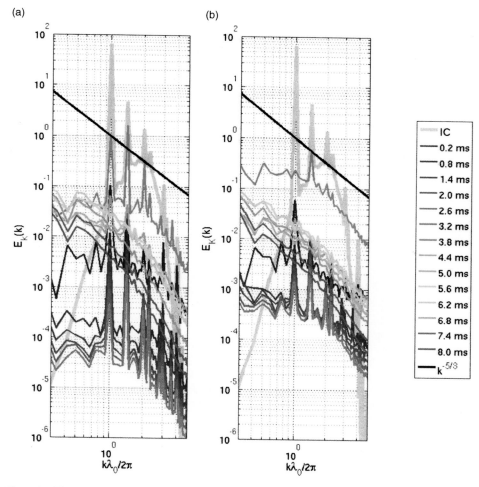

Figure 8.7 Time-dependent spectra of the turbulent kinetic energy K versus resolution for the weighted short case. (a) 0.1 cm grid; (b) 0.05 cm grid.

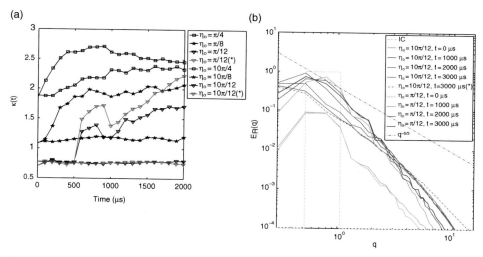

Figure 8.12 (a) Zero crossing frequency evolution; (b) representative R spectra; (*) finer resolution.

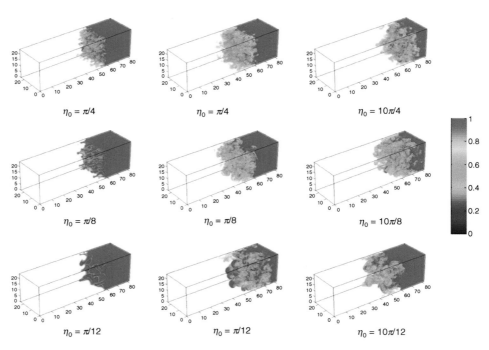

$\eta_0 = \pi/4$ $\eta_0 = \pi/4$ $\eta_0 = 10\pi/4$

$\eta_0 = \pi/8$ $\eta_0 = \pi/8$ $\eta_0 = 10\pi/8$

$\eta_0 = \pi/12$ $\eta_0 = \pi/12$ $\eta_0 = 10\pi/12$

Figure 8.16 (a) Mixing width evolution. Left: low η_o (shocked and reshocked); right: high η_o (first shocked only). (b) Mixing visualizations. Left: low η_o (first shocked only); center: low η_o (shocked and reshocked); right: high η_o (first shocked only). Selected times are $t=3000$ μs after first shock, or $t=3000$ μs after reshock. (c) PDFs of Y_{SF_6}. Left: first shocked only (high η_o vs. low η_o); right: first shocked only (high η_o) vs. shocked and reshocked (low η_o); selected representative times as in Fig. 8.15.

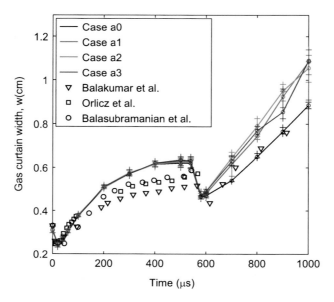

Figure 8.23 IC effects on shocked and reshocked gas curtains; comparison between simulated and laboratory [15, 34, 37] mixing widths.

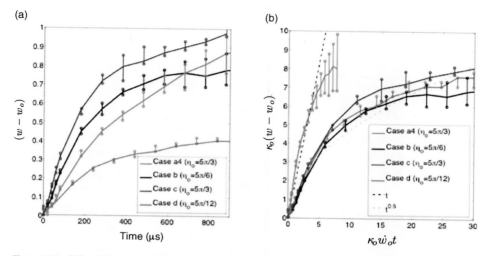

Figure 8.28. GC width as function of time for simulated (first-shocked-only) cases *a4*, *b*, *c*, and *d*; (a) dimensional; (b) nondimensional. Estimated uncertainty ranges are prescribed as in Fig. 8.23.

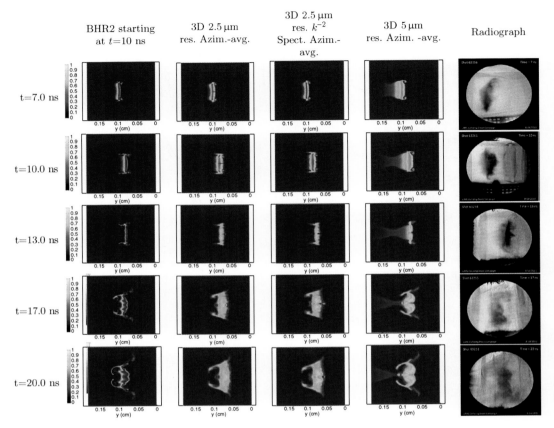

Figure 9.7 Azimuthally averaged aluminum mass concentration (c_{Al}) distributions for the reshock experiment along with radiographic data from the experiment.

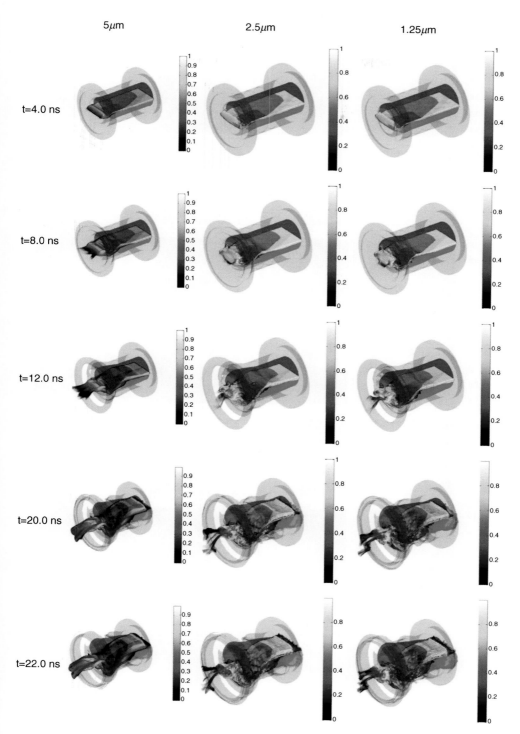

Figure 9.8 Visualization of c_{Al} from simulations of the shear experiment with a 1 g/cm^3 density isosurface overlayed in blue (providing the outline of the Be tube). The green outlines are 0.7 isosurfaces of the mass fractions for the two different foams.

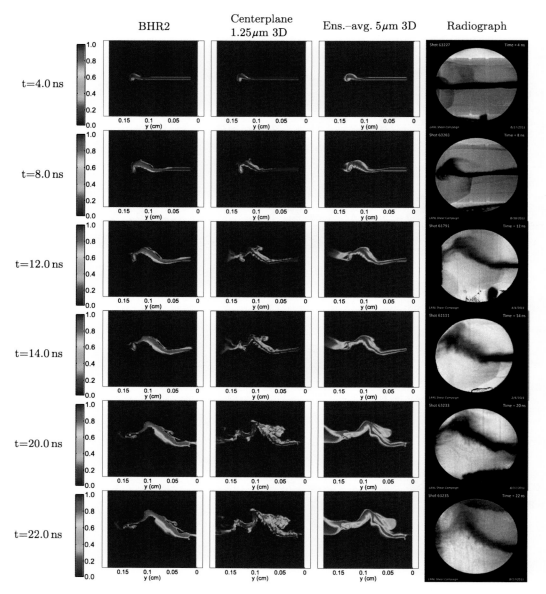

Figure 9.9 Aluminum mass concentration (c_{Al}) distributions for the shear experiment along with radiographic data (also from experiment). The 3D data are taken from the center plane (in the z direction).

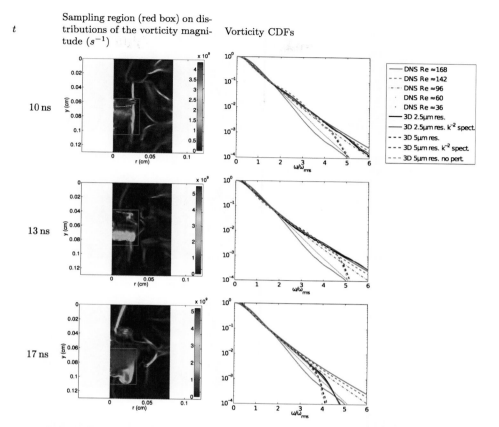

Figure 9.15 Cumulative distribution functions (CDFs) of vorticity ω/ω_{RMS} for the reshock experiment sampled from a uniform $5\mu m$ grid. Sampling regions (within red lines) on distributions of vorticity magnitude (s^{-1}). The DNS data were taken from [26].

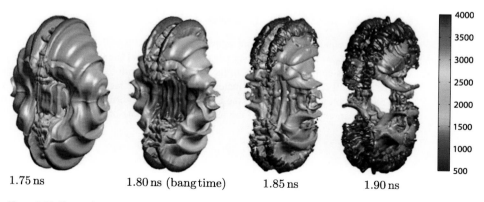

Figure 9.36 Deuterium mass concentration $c_D = 0.1$ isosurface colored with T_{ion} (in eV) calculated from the $1/8\mu m$ simulation data with $\chi = 10$ and no drive asymmetry. In order to maximize displayed detail for a converging problem, these pictures do not show the same spatial extent.

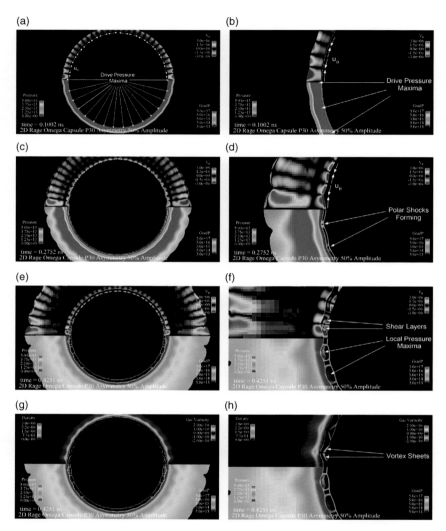

Figure 10.2 A sequence of early time snapshots from a 2D RAGE simulation of the P30 Omega capsule with $a_{30} = 0.50$. In the lower portion of each panel the CH shell is colored by pressure while the gas is colored by grad P. In the upper portion both the CH shell and the gas are colored by the θ component of the material velocity. In the last row of panels the θ velocity has been replaced on top by density in the CH and azimuthal vorticity in the gas. Each right hand panel shows an expanded view of the region at the left pole of the capsule near the CH/gas interface.

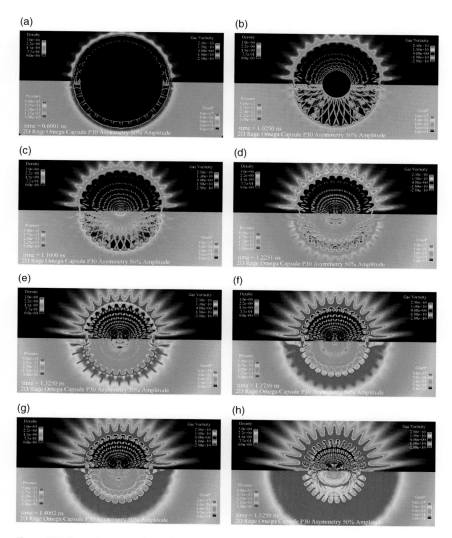

Figure 10.3 Late time snapshots from 2D RAGE simulation of the P30 Omega capsule with $a_{30} = 0.50$. In the lower portion of each panel the CH shell is colored by pressure while the gas is colored by grad P. In the upper portion the CH shell is colored by density while the gas is colored by azimuthal vorticity.

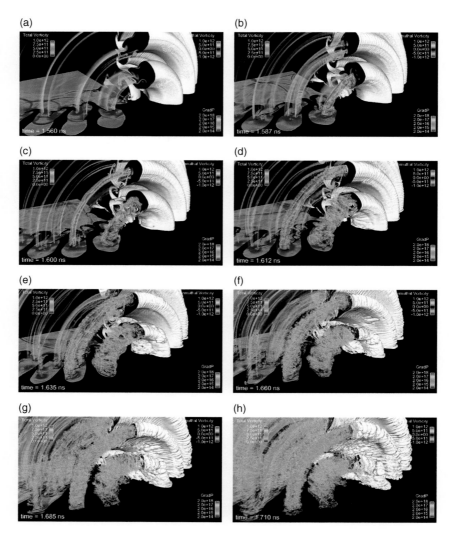

Figure 10.14 Eight time snapshots from the 0.05 μm 3D RAGE simulation of the P30 OMEGA capsule. Close up view of the development of the turbulence in the two bubbles nearest the polar axis.

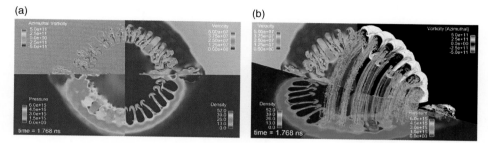

Figure 10.19 (a) Four panels showing azimuthal vorticity, velocity, pressure and density on a symmetry plane of the 5% simulation of Fig. 10.18. Note the 6 gigabar pressure in the capsule center which drives turbulent gas jets outward with velocities of 500 km/sec. (b) Another view of the four hydrodynamic quantities of panel (a) in the 5% 3D RAGE simulation of the P30 capsule with the developing turbulent vorticity shown in a volume rendered representation. The white surface is, as usual, the CH/gas interface.

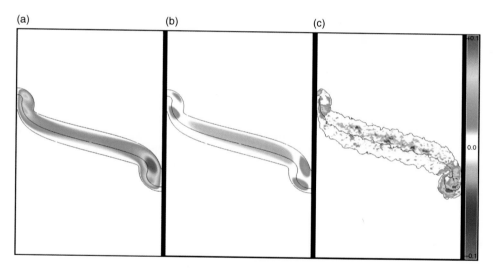

Figure 11.12 Comparison of gradient diffusion model (a) for the horizontal turbulent mass-flux velocity (a_x) and the transport model of BHR (b) and ILES (c) at $t = 45$ ms. Colors are values of a_x and contour lines are [0.025, 0.5, 0.975] of the heavy fluid volume fraction to show the center and edges of the mixing layer. Note the countergradient flux in the top half of the mix layer.

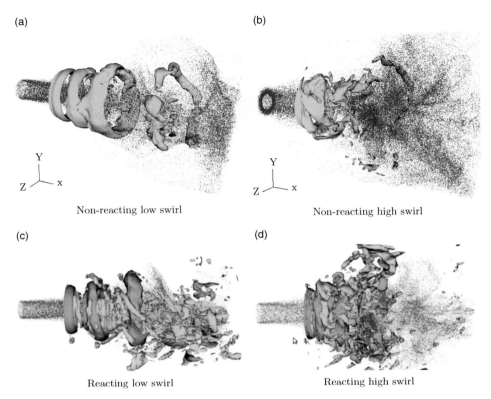

(a)

Y
Z — X

Non-reacting low swirl

(b)

Y
Z — X

Non-reacting high swirl

(c)

Reacting low swirl

(d)

Reacting high swirl

Figure 12.12 Iso-surface of the instantaneous azimuthal vorticity and the droplet distribution for nonreacting (a, b) and reacting (c, d) cases at low ($S_w = 0.5$) and high swirl ($S_w = 0.8$) numbers in the GE-DACRS combustor [20]. The yellow and blue iso-surfaces denote positive ($\widetilde{\omega}_\theta = 15000s^{-1}$) and negative ($\widetilde{\omega}_\theta = -15000s^{-1}$) values of azimuthal vorticity, respectively.

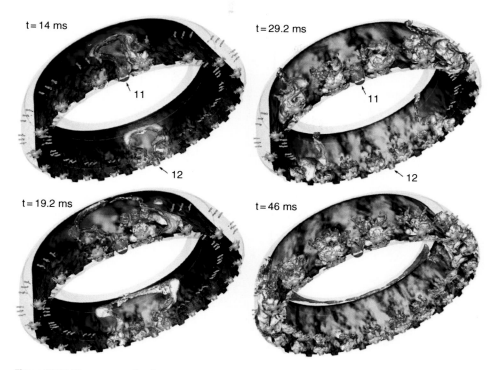

Figure 12.22 Four successive instants (t = 14, 19.2, 29.2, and 46 ms) of the ignition sequence. The cutting surface C is colored by axial velocity (light blue: -20 m/s $\rightarrow$ 20 m/s, iso-surface of the axial velocity $U = 38$ m/s colored by temperature (turquoise blue: 273 K $\rightarrow$ red: 2400 K), and iso-surface of progress rate $Q = 200$ mol/s (shiny light blue) representing the flame front. The two high speed hot jets used for ignition appear as red zones in the pictures (marked I1 and I2) [80].

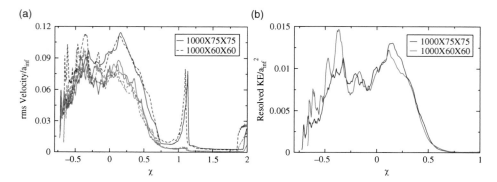

Figure 13.3 Grid independence study: (a) rms of velocity profiles (blue: u_r, red: u_θ, green: u_ϕ); (b) resolved turbulent KE at 4 ms. Note that the quantities are normalized using ambient speed of sound, a_{inf}.

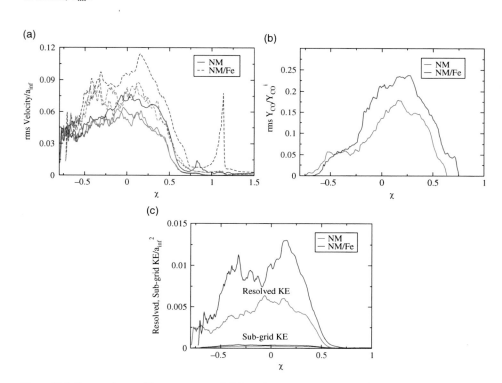

Figure 13.7 Comparison of homogeneous (NM) and heterogeneous (NM/Fe) explosives at 4 ms: (a) rms of velocity profiles (blue: u_r, red: u_θ, green: u_ϕ); (b) rms of Y_{CO}; (c) resolved and subgrid turbulent KE.

where $(\ ),_i \equiv \partial/\partial x_i$, P is the mean pressure, s_{ij}, S_{ij} are fluctuating and mean strain rate, $\bar{\rho}$ is the mean mass density, and $<\ >$ denotes ensemble averaging. The terms on the right hand side can represent several instability mechanisms.

I.   Represents 3D vortex stretching by nonlinear mode coupling.

II.   Accounts for the mean shear driven Kelvin–Helmholtz (KH) instability, mean dilatation by shock, and the Bell–Plesset (convergence) instability.

III.   Represents two baroclinic instabilities (proportional to pressure gradients) – of which the RMI is one.

IV.   Represents two baroclinic instabilities (proportional to mass density gradients).

In RMI, vorticity generation is primarily and only baroclinic (pressure and density gradients torqueing the fluid). Beyond stretching, the RGI can include as many as seven instabilities.

Parameters describing the balance of instabilities come from nondimensionalization of relevant equations (e.g., as mentioned earlier). We have recognized the RGI as occurring when a high η_o interface is first shocked [9] and in reshocked situations [25]. As noted earlier, both configurations involve high initial interfacial η_o, and both exhibit the same scalings and collapse of data [9, 25], and are also very different from the (low η_o) RMI. A simple way to demonstrate the very large physical differences is with the first shocked problem (Fig. 8.13 discussed earlier) and its trends with initial η_o.[2] We can thus characterize the RGI as combination of several different instabilities about which very little is known. There is no set of nondimensional statistical parameters describing the balance of the different instability mechanism as function of the statistics of the interface, material properties, Ma and Re, differential material impedances, density differences, mean post shock velocity, and pressure jumps. Moreover, current engineering turbulence models IC cannot distinguish between the RMI and the RGI trends and scalings.

Turbulent flows are, in general, known to have a sensitive dependence on IC [35]. The shock accelerated interface with high initial η_o is a challenging example of acute sensitivity to IC. This is due to the large amplification a shock has on the various parameters of the initial interface and the fact that the reshock problem is comprised of a group of instabilities, as listed earlier. Our work [10, 25] consistently exhibits predictive robustness to IC at first shock (t=0) and extreme sensitivity to interfacial conditions at reshock time (t=trs). Figure 8.17, from [25], shows the results of shocking a planar interface twice: first shocked at t=0, at which time η_o is in the RMI range, and reshocked at t~3700 ms, for which the initial interface has high η_o. We plot, as functions of time, mix layer width, bulk Reynolds number, and turbulent eddy viscosity [25] – used to scale nonlinear to linear interactions of the flow. These turbulence metrics commonly used in simple turbulence models are robust and approximately the same (despite different interfacial conditions) during the RMI phase up to the reshock time t~3700 ms. The mixed layer response (after reshock) and growth trends are hugely

---

[2] Different shock interface interactions are involved depending on having heavy to light (H–L) or light to heavy (L–H) conditions at shock time [36]. For the sake of the discussion, the focus here is on the first shock L–H case.

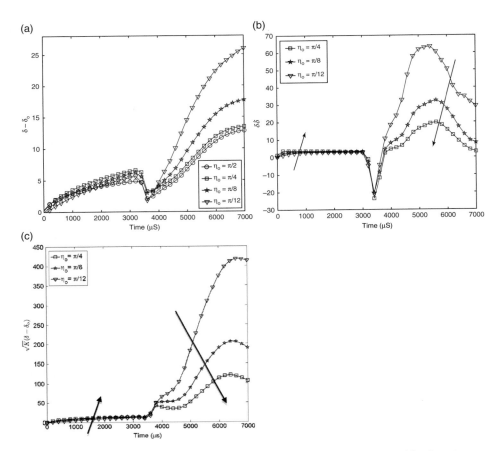

Figure 8.17 Mix width thickness (a), bulk Reynolds number (b), and turbulent eddy viscosity (c) for shocked/reshocked planar shocktube ILES (reshocked at t~3,700 μs) as function of initial rms interface slope η_o [13]. Arrows indicate the direction of increasing η_o.

different. Note that all the flows have similar values of layer thickness and eddy viscosity at reshock time and note the rapid divergence of the trajectories of these metrics on reshock. The predictability problem manifests itself as a rapid divergence of the trajectories and depends on subtleties of the enstrophy deposition and generation in high η_o interfaces in ways that are currently unknown. Such sensitivity to IC at reshock has been also observed in other shock tube experiments [10, 34]. This leads to very important questions regarding the nature of the RGI and late time predictability when using an incomplete set of IC metrics, and, of course, leads to the question of what a complete set of metrics might be to assure predictability.

8.3   Shocked Gas Curtain

While qualifying the accuracy of coarse grained simulation (CGS) is mainly thought to involve statements regarding resolution (ability to capture outer flow while emulating SGS effects), an equally important and frequently underrated issue for verification, validation, and uncertainty quantification (VVUQ) frameworks relates to IC characterization and modeling issues – alluded to earlier in the context of late time predictability. The quality of IC characterization and modeling limits the ability to reproduce observations carried out in different (computational or laboratory) experiments, and thus plays a crucial role in the simulation model evaluation and validation processes. Last but not least, establishing appropriate metrics and data reduction procedures to quantify the level of agreement between different experiments is also a crucial aspect in the model predictability assessment process.

In this section we demonstrate practical challenges encountered in CGS predictability evaluations in the complex 3D shocked GC problem for which state of the art laboratory data and diagnostics were available. First, uncertainties in characterizing initial GC conditions in the laboratory experiments are identified and used to motivate our IC modeling and simulation strategy. Analysis of the simulation results that follows addresses the impact of IC specifics on the shocked GC dynamics and transition, integral width measures, and grid resolution effects. Finally, appropriate analysis and scalings are used to recognize observational aspects of the shocked GC, which can be corroborated by both laboratory and computational experiments – despite the inherent uncertainty issues of IC characterization and modeling.

GC experiments to study RMI driven turbulence and mixing were designed at LANL to validate simulation code capabilities [15, 37]. The SF_6 GC (Fig. 8.18) is formed by forcing SF_6 through a linear arrangement of round nozzles into the shock tube test section; the GC is stabilized using coflowing air and suction at the bottom of the curtain [37]. Once a steady state is achieved, the GC is shocked ($Ma=1.2$), and its later evolution subject to RMI, transition, and nonequilibrium turbulence phenomena is investigated. Analysis in the laboratory experiments is based on particle image velocimetry (PIV) and planar laser induced fluorescence (PLIF) data acquired at the selected horizontal plane, 2 cm below the beginning of the GC (Fig. 8.18). Typically available data involve intensities (relative SF_6 concentration), $I=C_{SF6}/C_{max}$ where C_{max} is the peak SF_6 concentration at the slice (Fig. 8.19(a)). The evolving structures can

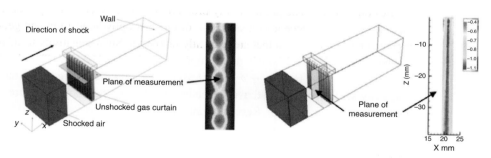

Figure 8.18 Computational domain for simulations of shocked gas curtain; the simulated gas curtain and planes where laboratory acquired data are available are indicated. (a) 2D relative concentration (C/C_{max}) of SF_6 in a horizontal plane 2 cm from nozzles; (b) vertical PIV in the center of the gas curtain gives estimates of nozzle exit velocity ~100 cm/s.

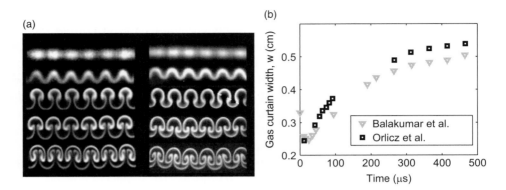

Figure 8.19 Laboratory visualizations of SF_6 intensities at measurement plane indicated in panel b; left from [15], right from [37]; times vary from shock time at $t=0$ (top frame) to $t=465$ μs (bottom frame).

be optionally reshocked at various times using a moveable reflecting wall to study IC effects on turbulence and mixing.

Even with sophisticated advanced measurement techniques, it has not been possible to adequately characterize the 3D nature of the SF_6 gas curtain concentration distributions. The impact of actual GC IC variability in the experiments is apparent in the reported results [15, 37]; it affects repeatability of observed instantaneous PLIF based flow patterns at the measuring plane (Fig. 8.19(a)), and also leads to significantly different (~10–20%) GC width measures (Fig. 8.19(b)) based on the evolution of mean thickness of 2D (digitized, ensemble averaged) PLIF intensity distributions.

The laminar flow rate of SF_6 through the nozzles is not well known. The more recent studies [34] have estimated the jet exit velocity to be $V_{in} \sim 100$ cm/s. In practice, the flow rate is adjusted slightly for each experimental realization to produce a stable GC for the given external condition pressure. Moreover, difficulty in positioning the PLIF camera used to characterize the initial GC images leads to significant optical distortion, and the associated data acquisition noise is also not adequately characterized. The precise composition of the mixture of SF_6, acetone (used as tracer for PLIF), and air at the jet exits is not known. Other uncharacterized aspects of the laboratory setup include specific features of the curtain stabilizing coflowing air and suction enforced at the bottom of the laboratory GC (used to remove excess SF_6 and control its spreading).

Depending on the method used (mass spectroscopy or PLIF), typical acetone volume fractions are estimated as high as 35%. The actual Atwood number, $A = (\rho_{mixture} - \rho_{air})/(\rho_{mixture} + \rho_{air})$, affecting the initial RMI growth rates in the laboratory studies is thus effectively smaller than that in the nominally pure SF_6 limit, and is not known accurately. Recent 2D simulations of the shocked (single) SF_6 column configuration [38] showed that such content of a third species (acetone) can have significant effects on the evolution of RMI.

Finally, although the possible relevance of 3D flow instabilities is clearly suggested by the (2D) visualizations in the shocked GC experiments [15, 37], and by their demonstrated important role in the closely related cited problem of the shocked

SF_6 column [39], their impact (and sensitivity to ICs) on transition and mixing remain to be understood. SF_6 concentration gradients developing in the vertical (z) and horizontal (x, y) directions due to diffusion of SF_6 in air and buoyancy effects can also give rise to considerable 3D effects on the GC (and consequent changes of C_{max} in the z direction). It is thus clear that knowledge of the detailed 3D distribution characteristics of the initial SF_6 concentration is really needed to realistically pose the shocked GC problem.

8.3.1   Modeling Strategy

Here, we first describe our GC simulation approach and IC modeling strategy. This is followed by analysis of the shocked GC 3D simulations [10], based on integral mixing measures and turbulent flow characterization. Data reduction using suitable IC parameterizations is shown to be useful in characterizing instability behaviors and establishing effective comparisons of integral GC width growth measures with available laboratory data.

The computational domain used in the RAGE simulations (Fig. 8.18) scales down to ~1/4th of the laboratory test section (factors of 2 shorter in the vertical (z) and spanwise (y) directions, while realizing the full scale in the shockwise direction). As in the planar singe interface case discussed earlier, the shock propagates in the (x) direction through the GC and reflects at the end of the simulation box on the right, where purely reflecting boundary conditions can be optionally enforced to address reshock effects. Periodic boundary conditions were used in the nozzle (y) direction. The simulations involved 5–6 level AMR [8], had computational cells being correspondingly reduced by factors of up to 32–64 (smallest grid sizes 0.0125 cm and 0.00625 cm, respectively), and used ~15–45 million control volumes.

Attempts have been made to model the initial 2D GC laboratory measured intensities (e.g., on top of Fig. 8.19(a)) based on Mikaelian's formula [40],

$$ I = \frac{A(1 + B\cos kx)}{1 + B} \exp\left[-y^2\alpha^2/(1 + \beta\cos kx)^2\right]. $$

To address possible effects of IC concentration uncertainties noted earlier, 2D RAGE simulations of the shocked SF_6 GC were carried out using idealized GC ICs based on this formula. We assumed a binary system involving SF_6 and air, different values of C_{max} in the range of 40 to 80% and we examined the mean growth of RMI triggered vortical structures in terms of a GC width modeling the corresponding laboratory quantity. The GC width is defined as the shockwise distance over which $\chi(x) > \chi_h$ – based on a selected threshold $\chi_h \sim 0.95$, where $\chi(x) = 4\xi(x)(1 - \xi(x))$,

$$ \xi(x) = \frac{1}{L_y}\int_0^{L_y}\left(\frac{\rho_{SF_6}(x, y, z = 2)}{\rho(x, y, z = 2)}\right)dy, $$

involves the mean mass fraction data at the measurement's horizontal slice, and L_i *denotes* domain length in i direction. The GC width is scaled to initially match corresponding laboratory measures of the unshocked GC (when available). The GC width was found to be very dependent on C_{max} with higher values of C_{max} (~ 75%), providing better match with the laboratory data at the measurements' plane – suggesting such a mixture of air and SF_6 at the jet exit.

Our strategy to produce a GC with realistic 3D characteristics involved carrying out a separate simulation to emulate the physics of a mixture of SF_6 and air falling through the shock tube test section constrained by the available laboratory GC IC information [10]. An SF_6 air mixture exits through nozzles of diameter 0.3 cm, and separation between nozzle centers $\lambda_o = 0.36$ cm, with inlet velocity $V_{in} = 100$ cm/s – typical value for the nozzle exit velocity [34]. SF_6 air mixture (C_{max}=75%) at the jet exit was assumed following our 2D analysis noted earlier. No attempt was made to model the (insufficiently characterized) coflowing air arrangement or the bottom suction in the laboratory setup; an advective outflow boundary condition (no suction) was enforced at the bottom of the simulated GC. The solution is allowed to evolve in time until a steady state is reached, and is then used as the initial 3D GC for the RAGE simulations. A fairly good match with laboratory GC features at the horizontal measurements' plane can be thus obtained (Fig. 8.20(a)). However, initial GC characteristics associated with using Mikaelian's formula [40] do not accurately represent the spectral content in the laboratory GCs (Fig. 8.20(b)): the dominant peak can be emulated, but not the additional (smaller and larger scale) modes apparent there. In the simulations that follow, the 3D fluctuations present in the laboratory experiments were simulated to address their potential impact on the shocked GC transition to turbulence.

Various ways of superimposing synthetic noise to the baseline varicose (nonperturbed) GC were additionally tested. Configurations with carefully controlled ICs were considered, differing from each other on the specifics of modeled low-amplitude multimode fluctuations. This was achieved through two strategies: (a) by slightly offsetting the nozzles in the shock (x) direction, and (b) by adding 3D perturbations to the initial (simulated) concentration field; by combining them, four different realistic IC cases were considered (Fig. 8.21):

(*a0*): Baseline varicose: nozzles are aligned along y (no added fluctuations);
(*a1*): Randomly offset nozzles in shock direction, with standard deviation 5% of nozzle radius;
(*a2*): Baseline (*a0*) with added 3D concentration fluctuations,

$$\Psi = \Lambda \sum_{l,m,n} a_{l,m,n} \sin\left(2\pi lx/L_x\right) \sin\left(2\pi my/L_y\right) \sin\left(2\pi nz/L_z\right),$$

where Λ fixes the target (2%) standard deviation, $-\frac{1}{2} < a_{l,m,n} < \frac{1}{2}$ are random numbers, and the smallest wavelengths involve at least eight computational cells on the baseline resolution.
(*a3*): Nozzles offset as in (*a1*) and added concentration fluctuations as in (*a2*) – Figure 8.21(b).

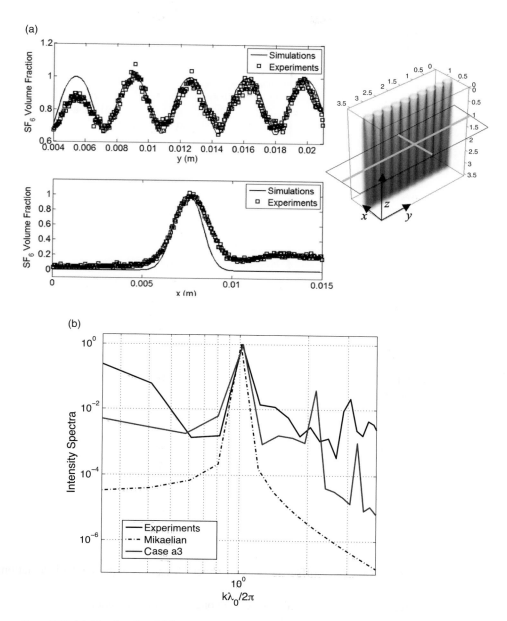

Figure 8.20 (a) Simulated and laboratory initial gas curtain characteristics at the measurements' plane and at indicated (gray) traces: *y* profiles through the GC center, and *x* profiles through the center of a typical gas cylinder; (b) typical GC IC intensity spectra of laboratory data show a dominant peak associated with Mikaelian's [24] intensity distribution, and also small and large scale fluctuations modeled in the simulated GCs (corresponding spectra for case *a3* is exemplified).

Table 8.3 Shocked GC Configurations (cf. Figs. 8.21 and 8.26)

Case #	Reshock	η_o	w_o	GC Fluctuations	Resolution	Laboratory Data
a0	yes	$5\pi/3$	$\sigma=0.3cm$	none	0.0125cm	n/a
a1	yes	$5\pi/3$	σ	long	0.0125cm	n/a
a2	yes	$5\pi/3$	σ	short	0.0125cm	n/a
a3	yes*	$5\pi/3$	σ	long+short	0.0125cm	[84*,65,89]
a4	no	$5\pi/3$	σ	as a3	0.0125cm	[89]
a5	yes*	$5\pi/3$	σ	as a3	0.00625cm	[84*,65,89]
b	no	$5\pi/6$	σ	as a3	0.0125cm	[89]
c	no	$5\pi/3$	2σ	as a3	0.0125cm	No
d	no	$5\pi/12$	$\sigma/2$	as a3	0.0125cm	No

* Reshocked GC laboratory data available.

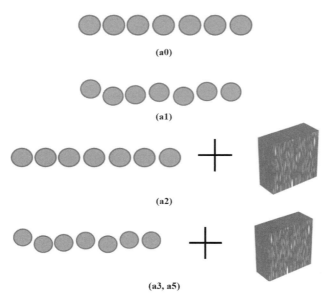

(a0)

(a1)

(a2)

(a3, a5)

Figure 8.21 Schematic of the imposed IC perturbations to emulate fluctuations in the laboratory experiments.

Various simulations varying ICs and grid resolution were carried out [10] (cf. Table 8.3). The Ma=1.2 shock impinges the GC at time t $>\sim$ 0, reflects on the wall, and eventually reshocks the GC at t $<\sim$ 600 μs. Figure 8.22 shows the 3D evolution of the shocked GC 3D based on visualizations of SF_6 mass fraction for the various ICs discussed earlier. The figure exemplifies significant 3D features of the shocked and reshocked GC in terms of SF_6 mass fraction distributions at times 200 μs before and after reshock, respectively. Since the concentration of SF_6 is higher and rapidly varying near the nozzle exits, bigger structures there reflect larger vorticity production and substantially 3D flow features. The structures near the SF_6 nozzles are more developed. Compared with the baseline (nonperturbed case a0), significantly more complex ICs in all

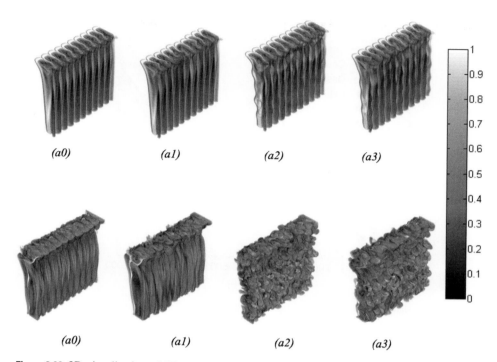

Figure 8.22 3D visualization of SF_6 mass fraction 200 μs before reshock (top) and 200 μs after reshock (bottom) for various GC ICs considered.

perturbed arrangements lead to enhanced vortex interactions and mixing after reshock; for IC cases *a2* and *a3*, where SF_6 concentration perturbations were also involved, more turbulent characteristics are suggested. Before reshock (top frames), small 3D structures are hardly visible in cases *a0* and *a1*, but their presence is noticeable in the other two (*a2* and *a3*) perturbed concentration cases. As noted, IC perturbations play a significant role in promoting transition to turbulence after reshock.

Analysis of instantaneous turbulence characteristics [10] was based on local and statistical flow quantities, mass weighted averaging over a (*y*, *z*) sampling plane domain including a relatively short vertical extent around the measurement plane (right of Fig. 8.18). Figure 8.23 shows GC width results for the various simulated reshocked cases at the baseline resolution. Estimated uncertainty ranges of simulation results reflect on variability associated with changing the data threshold χ_h (within $0.92 < \chi_h < 0.98$) in the GC width definition given earlier; corresponding uncertainties for the plotted laboratory values before reshock are estimated to be within 10%. The results are compared with the available published [15, 34, 37] laboratory counterparts – only one of which [37] involves reshock.

Before reshock, Figure 8.23 shows very robust simulated GC widths (fairly insensitive to varying ICs and resolution), and in fair agreement with the laboratory results. Scatter of the laboratory data reflects the noted IC repeatability difficulties in the experiments. Systematic overprediction of the simulated GC width reflects the

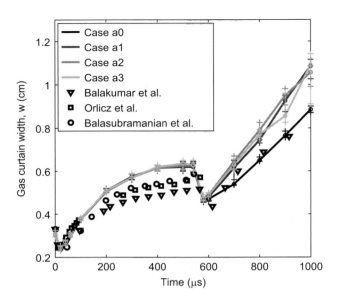

Figure 8.23 IC effects on shocked and reshocked gas curtains; comparison between simulated and laboratory [15, 34, 37] mixing widths. A black and white version of this figure will appear in some formats. For the color version, please refer to the plate section.

difference between initial (computational and laboratory) GCs, that is, acetone mixed with the SF_6 used as dye in conjunction with the laboratory PLIF visualization technique. The actual value of A affecting the initial RMI growth rate in the laboratory studies is effectively smaller than that in the nominally pure SF_6 limit, and this explains the simulation overpredictions. Additional IC issues are associated with the uncertainties in knowing C_{max} and precise IC fluctuation characteristics.

After reshock (t > 600 ms), Figure 8.23 shows consistency of the computed results associated with nonperturbed IC case *a0* with the only available reshocked GC laboratory data [37]. The latter apparent agreement between computed and measured data is regarded as fortuitous given the noted discrepancies before reshock and the fact that uncertainties of the laboratory data after reshock have not been reported. Predicted GC widths associated with the perturbed initial GC (cases *a1*, *a2*, and *a3*) are consistent with each other and significantly larger than that for the nonperturbed IC (case *a0*).

Actual values of *Re* characterizing the flow at the smallest resolved scales are not a priori available in CGS, where the smallest characteristic resolved scale is determined by the resolution cutoff wavelength prescribed by an explicit or implicit spatial filtering process. Using grid-dependent spatial filtering is the common feature of state of the art CGS used to simulate the RMI (see discussion in [7]). Since the small scale cutoff is determined by grid size, the implication is that grid refinement will be associated with "observing" smaller simulated physical structures – associated with a somewhat different problem with higher effective *Re*.

The noted inherent sensitivity of shock driven turbulent mixing experiments to ICs (and their resolution) is also a crucial aspect to be addressed in this context [7].

Representative cases without reshock were analyzed based on the higher resolution in planar RMI simulations in [9]. Integral measures such as integral mixing widths were found to be virtually insensitive to grid resolution; as the flow becomes nonlinear, finer resolution resulted in increased small-scale production as indicated and longer self-similar ranges in the spectra of R – which has been associated with higher effective Re [7, 30]. Our shocked/reshocked planar RMI studies [7] demonstrated mixing layer growth rates fairly insensitive to ICs before reshock, and very sensitive late time growth after reshock – indicating, in particular, the significant role of resolved details in the thicker, more complex mixing layers at reshock time.

Selected comparative results from the higher resolution shocked GC simulations are shown in Figures 8.24 and 8.25 for corresponding cases $a3$ and $a5$. Figure 8.24 compares 3D volume visualizations of SF_6 mass fraction. Fairly similar structures are predicted with both resolutions at $t = 400$ μs (before reshock), but some nonnegligible differences are apparent – attributed to sensitivity of GC dynamics and the RMI growth rate to more resolved small scale content and higher effective Re [7]; the latter differences are capable of producing significantly different spectral and morphological conditions (i.e., ICs) at reshock time. Smaller structures are apparent after reshock ($t = 800$ μs) for the higher resolution case – and more so for the late time frame ($t = 1200$ μs).

Comparison of computed GC widths in Figure 8.25(a) shows results essentially resolution independent before reshock, but there is larger predicted GC width for the finer resolution case after reshock. Following our previous analysis of IC resolution [7] and interfacial morphology effects on planar RMI evolution [25], we attribute GC width differences after reshock as reflecting higher (because better resolved) small scale content in the GC mixing layers at the more sensitive *reshock time* for the finer resolution simulation.

Comparison of K and Ω in Figure 8.25(b), (c) shows that higher resolution leads to substantially higher mean enstrophy, but only slightly higher K. Higher enstrophy reflects larger small scale production and dissipates K faster (i.e., following $-dK/dt \sim 2v_{eff}\Omega$). Resolution effects on turbulent spectra and late time velocity fluctuation variances were also reported in [10]. Longer inertial ranges are observed for the higher resolution spectra of K and R – which is consistent with the suggested impact of higher (effective) Re noted earlier; finally, the more similar (late time) variances for finer resolution depict more isotropic features associated with increased small scale production after reshock.

Following our previous analysis of IC resolution [7] – see also Chapter 5 – and interfacial morphology effects on planar RMI evolution [9], we attribute GC width differences after reshock as reflecting higher (because better resolved) small scale content in the GC mixing layers at the more sensitive reshock time for the finer resolution simulation.

8.3.2   Initial Condition Parameterization and Data Reduction

Because of its specifically different features – that is, two separate (and diffused) material interfaces – it was not clear a priori whether bipolar RMI behavior such as

Figure 8.24 3D visualization of SF$_6$ mass fraction for GC IC cases *a3* (left) and its more finely resolved version *a5* (right); top: before reshock ($t = 400$ μs), middle and bottom: after reshock ($t = 800$ μs, and $t = 1200$ μs, respectively).

(a)

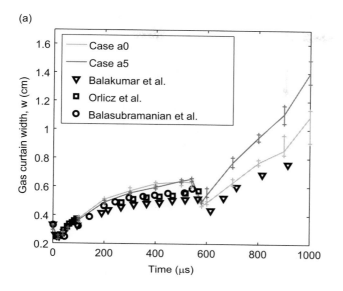

(b)

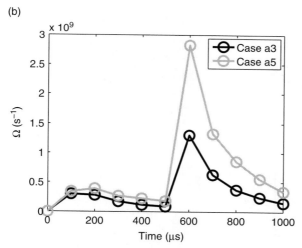

Figure 8.25 Simulation integral predictions for cases *a3* and its more finely resolved version *a5*. (a) GC width predictions with estimated uncertainty ranges as in Fig. 8.23 compared with laboratory experiments [15, 34, 37]; (b) mean enstrophy; (c) mean turbulent specific kinetic energy.

that discussed earlier can be realized also in the GC shocktube, or even whether such a IC parameterization (in terms of an appropriate single parameter η_o) can be useful to characterize the flow instability dynamics. To test the concept in the shocked GC experiments, we defined η_o by $\eta_o = w_o \times 2\pi/\kappa_o$, where $w_o = w(t{=}0)$ is the GC width, and $2\pi/\kappa_o$ is the distance between nozzles.

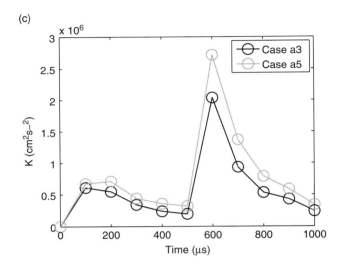

Figure 8.25 (*cont.*)

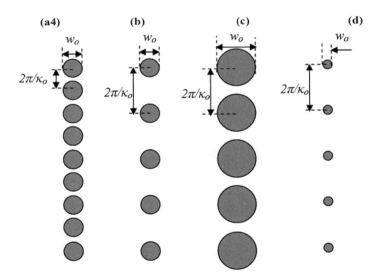

Figure 8.26 Nozzle configurations for various GC ICs; (a4) $\eta_o=5\pi/3$; (b) $\eta_o=5\pi/6$; (c) $\eta_o=5\pi/3$; (d) $\eta_o=5\pi/12$.

To study the implications of particular η_o selections, results of GC width for several nozzle configurations were considered and the GC evolution after first shock (no reshock) were investigated (Fig. 8.26, cf. Table 8.3). Two laboratory realizations from [34] were first examined: case *a4* with all nozzles open and $\eta_o = 5\pi/3$, and case *b* where alternate nozzles were closed, thus increasing the distance between nozzles,

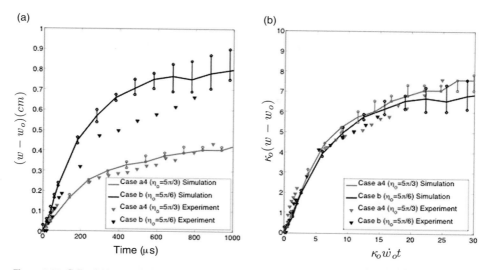

Figure 8.27 GC width as a function of time for laboratory and computational (first shocked only) cases *a4* and *b*; (a) dimensional; (b) nondimensional. Estimated uncertainty ranges of simulation data are prescribed as in Fig. 8.23; estimated uncertainty of reported laboratory data is less than 10%.

effectively reducing η_o by half ($\eta_o = 5\pi/6$). As it turned out, cases *a4* and *b* appeared as *nonlinear* regimes (GC width growth ~$t^{1/2}$). To examine whether bipolar RMI behavior could be also identified in the shocked GC problem, two additional simulation cases were considered by further varying the GC width (w_o) and keeping the distance between nozzles ($2\pi/\kappa_o$) constant (cases *c* and *d*); case *d*, with $\eta_o=5\pi/12$, was selected for its potential of falling in the "low η_o" (linear) regime.

GC width results for cases *a4* and *b* (without reshock) are compared in Figure 8.27(a). As already noted, simulations overpredict the laboratory GC width results. This again is attributed to the difficulty in accurately characterizing *A*, and to its effectively smaller value for the laboratory cases. For cases *a4* and *b*, we find that the higher the initial η_o, the lower the GC width, which is consistent with the previous findings for the *nonlinear* (high η_o) regime in the planar RMI case [9]. When nondimensional GC width curves for simulation and laboratory shocktube data are plotted against nondimensional time as in our previous planar RMI work [9] – using the time scaling noted earlier in connection with Figure 8.13(b), all curves tend to collapse in good agreement (Fig. 8.27(b)). This indicates that discrepancies due to *A* effects in the shocked GC can be largely reduced by appropriate scaling of time using *actual* initial RMI growth rate computed from the laboratory and simulation data.

Finally, on plotting the nondimensional GC width curves for all simulated GC cases in Figure 8.28(b), the results are found to fall into distinctly different linear (low η_o) and nonlinear (high η_o) correlation groups – with the presently available laboratory data falling in the latter. We were thus able to recognize the *bipolar RMI behavior* first reported in connection with the planar single material interface case [9] also in the more complex GC shocktube configuration [10].

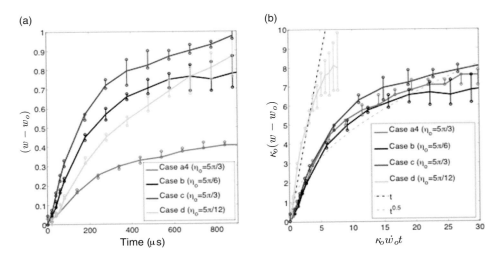

Figure 8.28 GC width as function of time for simulated (first-shocked-only) cases $a4$, b, c, and d; (a) dimensional; (b) nondimensional. Estimated uncertainty ranges are prescribed as in Fig. 8.23. A black and white version of this figure will appear in some formats. For the color version, please refer to the plate section.

8.4   Conclusions

A particular focus here has been on addressing some of the challenging issues of under-resolved RMI driven turbulent material mixing simulation. Because RMI is shock driven, resolving all physical space/time scales in numerical simulations is prohibitively expensive even on the largest supercomputers. By combining shock and turbulence emulation capabilities on a single model, ILES has the potential for providing a natural effective simulation strategy for the study of RMI. The noted inherent sensitivity of laboratory and computational experiments to initial and other boundary conditions is also a crucial aspect to be addressed in this context. Our work addressed the importance of IC characterization and modelling and their effects on 3D transition and mixing in RMI.

Shocked driven mixing is fairly insensitive to ICs before reshock, but very sensitive after reshock – when the shock accelerated interface is not flat but highly corrugated and stochastic and the complex balance of various different instability mechanisms (the RGI) drives the flow dynamics.

Vorticity production at shock time and eventual mode coupling (and transition) thereafter will generally depend on the initial interfacial characteristics, as well as other ICs, such as Ma, Re, and magnitude and sign of A. The presence of small scale material concentration fluctuations in the ICs – and their consequences on the morphology of thicker (high RMS slope η_o) mixing layers at reshock time, promote disorganization, larger spectral bandwidth, and late time features typically associated with transition to turbulence. A single IC parameter characterizing the initial RMS material interface slopes can be usefully identified as relevant in determining whether the RM driven flow is in a linear ballistic regime or nonlinear mode coupling regimes.

8.5   Acknowledgments

LANL is operated by the Los Alamos National Security, LLC, for the U.S. Department of Energy NNSA under Contract No. DE-AC52-06NA25396. The authors thank K. Prestridge, S. Balasubramanian, B.J. Balakumar, and C.A. Zoldi-Sood for stimulating discussions and for sharing information and data from their experiments and simulations. This work was made possible by funding from the LANL Laboratory Directed Research and Development Program on "Turbulence by Design" through directed research project number 20090058DR, and from the LDRD Program on "LES Modeling for Predictive Simulations of Material Mixing" through exploratory research project number 20100441ER.

References

1. Brouillete, M., "The Ritchmyer-Meshkov instability," *Annu. Rev. Fluid Mech.*, **34**, 445–468, 2002.
2. Sagaut P., *Large Eddy Simulation for Incompressible Flows*, 3rd edition, Springer, 2006.
3. Hill, D.J., Pantano, C., and Pullin, D.I., "Large-eddy simulation and multiscale modeling of a Richtmyer–Meshkov instability with reshock," *J. Fluid Mech.*, **557**, 29–61, 2006.
4. Grinstein, F.F, Margolin, L.G., and W.J. Rider, editors, *Implicit Large Eddy Simulation: Computing Turbulent Fluid Dynamics*, 2nd printing, Cambridge University Press, 2010.
5. Drikakis, D.,Grinstein, F.F., and Youngs, D., "On the computation of instabilities and symmetry-breaking in fluid mechanics," *Progress in Aerospace Sciences*, **41**(8), 609–641, 2005.
6. Dimonte, G., "Nonlinear evolution of the Rayleigh–Taylor and Richtmyer–Meshkov instabilities," *Phys. Plasma*, **6**(5), 2009–2015, 1999.
7. Grinstein, F.F., Gowardhan, A.A., and Wachtor, A.J., "Simulations of Richtmyer–Meshkov instabilities in planar shock-tube experiments," *Physics of Fluids*, **23**, 034106, 2011.
8. Gittings, M., Weaver, R., Clover, M., Betlach, T., Byrne, N., Coker, R., Dendy, E., Hueckstaedt, R., New, K., Oakes, W. R., Ranta, D., and Stefan, R., "The RAGE radiation-hydrodynamic code," *Comput. Science and Discovery*, **1**, 015005, 2008.
9. Gowardhan, A.A., Ristorcelli, J.R., and Grinstein, F.F., "The bipolar behavior of the Richtmyer–Meshkov instability," *Physics of Fluids*, **23** (Letters), 071701, 2011.
10. Gowardhan, A.A. and Grinstein, F.F., "Numerical simulation of Richtmyer–Meshkov instabilities in shocked gas curtains," *Journal of Turbulence*, **12**(43), 1–24, 2011.
11. Richtmyer, R. D., "Taylor instability in shock acceleration of compressible fluids," *Commun. Pure Appl. Maths.*, **13**, 297–319, 1960.
12. Vetter, M. and Surtevant, B., "Experiments on the Richtmyer–Meshkov instability of an air/SF6 interface," *Shock Waves*, **4**, 247–252, 1995.
13. Jacobs, J.W. and Sheeley, J.M., "Experimental study of incompressible Richtmyer–Meshkov instability," *Physics of Fluids*, **8**, 405–415, 1996.
14. Poggi, F., Thorembey, M.H., and Rodriguez, G., "Velocity measurements in turbulent gaseous mixtures induced by Richtmyer–Meshkov instability," *Phys. Fluids*, **10**, 2698, 1998.

15. Orlicz, G.C., Balakumar, B.J., Tomkins, C.D., and Prestridge, K.P., "A Mach number study of the Richtmyer–Meshkov instability in a varicose, heavy-gas curtain," *Phys. Fluids* **21**, 064102, 2009.

16. Leinov, E.,Malamud, G., Elbaz, Y., Levin, A., Ben-dor, G., Shvarts, D., and Sadot, O., "Experimental and numerical investigation of the Richtmyer–Meshkov instability under re-shock conditions," *J. Fluid Mech.*, **626**, pp. 449–475, 2009.

17. Schilling, O. and Latini, M., "High-order WENO simulations of three-dimensional reshocked Richtmyer–Meshkov instability to late times: dynamics, dependence on initial conditions, and comparisons to experimental data," *Acta Mathematica Scientia*, **30B**(2), 595–620, 2010.

18. Cohen, R.H., Dannevik, W.P., Dimits, A.M., Eliason, D.E., Mirin, A.A., Zhou, Y., Porter, D.H., and Woodward, P.R., "Three-dimensional simulation of a Richtmyer–Meshkov instability with a two-scale initial perturbation," *Physics of Fluids*, **14**, 3692–3709, 2002.

19. Youngs, D.L., "Rayleigh–Taylor and Richtmyer–Meshkov mixing," ch. 13 in *Implicit Large Eddy Simulation: Computing Turbulent Fluid Dynamics*, ed. by Grinstein, F.F, Margolin, L.G., and Rider, W.J., 2nd printing, Cambridge University Press, 2010.

20. Hahn, M.,Drikakis, D., Youngs, D.L., and Williams, R.J.R., "Richtmyer–Meshkov turbulent mixing arising from an inclined material interface with realistic surface perturbations and reshocked flow," *Physics of Fluids* **23**, 046101, 2011.

21. Thornber, B., Drikakis, D., Williams, R.J.R., and Youngs, D.L., "The influence of initial conditions on turbulent mixing due to Richtmyer-Meshkov instability," *J. Fluid Mech.* **654**, 99–139, 2010.

22. Ukai, S.,Balakrishnan, K., and Menon, S., "Growth rate predictions of single- and multi-mode Richtmyer–Meshkov instability with reshock," *Shock Waves*, **21**, 533–546, 2011.

23. Jimenez, J., Wray, A.A., Saffman, P.G., and Rogallo, R.S., "The structure of intense vorticity in isotropic turbulence," *J. Fluid Mech.*, **255**, 65–90, 832, 1993.

24. Mikaelian, K.O., "Extended model for Richtmyer-Meshkov mix," *Physica D*, **240**, 935–942, 2010.

25. Ristorcelli, J.R., Gowardhan, A.A., and Grinstein, F.F., "Two classes of Richtmyer-Meshkov Instabilities: a detailed statistical look," *Phys. Fluids*, **25**, 044106, 2013.

26. Porter, D.H., Pouquet, A., and Woodward, P.R., "Kolmogorov-like spectra in decaying three-dimensional supersonic flows," *Phys. Fluids*, **6**, 2133, 1994.

27. Grinstein, F.F., "Vortex dynamics and entrainment in regular free jets," *J. Fluid Mechanics*, **437**, 69–101, 2001.

28. Drikakis, D., Fureby, C., Grinstein, F.F., and Youngs, D., "Simulation of transition and turbulence decay in the Taylor-Green vortex," *Journal of Turbulence*, **8**, 020, 2007.

29. Sreenivasan, K.R., Prabhu, A., and Narasimha, R., "Zero-crossings in turbulent signals," *J. Fluid Mechanics*, **137**, 251–272, 1983.

30. Zhou, Y., Grinstein, F.F., Wachtor, A.J., and Haines, B.M., "Estimating the effective Reynolds number in implicit large eddy simulation," *Physical Review E*, **89**, 013303, 2014.

31. Dimotakis, P.E., "The mixing transition in turbulent flows," *J. Fluid Mech.* **409**, 69, 2000.

32. Zhou, Y., "Unification and extension of the similarity scaling criteria and mixing transition for studying astrophysics using high energy density laboratory experiments or numerical simulations," *Physics of Plasmas*, **14**, 082701, 2007.

33. Tennekes, H. and Lumley, J.L., *A First Course in Turbulence*, MIT Press, 1972.

34. Balasubramanian, S., Orlicz, G.C., Prestridge, K.P., and Balakumar, B.J., "Experimental study of initial condition dependence on Richtmyer-Meshkov instability in the presence of reshock," *Physics of Fluids*, **24**, 034103, 2012.

35. George, W.K. and Davidson, L., "Role of initial conditions in establishing asymptotic flow behavior," *AIAA Journal*, **42**, 438–446, 2004.

36. Lombardini, M.,Hill, D.J., Pullin, D.I., and Meiron, D.I., "Atwood ratio dependence of Richtmyer–Meshkov flows under reshock conditions using large-eddy simulations," *J. Fluid Mech.*, **670**, 439–480, 2011.

37. Balakumar, B.J., Orlicz, G., Tomkins, C., and Prestridge, K., "Simultaneous particle-image velocimetry–planar laser-induced fluorescence measurements of Richtmyer–Meshkov instability growth in a gas curtain with and without reshock," *Phys. Fluids*, **20**, 124103, 2008.

38. Shankar, S.K., Kawai, S., and Lele, S.K., "Two-dimensional viscous flow simulation of a shock accelerated heavy gas cylinder," *Physics of Fluids*, **23**, 024102, 2011.

39. Weirs, V.G., Dupont, T., and Plewa, T., "Three-dimensional effects in shock-cylinder interactions," *Physics of Fluids*, **20**, 044102, 2008.

40. Mikaelian, K.O., "Numerical simulations of Richtmyer–Meshkov instabilities in finite-thickness fluids layers," *Physics of Fluids*, **8**, 1269, 1996.

9   Laser Driven Turbulence in High Energy Density Physics and Inertial Confinement Fusion Experiments

Brian M. Haines, Fernando F. Grinstein, Leslie Welser–Sherrill, and James R. Fincke

9.1   Introduction

The mixing of initially separate materials in a turbulent flow by the small scales of turbulent motion is a critical and often poorly understood element of many research programs, such as inertial confinement fusion (ICF), supernova implosions and explosions, and combustion, as well as many other applications in engineering, geophysics, and astrophysics. In typical contexts of interest, we are interested in achieving detailed understanding of the consequences of material interpenetration, hydrodynamical instabilities, and mixing arising from perturbations at the material interfaces, that is, driven by Rayleigh–Taylor (RT), Richtmyer–Meshkov (RM), and Kelvin–Helmholtz (KH) instabilities (buoyancy, shock, and shear induced instabilities, respectively). Laboratory observations typically provide only limited integrated measures of complex nonlinear three-dimensional physical processes, leaving many details and mechanisms unresolved. Carefully controlled computational experiments based on the numerical simulations play a crucial complementary role, providing insight into the underlying dynamics. Collaborative laboratory/computational studies are used to establish predictability of the models in conjunction with the development of frameworks for analysis, metrics for verification and validation, and uncertainty quantification.

The broad goal of this work is to address how well we can predict turbulent mixing in high energy density physics (HEDP) and inertial confinement fusion (ICF) regimes at both macroscopic and subgrid scales. To this end, we perform simulations of laser driven reshock and shear experiments [1–3] as well as an ICF capsule implosion [4] in order to better understand material mixing driven by turbulent instabilities and, in the ICF case, its effects on capsule performance. KH driven shear flows such as mixing layers [5], wakes [6], and jets [7] are of great interest because of their fundamental roles in the practical applications. Crucial aspects of the RM and KH development to be addressed are global instabilities, flow self-organization, complex three-dimensional vortex dynamics, and transition to turbulence (e.g., [8]). The RM instability is induced by the misalignment of the pressure and density gradients when a shock passes a perturbed material interface. The KH instability is associated with the presence of a velocity gradient (shear) between fluids at their interface. In our case, these instabilities are induced by shocking targets with appropriate geometries. In contrast to forced isotropic homogeneous turbulence, where stable statistics are achieved once the flow is developed, shock induced turbulence is inherently unsteady, transitional, and decaying. Suitable metrics for determining transition to turbulence in this kind of unsteady flow are not well established, and the state of the art analysis largely relies on using diagnostics designed for fully developed homogeneous isotropic turbulent flow.

Small scale resolution requirements for simulations typically focus on those of continuum fluid mechanics described by the Navier–Stokes equations; different requirements are involved depending on the regime considered and on the relative importance of coupled physics such as multispecies diffusion and combustion as determined by Knudsen, Reynolds, Schmidt, Damkohler, and other characteristic nondimensional numbers; on the other hand, the longest wavelengths that can be resolved are constrained by the size of the computational domain. Ideally, we would like to resolve all relevant space/time scales and material interfaces in our simulations, the so-called direct numerical simulation (DNS) strategy. Resolving all such physical space/time scales in DNS is prohibitively expensive in the foreseeable future for most practical flows and regimes of interest at moderate-to-high Reynolds number (Re). On the other end of the simulation spectrum are the Reynolds averaged Navier–Stokes (RANS) approaches, which are the preferred industrial standard. In coarse grained simulation (CGS) strategies, large energy containing structures are resolved, smaller structures are filtered out, and unresolved subgrid scale (SGS) effects are modeled; this includes classical LES strategies [9] with explicit use of SGS models, and implicit LES (ILES) [10], relying on SGS modeling implicitly provided by physics capturing numerical algorithms. The CGS strategy of separating resolved and SGS physics effectively becomes the intermediate approach between DNS and RANS.

Transition to turbulence involves unsteady large scale coherent structure dynamics, which can be captured by CGS but not by single point closure turbulence modeling [11]. This brings up a crucial distinguishing aspect between CGS and RANS: because of noncommuting of averaging operations and the transport operator, the closures that define RANS strategies are not unique and work best for self-similar equilibrium problems for which they are designed. Weakly turbulent near equilibrium regimes are involved in many engineering flows of interest. In such cases, CGS may be prohibitive for flows involving complex geometries and multiphysics, but RANS could be a reasonable alternative to obtaining reasonable predictions with reduced computational resources. When trying to simulate a flow where results are strongly dependent on initialization, CGS methods are ideal to determine *how and when* to initialize RANS models [12].

Turbulent material mixing can be usefully characterized by the fluid physics involved: large scale entrainment, stirring due to velocity gradient fluctuations, and molecular diffusion. At moderately high Re, when convective time scales are much smaller than those associated with molecular diffusion, we are primarily concerned with the numerical simulation of the first two convectively driven processes. These processes can be captured with sufficiently resolved implicit large eddy simulation (ILES) [10]. Therefore, ILES serves as our primary simulation strategy. We validate these results using available experimental data.

The chapter proceeds as follows. In Section 9.2, we discuss simulations of two HEDP experiments [1, 2, 3]. Details of the experiments are outlined in Section 9.2.1 and our simulation strategies are presented in Section 9.2.2. Next, in Section 9.2.4, we compare our simulations to available experimental data. We compare feature shapes by showing relevant material concentration distributions for simulations and radiographic data from the experiments. These have comparable distributions of feature sizes, which is demonstrated by taking their discrete Fourier transforms. In Section 9.2.5, we test turbulence metrics designed for homogeneous isotropic turbulence in order to explore their applicability for the type of unsteady, transitional flow observed in our simulations. We demonstrate that several turbulent features, such as characteristic vorticity distributions and short correlation lengths, develop quickly, while others, such as self-similarity and isotropy, do not develop until the turbulent flow has decayed significantly. In Section 9.3, we discuss simulations of an OMEGA-type ICF capsule [4]. In Section 9.3.1, we discuss the simulation strategies employed in the ICF simulations. Next, in Section 9.3.2, we review the reshock experiment, discuss a novel 2D–3D simulation strategy employed to simulate it, and compare the results from this and standard 2D–3D strategies to our fully 3D simulation data. In Section 9.3.3, we detail application of our new 2D–3D strategy to an OMEGA–type ICF capsule and discuss results, comparing them to available experimental data. Finally, in Section 9.4, we present our conclusions.

9.2   HEDP Simulations

9.2.1   Description of Experiments

The experiments, outlined in detail in [1], were performed using the University of Rochester's OMEGA laser. The target geometries and other experimental details for both experiments are given in the following. The primary diagnostic for the experiments is x-ray radiography with scandium, chromium, or chlorine backlighters. Radiographs are obtained from two quasi-orthogonal views for each shot, one of which uses a target mounted backlighter and the other of which uses a TIM mounted backlighter. They are recorded on either a single strip CCD camera or on single strip film. Therefore, a maximum of two images can be acquired for each shot. Images for a range of times are thus obtained by repeating the experiment with identically constructed targets.

9.2.1.1   Reshock Experiment

The reshock target consists of a cylindrical beryllium (Be) tube ≈ 1.4mm in length and ≈ 0.5mm in diameter with a $\approx 100\mu m$ wall thickness (see Figure 9.1). The progression of events in the experiment can be visualized in terms of the schematic shock interface diagram in Figure 9.2. The target is successively hit from both sides by two laser pulses whose interactions with the ablators on each side generate shocks. The first pulse, ≈ 5 kJ, at $t = 0$ ns impacts the plastic ablator on the left, driving a Mach ≈ 5 shock through the $20\mu m$ aluminum tracer disk adjoining the ablator. The tracer disk is thus propelled to the right down the center of the cylinder, which is filled with a low density (60 mg/cc) CH foam. The second pulse, $\approx 4kJ$ at 5 ns, impacts a plastic ablator at the

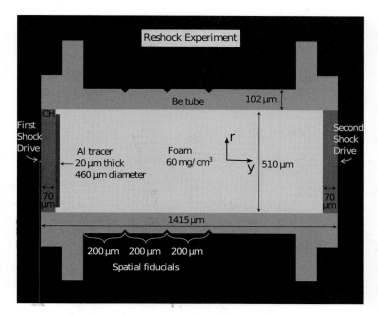

Figure 9.1 Reshock target geometry.

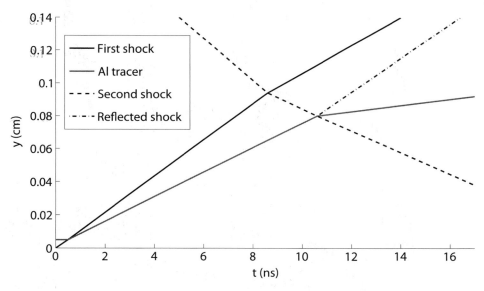

Figure 9.2 Schematic shock interface diagram depicting progression of events in the reshock experiment. Crossing locations are approximate.

right end of the tube. The shocks collide at approximately 8 ns to the right of the mixing layer and the second shock interacts with (reshocks) the mixing layer at approximately 10 ns, causing it to compress until approximately 13 ns. At approximately 17 ns, the second shock exits the mixing layer.

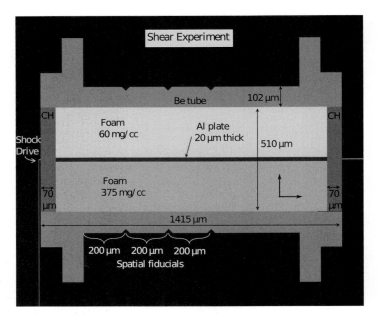

Figure 9.3 Shear experiment target geometry.

Aided by simulations and plasma viscosity formulae, we estimate that the experiment reaches a peak Reynolds number $Re = \frac{\delta_{MZ}\dot{\delta}_{MZ}}{\nu} \approx 1.6 \times 10^5$. Here, δ_{MZ} is the approximate width of the mixing region and ν is the viscosity, for which we use $\nu = 0.01\text{cm}^2/\text{s}$. This corresponds to a Taylor microscale based Re of $Re_\lambda \approx \sqrt{\frac{20}{3}Re} \approx 1000$, which is typical for laser driven experiments.

9.2.1.2   Shear Experiment

The shear target consists of a cylindrical Be tube ≈ 1.4mm in length and ≈ 0.5mm in diameter with a $\approx 100\mu m$ wall thickness (see Fig. 9.3). Gold cones (not shown) are attached to both ends of the tube in order to minimize stray laser light. One half of the cylinder is filled with a low density ($\rho = 60\text{mg/cm}^3$) CH foam, while the other is filled with a high density ($\rho = 375\text{mg/cm}^3$) CH foam. These are separated by a $20\mu m$ thick aluminum ($\rho = 2.43\text{g/cm}^3$) plate. The ablator is hit from the left by a single laser pulse (≈ 5 kJ), which generates a shock, producing a shear layer as the shock passes through the foams at different speeds.

Aided by simulations and plasma viscosity formulae, we estimate that the experiment reaches a peak Re of $Re = \frac{\delta_{MZ}\dot{\delta}_{MZ}}{\nu} \approx 2.0 \times 10^4$. Once again, we use $\nu = 0.01\text{cm}^2/\text{s}$. This corresponds to a Taylor microscale based Re of $Re_\lambda \approx \sqrt{\frac{20}{3}Re} \approx 370$.

9.2.2   Simulation Strategies

Three-dimensional simulations were performed using an ILES strategy [10] based on the radiation adaptive grid Eulerian (RAGE) [13] radiation hydrodynamics code. RAGE solves the Euler form of the multimaterial compressible conservation equations for mass density (ρ), momenta ($\rho \, \vec{u}$), and total specific energy (E), given by

$$\frac{\partial \rho}{\partial t} + \nabla \cdot \left(\rho \vec{u} \right) = 0$$

$$\frac{\partial \rho \vec{u}}{\partial t} + \nabla \cdot \left(\rho \vec{u} \vec{u} + pI \right) = 0$$

$$\frac{\partial \rho E}{\partial t} + \nabla \cdot \left(\rho \vec{u} E + p\vec{u} \right) = 0 \qquad (9.2.1)$$

$$\frac{\partial c_i \rho}{\partial t} + \nabla \cdot \left(c_i \rho \vec{u} \right) = 0,$$

where ρ is the material density, $\vec{u}$ is the fluid velocity, P is the pressure, c_i is the mass fraction of the ith species, and radiation transport, body forces, molecular shear, and heat conduction have been neglected. These equations are supplemented with SESAME tabular equations of state [14]. Despite the fact that the experiment rapidly ionizes and attains a plasma state, applying this hydrodynamic approximation remains valid due to the high densities of the experiments (200+ times the density of air). In addition, neglecting the effects of radiation transport on the flow is justified since the ratio of typical radiative to hydrodynamic fluxes can be estimated as $\sigma T^4 / \left(\rho U^3 \right) \approx 10^{-4}$, where σ is the Stefan–Boltzmann constant and $T = 10$ eV, $U = 80$km/s, and $\rho = 0.2$g/cc are characteristic of the experiment. To test this assumption, lower resolution simulations were performed with radiation effects turned on, and no substantial differences were observed.

RAGE uses a second-order Godunov finite volume scheme and adaptive mesh refinement (AMR) using cubic cells, as well as a variety of numerical options. In the present simulations, a van Leer limiter was used (see [2] for other details). As used in the present work, RAGE models miscible material interfaces (Schmidt number $Sc \approx 1$) and high Re convection driven flow with an effective (numerical) viscosity determined by the small scale cutoff. RAGE does not use fixed timesteps; rather, the timestep is set as the minimum of a Courant condition on the hydrodynamics and stability criteria determined by other physics in the code.

Three-dimensional simulations of the reshock experiment were performed at three maximum grid resolutions (1.25μm, 2.5μm, and 5μm), and all simulations used AMR. For each simulation, the total computational domain size was 0.32cm $\times$ 0.32cm $\times$ 0.32cm. The more computationally intensive 1.25μm simulations involved only a quadrant of the full domain, covered $t = 0 - 20$ ns, and ran for 116 hours with 32,000 timesteps using 4096 processors on Los Alamos National Laboratory's (LANL's) Cielo supercomputer, achieving a maximum of approximately 100 million cells. The 2.5μm and 5μm simulations covered $t = 0 - 30$ ns and ran for 159 and 48 hours with 25,000 and 14,000 timesteps, respectively, using 4096 processors. These employed a maximum of approximately 500 million and 23 million total cells, respectively.

Three-dimensional simulations of the Shear experiment were performed at three maximum grid resolutions (all simulations employed AMR). For each simulation, the total computational domain size was once again 0.32cm $\times$ 0.32cm $\times$ 0.32cm.

The high, intermediate, and low resolution simulations had maximum resolutions of 1.25μm, 2.5μm, and 5μm, respectively. The simulations employed approximately 388 million, 212 million, and 21 million total cells, respectively, at completion, and approximately 32,000, 19,000, and 10,000 timesteps. The 1.25μm resolution simulation ran for approximately 266 hours using 4096 processors on LANL's Cielo, while the 2.5μm resolution ran for approximately 538 hours using 1024 processors, and each 5μm simulation ran for approximately 89 hours using 512 processors. The low resolution simulations were performed 20 times with different realizations of the random interface perturbations (discussed in the following) in order to perform ensemble averaging. Differences between integrated velocity variances varied by at most 2% when comparing an ensemble of 15 simulations to an ensemble of 20 simulations, suggesting that a sufficient number of simulations has been performed to build an ensemble.

9.2.3   Interface Perturbation Spectra

As we consider simulating the growth of hydrodynamic instabilities in laboratory experiments, we must consider the effects of modeling initial conditions (ICs), for which limited experimental characterizations are available. The difficulties with the open problem of predictability of material stirring and mixing by underresolved multi-scale turbulent velocity fields in shock driven turbulent flows are compounded with the inherent sensitivity of turbulent flows to ICs [11, 22]. Flow instabilities driven by accelerated material interfaces are particularly sensitive to their ICs: small variations in the initial state of the interface can result in significant changes to the integral character of a mixing layer at late times [23–25].

Therefore, in order to properly simulate the growth of hydrodynamic instabilities, it is necessary to accurately model perturbations on the interfaces where such instabilities are expected to grow. For simulations of both the reshock and shear problems, we perturb the interfaces between the aluminum tracer and the foams. These perturbations are specified using a spectrum measured from a sample used in experiments. Specifically, the top and bottom of the aluminum plate were perturbed using the function

$$\zeta(x, y) = \Gamma \sum_{i,j=0}^{N} c_{ij} \cos\left(2\pi i x/l_x + \theta_i\right) \cos\left(2\pi j y/l_y + \phi_j\right), \qquad (9.2.2)$$

where $\theta_i, \phi_j \in [0, 2\pi]$ and $c_{ij} \in [0, 1]$ were chosen randomly using a uniform distribution and weighted appropriately to achieve the desired spectrum. N was set so that the spectra were cut off at $\lambda = 2.5\mu$m, the size of the smallest cell in the high resolution simulation, for the reshock experiment and $\lambda = 10\mu$m, the length of two cells in the coarse grid, for the shear experiment. Γ was set in order that the perturbations have a standard deviation of 0.33μm, corresponding to the experimentally obtained data. The spectra are shown in Figure 9.4 for the reshock experiment and Figure 9.5 for the shear experiment. Note that the spectra for the different grids do not line up

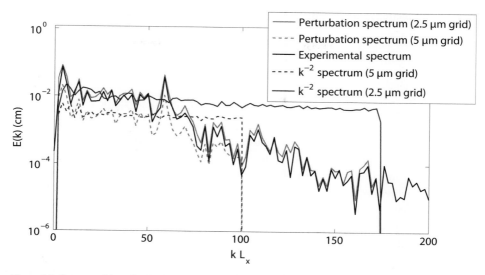

Figure 9.4 Spectra of interface perturbations. Here, $L_x = 460\mu$m, the width of the aluminum tracer.

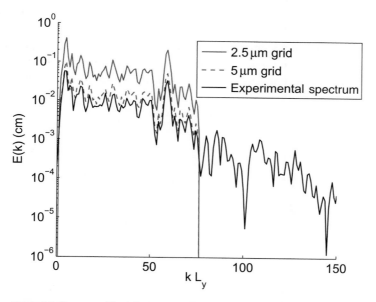

Figure 9.5 Spectra of interface perturbations for the shear experiment. Here, $L_y = 1275\mu$m, the length of the aluminum plate. The computational spectra are generated using fast Fourier transforms (FFT) over the respective computational grids (differences arise because the standard deviation was specified over the coarse computational grid and both grids use the same coefficients).

because the standard deviation is fixed on one grid and the same coefficients are used to specify ζ on each grid. For the low resolution (5μm) 3D simulations, 20 simulations were performed for which the coefficients $\{c_{ij}\}$ were varied maintaining the desired spectrum and standard deviation.

9.2.4   Visualizations and Comparison with Experimental Data

Visualizations of the aluminum concentration for the reshock experiment are shown in Figure 9.6. Here, the aluminum mass concentration from the 2.5μm resolution 3D data is visualized corresponding to the colorbars with a 1g/cm^3 density isosurface overlaid in blue (providing the outline of the Be tube). The presence of small scale features in the data suggests turbulent features in the mixing layers that become more pronounced after reshock ($t \approx 10$ ns). The 5μm resolution data (not shown) appears more diffuse and does not exhibit the small scale features.

In Figure 9.7, we compare azimuthally averaged 3D aluminum concentrations for the reshock experiment radiographic data. The inconsistent radiograph shapes are due to target imperfections that were not modeled in simulations. The 5μm resolution 3D data have a much wider mixing layer than the 2.5μm resolution 3D data and experiment, suggesting significant numerical diffusion effects. Small differences are noticeable among the 2.5μm resolution 3D simulations with different IC spectra. However, the mixing layer size and shape are largely consistent.

In Figure 9.8, the aluminum mass concentration is visualized for the shear experiment corresponding to the colorbars with a 1 g/cm^3 density isosurface overlayed in blue (providing the outline of the Be tube). The green outlines are 0.7 isosurfaces of the mass fractions for the two different foams. In Figure 9.9, we compare ensemble averaged low resolution (5μm) 3D aluminum concentrations at the (y, z) center plane, which are ensemble averaged over 20 performed simulations, to corresponding laboratory radiographic data. The shapes and feature sizes of the simulations compare well to the radiographic data.

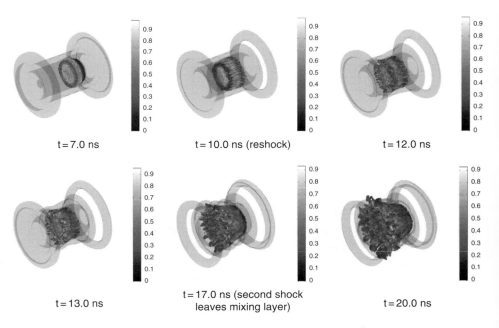

Figure 9.6 Visualization of c_{Al} for the reshock experiment with a 1g/cm^3 density isosurface overlaid in gray (providing the outline of the Be tube).

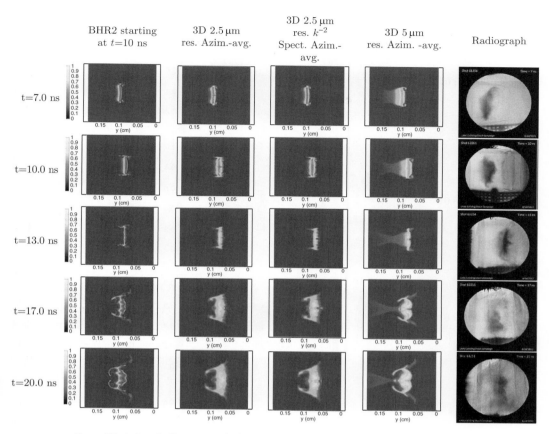

Figure 9.7 Azimuthally averaged aluminum mass concentration (c_{Al}) distributions for the reshock experiment along with radiographic data from the experiment. A black and white version of this figure will appear in some formats. For the color version, please refer to the plate section.

The visualizations, while useful for providing an overview of the progress of the experiment, are entirely qualitative in nature. It is desirable to obtain more quantitative comparisons as to how well simulations compare to the experimental data. To demonstrate such a quantitative comparison, in Figures 9.10 and 9.11, we show the radially averaged FFT of the simulated aluminum concentrations for the reshock and shear experiments, respectively, compared to the radiographic data (the 3D c_{Al} shear data is restricted to the center plane, while the reshock data is spatially averaged). By looking at the FFTs, we can compare the distribution of feature sizes in simulations to radiographic data rather than just shape. Since the magnitudes of the concentration are not directly comparable to the radiograph intensities, the spectra are rescaled individually to have unit maxima. The small scale cutoff on the plots is set at $20\mu m$, which is approximately the resolution of the radiographic data.

For the reshock experiment, a good match is obtained for the $2.5\mu m$ resolution 3D data for all displayed length scales (the cutoff, $20\mu m$, corresponds to 8 grid cells for the $2.5\mu m$ resolution data) through 17 ns, well after reshock (approx. 10.4 ns). The low resolution 3D data show good agreement for both spectra only to $50\mu m$, corresponding

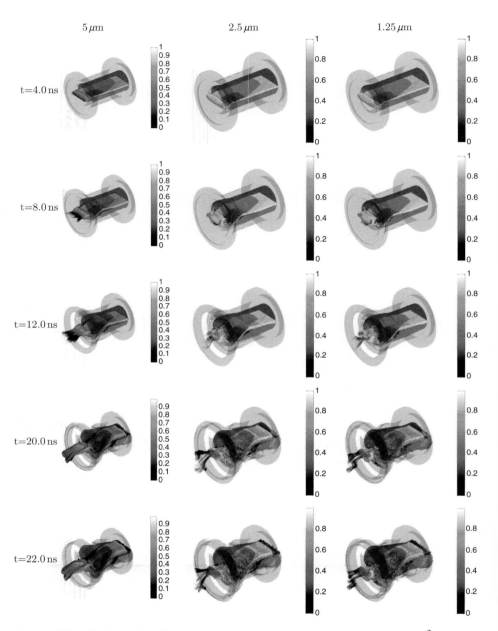

Figure 9.8 Visualization of c_{Al} from simulations of the shear experiment with a 1 g/cm^3 density isosurface overlayed in blue (providing the outline of the Be tube). The green outlines are 0.7 isosurfaces of the mass fractions for the two different foams. A black and white version of this figure will appear in some formats. For the color version, please refer to the plate section.

to 10 grid cells. Therefore, the 2.5μm resolution data appear to be more appropriate for studying the flow in the experiments. While there is not much sensitivity to the detailed spectral content of the aluminum surfaces, the simulations using actual IC spectral content from the experiments show better agreement with the laboratory data.

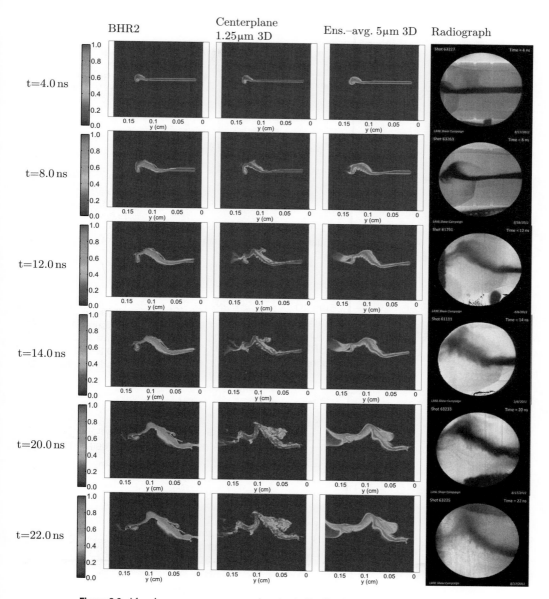

Figure 9.9 Aluminum mass concentration (c_{Al}) distributions for the shear experiment along with radiographic data (also from experiment). The 3D data are taken from the center plane (in the z direction). A black and white version of this figure will appear in some formats. For the color version, please refer to the plate section.

For the shear experiment, a good match is obtained for the 1.25μm and 2.5μm 3D data for all displayed length scales (the cutoff, 20μm, corresponds to 8 grid cells for the 2.5μm data) for all simulated times. The low resolution 3D data shows good agreement for both spectra only to 50μm, corresponding to 10 grid cells.

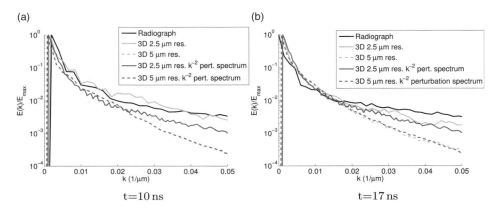

Figure 9.10 Radially averaged FFT of averaged aluminum mass concentrations (c_{Al}) for 3D simulations of the reshock experiment compared with radially averaged FFT of comparable regions of radiographic data.

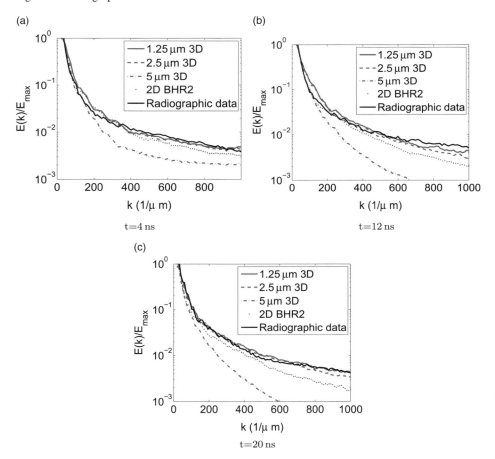

Figure 9.11 Radially averaged FFT of aluminum mass concentrations (c_{Al}) for 3D simulations of the shear experiment (taken from center planes) compared with radially averaged FFT of comparable regions of radiographic data.

9.2.5 Comparison with Results for Homogeneous Isotropic Turbulence

In this section, we analyze which features of isotropic homogeneous turbulence appear in the 3D simulations and how they develop. The vorticity field in equilibrium turbulent flows tends to be highly organized with intense vorticity concentrated in elongated filaments, characterizing the smallest coherent structures of turbulent flows [26]. Achieving self-similarity in turbulence has precise consequences in addition to the existence of an inertial subrange in the turbulent kinetic energy spectra – specifically, in terms of characteristic probability density functions (PDFs) of various velocity functions (e.g., vorticity), as well as autocorrelation and structure functions. Such information associated with small scale flow features is available for isotropic steady turbulence (e.g., [26]), and provides a useful basis for verification and validation of LES (ILES) strategies in the present context.

Actual values of Re characterizing the flow at the smallest resolved scales are not a priori available in CGS such as LES or ILES, where the smallest characteristic resolved scale is determined by the resolution cutoff wavelength prescribed by an explicit or implicit spatial filtering process. Since the small scale cutoff is determined by grid size, the implication is that grid refinement is associated with observing smaller simulated physical structures associated with a somewhat different problem with higher effective Re, Re_{eff}. The concept of effective (resolution dependent) turbulent Re for ILES has been invoked in the past for statistically steady flows (e.g., [27, 28]); practical convergence of turbulence characteristics can be achieved once resolution is fine enough to ensure that a computed associated effective Re [27] is higher than a critical value such as the mixing transition threshold Re, Re_{mix} [29]. Here, we will demonstrate that the effective Re is comparable to or larger than the mixing transition threshold for the high resolution Reshock simulations, but has not been achieved for the shear simulations.

We seek to demonstrate self-similarity and other turbulent characteristics through established metrics, such as probability density functions (PDFs) of various velocity functions (e.g., vorticity magnitude) and characteristic behaviors (e.g., of structure functions). Such information associated with small scale flow features is available for isotropic steady turbulence (e.g., [26]) and has been extensively used as a useful basis for verification and validation of ILES (e.g., [25, 27, 28, 30]). Shock induced turbulent flow is inherently unsteady and involves transitional and decaying turbulence aspects. We mined our simulation database using suitably adapted diagnostics originally designed for statistically steady homogeneous isotropic flow. For the reshock experiment, we take advantage of the cylindrical geometry of the problem and azimuthal spatial averaging is carried out to evaluate mean quantities. For the shear problem, we use the ensemble of low resolution simulations to evaluate mean quantities. Deviations around such mean values are used to characterize relevant statistical fluctuations.

Despite the fact that the unsteady flow in the experiments is not homogeneous, nor isotropic, nor fully turbulent, there are small regions in which the flow does demonstrate characteristics of homogeneous isotropic turbulence. We identify these regions by the presence of high levels of TKE and vorticity. In this section, we isolate these regions and perform turbulence analysis and compare the results to those expected for homogeneous isotropic turbulence. Due to the cylindrical geometry of the reshock target, our analysis takes advantage of the spatial homogeneity in the azimuthal direction, in which spatial averaging is carried out in order to evaluate mean quantities. Since the shear target has no such spatial symmetry, we cannot take advantage of spatial homogeneity to perform these analyses for this experiment, so we use an ensemble of simulations for the 5.0μm data and perform analysis within a sufficiently spatially homogeneous region inside the Be tube for the high resolution 3D data.

9.2.5.1   Integrated Flow Quantities

In Figure 9.12, we show typical distributions of the TKE for the reshock experiment. The TKE measures the amount of energy in the portion of the flow that deviates from the radially averaged flow. We define the turbulent kinetic energy as

$$K(y, r) := \left\langle \frac{1}{2}\rho\left((u_r - \widetilde{u}_r)^2 + (u_y - \widetilde{u}_y)^2 + u_\theta^2 \right) \right\rangle_\theta \Big/ \langle\rho\rangle_\theta,$$

where $\langle\cdot\rangle_\theta$ denotes azimuthal averaging and $\widetilde{u}_i := \langle u_i\rho\rangle_\theta/\langle\rho\rangle_\theta$. Obtaining an equivalent measure for the shear experiment is more delicate, since it lacks the spatial symmetry of the reshock target. One method of doing this is by using the ensemble performed at low resolution and defining the TKE using ensenble averages as

$$\text{TKE}_{\text{ens}} := \left\langle \frac{1}{2}\rho\left((u_x - \widetilde{u}_x)^2 + (u_y - \widetilde{u}_y)^2 + (u_z - \widetilde{u}_z)^2 \right) \right\rangle_{\text{ens}} \Big/ \langle\rho\rangle_{\text{ens}}, \qquad (9.2.3)$$

where $\langle\cdot\rangle_{\text{ens}}$ denotes ensemble averaging, $\widetilde{u}_i := \langle u_i\rho\rangle_{\text{ens}}/\langle\rho\rangle_{\text{ens}}$, and the TKE is restricted to the plane $x = 0$. This definition, however, does not allow us to study convergence as a function of grid resolution since it can only be applied when an ensemble of simulations is available. Such an ensemble is too computationally expensive to obtain at the higher resolutions, so for these we define the TKE as

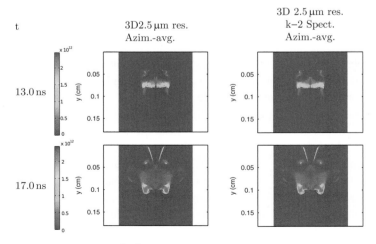

Figure 9.12 TKE (g cm^2/s^2) for the reshock experiment.

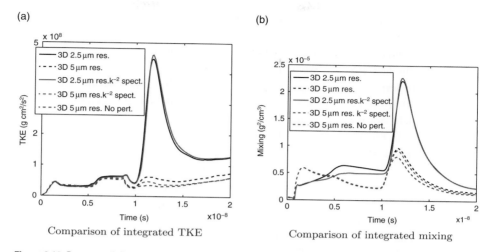

Figure 9.13 Integrated flow quantities for the 3D simulations of the reshock experiment.

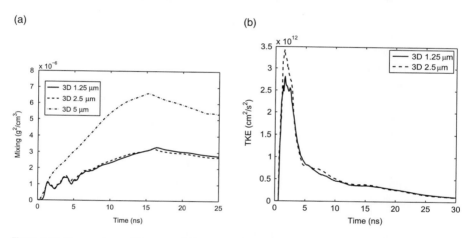

Figure 9.14 Comparison of integrated flow quantities for the 3D simulations of the shear experiment. (a) Integrated mixing; (b) volumetrically averaged TKE. The mixing measure is defined in Equation (9.2.5) and the TKE is the density weighted volumetric integral of the quantity in Equation (9.2.4).

$$\text{TKE}_{\text{vol}} := \left\langle \frac{1}{2}\rho\left((ux - \widetilde{u}x)^2 + (u_y - \widetilde{u}_y)^2 + (u_z - \widetilde{u}_z)\,2\right)\right\rangle_z \Big/ \langle\rho\rangle_z, \qquad (9.2.4)$$

where the spatial averaging is performed in the z direction over a homogeneous region excluding the Be tube. Note that, due to the presence of the Be tube, this region is only approximately homogeneous.

The density weighted volumetrically integrated TKE is plotted in Figure 9.13 for the reshock experiment and in Figure 9.14 for the shear experiment. The 2.5μm resolution 3D reshock simulation becomes much more turbulent after reshock, with a peak in TKE at approximately $t = 12$ns, during which time the mixing layer is being compressed by the second shock. This spike is entirely absent from the $5\mu m$ resolution data, indicating that small scales are not being captured well enough to allow for the development of turbulence. For the shear simulations, the average TKE peaks at about 1.6 ns, despite the fact that, as we will demonstrate, many other turbulent features have yet to materialize. The $1.25\mu m$ simulation is able to better resolve the initial interface perturbations and turbulence generation mechanisms and thus demonstrates more complex behavior near the peak.

In Figures 9.13 and 9.14, we also show a volumetrically integrated mixing measure for the reshock and shear experiments, respectively, given by

$$\text{mixing} := \int \rho^2 c_{Al} \, c_{\text{foam}} \, dV. \qquad (9.2.5)$$

Comparing the mixing and TKE for the reshock TKE in Figure 9.13, we can infer that the difference between mixing in the $2.5\mu m$ and $5\mu m$ resolution data represents turbulent mixing. The bulk of the turbulent mixing thus occurs post-reshock ($\approx 10 - 17$ ns). It is important to note that the reduction in the value of the mixing measure after $t \approx 12$ ns largely reflects a decrease in the density of the material due to post-reshock expansion (rather than demixing).

The mixing analysis for the shear problem in Figure 9.14 demonstrates reasonably grid independent properties for the $2.5\mu m$ and $1.25\mu m$ data. The more complex variability of the mixing behavior at early times for the $1.25\mu m$ and $2.5\mu m$ data reflects compression and expansion due to shock wave interactions with the aluminum tracer which are not adequately resolved with the $5\mu m$ resolution.

We demonstrate effects of interface perturbations for the reshock simulations in Figure 9.13. For the $2.5\mu m$ resolution data, the specific choice of initial interface perturbations appears to have little effect. For the $5\mu m$ resolution data, we also compare with results where no interface perturbations were imposed. The case with no perturbations shows the smallest growth in TKE after reshock, while the case closely emulating the experimental spectra shows the largest growth. This suggests that while the presence of initial interface perturbations is necessary for the development of turbulence after reshock, their precise nature may not be important; however, as previously noted [25, 30], the flow after reshock is found to be very sensitive to actual mixing layer conditions at reshock time.

9.2.5.2   Vorticity Distributions

Cumulative distribution functions (CDFs) of vorticity magnitude are characteristic of isotropic homogeneous turbulence [26]. CDF(ω) – the integral of PDF(ω) – represents the volume fraction with vorticity magnitude greater than ω. The CDFs associated with the DNS data tend to approach a high Re_λ limit when Re_λ is above the mixing transition threshold $Re_\lambda \approx 100 - 140$ [29], where Re_λ is based on the Taylor microscale λ. Most of the volume fraction is occupied by relatively weak vorticity $\omega/\omega_{\text{rms}} \lesssim 1$, and intense (worm) vortices with $\omega/\omega_{\text{rms}} \gtrsim 3$ fill a small volume fraction $\approx 1\%$ for the higher Re_λ cases. Here, ω_{rms} denotes the volumetric root mean square (RMS) value. Similar analysis of RAGE shock tube simulation data has also

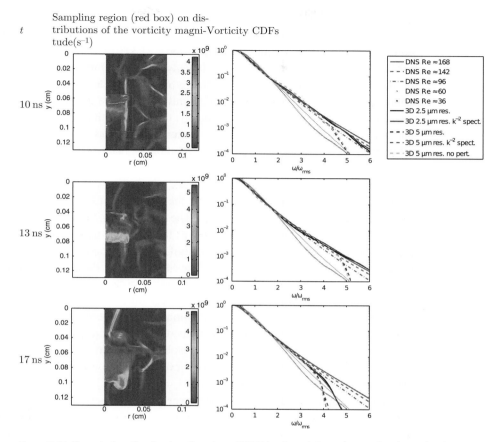

Figure 9.15 Cumulative distribution functions (CDFs) of vorticity ω/ω_{RMS} for the reshock experiment sampled from a uniform $5\mu m$ grid. Sampling regions (within red lines) on distributions of vorticity magnitude (s^{-1}). The DNS data were taken from [26]. A black and white version of this figure will appear in some formats. For the color version, please refer to the plate section.

been carried out based on PDFs of other fundamental magnitudes of interest such as longitudinal and transverse velocity derivatives [30].

Figure 9.15 shows CDFs of the vorticity magnitude for simulations of the reshock experiment at various times and compares them to those corresponding to DNS of isotropic turbulence data from Jiménez et al. [26]. It is important to note that the DNS data involve a periodic box sampling of homogeneous isotropic turbulence. In contrast, the sampling domains of reshock data (frames on the left of Figure 9.15) focus on flow regions selected because they contain high levels of TKE and vorticity, but which (as also noted earlier) are not homogeneous, not isotropic, nor fully turbulent. In particular, the vorticity distributions associated with the reshock simulation data are very different for small vorticity values from those of the DNS isotropic turbulence simulations. Specifically, PDFs of the vorticity magnitude for the homogeneous DNS data tend to vanish for $\omega \approx 0$, whereas the corresponding PDF based on the reshock data tend to peak in that limit due to the irrotational flow content in the sampling domains. This has a direct effect on the ω_{rms} value used as scale the vorticity in a figure such as Figure 9.15. In order to have meaningful comparative PDF analysis, vorticity magnitude thresholds were chosen such that we set $\text{PDF}(\omega) = 0$ (i.e., $\text{CDF}(\omega) = 1$) for $\omega \lesssim \omega_{\text{thr}}$. For the reshock data, the threshold value $\omega_{\text{thr}} = 0.5\omega_{\text{rms}}$ was used. Importantly, with this choice, the ILES data pass through the point that is Re_λ independent for the DNS data. The 2.5μm resolution 3D data match the (self-similar) DNS data the best at $t = 13$ ns, with the CDF tail decaying for later times. Notably, the match with DNS at $t = 13$ ns is better than at $t = 12$ ns (not shown), where the integrated TKE has its maximum (Fig. 9.14).

Figure 9.15 suggests that the 5μm resolution is inadequate to capture turbulent features in the reshock problem – that is, it is associated to very low effective Re_λ whereas the 2.5μm resolution CDFs are consistent with the higher Re_λ ("self-similar") DNS results. Such comparisons are used here to characterize Re_λ for the (nominally inviscid) RAGE simulations. The effective Re_λ associated with the finer (2.5μm) resolution appears to be above the mixing transition threshold value ($Re_\lambda \approx 100 - 140$) proposed by Dimotakis [29] for sustained turbulent flows. The higher effective Re results are also the ones that provide the best match in the FFT comparisons with laboratory experiments discussed in Section 9.2.4.

It is notable that the vorticity CDFs show very little sensitivity to the nature of the interface perturbations. For the low resolution simulations, the case with no perturbations matches very well with the simulations with interface perturbations. This is likely due to the fact that the flows in the 5μm resolution case do not appear turbulent – an indication that either the effective Re is low or the resolution is insufficient to resolve the interface perturbations satisfactorily.

Figure 9.16, showing data for the shear experiment, suggests that the 5μm resolution is inadequate to capture turbulent features downstream from the shock: these data match the DNS data best at $t \approx 12$ ns, but the tail of the CDF decays significantly by $t = 22$ ns. The 5μm data are thus associated with very low effective Re_λ whereas the 1.25μm and 2.5μm resolution CDFs are consistent with the higher Re_λ ("self-similar") DNS results. Such comparisons are used here to characterize Re_λ for the (nominally inviscid) RAGE simulations. Once again, a vorticity threshold of $\omega_{\text{thr}} = 0.5\omega_{\text{rms}}$ was used to filter the vorticity data.

9.2.5.3   Autocorrelation Analysis

Velocity autocorrelation functions indicate the level of correlation in a flow. In isotropic homogeneous turbulence, the spatial autocorrelation functions decay rapidly to zero, indicating that the correlation length in the flow is very small and there are no long range correlations. For the reshock experiment, we define

$$\text{ACF}(r, d) := \left\langle \frac{\left\langle (u_\theta(r, \phi, y) - \overline{u}_\theta)\cdot(u_\theta(r, \theta + \phi, y) - \overline{u}_\theta) \right\rangle_\phi}{\left\langle (u_\theta(r, \phi, y) - \overline{u}_\theta)^2 \right\rangle_\phi} \right\rangle_y,$$

where $d = \frac{\sin\theta}{\cos\frac{\theta}{2}} r$ and $\overline{u}$ is the average of u as a cylindrical vector. These are plotted in Figure 9.18 for the 2.5μm resolution 3D data. These are calculated in the spatially homogeneous azimuthal direction. The flow demonstrates short correlation lengths for

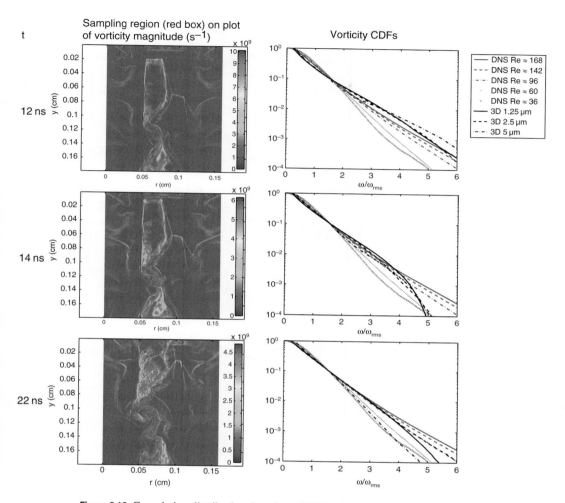

Figure 9.16 Cumulative distribution functions (CDFs) of vorticity ω/ω_{RMS} for the shear experiment inside sampling regions compared with expected CDFs for isotropic homogeneous turbulence (red curves) at various Taylor Re. Sampling regions (areas inside red lines) are shown on distributions of vorticity magnitude (s^{-1}). The DNS data were taken from [26].

all times (ACF$(r, d) = 0$ indicates no correlation). However, weak long range correlations (corresponding to $|$ACF$(r, d)| > 0$) become more prominent from $t \approx 13$ ns onwards. Analysis of the 5μm resolution data (not shown) demonstrates longer correlation lengths and stronger long range correlations, consistent with previous analyses indicating that turbulence does not fully develop in the 5μm resolution simulations.

For the shear experiment, we define

$$\text{ACF}(r) := \frac{\langle (u_x(x, y, z) - \overline{u_x}(y, z)) \cdot (u_x(x + r, y, z) - \overline{u_x}(y, z)) \rangle}{\left\langle (u_x(x, y, z) - \overline{u_x}(y, z))^2 \right\rangle}, \tag{9.2.6}$$

where $\overline{}$ denotes x averaging and $\langle \cdot \rangle$ denotes averaging over the volumes outlined in red in Figure 9.16 extending 50μm above and below the displayed plane. These are plotted

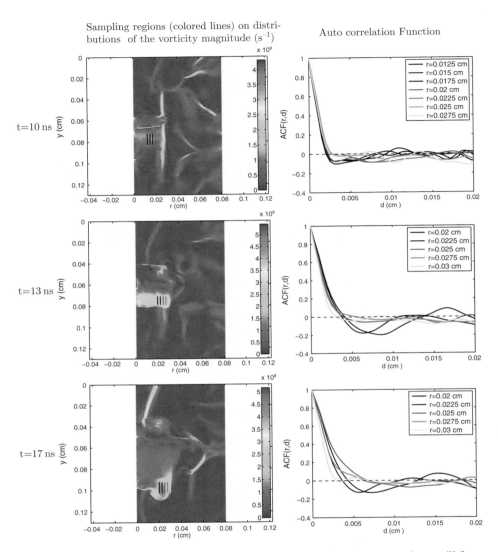

Figure 9.17 Autocorrelation functions $ACF(r, d)$ for the reshock experiment at various radii for 2.5μm resolution 3D data. The vertical colored bars in the vorticity distributions are cross-sections of the annular sampling regions.

in Figure 9.18 for the 1.25μm, 2.5μm, and 5μm resolution 3D data along with sampling regions. The flow demonstrates short correlation lengths for $t \approx 10 - 12$ns ($\mathrm{ACF}(r, d) = 0$ indicates no correlation) near the shock front as well as for $t \approx 20 - 22$ns downstream from the shock for the 1.25μm and 2.5μm resolution data. However, weak long range correlations (corresponding to $|\mathrm{ACF}(r, d)| > 0$) become more prominent near the shock front from $t \approx 14$ns onwards. Analysis of the 5μm resolution data demonstrates longer correlation lengths and stronger long range correlations.

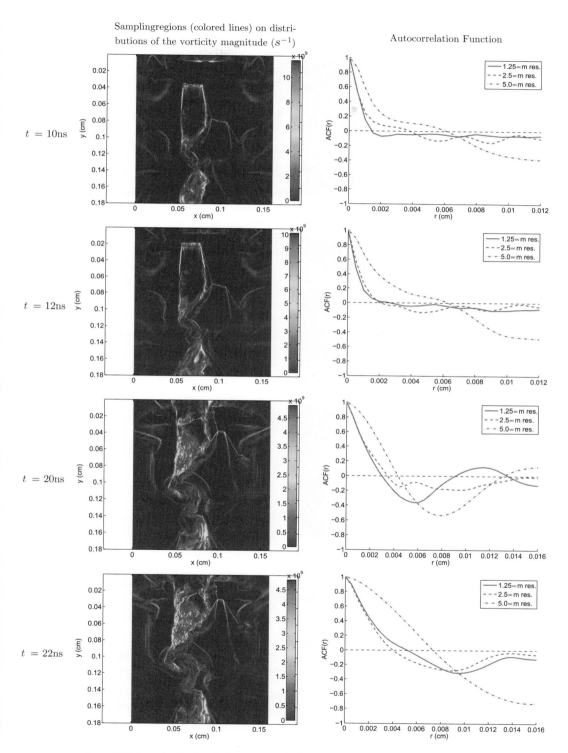

Figure 9.18 Longitudinal velocity autocorrelation functions $ACF(r)$ for the shear experiment, defined in Equation (9.2.6), for 3D simulation data. Sampling regions are shown inside red boxes.

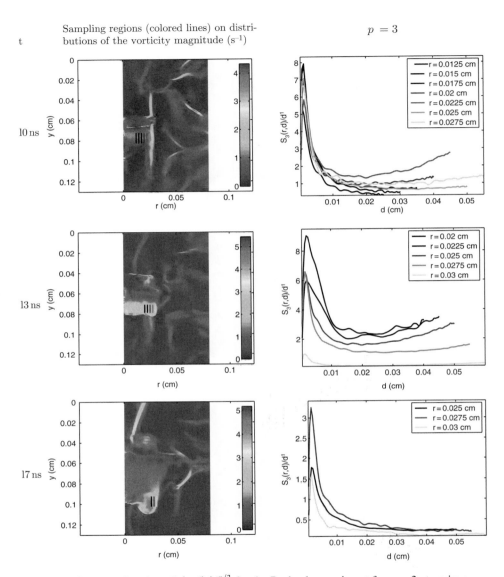

Figure 9.19 Structure functions $S_p(r,d)/d^{p/3}$ for the Reshock experiment for $p = 3$ at various radii for the 2.5μm resolution 3D data. The colored bars are cross-sections of the annular sampling regions.

9.2.5.4   Velocity Structure Functions

Velocity structure functions are used next to determine length scales over which a flow is self-similar. For the reshock simulations, we define the velocity structure functions as

$$S_p(r,d) := \frac{1}{2\pi|y_{\max} - y_{\min}|} \int_{y_{\min}}^{y_{\max}} \int_0^{2\pi} |u(r, \theta + \phi, y) - u(r, \theta, y)|^p \, d\phi dy,$$

where $d = \frac{\sin\theta}{\cos\frac{\theta}{2}} r$ and u takes polar arguments (r, θ, y). These are shown in Figure 9.19 for $p = 3$ using the 2.5μm resolution 3D data, plotted as $S_p(r,d)/d^{p/3}$. For length scales over which a flow is self-similar, the structure functions should exhibit the power law behavior

$S_p \propto s^{\sigma_p}$ where $\sigma_p = hp$. The Kolmogorov theory predicts $h = \frac{1}{3}$ [31], though laboratory experiments indicate this starts to break down for large p (see, e.g., [32] for a review).

At $t = 10$ns (immediately before reshock), a short range of self-similarity has developed, indicating that turbulent kinetic energy is only beginning to cascade to smaller scales. This range then grows, and by $t = 17$ ns, the 2.5μm resolution data indicates a full range of self-similarity, starting from ≈ 0.005cm (20 cells). For the 5μm resolution data (not shown), this threshold occurs at ≈ 0.01cm (also 20 cells).

We do not calculate velocity structure functions for the shear simulations since the turbulent regions in the experiment do not have a large enough spatial extent to observe self-similar behavior given the resolution of the simulations. This is consistent with the lack of convincing evidence, using other presented metrics, that the mixing transition has been achieved for this experiment.

9.2.5.5   Velocity Variances

Figure 9.20 shows the variance of each component of the velocity as a function of y for the 3D reshock data. For isotropic flows, each component of the velocity will have a similar variance. We calculate the variances as

$$\mathrm{var}[u_i](y) := \frac{r_i}{\sum_j \langle r_j \rangle_y},$$

where

$$r_i := \left\langle \left\langle (u_i - <<u_i>_\theta>_r)^2 \right\rangle_\theta \right\rangle_r,$$

and $\langle \cdot \rangle_i$ denotes averaging over the ith coordinate. The reshock can be observed as a spike in the variance of u_r. The flow immediately becomes more isotropic in the wake of the shock, which passes through the mixing layer from $t \approx 10 - 17$ ns. The flow is most isotropic at $t \approx 17$ ns, when the shock has left the mixing layer.

It is notable that the variance and structure function analyses indicate that the flow is most isotropic and self-similar at $t = 17$ ns. Nevertheless, as measured by the integrated TKE and vorticity CDFs, the turbulence has decayed significantly by this point. This suggests that isotropy takes longer to develop than other turbulent characteristics such as self-similar vorticity distributions and decorrelation with strong fluctuations.

Figure 9.21 shows the variance of each component of the velocity as a function of y for the 5μm resolution 3D shear data (variances for the 1.25μm and 2.5μm data are unavailable since we performed an ensemble of simulations only for the 5μm resolution data). For isotropic flows, each component of the velocity will have a similar variance. We calculate the variances as

$$\mathrm{var}[u_i](y) := \frac{\langle r_i \rangle_{x,z}}{\sum_j \langle r_j \rangle_{x,z}} \qquad (9.2.7)$$

where

$$r_i := \left\langle (u_i - \langle u_i \rangle_{\mathrm{ens}})^2 \right\rangle_{\mathrm{ens}},$$

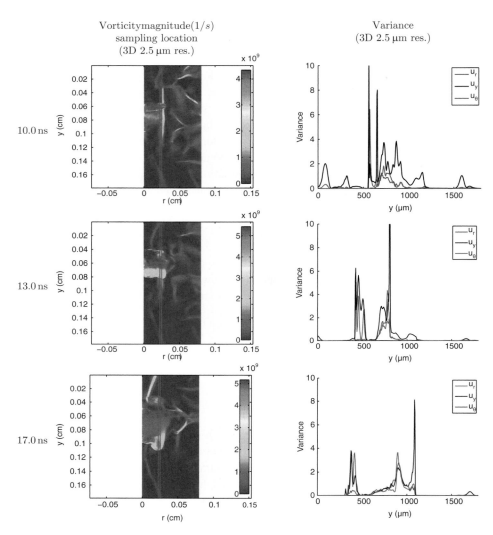

Figure 9.20 Velocity variances for the reshock experiment as a function of y for the three velocity components. The red lines outline cross-sections of the annular sampling regions.

$\langle \cdot \rangle_{ens}$ denotes ensemble averaging, and $\langle \cdot \rangle_{x,z}$ denotes averaging over the x and z coordinates. The flow is highly anisotropic at $t = 8ns$ and slowly becomes more isotropic through $t = 22ns$. In the reshock experiment, the flow does not become appreciably isotropic until after it has been reshocked. In the shear experiment, the flow is not reshocked, and hence isotropy takes much longer to develop.

9.2.5.6   Effective Reynolds Number Analyses

The spectral bandwidth of fully developed turbulence can be scaled by a turbulence Re, usually taken as a ratio of an outer (integral) scale and the Taylor microscale [33].

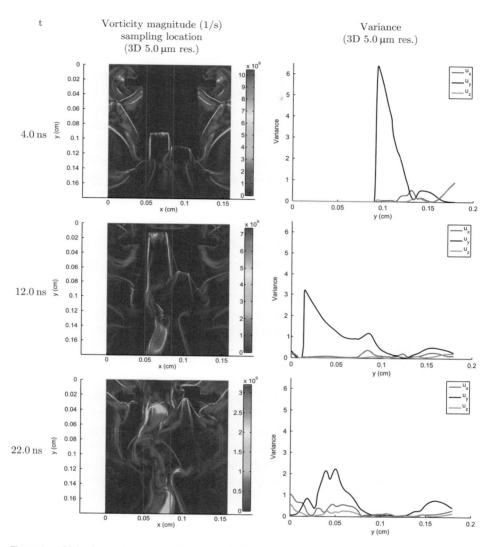

Figure 9.21 Velocity variances as a function of y for the three velocity components for the shear experiment, as defined in Equation (9.2.7). The red lines contain cross-sections of the sampling regions.

Transition to turbulence is traditionally viewed in terms of rapid increase in population of small scale motions and more isotropic features (e.g., [34], and references therein). The Taylor microscale characterizes the largest length scale at which viscosity affects the dynamics of turbulent eddies [33]. A quantitative estimate of an effective Taylor microscale Reynolds number $\mathrm{Re}^{\lambda}_{\mathrm{eff}} = U\lambda/\nu_{\mathrm{eff}}$ can be obtained for the reshock simulation data at times in the approximate interval 18–19 ns as follows [35] – see also Chapters 1 (Sec. 1.3) and 2 (Sec. 2.5.2). A mean velocity fluctuation can be evaluated as $U = \sqrt{2\langle k \rangle_V/3}$, where volumetric averages are denoted by $\langle \cdot \rangle_V$. We can compute a mean Taylor microscale λ by averaging its value over spatial directions $\lambda_i = \sqrt{U^2/\langle(u_{i,i})^2\rangle_V}$, estimate the dissipation as $\varepsilon \sim \langle -dK/dt\rangle_V$, where K is the kinetic energy, and formally compute an effective viscosity ν_{eff} in terms of this measure of dissipation and the volumetric averaged strain rate magnitude as $\nu_{\text{eff}} = \varepsilon/2\langle S_{ij}S_{ij}\rangle_V$, where $S_{ij} = (u_{i,j} + u_{j,i})/2$ is the strain rate tensor, as in [27]. In order to actually carry out these computations we recall the noted basic differences with homogeneous isotropic turbulence. Sampling of reshock simulation data focuses on flow regions containing high TKE levels and/or vorticity (the sampling region is chosen manually as a region that includes the most turbulent flow and then filtered by vorticity content), which (as noted) are neither homogeneous nor isotropic. Weak vorticity distributions associated with the sampled transitional data are very distinct from the isotropic homogeneous data due to their irrotational flow content. Data filtering based on vorticity content was used in order to have meaningful comparative analysis, and was similarly used here for all volumetric averages; we thus obtain $\lambda \approx$ 3–4 cells and $\text{Re}^\lambda_{\text{eff}} \sim 80$ for the 2.5μm resolution, and $\lambda \approx$ 5–6 cells and $\text{Re}^\lambda_{\text{eff}} \sim 100$ for the 1.25μm resolution. Due to the lack of effective means for obtaining averaged quantities for the high resolution shear data, this analysis was not carried out for that experiment.

Asymptotic estimates for an outer scale effective Re can be generated as in [35]. Based on dimensional grounds, a dimensionless ratio, D, can be introduced as $D := \frac{\varepsilon L}{U^3}$, where L and U are characteristic outer length and velocity scales, respectively (as used in the definition of Re). Recently, a significant amount of work has been devoted to investigate the behavior of D as a function of Re (see [36] and references therein). A large body of experimentally and numerically generated data suggested that D approaches a constant $D \to \frac{1}{2}$ when Re becomes sufficiently large – specifically, for $Re_\lambda \gtrsim 100$ near the mixing transition [29]. In our present context, we thus assume that high Re isotropic turbulence regimes for which $D = \frac{1}{2}$ have been achieved for the 1.25μm and 2.5μm 3D shear data, and define a time dependent L to be the outer scale prescribed by a mixing width δ_{MZ}. As in [37], the material mixing zone thickness δ_{MZ} is defined in terms of the mixedness $M(z)$ by

$$\delta_{\text{MZ}} = \int M(z)dz, \qquad (9.2.8)$$

where

$$M(z) = 4\psi(z)(1 - \psi(z)),$$
$$\psi(z) = \frac{1}{A}\int c_{\text{Al}}\, dxdy,$$

and $A = 6.5025 \times 10^{-3}\text{cm}^2$ is the area of the top of the Al tracer. We thus formulate an asymptotic model for the dissipation $\varepsilon = \frac{U^3}{2L}$, where we use $U = \sqrt{2\langle\text{TKE}\rangle_V/3}$, where $\langle\cdot\rangle_V$ denotes volume averaging (with the volume averaged TKE taken from the data in Fig. 9.14), and data filtering based on vorticity content; in this context we can evaluate outer scale (asymptotic) measures of $\nu_{\text{eff}} = \varepsilon/2\langle S_{ij}S_{ij}\rangle_V$, where $S_{ij} := (u_{i,j} + u_{j,i})/2$ is the rate of strain, and $\text{Re}_{\text{eff}} = \frac{UL}{\nu_{\text{eff}}}$. The outer scale based asymptotic measures of mixing width and effective outer scale Re are plotted in Figure 9.22 for the 1.25μm and 2.5μm

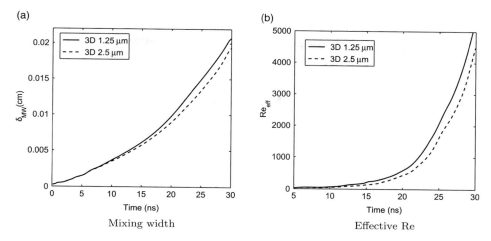

Figure 9.22 Reshock experiment effective Reynolds number analysis. (a) Mixing width measure δ_{MZ}, as defined in Equation (9.2.8), for the 3D data. This is used as an outer scale length L. (b) Asymptotic outer scale $Re_{eff} := \frac{UL}{\nu_{eff}}$, where L is taken from (a), $U := \sqrt{2\langle TKE \rangle_V / 3}$, where TKE is defined in Equation (9.2.4), $\nu_{eff} := \varepsilon / 2\langle S_{ij}S_{ij} \rangle_V$, S_{ij} is the strain tensor, and $\varepsilon = U^3 / 2L$ is the asymptotic dissipation.

data (the $5\mu m$ data is omitted since the assumption that $D = 1/2$ is incompatible with the turbulence metrics that suggest the mixing transition is far from being achieved). On the other hand, values for Re_{eff} between 10^3 and 5×10^3 are consistently obtained for $1.25\mu m$ and $2.5\mu m$ after ≈ 20ns, approaching the (integral scale based Re) mixing transition threshold values (between 10^4 and 2×10^4) from [29]. While below the value of 2×10^4 estimated for experiments, these numbers are likely underestimates due to the fact that turbulence is only found in small, isolated regions of the flow.

Using the aforementioned analysis for the reshock problem, values for Re_{eff} between 10^4 and 5×10^4 are consistently obtained between $\approx 14 - 20$ ns, above the (integral scale based Re) mixing transition threshold values (between 10^4 and 2×10^4) from [29] and roughly consistent with simulation aided[1] experimental estimates of $Re = \frac{\delta_{MZ}\dot{\delta}_{MZ}}{\nu} \approx 1.6 \times 10^5$, which is typical for laser driven experiments such as these.

9.3 Inertial Confinement Fusion Simulations

Inertial confinement fusion (ICF) uses a uniform laser or x-ray drive to heat a spherical shell, which then ablates, generating a reactive force that implodes fuel, typically a mixture of deuterium and tritium, inside the shell. The convergence of the fuel raises its pressure, causing a deceleration of the shell and converting its kinetic energy into internal energy in the fuel and thus initiating fusion reactions [38]. Several phases of the implosion are

---

[1] While the mix width δ_{MZ} can be calculated in experiment and matched to simulation [1], ν must be determined from plasma viscosity formulae using state information from simulation.

known to be subject to hydrodynamic instabilities. The interface between the shell and surrounding gas is RT unstable as the shell material ablates. In addition, the interface between the shell and fuel is RM unstable as a shock generated by the ablasion exits the shell, as well as during subsequent interactions between the shell and reflected shocks. This interface is also RT unstable when the shell decelerates. These turbulent instabilities are seeded by asymmetries in the drive and on the surfaces of material interfaces. Understanding turbulent instability growth and corresponding material mixing during the implosion and burn of ICF capsules is important for determining their performance. Indeed, turbulence development may displace fuel in the hot spot (the hot, dense core of the implosion where a majority of the fusion reactions occur) and cause mixing of cold shell material with the fuel; these effects may degrade capsule yield significantly. Nevertheless, accurate prediction of turbulence development and the amount of material mixing that occurs during an implosion is hindered by our present computational capabilities.

Performing simulations of ICF capsules is a particularly difficult task due to the range of physics and scales involved. Using present computational power, it is necessary to either ignore relevant physics in the problem or reduce the dimensionality in order to successfully complete a simulation of an ICF implosion. It is well known, however, that reduced dimension (1D and 2D) hydrodynamic simulations significantly underpredict their growth of various hydrodynamic instabilities due to the absence of vortex stretching effects (see, e.g., [12]). An intermediate approach, pioneered by Thomas and Kares in [39], involves starting the simulation in 2D then rotating it to 3D once sufficient convergence of the ICF capsule has been achieved. Using this approach, Thomas and Kares demonstrate the effects of drive asymmetry on the development of turbulence in the implosion through Bell–Plesset related convergence effects [60], in which perturbations of an incompressible fluid interface grow during an implosion as a consequence of mass conservation. In the present chapter we qualify and generalize the use of such an approach in order to study the impact of surface defects and drive asymmetry in promoting the development of turbulent instabilities and associated material mixing, as well as their effects on capsule yield.

In order to address the possible limitations of the use of a 2D–3D rotation strategy, we employ this approach to perform simulations of the laser driven reshock experiment outlined in Section 9.2.1 for which fully 3D simulation data are available (see Section 9.2.2). The reshock experiment involves similar physics to that in an ICF implosion but does not involve the convergence (implosion) physics, which makes the problem tractable for performing a fully 3D simulation. We show that, when (nominally) 2D initial conditions are used as in the pure 2D–3D data rotation, the growth of physical 3D modes is significantly inhibited and the simulation does not develop sufficient turbulence nor turbulent mixing to match the fully 3D simulations – despite the presence of grid scale azimuthal perturbations of the 2D rotated data. In contrast, we find that we can generate turbulence and mixing that satisfactorily compare to that in the fully 3D simulation when additional (azimuthal) perturbations are applied to the 2D–3D mapping function at the time of rotation. These perturbations compensate for the reduced growth of turbulent instabilities during the 2D phase of the simulation as well as the artificial smoothing of the data in the azimuthal direction caused by the 2D–3D rotation.

Reduced dimension simulations are known to overpredict the yield of ICF capsules compared to experiment (see, e.g., [40–43], among many others). Many explanations have been suggested – for example, turbulent mixing seeded by drive asymmetry [39, 44–46], the depletion of high energy ions due to their long mean free paths [47], and the presence of self-generated electric and magnetic fields that are not typically modeled in simulations [48]. While reduced dimension simulations can be tuned to achieve reasonable results [44, 45, 49, 50], they do not provide predictive capability. Furthermore, three-dimensional ICF simulations that have been performed [39, 51] have not considered or adequately resolved the effects of long wavelength asymmetries, which have the largest amplitudes [43, 45]. Here, we use our newly developed simulation strategy to demonstrate that turbulent instabilities seeded by long wavelength surface defects and short wavelength drive asymmetries are sufficient to explain the discrepancies between the yield predicted by previous reduced dimension simulations of ICF implosions and experiments. Our simulations indicate that this is achieved primarily through the displacement of fuel away from the hot spot by shell material. This provides evidence for a recent hypothesis in [50], which suggests that long wavelength asymmetries in the x-ray drive of a National Ignition Facility (NIF) capsule may be a significant source of distorted hot spot shapes and yield degradation. It also supports the supposition in [52] that escaping fuel mass may be the cause of discrepancies between the size of the fuel assembly as predicted by 2D simulations and observed experimentally.

9.3.1   Simulation Strategies

The present simulations were performed using an ILES strategy [10] based on LANL's RAGE hydrodynamics code [13] discussed earlier. Radiation transport is neglected in the reshock problem, as the ratio of typical radiative to hydrodynamic fluxes can be estimated as $\sigma T^4/(\rho U^3) \approx 10^{-4}$, where σ is the Stefan–Boltzmann constant and $T = 10\text{eV}$, $U = 80\text{ km/s}$, and $\rho = 0.2\text{ g/cc}$ are characteristic of the experiment. For the ICF problem, Gray radiation diffusion is used along with a three temperature (ion, electron, and radiation) treatment as well as electron and ion heat conduction.

RAGE uses a second-order Godunov finite volume scheme and AMR using cubic cells, as well as a variety of numerical options. In the present simulations, we use a van Leer limiter, no material interface treatment, and, as in our recent work [12], a particular artificial diffusion option meant to compensate for anisotropic errors due to directional splitting was not activated. As used in the present work (based on nominally inviscid equations) RAGE models miscible material interfaces and high Reynolds number convection driven flow with an effective (numerical) viscosity determined by the small scale cutoff and residual numerical diffusion of the simulation algorithm. Details of the 2D–3D simulation procedure varied between the reshock and ICF problems due to the differing geometries of the problems, and are provided in the following relevant sections. The ILES simulation strategy as implemented in RAGE has been used extensively to model relevant high energy density physics experiments and has been validated in this context through both code comparison work [53], involving codes with a variety of simulation strategies, and direct comparison to experiment [2, 3, 54].

For the ICF simulations, the new Singe package in RAGE was used to simulate thermonuclear reactions. This package tracks material isotopes and calculates thermonuclear reaction rates. In each computational cell, these obey the formula $f = \frac{n_1 n_2 \langle \sigma v \rangle}{V_{\text{eff}}}$, where f is the reaction rate, n_i is the number of reactants of the ith species, v is the relative velocity of two reactants, $\langle \cdot \rangle$ indicates averaging over an effective velocity distribution σ as a function of T_{ion}, and V_{eff} is the effective volume. The value of f is halved for reactions involving a species of nuclei reacting with itself. The effective volume V_{eff} can be calculated in two limits to obtain bounds on the reaction rate: a "clean" limit and a fully atomically mixed limit. In the former limit, V_{eff} is set to be the volume of the reactants in each cell. In the atomically mixed limit, V_{eff} is set to the volume of the computational cell. The true burn rate lies somewhere between these limits and varies between cells; we present these limits in the absence of any information about the distribution of materials at a subgrid scale level. Singe assumes that all charged particle reaction products are deposited locally and that neutrons are lost from the system.

ICF simulations were performed three different maximum resolutions: $1/2\mu$m, $1/4\mu$m, and $1/8\mu$m. The $1/2\mu$m and $1/4\mu$m simulations were run on LANL's Cielo supercomputer, employing a maximum of 1.1 billion and 6.4 billion cells, respectively, and a total of 0.01 and 0.2 Cielo days. The number of processors used was varied as the simulations progressed; the $1/2\mu$m simulations used a maximum of 4,096 processors and the $1/4\mu$m simulations used a maximum of 16,384 processors. The $1/8\mu$m simulation was run in the 2D phase on Cielo, using 0.14 Cielo days, then completed in 3D on LLNL's Sequoia supercomputer, using a further 1.1 Sequoia days with as many as 131,072 processors and a maximum of 36 billion cells.

9.3.2   Testing and Validating the 2D–3D Strategy: Reshock Experiment

We validate our 2D–3D strategy using the previously presented 3D simulations of the laser driven reshock experiment.

9.3.2.1   2D–3D Mapping and Flow Initialization Strategies

In all cases we performed the 2D–3D mapping at 10 ns, immediately before reshock. First, we examined the impact of changing the surface roughness amount specified by the function $\zeta(x, z)$ defined in Equation (9.2.2) and multiplying it by of χ, in the context of the unperturbed mapping – that is, pure data rotation, in which all state and velocity variables were set using the mapping

$$\begin{Bmatrix} x \\ y \end{Bmatrix} = \begin{Bmatrix} \sqrt{x'^2 + z'^2} \\ y' \end{Bmatrix},$$

where (x, y) are the coordinates for the 2D problem and (x', y', z') are the coordinates for the 3D problem. Vector valued quantities, such as the velocity, were also appropriately rotated. Next, we also considered simulations where the 2D–3D mapping itself was modified using the perturbation function $\zeta(x,z)$ defined in Equation (9.2.2) and multiplying it by χ. Specifically, the perturbed mapping was defined as

$$\begin{Bmatrix} x \\ y \end{Bmatrix} = \begin{Bmatrix} \sqrt{x'^2 + z'^2} \\ y' + \zeta(x', z') \end{Bmatrix}.$$

9.3.2.2 Hydrodynamic Instability Growth in the Reshock Experiment

Turbulence generation in the reshock experiment is associated primarily with the RM instability (RMI). Richtmyer [55] derived an expression for the growth of a perturbation subject to an incident shock wave:

$$\frac{da}{dt} = kAa_0 u_c, \tag{9.3.1}$$

where k is the wavenumber, $A = \frac{\rho_0 - \rho_1}{\rho_0 + \rho_1}$ is the post shock Atwood number, $a_0 = \frac{1}{2}(a_{0-} + a_{0+})$ is the initial amplitude (a_{0-} is preshock and a_{0+} is postshock), and u_c is the change in interface velocity due to the shock. Using values appropriate to the reshock experiment ($k = 4\pi/460\mu m$ corresponding to the longest wavelength IC perturbation, $A \approx 0.97$ and $u_c \approx 10^7 \text{cm/s}$), we can obtain an estimate of $a(10\text{ns})/a_0 \approx 30$, which provides a rough estimate of the growth factor of the interface perturbations at the time of 2D–3D rotation. Formula (9.3.1) is valid in the linear growth regime for RMI – in particular, experimental and numerical comparisons in [56] suggest it is valid for $ka_0 < 1$. For the values used here, we have $ka_0 \approx 0.007$.

The analysis considers the growth of only the longest wavelength perturbation. It is likely that since the longer wavelengths have the largest amplitude (see Fig. 9.4), these dominate the hydrodynamic development of the experiment, so it is most important to capture their growth correctly. Nevertheless, our initial perturbation spectra includes a wide range of modes, many of which are outside of the linear regime $ka_0 < 1$. As a further complication, the growth of these modes is, in reality, coupled. A rationale for justifying the relevance of this estimate comes from our recent RMI simulation work, [30, 37, 57] (see, e.g., fig. 9 in [37] and fig. 5 in [57]). Shock driven flows are typically investigated using the early time relevant RMI model for a first shocked relatively flat and thin material interface modeled by the Richtmyer formula [55]. However, for most classes of shock induced mixing, reshocks, which happen at a later time and involve morphologically more complex interfacial layers and fluid instability mechanisms distinctly different from the classical RMI, are more important phenomena [57]. Before reshock, growth tends to follow Richtmyer's formula – shorter wavelengths grow faster, whereas after reshock actual growth features get inverted and the dominant modes are the longer wavelengths reflecting what we have called the bipolar behavior of the RMI [30].

9.3.2.3 Augmenting Initial Perturbation Level in 2D–3D Pure Rotation Context

In this section, we will show results from performing 2D–3D simulations where the initial 2D surface perturbations are augmented by a factor χ. We will demonstrate that this method does not produce satisfactory results in comparison with purely 3D

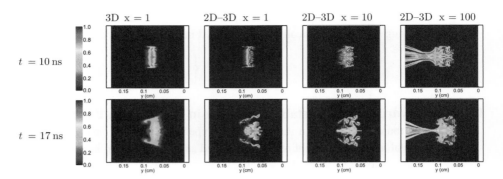

Figure 9.23 c_{Al} plots for simulations of the reshock experiment; here, 2D–3D perturbations are applied in 2D at $t = 0$.

simulation data. Therefore, this method *will not* be used subsequently for ICF simulations. However, we present these results for completeness and as a caveat to others who may want to try it.

In Figure 9.23, we show 2D plots of the aluminum concentrations c_{Al} for the 2D–3D simulations (azimuthally averaged for $t \geq 10$ns) compared with azimuthally averaged values for the purely 3D simulation. The magnitude of the initial interface perturbations was kept fixed ($\chi = 1$ in Equation (9.2.2)) for the 3D simulation and varied for the 2D–3D simulations ($\chi = 1, 10, 100$). Note that in the $\chi = 100$ case (experimentally observed perturbations enhanced by a factor of 100), the aluminum tracer is fragmented from the perturbations at $t = 0$. In all of the 2D–3D cases presented here, the aluminum is significantly more concentrated and fragmented than in the 3D case, and no satisfactory comparison is obtained.

In Figure 9.24, we show a volumetrically integrated measure of the density weighted velocity variance – turbulent kinetic energy TKE – as a function of time. We define the local velocity variance TKE as

$$K(y, r) := \frac{\left\langle \frac{1}{2}\rho \left(\left(u_r - \widetilde{u}_r\right)^2 + \left(u_y - \widetilde{u}_y\right)^2 + u_\theta^2 \right) \right\rangle_\theta}{\langle\rho\rangle_\theta}, \qquad (9.3.2)$$

where $\langle\cdot\rangle_\theta$ denotes azimuthal averaging and $\widetilde{u}_i := \langle u_i\rho\rangle_\theta / \langle\rho\rangle_\theta$. The 2D–3D simulations presented all significantly underpredict the integrated TKE when compared with the purely 3D simulation.

In Figure 9.24, we also show an integrated mixing measure as a function of time, defined by Equation (9.2.5), where azimuthally averaged fields are used when the data are 3D and the integration is performed over the entire 2D computational domain (thus allowing us to compare the results even when the 2D–3D data has not yet been rotated to 3D). The 2D–3D simulations all significantly underpredict the amount of mixing when compared to the 3D data.

By all metrics considered, the 2D–3D simulation strategy is insufficient to obtain reasonable comparisons to 3D simulation data when augmenting initial perturbation

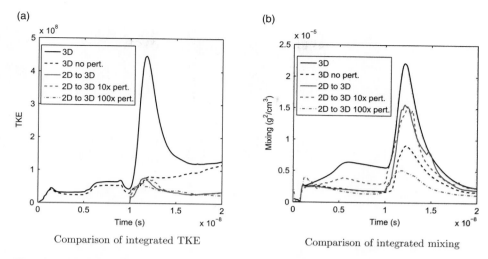

Figure 9.24 Integrated flow quantities for the reshock simulations; here, 2D–3D perturbations are applied in 2D at $t = 0$.

level of 2D ICs in 2D–3D pure rotation context. This is likely due to the fact that increased turbulence generation during the 2D phase of the simulation is offset by the artificial azimuthal smoothing of the 2D–3D rotation procedure.

9.3.2.4 Generalized 2D–3D Mapping with Perturbations

In this section, we show results for 2D–3D simulations where the 2D–3D mapping is additionally perturbed (azimuthally) based on the initial surface roughness. We will show that this method produces satisfactory results compared with purely 3D simulation data when the mapping perturbations are set at a factor of $\chi = 30$ times the initial surface roughness level. This corresponds to the approximate RMI growth rate of the longest wavelength mode between $t = 0$ and $t = 10$ ns, the time at which the simulation is rotated to 3D.

In Figure 9.25, we show 2D plots of the aluminum concentrations c_{Al} for the 2D–3D simulations (azimuthally averaged for $t \geq 10$ ns) compared to azimuthally averaged values for the purely 3D simulation. In all cases, the initial aluminum tracer interfaces were perturbed using experimental data (i.e., $\chi = 1$). In addition, for the 2D–3D data, the state and velocity variables were perturbed upon rotating the data to 3D at $t = 10$ ns, using the method outlined in Section 9.3.2.1 with $\chi = 1, 10, 30$. A satisfactory comparison is obtained for the case $\chi = 30$, corresponding to the long wavelength RMI growth rate up to the time of rotation. For the cases $\chi = 1, 10$, the aluminum concentration is significantly fragmented and overly concentrated compared to the 3D data.

In Figure 9.26, we show a comparison of a density weighted volumetrically integrated measure of the velocity variance, defined in Equation (9.3.2). The peak TKE value increases nonlinearly with the magnitude of the interface perturbations (χ). For the case $\chi = 30$, a reasonable comparison is obtained with the 3D data.

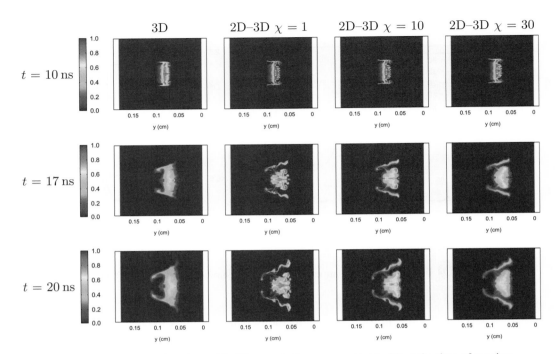

Figure 9.25 c_{Al} plots; here, 2D–3D perturbations are applied in 3D at the time of rotation at $t = 10$ ns.

In Figure 9.26, we also show a comparison of an integrated mixing measure defined in Equation (9.2.5). The peak mixing value increases nonlinearly with the magnitude of the interface perturbations (χ). For the case $\chi = 30$, a good comparison is obtained with the purely 3D data.

In Figure 9.27, we show CDFs of the vorticity magnitude ω/ω_{RMS}, where ω_{RMS} is the root mean square vorticity, for 2D–3D simulations and compare them with those corresponding to DNS of isotropic turbulence data from Jiménez et al. [26] and our purely 3D simulations [2]. CDF(ω) – the integral of PDF(ω) – represents the volume fraction with vorticity magnitude greater than ω. The CDFs associated with the DNS data tend to approach a high Re_λ limit when Re_λ is above the mixing transition threshold $Re_\lambda \approx 100 - 140$ [29, 34], where Re_λ is based on the Taylor microscale λ. Of the 2D–3D simulation data, the case with $\chi = 30$ compares best with the purely 3D data, with almost negligible differences at $t = 12$ns. The quality of the comparison degrades only slightly by $t = 17$ns.

By all measures considered, a good comparison can be obtained with purely 3D data by performing a 2D–3D simulation where the 2D–3D mapping is perturbed using the experimentally observed initial condition perturbations enhanced by a factor of $\chi = 30$, corresponding to the RMI growth rate of the longest wavelength mode up to the time of rotation.

9.3.2.5   Effects of Simulating a Quadrant of the Reshock Problem

Here, we assess the possible effects of simulating only a quadrant of the reshock problem – which is a reasonable approximation given the rotational symmetry of

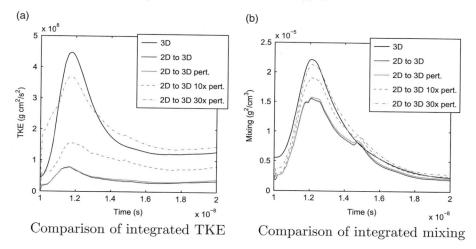

Comparison of integrated TKE   Comparison of integrated mixing

Figure 9.26 Integrated flow quantities for the reshock simulations; here, 2D–3D perturbations are applied in 3D at the time of rotation at $t = 10$ ns.

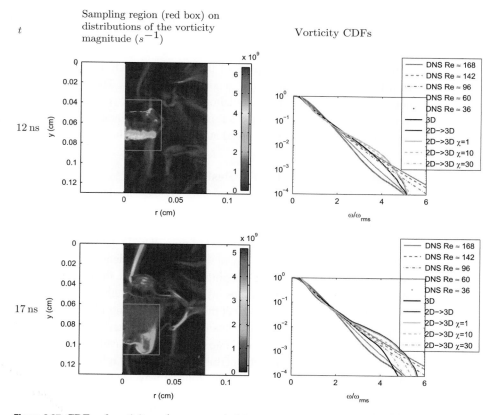

Figure 9.27 CDFs of vorticity $\omega/\omega_{\mathrm{RMS}}$ sampled from a uniform 5μm grid. Sampling regions (within red lines) on distributions of vorticity magnitude (s^{-1}). The DNS data were taken from [26].

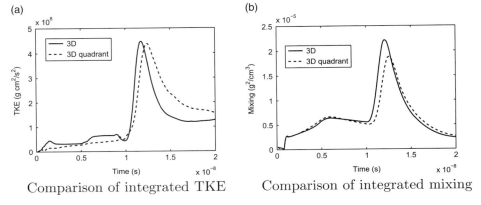

Figure 9.28 Integrated flow quantities for the reshock simulations.

the problem. Nevertheless, this can introduce inaccuracies due to the fact that there are differently allowed azimuthal constraints and nonlinear mode couplings when only a single quadrant of the reshock problem is involved. In order to assess the possible limitations effects of simulating a quadrant of the problem, we performed both single quadrant and full 3D simulations at $2.5\mu m$ and compared the results.

In Figure 9.28, we show plots of the integrated TKE and mixing, respectively, for both the full 3D and 3D single quadrant simulations. The differences are minor but noticeable. For the quadrant simulation, the TKE peak is delayed and the mixing peak diminished slightly. Moreover, TKE peak values (and hence TKE production mechanisms) are found to be well resolved at this $2.5\mu m$ resolution [35].

9.3.2.6   Summary of Results for the Reshock Problem

We considered two methods of perturbing 2D–3D simulation data in order to enhance turbulence generation and turbulent material mixing seeking to better match the fully 3D simulation data. For the first method, we enhanced the initial interface roughness by factors of $\chi = 1, 10, 100$ in the context of pure 2D–3D data rotation, but found that this did not improve our results by any of the metrics considered. For the second method, we perturbed the 2D–3D rotation mapping used to initialize the state and velocity variables. We used the same perturbations as in the first method, enhanced by factors of $\chi = 1, 10, 30$, to also directly affect the 2D–3D mapping function. We found that good matches to the 3D data can be obtained with the latter strategy when using $\chi = 30$, corresponding to the RMI growth rate of the longest wavelength mode up to the time of rotation. It is likely because the perturbation spectrum relevant to instability growth in the problem (see Fig. 9.4) peaks at the longest wavelengths. This is also the case in the ICF experiments we consider in the following (see Section 9.3.3.1). Nevertheless, for cases in which long wavelength perturbations are not dominant, it remains an open question as to whether this method is adequate.

Figure 9.29 Diagram of the ICF capsule.

20 μm CH

420 μm D
0.27 μg

9.3.3   ICF Simulations

We perform simulations of the implosion of a simple ICF capsule with a polystyrene shell of width $d = 20\mu m$ and outer radius $r = 437.5\mu m$ filled with $0.27\mu g$ of deuterium gas (see Fig. 9.29). This models a number of capsules that have been imploded at the University of Rochester's OMEGA laser facility. These capsules are driven by a direct drive approach using all 60 beams of the OMEGA laser. In order to study the relative importance of drive asymmetry and surface defects to the degradation of yield, we perform simulations with both a uniform drive and drive asymmetries modeling those observed when using the OMEGA laser [58]. In both cases, a total of 24 kJ is deposited directly into the outer polystyrene shell over a period of approximately 1.35 ns.

In what follows, we discuss surface roughness and drive asymmetries in Section 9.3.3.1. Next, in Section 9.3.3.2, we discuss the growth of hydrodynamic instabilities seeded by these sources of asymmetry. We use this to inform a 2D–3D perturbation strategy similar to that used in the reshock simulations, which is described in detail in Section 9.3.3.3. Results of simulations using our 2D–3D strategy are discussed and compared to standard 2D–3D strategies, reduced dimension simulations, and experimental data in Section 9.3.3.4.

9.3.3.1   Initial Material Interface Conditions and Drive Asymmetries

Interface perturbations are applied initially on the inner and outer surfaces of the polystyrene shell. The interface perturbations are applied radially as a sum of spherical harmonics:

$$\zeta(\theta, \phi) = \chi \Gamma \sum_{l=1}^{5} \sum_{\substack{-l \leq m \leq l \\ \sqrt{l^2 + m^2} \leq 5}} a_{l,m} e^{im\phi} P_l^m(cos(\theta)), \qquad (9.3.3)$$

where P_l^m is an associated Legendre polynomial, $\{a_{l,m}\}$ are set so the perturbations have the outer surface spectrum in Figure 9.30 (defined in [43]), and Γ is set so that the standard deviation of the perturbations is $1\mu m$. Note that we only include modes with $l \leq 5$ due to the sharp drop in the perturbation spectrum beginning at $l \approx 5$. The constant $\chi = 1$ for interface perturbations and $\chi = 10, 20$ for 2D–3D mapping perturbations, discussed in the following.

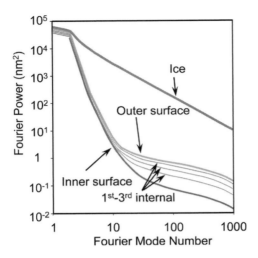

Figure 9.30 Approximate surface roughness power spectra for the various surfaces in an ICF capsule. From [43].

We model the drive asymmetries in the implosion on work done by Marshall et al. [58] profiling the drive of the OMEGA laser. Specifically, we apply a P_{30} perturbation with 1.2% standard deviation to the ablator layer where energy is deposited to model the drive at $t = 0$. We do not model beam imbalances in the present simulations. However, the spectrum of beam imbalances is similar to that of surface defects [45], so some of the effects of long wavelength asymmetries that we attribute to surface defects may be caused by beam imbalances in experiment.

9.3.3.2   Hydrodynamic Instability Growth in an ICF Capsule

The dominant hydrodynamic instability during an ICF implosion is the RT instability (RTI), which occurs at early times as ablator material accelerates into the surrounding gas and at "stagnation," when the shell decelerates due to increased fuel pressure. The fuel ablator interface is also subject to the RMI as the shock exits the shell at early times and at later times when reflected shocks interact with the fuel/shell interface. In addition, perturbations grow due to Bell–Plesset effects (see, e.g., [60]), in which incompressible fluid perturbations grow in a converging geometry as a consequence of mass conservation.

For the classical RTI in planar geometries, perturbations with wavenumber k grow exponentially [59] as $a(t) = a_0 \exp(\sqrt{Agk}t)$, where $A = \frac{\rho_0 - \rho_l}{\rho_0 + \rho_l}$ is the Atwood number and g is the acceleration. In a converging spherical geometry, this equation must be modified to include Bell–Plesset convergence effects, whose growth is coupled to the RTI growth [60], yielding

$$a(t) = a_0 \exp\left(\int_0^t \gamma_0(s)\, ds \right), \tag{9.3.4}$$

where

$$\gamma_0^2(t) = \frac{l(l+1)}{R} \frac{\rho_0 - \rho_1}{l\rho_0 + (l+1)\rho_1} g. \tag{9.3.5}$$

Here, l is the Legendre mode number, and R, ρ_0, ρ_1, and g are time dependent quantities. Aided by 2D calculations, we can estimate, taking the maximum over the dominant modes $1 \le l \le 5$,

$$\chi = \frac{a(1.4\text{ns})}{a_0} \approx 15.$$

Shorter wavelength modes are neglected due to their vanishingly small amplitudes, as seen in Figure 9.30.

This estimate neglects feed through effects from the ablative RTI, RMI effects, mode coupling, and the spatial nonuniformity of instability growth in an ICF implosion. Without purely 3D data to compare, there is significant uncertainty in the estimate. Therefore, we perform simulations using $\chi = 10$ and $\chi = 20$ in order to obtain approximate upper and lower bounds on instability growth rates for the long wavelength modes and their effects on capsule performance.

9.3.3.3 2D–3D Mapping for ICF Capsule Simulations

For our ICF capsule simulations, 2D–3D mapping is performed at 1.4 ns. This is the approximate time at which the shock wave converges at $r = 0$. The unperturbed mapping is the same as for the reshock problem – namely, all state and velocity variables were set using the mapping

$$\begin{Bmatrix} x \\ y \end{Bmatrix} = \begin{Bmatrix} \sqrt{x'^2 + z'^2} \\ y' \end{Bmatrix},$$

where (x, y) are the coordinates for the 2D problem and (x', y', z') are the coordinates for the 3D problem. Vector valued quantities, such as the velocity, were also appropriately rotated.

We performed simulations where the 2D–3D map was perturbed using the perturbation function $\zeta(x, z)$ defined in Equation (9.3.3) using $\chi = 10$ and $\chi = 20$ to obtain upper and lower bounds on perturbation growth for comparison with experimental results. Specifically, the perturbed mapping was defined as

$$\begin{Bmatrix} x \\ y \end{Bmatrix} = \begin{Bmatrix} r' + \zeta\left(\arctan\left(\frac{z'}{x'}\right), \arctan\left(\frac{y'}{r'}\right)\right) \\ y' \end{Bmatrix},$$

where $r' := \sqrt{x'^2 + z'^2}$.

9.3.3.4 Results

In Figure 9.31(a), we show the shock and shell position versus time from a 1D calculation. This shows the shock colliding with itself at the center of the capsule at 1.4 ns proceeded by a series of reflections off the gas/shell interface and the center. In Figure 9.31(b), (c), we show shock profiles from the 2D phase of our 2D–3D

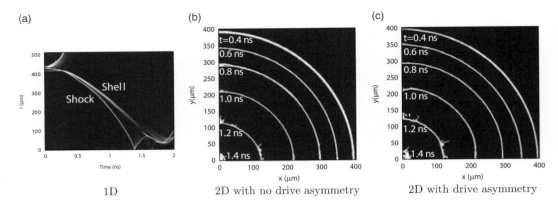

Figure 9.31 (a) Shock and shell position versus time from a 1D calculation (b, c) Shock profiles in 0.2 ns increments from 0.4 ns to 1.4 ns (time of rotation) for 2D phase (b) without and (c) with drive asymmetry.

simulations without and with drive asymmetry, respectively. This covers the period of the implosion from 0.4 to 1.4 ns as the shock is converging for the first time.

In Figure 9.32, we show visualizations of the main shock as it traverses the gas from 1.52 ns through 1.88 ns for the 3D phase of the $1/8\mu$m resolution 2D–3D simulation with no drive asymmetry and $\chi = 10$. During this period, having reflected off the center at approximately 1.40 ns, the main shock reflects off the shell at approximately 1.60 ns. This interaction distorts the shock shape significantly due to the deformed shell shape. As the shock converges again at approximately 1.65 ns, it does not converge to a point due to its distorted shape; rather, the shock becomes oblate then breaks in two as it reflects off the center again. When these two resulting shocks proceed to interact with the now highly distorted shell at approximately 1.69 ns, they break apart further into many smaller shocks traveling in all directions. Each time a shock interacts with shell material, it induces an RMI, which leads to turbulent mixing of gas and shell material.

The evolution of the main shock as it traverses the gas is largely the same for the simulations with drive asymmetry, though the shock breaks up faster due to the presence of shorter wavelength (P_{30}) deformations of the shell. In Figure 9.33, we show visualizations of the main shock as it reflects inside the gas from 1.50 ns through 1.72 ns for the 3D phase of the $1/2\mu$m resolution 2D–3D simulation with drive asymmetry and $\chi = 20$. During this period, having reflected off the center at approximately 1.40 ns, the main shock reflects off the shell at approximately 1.54 ns. This distorts the shock shape significantly due to the deformed shell shape. As the shock converges once more, it breaks up into many smaller shocks as the shock interacts with itself and protruding pieces of shell, and it is no longer possible to distinguish a main shock.

In Figure 9.34, we show a visualization of the deuterium mass concentration for the $1/8\mu$m resolution simulation. Long wavelength surface defects seed vortex rings that pull shell material into the gas and fragment it. Significant mixing develops near the fuel/shell interface at bang time and quenches the burn as shell material protrudes

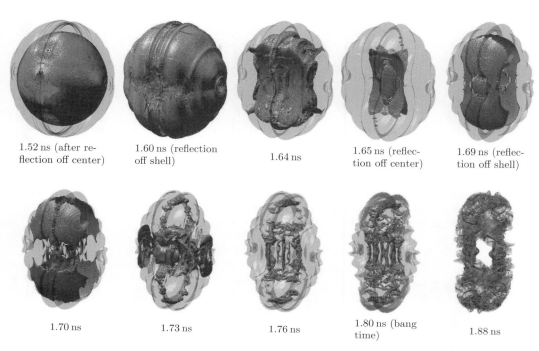

Figure 9.32 Visualization of the main shock (in red) and its interactions with the fuel/gas interface (in blue) for the 3D phase of the $1/8\mu$m resolution 2D–3D simulation with no drive asymmetry and $\chi = 10$. In order to maximize displayed detail for a converging problem, these pictures do not show the same spatial extent.

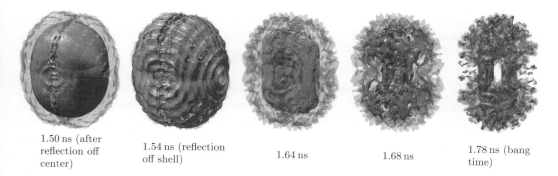

Figure 9.33 Visualization of the main shock (in red) and its interactions with the fuel /gas interface (in blue) for the 3D phase of the $1/2\mu$m resolution 2D–3D simulation with drive asymmetry and $\chi = 20$. In order to maximize displayed detail for a converging problem, these pictures do not show the same spatial extent.

through the center of the gas, even as the gas retains an overall shape that is fairly spherical. While no imaging was performed for relevant OMEGA experiments, ICF experiments at the NIF are typically oblate [52] and have been observed to develop a similar toroidal shape near bang time [61].

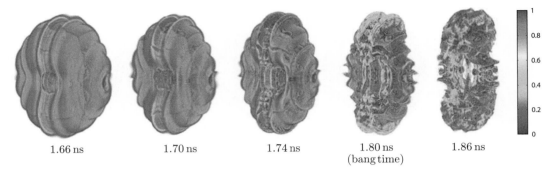

1.66 ns   1.70 ns   1.74 ns   1.80 ns   1.86 ns
                                                  (bang time)

Figure 9.34 Visualization of the deuterium mass concentration for the $1/8\mu$m resolution simulation with $\chi = 10$ and no drive asymmetry. In order to maximize displayed detail for a converging problem, these pictures do not show the same spatial extent.

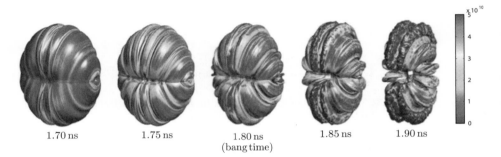

1.70 ns   1.75 ns   1.80 ns   1.85 ns   1.90 ns
                                  (bang time)

Figure 9.35 Visualization of hot spot shape, defined as the $T_{ion} = 1$keV isosurface, colored with vorticity magnitude (in s^{-1}) for the $1/8\mu$m resolution simulation with $\chi = 10$ and no drive asymmetry. In order to maximize displayed detail for a converging problem, these pictures do not show the same spatial extent.

The somewhat "squarish" shape of the gas in Figure 9.34 reflects the directional splitting strategy used by the Riemann solver in the RAGE code. Nevertheless, the shape distorting effects of this feature are dominated by the instability growth due to long wavelength surface defects. Therefore, the negative effects are largely cosmetic and do not have a significant impact on the quantitative results presented in the following.

In Figure 9.35, we show a visualization of the hot spot shape, defined as the $T_{ion} = 1$keV isosurface. This visualization more clearly shows the hole that develops in the center of the hot spot that quenches the burn, while the hot spot maintains a fairly spherical shape until after bang time.

9.3.3.5 Capsule Performance and Comparison to Experiment

Many thermonuclear reactions take place in the capsule during its implosion. We focus our attention on the D+D$\rightarrow$n+He^3 reaction, for which experimental data are available for comparison. While experimental data about the D+T$\rightarrow$n+α reaction are also

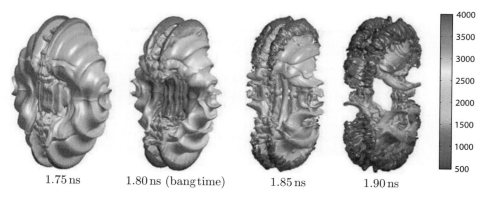

| 1.75 ns | 1.80 ns (bang time) | 1.85 ns | 1.90 ns |

Figure 9.36 Deuterium mass concentration $c_D = 0.1$ isosurface colored with T_{ion} (in eV) calculated from the $1/8\mu$m simulation data with $\chi = 10$ and no drive asymmetry. In order to maximize displayed detail for a converging problem, these pictures do not show the same spatial extent. A black and white version of this figure will appear in some formats. For the color version, please refer to the plate section.

available, we cannot accurately simulate this reaction with the Singe package since it relies on the presence of T produced by the D+D→p+T reaction, whose products are deposited locally. In experiments, the majority of these reactions take place after the tritons, generated with energies of 1MeV, have slowed down to ≈ 50keV, where the D+T→n+α reaction rate peaks. By the time this has occurred, the tritons have traveled a sufficient distance to invalidate the assumption of local deposition. For this reason, the number of neutrons produced by this reaction is significantly underpredicted, and so we do not report these results. In the absence of any data about how deuterium and shell material are distributed at a subgrid scale, we perform calculations in two limiting cases: "clean," in which the deuterium is assumed to burn in "chunk" form (i.e., it is atomically isolated from the shell material), and atomically mixed, in which the shell material in a cell is assumed to be uniformly atomically mixed with the deuterium.

In Figure 9.36, we show a deuterium mass concentration $c_D = 0.1$ isosurface colored with the ion temperature T_{ion} (in eV) for the $1/8\mu$m resolution simulation data. This gives an indication of the approximate fuel/shell interface and the temperature there. The annular region that contains most of the fuel mass has a lower temperature than the thin fuel region in the center, where the burn rates are highest. It is notable that the hot spot, defined as the region where $T_{ion} \geq 1$keV, extends beyond this surface and contains part of the shell.

Results are presented in Table 9.1 for $1/2\mu$m resolution simulation data. For the 2D–3D simulations, data are averaged over two simulations, with $\chi = 10, 20$, and using two techniques for performing neutron production rate calculations: "clean" and atomically mixed, as defined earlier. For the other simulations, data is averaged over the two neutron production rate calculations. Presented uncertainties are the standard deviations. Here, we show the yield in terms of the total number of neutrons produced by the D+D→n+He^3 reaction, as well as bang time (the time at which the DD neutron

Table 9.1 Integrated Quantities for $1/2\mu$m ICF Capsule Simulations*

	DD Neutron Yield (10^{10} Neutrons)	Bang Time (ns)	Burn weighted T_{ion} (KeV)	Burn Width (ns)
2D	36.1±4.9	1.79±.01	3.2±.3	0.11±.01
2D–3D with no mapping perturbations and no drive asymmetry	28.7±2.9	1.74±.02	3.4±.1	0.09±.02
2D–3D with surface defects	14.6±3.6	1.75±.01	2.8±.2	0.11±.01
2D–3D with surface defects and drive asymmetry	9.7±2.6	1.78±.01	2.7±.2	0.15±.02

* For the 2D–3D simulations with mapping perturbations, data are averaged over four simulations, with $\chi = 10, 20$ and using two techniques for performing neutron production rate calculations: "clean" and atomically mixed. For the other simulations, data is averaged over the two neutron production rate calculations.

production rate reaches its peak), the burn weighted ion temperature, and the burn width (the temporal extent of the burn, calculated as the time during which the neutron production rate exceeds half of its maximum). For the atomically mixed calculations, the DD neutron yield was typically $\approx 15\%$ lower and the ion temperatures $\approx 3\%$ higher than for the "clean" calculations. For comparison, we show results from a 2D simulation and a 2D–3D simulation with no mapping perturbations. Comparing 2D–3D results with mapping perturbations, we see that a majority of yield degradation results from long wavelength surface defects. It is notable that drive asymmetry extends the burn width. This is due to the fact that the perturbations associated with drive asymmetry lead to the presence of several burn regions that are separated both spatially and temporally.

In order to perform a comparison with experimental data, we performed a $1/2\mu$m simulation adapted to OMEGA shot 65036 [62]. This shot differs from the capsule considered above by the inclusion of a 1.8% titanium dopant in a 2μm-wide layer adjacent to the fuel. While this dopant does not affect the hydrodynamic development of the capsule appreciably, it raises the ion temperature in the fuel by absorbing some of the energy radiated by the hot spot and changing the heat conductivity of the doped portion of the shell. In Table 9.2, we show a comparison of our simulation of shot 65036 to available experimental data from [62]. The simulation uses a growth factor of $\chi = 10$ for the perturbations in the 2D to 3D mapping and neutron production rates are calculated in the "clean" limit. The comparisons to all available experimental data are well within the experimental uncertainties. We are unaware of any previous simulations that have been able to match the neutron yield, bang time, burn width, and burn weighted ion temperatures from experiment simultaneously without tuning.

Grid resolution studies for integrated quantities are given in Table 9.3. While the bang time, burn weighted ion temperature, and burn width do not show appreciable variation, the neutron yield varies by approximately 20%. This is a reflection of how sensitive the reaction rate is to the distribution of gas and shell material, which is in turn highly sensitive to resolution and initial conditions in imploding problems.

Table 9.2 Comparison to Experiment (OMEGA shot 65036 [62]) for $1/2\mu$m ICF Capsule Simulation*

	DD Neutron Yield (10^{10} Neutrons)	Bang Time (ns)	Burn Weighted T_{ion} (KeV)	Burn Width (ns)
Experiment	$3.39 \pm .09$	$1.71 \pm .03$	$3.7 \pm .5$	$0.15 \pm .03$
Simulation	3.33	1.72	3.7	0.13

* The simulation uses a growth factor $\chi = 10$, neutron production rates are calculated in the "clean" limit, and a titanium dopant was added to the shell as in experiment.

Table 9.3 Integrated Quantities as a Function of Grid Resolution for 2D–3D ICF Simulations with No Drive Asymmetry and $\chi = 10$*

Simulation resolution	DD Neutron Yield (10^{10} Neutrons)	Bang Time (ns)	Burn–Weighted T_{ion} (KeV)	Burn Width (ns)
$1/2\mu$m	19.0	1.74	3.0	0.10
$1/4\mu$m	23.4	1.74	3.2	0.09
$1/8\mu$m	15.4	1.80	3.0	0.09

* Neutron production rates are calculated using the "clean" method.

9.4 Conclusions

We performed simulations of laser driven reshock and shear experiments [1–3] in the strong shock high energy density regime to better understand material mixing driven by the RM and KH instabilities. Validation of the simulations was based on direct comparisons of simulation and experimental radiographic data supplemented with spectral analysis. Simulations were also compared with published DNS and the theory of homogeneous isotropic turbulence, seeking to characterize the small scale flow features and their turbulence statistical measures.

While the flow is not isotropic or homogeneous, we have shown that the 3D ILES simulations have flow regions that exhibit features of isotropic homogeneous turbulence. For the reshock problem, after reshock, our analysis shows self-similarity characteristics and effective Re assessments suggesting that the mixing transition is achieved and that the simulations are reasonably converged at the finest resolution. Our results also show that some turbulent features, such as self-similarity and isotropy, only fully develop once others, such as decorrelation, characteristic vorticity distributions, and integrated turbulent kinetic energy, have decayed significantly.

Our analysis of the shear experiment shows that many turbulent features, such as self-similarity, decorrelation, and characteristic vorticity distributions are present in the flow in two regions: near the shock front and in a separated region in the wake of the shock. Turbulent features develop faster near the shock front, peaking around $t = 10-12$ ns, whereas these features begin to appear around $t = 20\,\text{ns}$ in the wake of the shock. Nevertheless, despite observing grid independence for integrated flow quantities, our effective *Re* analysis indicates that the highest resolution simulations are near the mixing transition, but have not reached it.

We have also presented an improved strategy for performing simulations that begin in 2D and are later rotated to 3D. This strategy compensates for the reduced turbulence development in the 2D phase of the simulation as well as the artificial smoothing effects of the 2D–3D rotation, and was validated using simulations of the reshock problem. We applied our method to simulate the implosion of a simple OMEGA type ICF capsule, consisting of a polystyrene shell surrounding deuterium gas.

Our ICF simulations compared well with available experimental data. We demonstrated that the dominant mechanism for yield degradation in the capsule compared with lower dimensional simulations is the displacement of fuel from the hot spot by shell material induced by turbulent instability growth generated by long wavelength surface defects. In our simulations, this resulted in yield degradation of approximately 60%. This effect is compounded by drive asymmetry, which breaks up the burn region both spatially and temporally, reducing the yield in our simulations by a further 30% and extending the burn width by approximately 40%. Thus, discrepancies between the yield predicted by previous simulations of ICF implosions can be explained by the reduced growth of turbulent instabilities inherent in reduced dimension hydrodynamic simulations and the absence of appropriate modeling of long wavelength asymmetries that are present due to surface defects and drive asymmetries.

References

[1] L. Welser-Sherrill, J. Fincke, F. Doss, E. Loomis, K. Flippo, D. Offermann, P. Keiter, B.M. Haines, and F.F. Grinstein. "Two laser-driven mix experiments to study reshock and shear." *High Energy Density Physics Journal* 9(3):496–499, 2013.

[2] B.M. Haines, F.F. Grinstein, L. Welser–Sherrill, and J. R. Fincke. "Simulations of material mixing in laser-driven reshock experiments." *Phys. Plasmas* 20:022309, 2013.

[3] B.M. Haines, F.F. Grinstein, L. Welser–Sherrill, J.R. Fincke, and F. W. Doss. "Simulation ensemble for a laser-driven shear experiment." *Phys. Plasmas* 20:092301, 2013.

[4] B.M. Haines, F.F. Grinstein, and J.R. Fincke. "Three-dimensional simulation strategy to determine the effects of turbulent mixing on inertial-confinement-fusion capsule performance." *Physical Review E*, 89:053302, 2014.

[5] C.-M. Ho and P. Huerre. "Perturbed free shear layers." *Ann. Rev. Fluid Mech.* 16:365–424, 1984.

[6] C.H.K. Williamson. "Vortex dynamics in the cylinder wake." *Annu. Rev. Fluid. Mech.* 28:477–539, 1996.

[7] E.J. Gutmark and F.F. Grinstein. "Flow control with noncircular jets." *Annu. Rev. Fluid Mech.* 31:239–272, 1999.

[8] F.F. Grinstein. "Vortex dynamics and transition to turbulence in free shear flows," in *Implicit Large Eddy Simulation: Computing Turbulent Flow Dynamics*, ed. by F. F. Grinstein, L. G. Margolin, and W. J. Rider. Cambridge University Press, New York, 2nd printing, 2010.

[9] P. Sagaut. *Large Eddy Simulation for Incompressible Flows*, 3rd ed. Springer, New York, 2006.

[10] F.F. Grinstein, L.G. Margolin, and W.J. Rider, eds. *Implicit Large Eddy Simulation: Computing Turbulent Flow Dynamics*. Cambridge University Press, New York, 2nd printing, 2010.

[11] W.K. George and L. Davidson. "Role of initial conditions in establishing asymptotic flow behavior." *AIAA Journal*. 42:438–446, 2004.

[12] B.M. Haines, F.F. Grinstein, and J.D. Schwarzkopf. "Reynolds-averaged Navier-Stokes anitialization and benchmarking in shock-driven turbulent mixing." *Journal of Turbulence* 14(2):46–70, 2013.

[13] M. Gittings et al. "The RAGE radiation-hydrodynamic code." *Comput. Science & Discovery* 1:015005, 2008.

[14] S.P. Lyon and J. D. Johnson. "SESAME: the Los Alamos National Laboratory Equation of State Database." *Los Alamos National Laboratory* LA-UR-92-3407, 1992.

[15] B.E. Launder, G.J. Reese, and W. Rodi. "Progress in the development of a Reynolds-stress turbulence closure." *J. Fluid Mech.*, 68:537, 1975.

[16] D. Besnard, F.H. Harlow, F.H. Rauenzahn, and C. Zemach. "Turbulence transport equations for variable-density turbulence and their relationship to two-field models." LA-UR-12303, Los Alamos National Laboratory, 1992.

[17] O. Grégoire, D. Souffland, and S. Gauthier. "A second-order turbulence model for gaseous mixtures induced by Richtmyer-Meshkov instability." *J. Turbul.* 6:1, 2005.

[18] D. Livescu, J.R. Ristorcelli, R.A. Gore, S.H. Dean, W.H. Cabot, and A.W. Cook. "High-Reynolds number Rayleigh-Taylor turbulence." *J. Turbul.* 10:1, 2009.

[19] J.D. Schwarzkopf, D. Livescu, R.A. Gore, R.M. Rauenzahn, and J.R. Ristorcelli. "Application of a second-moment closure model to mixing processes involving multi-component miscible fluids." *J. Turbul.* 12:1–35, 2011.

[20] A. Banerjee, R.A. Gore, and M. J. Andrews. "Development and validation of a turbulent mix model for variable-density and compressible flows." *Phys. Rev. E* 82:046309, 2010.

[21] A.W. Cook. "Enthalpy diffusion in multicomponent flows." *Phys. Fluids* 21:055109, 2009.

[22] W.K. George. "The Self-preservation of turbulent flows and its relation to initial conditions and coherent structures," in *Advances in Turbulence*, ed. by W.K. George and R.E.A. Arndt, New York, 1989.

[23] G. Dimonte. "Nonlinear evolution of the Rayleigh-Taylor and Richtmyer-Meshkov instabilities." *Phys. Plasma.*, 6:2009–15, 1999.

[24] D. Youngs. "Rayleigh-Taylor and Richtmyer-Meshkov mixing," in *Implicit Large Eddy Simulation: Computing Turbulent Flow Dynamics*, ed. by F.F. Grinstein, L.G. Margolin, and W.J. Rider. Cambridge University Press, 2nd printing, 2010.

[25] A.A. Gowardhan and F.F. Grinstein. "Numerical simulation of Richtmyer–Meshkov instabilities in shocked gas curtains." *Journal of Turbulence* 12(43):1–24, 2011.

[26] J. Jiménez, A.A. Wray, P.G. Saffman, and R.S. Rogallo. "The structure of intense vorticity in isotropic turbulence." *J. Fluid Mech.*, 255:65–90, 1993.

[27] A.J. Wachtor, F.F. Grinstein, C.R. DeVore, J.R. Ristorcelli, and L.G. Margolin. "Mixing in implicit large-eddy simulation of statistically stationary isotropic turbulence." *Physics of Fluids*, 25:025101, 2013.

[28] C. Fureby and F.F. Grinstein. "Monotonically integrated large eddy simulation of free shear flows." *AIAA J.*, 37:544, 1999.

[29] P.E. Dimotakis. "The mixing transition in turbulent flows." *J. Fluid Mech.*, 409:69, 2000.

[30] A.A. Gowardhan, J.R. Ristorcelli, and F.F. Grinstein. "The bipolar behavior of the Richtmyer–Meshkov instability." *Physics of Fluids* 23:071701, 2011.

[31] A.N. Kolmogorov. *C. R. Acad. Sci. URSS* 30:301, 1941.

[32] G. Falcovich, K. Gawedzki, and M. Vergassola. "Particles and fields in fluid turbulence." *Reviews of Modern Physics*, 73:913–75, 2001.

[33] H. Tennekes and J.L. Lumley. *A First Course in Turbulence*. MIT Press, Cambridge, MA, 1972.

[34] Y. Zhou. "Unification and extension of the similarity scaling criteria and mixing transition for studying astrophysics using high energy density laboratory experiments or numerical simulations." *Physics of Plasmas*, 14:082701, 2007.

[35] Y. Zhou, F.F. Grinstein, A.J. Wachtor, and B.M. Haines. "Estimating the effective Reynolds number in implicit large eddy simulation." Phys. Rev. E 89:013303, 2014.

[36] Y. Kaneda, et al. "Energy dissipation rate and energy spectrum in high resolution direct numerical simulations of turbulence in a periodic box." *Phys. Fluids* 15, L21. 2003.

[37] F.F. Grinstein, A.A. Gowardhan, and A.J. Wachtor. "Simulations of Richtmyer-Meshkov instabilities in planar shock-tube experiments." *Physics of Fluids*, 23:034106, 2011.

[38] J. Nuckolls, L. Wood, A. Thiessen, and G. Zimmerman. "Laser compression of matter to super-high densities: Thermonuclear (CTR) applications." *Nature* 239, 139, 1972.

[39] V.A. Thomas and R.J. Kares. "Drive asymmetry and the origin of turbulence in an ICF implosion." *Phys. Rev. Letters* 109, 075004, 2012.

[40] S.W. Haan et al. "Design and modeling of ignition targets for the National Ignition Facility." *Phys. Plasma.*, 2:2480–2487, 1995.

[41] D.A. Callahan et al. "Optimization of the NIF ignition point design hohlraum." *J. Phy.s. Conf. Ser.*, 112:022021, 2008.

[42] D.S. Clark, S.W. Haan, B.A. Hammel, J.D. Salmonson, D.A. Callahan, and R.P.J. Town. "Plastic ablator ignition capsule design for the National Ignition Facility." *Phys. Plasma.* 17:052703, 2010.

[43] S.W. Haan et al. "Point design targets, specifications, and requirements for the 2010 ignition campaign on the National Ignition Facility." *Phys. Plasma.* 18:051001, 2011.

[44] P.B. Radha et al. "Two-dimensional simulations of plastic-shell, direct-drive implosions on OMEGA." *Physics of Plasmas* 12:032702, 2005.

[45] P.B. Radha et al. "Multidimensional analysis of direct-drive, plastic-shell implosions on OMEGA." *Physics of Plasmas* 12:056307, 2005.

[46] P.B. Radha et al. "Triple-picket warm plastic shell implosions on OMEGA." *Physics of Plasmas* 18:012705, 2011.

[47] K. Molvig, N.M. Hoffman, B.J. Albright, E.M. Nelson, and R.B. Webster. "Knudsen layer reduction of fusion reactivity." *Phys. Rev. Lett.*, 109:095001, 2012.

[48] P. Amendt, O.L. Landen, H.F. Robey, C.K. Li, and R.D. Petrasso. "Plasma barrodiffusion in inertial-confinement-fusion implosions: application to observed yield anomalies in thermo-nuclear fuel mixtures." *Phys. Rev. Lett.*, 105:115005, 2010.

[49] E.S. Dodd et al. "The effects of laser absorption on direct-drive capsule experiments at OMEGA." *Phys. Plasma.* 19:042703, 2012.

[50] R.H.H. Scott et al. "Numerical modeling of the sensitivity of x-ray driven implosions to low-mode flux asymmetries." *Phys. Rev. Letters* 110:075001, 2013.

[51] D.S. Clark et al. "Short-wavelength and three-dimensional instability evolution in National Ignition Facility ignition capsule designs." *Physics of Plasmas* 18:082701, 2011.

[52] G.P. Grim et al. "Nuclear imaging of the fuel assembly in ignition experiments." *Physics of Plasmas* 20:056320, 2013.

[53] C.C. Joggerst et al. "Cross-code comparison of mixing during the implosion of dense cylindrical and spherical shells." *J. Comput. Phys.* 275:154–173, 2014.

[54] B.M. Haines, F.F. Grinstein, L. Welser–Sherrill, J.R. Fincke, and F.W. Doss. "Analysis of the effects of energy deposition on shock–driven turbulent mixing." *Physics of Plasmas* 20:072306, 2013.

[55] R.D. Richtmyer. "Taylor instability in shock acceleration of compressible fluids." *Comm. Pure Appl. Math* 13:297–319, 1960.

[56] R.L. Holmes et al. "Richtmyer–Meshkov instability growth: Experiment, simulation, and theory." *J. Fluid Mech.* 389:55–79, 1999.

[57] J.R. Ristorcelli, A.A. Gowardhan, and F.F. Grinstein. "Two classes of Richtmyer-Meshkov instabilities: A detailed statistical look." *Physics of Fluids* 25:044106, 2013.

[58] F.J. Marshall et al. "Direct-drive-implosion experiments with enhanced fluence balance on OMEGA." *Physics of Plasmas* 11(1):251–259, 2004.

[59] P.G. Drazin. *Introduction to Hydrodynamic Stability*, Cambridge University Press, 2002.

[60] R. Epstein. "On the Bell–Plesset effects: the effects of uniform compression and geometrical convergence on the classical Rayleigh–Taylor instability." *Physics of Plasmas* 11(11): 5114–5124, 2004.

[61] H.S. Park et al. "High-adiabat high-foot inertial confinement fusion implosion experiments on the National Ignition Facility." *Phys. Rev. Lett.* 112:055001, 2014.

[62] J.A. Baumgaertel et al. "Observation of early shell-dopant mix in OMEGA direct-drive implosions and comparisons with radiation-hydrodynamic simulations." *Physics of Plasmas* 21:052706, 2014.

10 Drive Asymmetry, Convergence, and the Origin of Turbulence in Inertial Confinement Fusion Implosions

Vincent A. Thomas and Robert J. Kares

10.1 Introduction

Deviations from perfect spherical symmetry in an inertial confinement fusion (ICF) capsule's pressure drive can arise from several different sources. Nonuniformities in laser illumination due to variations in individual beam profiles and time integrated energies, beam pointing errors, and target offsets from the nominal position of the beam focus in the target chamber can all lead to asymmetry in the pressure drive on the capsule. The impact of this asymmetry on capsule performance has been studied for a number of years because of its potential importance for ICF ignition attempts. Two recent experimental studies are of particular interest.

In Rygg et al.[1] the authors performed experiments with spherical plastic (CH) capsules with a D^3He gas fill on the OMEGA laser system. Drive asymmetries were induced by offsetting the capsule from the nominal beam focus and the resulting capsule performance was measured. In this study the authors make the important distinction between the "shock burn" induced by the heating of the gas as a result of the initial collapse of the spherically converging gas shock on the capsule center and the subsequent "compression burn" induced by the compression and heating of the gas by the imploding shell. Results of this study demonstrate that the "shock burn" is relatively insensitive to the drive asymmetry. In contrast the "compression burn," which is the main component of the burn yield, is substantially diminished by increasing drive asymmetry.

In Li et al.[2] the authors again performed experiments with CH capsules on OMEGA using charged particle spectrometry techniques to measure the amplitude of low mode number ρR asymmetries in the imploded shell at the time of fusion burn and to show quantitatively how these low mode ρR asymmetries are directly correlated with the amplitude of asymmetries in the time averaged, on target laser intensity. From these and related studies the authors concluded that the observed growth of low mode ρR asymmetries in the imploded shell are initially seeded by laser illumination asymmetries and grow predominantly by Bell–Plesset[3] related convergence effects while capsule imperfections do not seem to play a dominant role in determining the observed ρR asymmetries.

In addition to results of these specific and very illuminating experiments, a closely related fundamental theoretical question arises: precisely how is turbulence generated in an ICF implosion and what, if any, is the relationship between drive asymmetry and the generation of turbulence in an ICF capsule? In this paper we present results from some very high resolution three-dimensional (3D) numerical simulations of the implosion of a highly idealized OMEGA capsule similar to the one used in the experiments of Rygg et al.[1], which address both the results of the aforementioned experiments and the more general question of how drive asymmetry can lead to turbulence generation in ICF implosions (cf. Thomas and Kares[4]).

Scaling studies are also presented which suggest the important role convergence may play in the hydrodynamic disruption of asymmetrically driven ICF implosions. These studies provide a qualitative understanding of the empirical observation that high convergence ICF implosions perform much worse in reality than would be expected from simulations based on the unstable Rayleigh–Taylor (RT) growth of capsule surface irregularities alone. The prime example of this discrepancy at high convergence is the recent failure of the National Ignition Campaign (NIC) at the National Ignition Facility (NIF).

The simulations described here were performed using the Los Alamos National Laboratory's Eulerian adaptive mesh refinement (AMR) radiation hydrodynamics code RAGE[5] and utilize spatial resolutions down to 0.05 μm in 3D that approach the order of the Kolmogorov length scale in this system. While these inviscid simulations do not explicitly resolve the Kolmogorov length scale and are thus not direct numerical simulations (DNS) of the capsule implosion, we will show that the resolutions achieved are sufficient to see important details of how the turbulence is generated in an asymmetrically driven ICF implosion.

In particular we will see that the asymmetric pressure drive creates initial density perturbations in the drive shell that are amplified by Bell–Plesset related convergence effects. When the incoming shell collides with the outgoing reflected gas shock these perturbations act as seeds for the growth of Richtmyer–Meshkov (RM) fingers of shell material that penetrate the DT gas. The growth of these fingers is further enhanced at late time by the RT instability. These same pressure drive asymmetries also create coherent vortical structures in the DT gas, counterrotating rings and sheets of azimuthal vorticity of opposite sign in close proximity to one another. These structures are strongly unstable to both short and long wavelength azimuthal instabilities of the Widnall[6] and Crow[7] type. They very rapidly evolve to a fully turbulent state by a process similar that described by Leweke and Williamson[8]. Thus, both the asymmetries in the shell and the turbulence in the gas are seen to have a common origin in the asymmetry of the pressure drive.

In Section 10.2 of this chapter we discuss high resolution 2D and 3D RAGE simulations of a simplified OMEGA capsule implosion that illustrate the central role played by drive asymmetry in the determination of both the penetration of fingers of shell material into the gas and generation of the gas turbulence. In this section we present linked 3D RAGE simulations of the turbulence generation process in the capsule implosion that use spatial resolutions as high as 0.05 μm and that require nearly 1×10^9 AMR cells in 3D. We demonstrate directly that the power spectra of the kinetic energy of the resulting turbulence observed in these simulations contains an inertial subrange whose scaling is consistent with a Kolmogorov $k^{-5/3}$ spectrum. Section 10.3 discusses possible implications of these results for NIF ignition. And finally, Section 10.4 summarizes our conclusions.

10.2   RAGE Simulations of an ICF Capsule with an Imposed Asymmetry

In this section we utilize high resolution 2D and 3D RAGE simulations to demonstrate the effect of imposing a well-defined drive asymmetry on an ICF capsule implosion. The example capsule chosen for our RAGE simulations is a plastic (CH) capsule similar to ones used in the experiments of Rygg et al.[1] on OMEGA. This capsule consisted of a round plastic shell with outer radius 425 μm and inner radius 400 μm containing an equimolar mixture of DT gas at an initial density of 2.5×10^{-3} g/cm^3.

The goal of these simulations is not to model the results of an actual OMEGA experiment, but instead to use the plastic OMEGA capsule as a simple example, a perfectly spherical CH shell with DT gas, with which to explicitly exhibit how an asymmetric pressure drive leads to both the penetration of fingers of shell material into the DT fuel and the generation of turbulence in the DT fuel as the capsule implodes.

For this purpose we need only purely hydrodynamic calculations with no radiation or nuclear burn physics included and even these calculations are highly idealized. Rather than attempting to drive the capsule with an asymmetric radiation field, the energy was imparted to the plastic shell in the following way. The shell was divided into two regions, an inner region (the pusher) between $r = 400$ μm and 405 μm and an outer region (the ablator) between $r = 405$ μm and 425 μm. Energy was then sourced into this outer ablator region as an energy source per unit mass S with a fixed spatial profile of the form,

$$S = At(1 + a_\ell P_\ell(\cos\theta))\tanh((r - r_0)/\Delta) \qquad (10.1)$$

for $r > r_0$ and $0 < t < 1$ ns

where $r_0 = 405$ μm and $\Delta = 20$ μm. Here the value of $A = 1 \times 10^{16}$ ergs/gm/ns was chosen to source in a total of 17.35 kJ in a 1 ns square power pulse. This form for the energy deposition was chosen to suppress grid induced RT instabilities at the ablator/pusher interface driven by the acceleration of the ablator on the denser pusher. This treatment allows the simulation to retain only the imprinting from the drive asymmetry without having to worry about possible feed-through of unwanted short wavelength RT instabilities. The pusher region moves under the influence of the drive pressure generated in the ablator. No energy was sourced directly into the pusher so that a high density pusher region was present late in the implosion.

The term proportional to $P_\ell(\cos\theta)$, the Legendre polynomial of order ℓ, provided a simple asymmetry of amplitude a_ℓ for the drive. The results presented here focus on the case of an energy source with an $\ell = 30$ asymmetry. A drive with P30 asymmetry was chosen so that a rather large number of features is created, which may have some relevance to a laser drive with a large number of beams. However, no attempt has been made to use real asymmetries from actual laser drives but rather an attempt has been made to choose a simple asymmetry that clearly illustrates the physical phenomena associated with asymmetric drive.

In this chapter we are interested in examining the fundamental mechanisms of turbulence production in our example ICF implosion, a study that requires very high resolution 3D numerical simulations of the implosion. However, such simulations are so computationally intensive that they cannot be practically carried out from $t = 0$ with the current generation of parallel supercomputers. To avoid this difficulty we have adopted the procedure of performing the initial phase of the implosion in 2D using the RAGE code. Then at a suitably chosen time late in the implosion the 2D RAGE simulation is rotated in 3D and used to initialize a 3D RAGE simulation of the late time behavior of the implosion. This 2D to 3D linking procedure allows us to carry out very high resolution simulations at late time of the turbulence generation while avoiding the practical limitations of existing compute platforms. However, it also restricts us to examining only axisymmetric rather than fully 3D asymmetries of the pressure drive. Despite this restriction we shall see that the basic mechanisms of turbulence generation are clearly exhibited. We now turn to results from 2D RAGE simulations of our idealized capsule implosion.

10.2.1   2D RAGE Simulations of the Capsule Implosion

Figure 10.1 shows an r-t plot of the motion of the pusher/DT gas interface from a 1D RAGE simulation of our idealized OMEGA capsule. This plot shows the abrupt acceleration, the roughly constant velocity implosion, and the stagnation phase with the multiple shock bounces. Also shown is the corresponding r-t motion for the main gas shock, which suffers its first collapse onto the capsule center at about $t = 1.08$ ns. We are primarily interested in the implosion up to the stagnation point at about $t = 1.75$ ns, since beyond this time in a real experiment the burn would have begun. The convergence ratio for the initial to final radius of the pusher/gas interface at $t = 1.75$ ns is 7.9 in this 1D simulation. Superimposed in Figure 10.1 on the r-t plot of the pusher/gas interface from the 1D RAGE simulation is the corresponding result from a 2D RAGE simulation (hollow squares) for the case of a spherically symmetric drive with $a_{30} = 0$. The interface motion in the symmetrically driven 2D RAGE simulation tracks the interface motion in the 1D simulation extremely well.

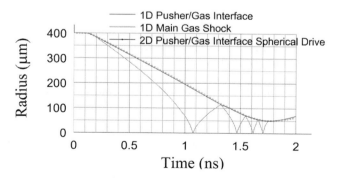

Figure 10.1 r–t plot of the motion of the pusher/gas interface and the main gas shock in a 1D RAGE simulation of the OMEGA capsule. Hollow squares show the pusher/gas interface motion in a 2D RAGE simulation using a spherically symmetric drive with $a_{30} = 0$. Effective interface radius in 2D is computed from the DT gas volume using $R_{eff} = \left(3 \cdot V_{gas} / 4\pi \right)^{1/3}$.

2D RAGE simulations were performed with a range of different values for the amplitude a_{30} of the P30 drive asymmetry from $a_{30} = 0$ (spherically symmetric drive) to $a_{30} = 0.50$ (50% P30 asymmetry) with the same total energy of 17.35 kJ. All of these 2D RAGE simulations had a nominal convergence of about 8 for the pusher/gas interface. The 2D simulations use AMR with a finest spatial resolution of 0.4 μm up until a time of $t = 1$ ns just before the first collapse of the main gas shock on the center of the capsule at around $t = 1.08$ ns. After that time a fixed spatial resolution of approximately 0.4 μm is forced throughout the entire region that encompasses the pusher and the gas in order to achieve the highest fidelity for the dynamics of the gas and the pusher shell during the later stages of the implosion.

Figure 10.2 shows a series of panels, each a time snapshot from the early portion of a 2D RAGE simulation of the example OMEGA capsule for the case with a 50% P30 asymmetry, which shows how the implosion progresses early in time as the asymmetric drive is applied. In the lower portion of each panel the CH shell, which includes both the ablator and pusher regions, is colored by pressure while the DT gas is colored by grad P. In the upper portion of each panel both the CH shell and the gas are colored by the θ component of the material velocity. The panels are arranged in rows with the left hand panel in each row showing a full view of the capsule at the particular snapshot time while the corresponding right hand panel shows a blowup of a small region at the left pole of the capsule near the CH/gas interface at that same time.

Figure 10.2(a) at $t = 0.1$ ns shows the angular position of the 16 pressure maxima associated with the P30 drive. The drive produces predominantly radial flow, but also produces flow in the θ direction away from the maxima as indicated in the figure because the drive is not perfectly spherically symmetric. At this early time of $t = 0.1$ ns, motion is confined to the ablator shell.

By $t = 0.275$ ns (Figs. 10.2(c) and (d)) the main shock has begun to penetrate into the gas. Behind this inward moving main shock, gas is flowing in the θ direction away from the angular positions of the original drive pressure maxima in exactly the same way as the adjacent shell material. While the main shock moves inward, the gas flow behind the main shock converges in the θ direction at the angular positions of the original drive pressure minima. Gas collides there forming pairs of shocks behind the main shock that spread out in the θ direction away from the angular position of the original drive minima. Figure 10.2(d) at $t = 0.275$ ns shows these polar shocks just beginning to form behind the main shock as gas collides at the angular position of the original drive minima.

As a result of this convergent flow in the θ direction, local pressure maxima form in the gas at the angular positions of the original drive minima. These local pressure maxima produce additional grad P accelerations directed away from the local pressure maxima, which drive gas inward into the unshocked region. By $t = 0.425$ ns (Figs. 10.2(e) and (f)) these local pressure maxima have produced convex perturbations on the main shock as well as regions of gas behind the main shock in which the θ component of the gas velocity has reversed sign. The result is the formation of shear layers in the gas, which can be clearly seen in the θ velocity field shown in the upper portion of Figure 10.2(f) at $t = 0.425$ ns.

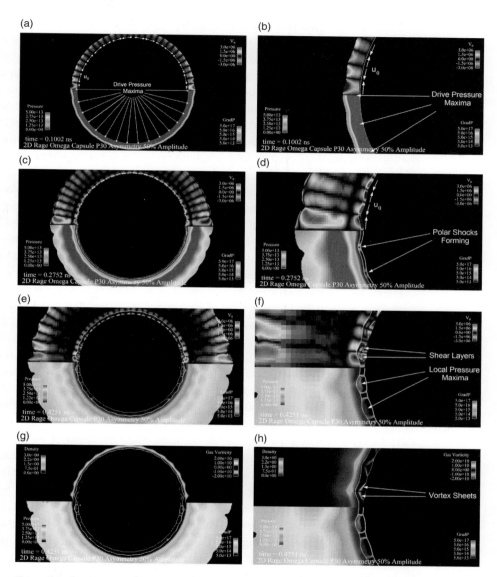

Figure 10.2 A sequence of early time snapshots from a 2D RAGE simulation of the P30 Omega capsule with $a_{30} = 0.50$. In the lower portion of each panel the CH shell is colored by pressure while the gas is colored by grad P. In the upper portion both the CH shell and the gas are colored by the θ component of the material velocity. In the last row of panels the θ velocity has been replaced on top by density in the CH and azimuthal vorticity in the gas. Each right hand panel shows an expanded view of the region at the left pole of the capsule near the CH/gas interface. A black and white version of this figure will appear in some formats. For the color version, please refer to the plate section.

In Figures 10.2(g) and (h), also at the same time of $t = 0.425$ ns, we have replaced the θ velocity in the upper portion of the panel by density in the CH shell and the azimuthal component of vorticity in the gas. This pair of panels illustrates that the shear layers seen in the θ velocity field of Figures 10.2(e) and (f) correspond to vortex sheets in the gas. As the main shock travels inward, the moving intersection points between the main shock and the polar shocks formed behind it trace out these vortex sheets in the gas. This process is particularly evident in the expanded view in Figure 10.2(h) at $t = 0.425$ ns. Notice that as a result of this mechanism, vortex sheets of opposite sign for the azimuthal vorticity, yellow for positive and blue for negative, always form in close proximity in the gas.

This same asymmetry of the pressure drive also produces convergent flow in the θ direction inside the shell material as well. The result are the 15 density maxima in the shell at the angular positions of the original drive minima, which are clearly visible in Figure 10.2(g) and (h) at $t = 0.425$ ns. As the capsule implodes these initial density perturbations seeded by the asymmetry of the pressure drive are amplified by Bell–Plesset related convergence effects.

Figure 10.3 shows a series of panels, each a time snapshot from the remainder of the implosion. Here in the lower portion of each panel the CH shell is colored by pressure while the gas is colored by grad P. In the upper portion of each panel the CH shell is colored by density while the gas is colored by the azimuthal component of the gas vorticity. Figure 10.3(a) at $t = 0.6$ ns shows that as the polar shocks pass through each other, new layers of vortex sheets are formed in the gas by the process described earlier.

This process continues as the main shock converges radially inward toward the center of the capsule so that by $t = 1.025$ ns (Fig. 10.3(b)), a time just before the first bounce of the main shock, several layers of vortex sheets have been formed within the gas. By $t = 1.1$ ns (Fig. 10.3(c)) the main shock has bounced off the center of the capsule and is moving radially outward. As this outbound main shock passes over the vortex sheets, it rolls them up into vortex rings, a process that is clearly visible in the $t = 1.225$ ns panel in Figure 10.3(d).

Simultaneous with the deposition and evolution of the vorticity in the gas is the development of pressure and density enhancements in the plastic shell as can been seen in the top four panels of Figure 10.3. Because the original pressure drive was not spherically symmetric, the shell material flows in the θ direction as shown in Figure 10.2(a) and deterministic pressure and density enhancements develop in the shell at the angular positions of the original drive minima. These perturbations grow in time due to convergence (the Bell–Plesset effect), becoming increasingly elongated in the radial direction until shell material protrudes out into the gas as shown in the $t = 1.225$ ns panel in Figure 10.3(d). At a slightly later time, $t = 1.325$ ns (Fig. 10.3(e)), the outgoing reflected main gas shock collides with the shell/gas interface and these perturbations are amplified by the RM instability to create large scale deterministic fingers of shell material in the gas. The continued growth of these deterministic RM fingers is seen in the last three panels of Figure 10.3.

The collision of the outgoing gas shock with the perturbed shell/gas interface also deposits significant vorticity on the interface as part of the RM development. For a

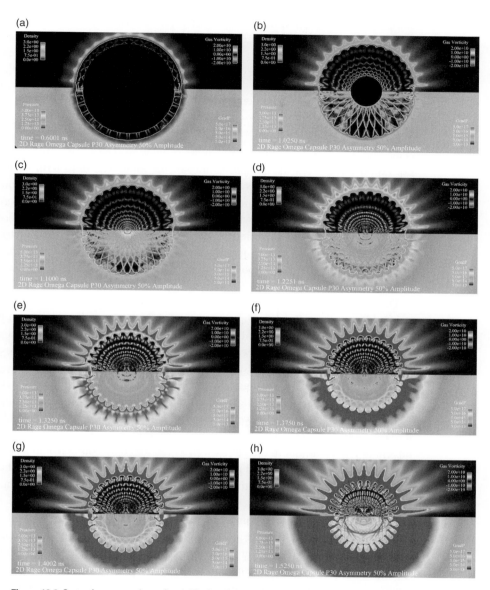

Figure 10.3 Late time snapshots from 2D RAGE simulation of the P30 Omega capsule with $a_{30} = 0.50$. In the lower portion of each panel the CH shell is colored by pressure while the gas is colored by grad P. In the upper portion the CH shell is colored by density while the gas is colored by azimuthal vorticity. A black and white version of this figure will appear in some formats. For the color version, please refer to the plate section.

compressible, inviscid simulation, vorticity production is controlled by the vorticity evolution equation,

$$\frac{\partial \vec{\omega}}{\partial t} + \left(\vec{v} \cdot \nabla \right) \vec{\omega} = \frac{1}{\rho^2} \nabla \rho \times \nabla P - \vec{\omega} \left(\nabla \cdot \vec{v} \right) + \left(\vec{\omega} \cdot \nabla \right) \vec{v} \qquad (10.2)$$

where ρ and P are the scalar density and pressure, and where

$$\vec{\omega} = \nabla \times \vec{v}$$

is the vorticity of the velocity field $\vec{v}$ in the simulation. The second term on the right in Eq. (10.2) arises from the compressibility of the fluid. The third term on the right in Eq. (10.2) is the vortex stretching term, which for axisymmetric 2D simulations is identically zero. The first term on the right in Eq. (10.2), the baroclinic torque term, results in vorticity production wherever pressure and density gradients are not aligned. In particular, where the outgoing reflected gas shock in our 2D asymmetric capsule simulation collides at an oblique angle with the perturbed surface of the shell/gas interface, then vorticity is deposited along the interface as a result of the baroclinic torque produced.

This process can be observed in the $t = 1.325$ ns snapshot of Figure 10.3(e) where red and blue vortex sheets of opposite sign for the azimuthal vorticity can been seen forming along opposite sides of the fingers of shell material as the outgoing reflected gas shock sweeps radially outward over them. These vortex sheets are a result of the baroclinic torque generated by the oblique collision of the gas shock with the perturbed surface of the shell/gas interface. As the large scale fingers develop, much of this vorticity is rolled up into vortex rings of opposite sign that are trapped in the developing bubbles of gas between the growing RM mushrooms. By the end of the 2D RAGE simulation, at $t = 1.525$ ns shown in Figure 10.3(h), there are large scale fingers of shell material in the gas as well as red and blue counterrotating vortex rings in both the body of the gas and in the developing bubbles of gas between the growing RM fingers. Such counterrotating ring pairs in close proximity are typically highly unstable in 3D.

Figure 10.4 illustrates the complex structure in the gas density associated with these counterrotating vortex rings. Figure 10.4(a) shows another snapshot taken from the time sequence in Figure 10.3 at a time $t = 1.50$ ns just after the second bounce of the main gas shock off the capsule center. In the lower portion of Figure 10.4(a) the gas is colored by the pressure gradient to reveal the complex reflected gas shock traveling radially outward. In the lower portion of Figure 10.4(b) the gas is colored by density instead to reveal the intricate structure of density mushrooms in the interior of the gas associated with the counterrotating vortex rings there. Figure 10.4(b) serves to emphasize the complex dynamical nature of the gas, which arises from the nonradial flow in the gas due to drive asymmetry.

Figure 10.5 compares results from six separate 2D RAGE simulations of the OMEGA capsule implosion in which the amplitude of the imposed P30 drive asymmetry is varied with a fixed total drive energy of 17.35 kJ. Simulation results for a_{30} values of 0.00 (spherically symmetric), 0.01, 0.05, 0.10, 0.25, and 0.50 are shown that

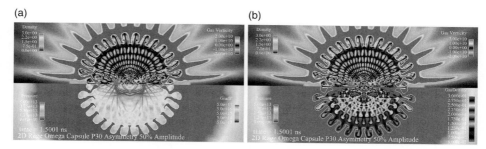

Figure 10.4 Snapshots at a time $t = 1.50$ ns from the 2D RAGE simulation of the idealized OMEGA capsule with $a_{30} = 0.50$. The gas region in the lower portion of the left hand panel is colored by the pressure gradient. The gas region in the lower portion of the right hand panel is colored by the gas density. Note the complex density structure in the interior of the gas, which corresponds to the gas azimuthal vorticity in the upper portion of the right hand panel. The time shown is just after the second bounce of the gas shock off the capsule center which occurred at $t = 1.485$ ns.

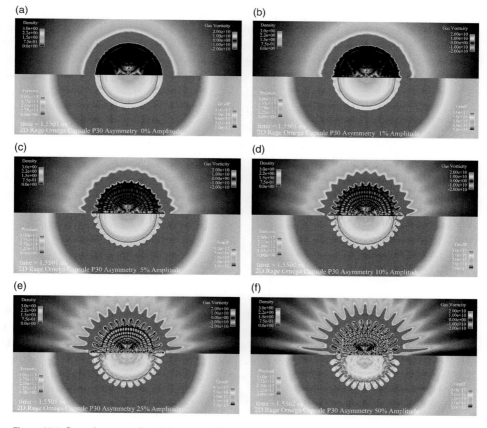

Figure 10.5 Snapshots at a fixed time $t = 1.55$ ns from six separate 2D RAGE simulations of the idealized OMEGA capsule illustrating the effect of varying the amplitude of the P30 asymmetry. Simulation results for a_{30} values of 0.00 (spherically symmetric), 0.01, 0.05, 0.10, 0.25, and 0.50 are shown. The time of these snapshots is chosen to be about 0.20 ns before minimum gas volume in the $a_{30} = 0$ simulation at $t = 1.75$ ns. In the lower portion of each panel the CH shell is colored by pressure while the gas is colored by grad P. In the upper portion the CH shell is colored by density while the gas is colored by azimuthal vorticity.

illustrate the effect of increasing the amplitude of the drive asymmetry. Figure 10.5 shows time snapshots from each of these six simulations at a time $t = 1.55$ ns shortly before stagnation. The convergence ratio of the pusher/gas interface is 5.6 for the spherically symmetric case at this time. Figure 10.5(a) clearly shows that for this choice of drive energy and a resulting convergence ratio of 5.6, the symmetrically driven RAGE simulation with $a_{30} = 0$ gives a round shell with only minor numerical artifacts apparent at the 45° diagonal directions as well as a round main gas shock, even after two bounces of the main shock off the center of the capsule. In contrast, the 2D RAGE simulation shown in Figure 10.5(f), which was driven asymmetrically with a P30 component of amplitude 50% ($a_{30} = 0.50$), shows a great deal of real structure that is due to the imposed drive asymmetry. Fifteen well-defined RM fingers of shell material are visible in this bottom right hand panel that are growing radially inward together with a pattern of strong counterrotating vortices which have been generated in the gas. The snapshots in Figure 10.5 illustrate how such structures continuously emerge as the amplitude of the P30 drive asymmetry is increased from 0% to 50%. Note, for example, how the 15 finger structure of the gas region which is the result of the P30 asymmetry is evident in all the implosions with a nonzero value of a_{30}, even in the 1% amplitude case shown in Figure 10.5(b) where a slight 15-fold perturbation visible on the CH/gas interface reveals the action of the P30 drive asymmetry.

Figure 10.6 shows the same six snapshots at a later time, $t = 1.75$ ns, corresponding to the stagnation of the implosion. Here the spherically symmetric case in Figure 10.6(a) shows the development of irregular RT structure at the CH/gas interface that grows from initial numerical perturbations provided by the grid as would be expected in this physically RT unstable situation. Compare this, however, with Figure 10.6(b), which shows the $a_{30} = 0.01$ case at this same time. Even at a 1% amplitude for the P30 asymmetry the 15 fingers of shell material which arise from the P30 asymmetry are clearly evident. For amplitudes of 5%, 10%, 25%, and 50% the P30 asymmetry overwhelmingly dominates the final structure of the gas region.

As Figure 10.7(a) illustrates, even in the 1% case, at stagnation the fingers of shell material extend inward about 14% of the 51.17 µm radius of the gas region obtained in the 1D RAGE simulation, which corresponds to about 35% of the gas volume being significantly perturbed. For the 5% asymmetric drive (Fig. 10.7(b)) the fingers penetrate more than 26% of the radius of the gas region and about 56% of the total gas volume is perturbed at stagnation. For larger asymmetries the vast majority of the gas volume is severely perturbed by the presence of the fingers at stagnation. These fingers are at well-defined angular positions since their geometry is basically determined by the drive asymmetry followed by growth due to convergence and, later in time, the Richtmeyer–Meshkov instability. Fingers of this type may be related to the angular asymmetries in ρR inferred in the proton spectrometry experiments of Li et al.[2]. With the proper choice of asymmetric drive the shell material fingers might be created in specific locations to test how well one might be able to predict such behavior.

Another numerical artifact that is most clearly visible for the spherically driven case with $a_{30} = 0$ in Figure 10.5(a) is the spurious vorticity generated at the center of the capsule by the spherical collapse of the gas shock. This spurious vorticity cannot

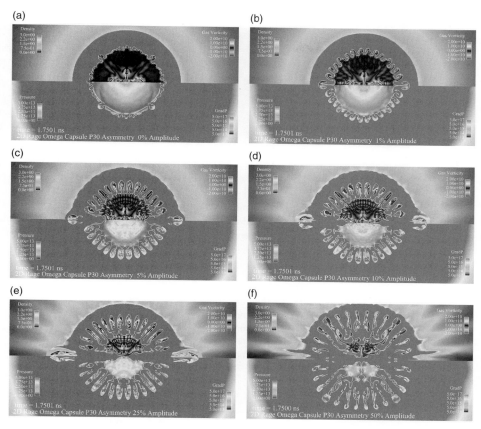

Figure 10.6 Snapshots at a later time $t = 1.75$ ns from the same six 2D RAGE simulations of the idealized OMEGA capsule. The time of these snapshots is chosen to be near minimum gas volume so that all of the implosions have stagnated at this time. Note the appearance, as expected, of irregular RT structure on the CH/gas interface for the case in panel (a) with spherically symmetric drive. Note also that even with an imposed P30 asymmetry of only 1%, a well-defined pattern of 15 fingers of shell material associated with the P30 asymmetry is clearly visible in panel (b).

be eliminated if a fixed Eulerian square grid is used to compute the spherical collapse of the gas shock onto the capsule center, as in these 2D RAGE simulations. As would be expected, a similar vorticity artifact is seen at the capsule center in all of the panels of Figure 10.5 for all values of a_{30} from 0 to 50%. However, because this spurious vorticity occurs at small radius it makes a negligible contribution to the total enstrophy of the simulation. The total enstrophy of one of the 2D simulations can be computed as

$$\int \frac{1}{2} |\omega_\phi|^2 2\pi r \, dr \, dz, \tag{10.3}$$

where ω_ϕ is the azimuthal component of vorticity in cylindrical coordinates (z, r, ϕ) and the integral is extended over the entire axisymmetric 2D simulation volume.

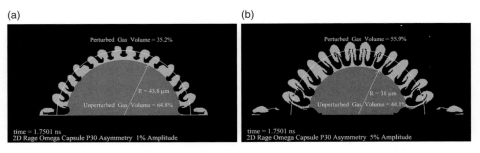

Figure 10.7 Snapshots at a time $t = 1.750$ ns from the two 2D RAGE simulations of the idealized OMEGA capsule with $a_{30} = 0.01$ and 0.05. The gas has been divided into two regions, a central unperturbed core and a region perturbed by the growth of fingers of shell material into the gas. For 1% imposed asymmetry in panel (a) the perturbed region contains 35.2% of the total gas volume. For 5% imposed asymmetry in panel (b) the perturbed region contains 55.9% of the total gas volume. The white circle of radius 51.17 μm represents the position of the CH/gas interface in the 1D RAGE simulation.

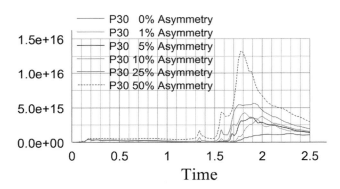

Figure 10.8 Time history for the total enstrophy for each of the six 2D OMEGA simulations of Fig. 10.5 with imposed P30 drive asymmetries of 0, 1, 5, 10, 25, and 50%. Enstrophy is in units of cm^3/sec^2.

Figure 10.8 shows time history plots of the total enstrophy for all six of the 2D RAGE simulations. Early in time before $t = 1$ ns there is a small but nonzero contribution to the total enstrophy from the region outside the capsule as a result of the expansion of the ablated plastic against the background gas. However, by about $t = 1$ ns this contribution has largely dissipated and for the calculations with low drive asymmetry little total enstrophy is generated prior to the first collision of the outgoing reflected gas shock with the incoming CH/gas interface at about $t = 1.34$ ns. The increased total enstrophy for the 50% asymmetry case in the time period prior to $t = 1.34$ ns is primarily a result of the enstrophy associated with the strong vortex rings produced in the body of the gas as seen in Figure 10.3. Sharp increases in total enstrophy are seen in all simulations with nonzero values of a_{30} at times near $t = 1.34$ ns, $t = 1.56$ ns, and $t = 1.68$ ns corresponding to the first, second, and third collisions of the outgoing reflected gas shock with the incoming CH/gas interface. Beyond this time the structure of the reflected shocks traversing the gas becomes increasingly complex and the observed increase in total enstrophy at late time arises from multiple complex gas shock interactions with the perturbed CH/gas interface combined with late time RT growth. The CH/gas interface is more perturbed for larger asymmetries and so the total vorticity production is larger for larger asymmetries.

Of particular note is that the spherically driven simulation in Figure 10.8 shows a very low total enstrophy until after stagnation when the CH/gas interface is grossly RT unstable. For the spherically driven case, no significant enstrophy generation is observed either from the collapse of the gas shock on axis or from the collision of the reflected gas shock at the CH/gas interface. This observation provides quantitative evidence that the spherically driven 2D simulation remains spherically symmetric to a high degree and that the spurious vorticity seen near the origin in these 2D simulations plays only a small role in the overall behavior of the system.

We now turn to a discussion of some detailed scaling studies of our idealized OMEGA capsule performed using 2D RAGE. In these studies we consider variations of the basic P30 implosion that include changing the mode number of the drive asymmetry, amplification of ρR nonuniformities of the CH shell, and the effect of changing the initial fill density of the DT gas and the consequent increase in the convergence ratio of the implosion. For these studies we have adopted a slightly different gridding strategy from the one used for the 2D simulations of the P30 capsule discussed earlier. Here we have utilized the AMR control of the mesh available in RAGE to force a very high spatial resolution of 0.05 μm in the region of the CH pusher shell at the expense of sharply reduced resolution in the body of the DT gas. The purpose of this gridding strategy is to focus computational resources on the dynamical behavior of the CH shell in 2D. We begin by considering the effect of varying the mode number of the asymmetric drive on the implosion.

10.2.1.1 Effect of Varying the Mode Number of the Imposed Asymmetry

Figure 10.9 shows a series of time snapshots at $t = 1.75$ ns, the time of stagnation in the 1D implosion, from eight different 2D RAGE simulations of our idealized OMEGA capsule implosion with varying ℓ values for the imposed drive asymmetry. Figure 10.9(a) shows the case with no imposed asymmetry, the spherically symmetric case. In Figure 10.9(a) only random short wavelength RT structure is visible on the inner surface of the shell as would be expected in this case. For the case of $\ell = 4$ shown in Figure 10.9(b) the long wavelength low order perturbation is visible with superimposed random short wavelength RT structure on the inner surface of the shell. Note, however, that as the order of the perturbation is increased from $\ell = 14$ to $\ell = 20$ (Fig. 10.9(e) though (g)), a coherent structure emerges on the inner surface of the shell containing $\ell/2$ large scale fingers of shell material that protrude into the DT gas. In the final snapshot of Figure 10.9(h) with $\ell = 30$ the structure with 15 large fingers observed in our previous P30 simulations is again seen.

This result is consistent with the well-known elementary estimate (cf. Yabe[9]) that the most dangerous mode for the disruption of an imploding spherical shell with initial radius R_0 and thickness ΔR_0 by RT growth has a mode number of order $\ell = R_0/\Delta R_0$. For the current case with $R_0 = 412.5$ μm and $\Delta R_0 = 25$ μm this simple estimate gives

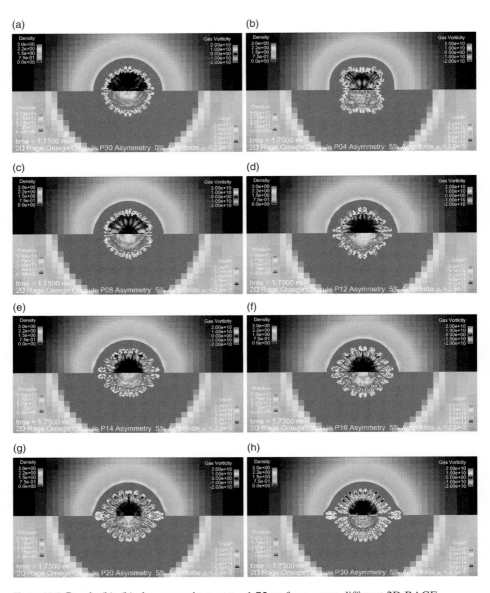

Figure 10.9 Panels (b)–(h) show snapshots at $t = 1.75$ ns from seven different 2D RAGE simulations of the idealized OMEGA capsule with $a_{30} = 0.05$ varying the mode number of the imposed drive asymmetry with $\ell = 4, 8, 12, 14, 16, 20,$ and 30. Panel (a) shows the case with no imposed asymmetry. Near $\ell = 16$ a clear mode structure emerges on the inner surface of the shell with $\ell/2$ fingers of shell material that reflects the symmetry of the imposed perturbation.

$\ell = 16.5$. Thus, from elementary considerations we might expect for modes numbers around $\ell = 16$ to observe the growth of large amplitude penetrations of the DT gas by fingers of capsule material and that is indeed what is observed in the sequence of Figure 10.9.

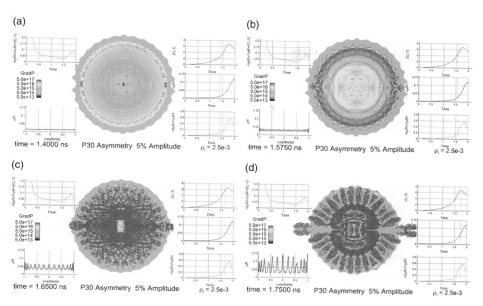

Figure 10.10 Time snapshots from a 2D RAGE simulation of the P30 capsule with $a_{30} = 0.05$ illustrating the rapid growth in ρR asymmetries of the CH shell.

10.2.1.2 Growth and Scaling in the ρR Asymmetries of the CH Shell

A quantity of particular interest experimentally is the angular distribution in the ρR of the imploding shell that has been measured for OMEGA capsules in the proton spectrometry experiments of Li et al.[2] Here we present results for the angular distribution of ρR for the CH shell obtained from a 2D RAGE simulation of our idealized P30 capsule with the nominal gas fill density of $\rho_i = 2.5 \times 10^{-3}$ g/cm^3 and an imposed drive asymmetry of 5%, $a_{30} = 0.05$. The effective ρR along any line of sight from the center of the capsule that passes completely through the imploding shell is defined as the line integral of the shell density along that line of sight. By varying the angular position of the line of sight at a fixed time in the simulation and computing the line integral of the shell density along that line of sight, we can generate the angular distribution for the ρR of the imploding shell. Once this distribution is obtained at each simulation time, both the angular average $\langle \rho R \rangle$ and the rms deviation from this average $\langle \delta \rho R \rangle$ can also be computed as a function of time. Figure 10.10(a)–(d) presents snapshots from four different times in the 5% 2D RAGE simulation. The center of each snapshot shows a view of the capsule with the pusher region colored in gray and the interior of the fill gas colored by the gradient of pressure to illustrate the position of the gas shocks. The plotter in the lower left corner of each snapshot shows the angular distribution of ρR plotted versus $\cos(\theta)$ at the corresponding simulation time. Also indicated on this plotter are the values of $\langle \rho R \rangle$ and $+/-\langle \delta \rho R \rangle$ at that simulation time. The plotter in the lower right corner of each snapshot shows a time history plot of $\langle \delta \rho R \rangle / \langle \rho R \rangle$ with a gray vertical line indicating the current simulation time. The plotter in the middle right region of each snapshot shows a time history plot of $\langle \rho R \rangle$ where, again, a gray vertical line indicates the current simulation time. The plotter in the upper right corner of each snapshot shows a time history of the convergence ratio of the CH/gas interface $C_r = R(t)/R_0$ where the effective interface radius in 2D is computed from the DT gas volume using $R(t) = (3 \cdot V_{gas}/4\pi)^{1/3}$ and where again the current simulation time is indicated by the gray vertical line. The plotter in the upper left corner of each snapshot is a time history of the scaled fractional ρR asymmetry which we will discuss in greater detail shortly.

Figure 10.10(a) shows a simulation time $t = 1.4$ ns slightly after the first collision of the reflected gas shock with the incoming CH shell. The angular distribution of ρR in the lower left plotter shows only low amplitude growth of the ρR asymmetry in the period prior to the first collision of the reflected gas shock. The time history plot of $\langle \delta \rho R \rangle / \langle \rho R \rangle$ in the lower right plotter shows a sharp increase in the slope of the $\langle \delta \rho R \rangle / \langle \rho R \rangle$ growth curve at the time of the first reflected shock collision with the CH shell. $\langle \delta \rho R \rangle / \langle \rho R \rangle = 0.086$ at this time.

Figure 10.10(b) shows a simulation time $t = 1.575$ ns slightly before the second collision of the reflected gas shock with the incoming CH shell. The angular distribution of ρR in the lower left plotter shows noticeable growth as a result of the first shock collision. The time history plot of $\langle \delta \rho R \rangle / \langle \rho R \rangle$ in the lower right plotter shows a much sharper increase in the slope of the $\langle \delta \rho R \rangle / \langle \rho R \rangle$ growth curve at the time of the second shock collision. $\langle \delta \rho R \rangle / \langle \rho R \rangle = 0.172$ at this time.

Figure 10.10(c) shows a simulation time $t = 1.65$ ns near the time of the third shock collision although by this time the shock reflections in the 2D simulation are so complex that a well-defined main gas shock is no longer clearly evident. Here the angular distribution of ρR in the lower left plotter shows substantial growth in the ρR asymmetry as a result of the second shock collision. $\langle \delta \rho R \rangle / \langle \rho R \rangle = 0.339$ at this time.

Finally, Figure 10.10(d) shows a snapshot at stagnation, $t = 1.75$ ns. The angular distribution of ρR in the lower left plotter shows that the ρR asymmetry has grown to large amplitude with $\langle \delta \rho R \rangle / \langle \rho R \rangle = 0.500$ at this time.

The sequence of snapshots in Figure 10.10 illustrates how repeated collisions of the reflected gas shock with the incoming shell amplify the ρR asymmetries until by stagnation fractional rms asymmetries of 50% are observed in the CH shell. Thus, a 5% asymmetry in the drive is seen to produce a 50% asymmetry in the ρR of the shell at stagnation for a convergence ratio of only 8.

In their proton spectrometry measurements of ρR asymmetries in OMEGA experiments, Li et al.[2] have attempted to correlate the measured fractional rms asymmetries $\langle \delta \rho R \rangle / \langle \rho R \rangle$ observed in the CH shell with fractional rms asymmetries $\langle \delta I \rangle / \langle I \rangle$ in the direct drive laser illumination. The relationship they found is given by

$$\frac{\langle \delta \rho R \rangle}{\langle \rho R \rangle} \approx \frac{1}{2}(C_r - 1)\frac{\langle \delta I \rangle}{\langle I \rangle}. \tag{10.4}$$

This relationship is based on a simple 1D scaling argument for the growth of ρR asymmetries from initial velocity perturbations in the shell. It suggests that the scaled fractional asymmetry $\langle \delta \rho R \rangle / \langle \rho R \rangle / (C_r - 1)$ should be of order one half the amplitude of the drive asymmetry. Figure 10.11 shows a time history plot of this quantity for five

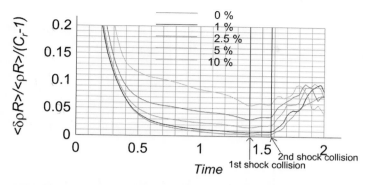

Figure 10.11 Time history of the scaled fractional ρR asymmetry of the CH shell in 2D RAGE simulations of the P30 capsule with a_{30} values of 0.00 (spherically symmetric), 0.01, 0.025, 0.05, and 0.10.

different 2D RAGE simulations with imposed asymmetries of 0, 1, 2.5, 5, and 10% amplitude. The time of the first reflected gas shock collision with the incoming shell is indicated in the figure. It can be readily seen from Figure 10.11 that the scaled fractional asymmetry in ρR is indeed approximately equal to one half the amplitude of the imposed drive asymmetry at the time of this first shock collision. Note, however, that effects of the multiple decelerating shocks destroy this simple scaling relationship and lead to fractional asymmetries at stagnation that are much larger than would be expected from the simple 1D scaling argument.

10.2.1.3 Effect of Varying the Initial Gas Fill Density and Increasing Convergence Ratio

In this section we consider the effect of varying the initial gas fill density on the behavior of the implosion. To study this effect we performed eight different 2D RAGE simulations of the implosion of our idealized OMEGA capsule with a 5% imposed drive asymmetry but with initial gas fill densities ranging from $\rho_i = 2.5 \times 10^{-2}$ g/cm^3, 10 times the nominal fill density of 2.5×10^{-3} g/cm^3, down to $\rho_i = 5.0 \times 10^{-4}$ g/cm^3, one-fifth the nominal fill density. Reducing the initial gas fill has the effect of increasing the final convergence ratio achieved at stagnation. However, this convergence ratio cannot be increased indefinitely because at some point in the process the capsule is completely disrupted and the integrity of the DT gas is totally compromised.

Figure 10.12 shows snapshots of the eight different 2D RAGE simulations with decreasing gas fill at the time of minimum gas volume in each of the simulations. The gas at minimum volume has been divided into two regions: a perturbed region that is heavily penetrated by fingers of shell material and a relatively unperturbed interior core of gas. This division is approximate, and is intended only to provide an estimate of the fraction of total gas volume, which is perturbed and unperturbed in each case. The percentage of the total gas volume that is unperturbed and perturbed is indicated for each simulation along with the approximate value of $(C_r - 1)$ computed by taking $C_r = R(t)/R_0$ where the effective interface radius in 2D is computed from the DT gas volume using

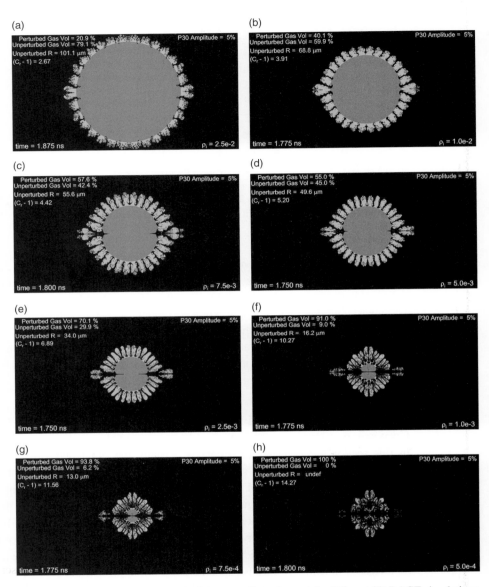

Figure 10.12 Snapshots at the time of minimum volume from eight different 2D RAGE simulations of the P30 capsule with $a_{30} = 0.05$ but with different values for the initial gas fill density ρ_i illustrating the disruption of the capsule for convergence ratios above roughly 13.

$R(t) = \left(3 \cdot V_{gas}/4\pi\right)^{1/3}$ in the usual manner. The radius of this unperturbed gas region is also indicted for each snapshot. Note that the same magnification is utilized in each of these snapshots so that the snapshots can be compared directly with one another.

Figure 10.12(a) shows the case with 10 times nominal fill density. In this case a convergence ratio of 3.7 is achieved at stagnation and only 21% of the total gas volume is perturbed by the intrusion of fingers of shell material. As the fill density is reduced the percentage of the total gas volume that is perturbed by fingers of shell material rises rapidly. Figure 10.12(e) shows the case with the nominal fill density of $\rho_i = 2.5 \times 10^{-3}$ g/cm^3 where 70% of the total gas volume is perturbed and the convergence ratio is 7.9. Figure 10.12(g) shows the simulation with 2/5 the nominal gas fill. Here the percentage of perturbed gas volume has risen to 94% and the convergence ratio is 12.6. Beyond this point it is no longer possible to meaningfully define an unperturbed core of gas, as the final snapshot of Figure 10.12(h) demonstrates. In the cases considered here it is not possible to exceed convergence ratios of 12 to 13 because the capsule is completely disrupted and the integrity of the gas is lost. It is interesting to note that in the experiments of Li et al.[2] no convergence ratios above 12 to 13 were experimentally achieved despite substantial reductions to the fill pressure of the CH capsules.

Figure 10.12 also illustrates the computational challenges of accurately simulating very high convergence flows with RAGE. The snapshots of Figure 10.12(f)–(h) show that for convergence ratios of 11.3 and above, unphysical flattening of the capsule along the 45° directions becomes increasingly pronounced. Accurate simulation of ICF implosions with convergence ratios in the range of 30 to 40 relevant for ignition targets without imposing an artificially high degree of implosion symmetry remains to date an unsolved problem of ICF simulation.

10.2.1.4 Summary of 2D RAGE Simulation Results for the Idealized OMEGA Capsule

We can briefly summarize the results of our 2D RAGE simulations of the idealized OMEGA capsule implosion as follows. For a spherically symmetric drive, the implosion remains to a high degree spherical until stagnation. At stagnation the CH/gas interface exhibits the RT structure one should expect when the system becomes physically RT unstable. For all amplitudes of asymmetric drive from 1% to 50% the density perturbations in the CH shell created by the drive asymmetry are amplified by Bell–Plesset related convergence effects. When acted upon by the outgoing reflected gas shock these perturbations act as seeds for the formation of RM fingers of shell material that penetrate the DT gas. The formation of these fingers is an essentially deterministic process with the number and angular position of the fingers being determined by the asymmetry of the drive. Even for 1% asymmetry and the convergence ratio of 8 achieved in these simulations, a clear pattern of 15 fingers of shell material is observed in the gas at stagnation. This process is very similar to the one described by Li et al.[2] in connection with the ρR asymmetries of imploded shells observed in their charged particle spectrometry experiments on OMEGA.

The 2D simulations also show that the same drive asymmetry that leads to the formation of the RM fingers also results in the formation of coherent vortical structures in the DT gas. Strong, counterrotating vortex rings are observed to form in close proximity both in the body of the DT gas and in the growing bubbles of DT gas which form between the RM fingers. It is well known that such vortical structures are subject to the unstable growth of azimuthal waves in 3D. There are two general classes of such azimuthal instabilities of counterrotating vortex rings. The first is the short wavelength Widnall[6] instability with wavelengths along the azimuthal direction of the rings that are of the order of twice the transverse size of a vortex core. The second is the long wavelength Crow[7] instability between a pair of counterrotating vortex rings with a wavelength along the azimuthal direction that is typically 6 to 8 times the separation distance between the rings. Generalizations of these simple idealized configurations involving many counterrotating vortex rings interacting simultaneously are also possible. In the next section we investigate the stability of these strong, coherent vortical structures directly by performing 3D simulations initialized from our 50% asymmetrically driven 2D RAGE simulation.

10.2.2   Linked 3D RAGE Simulations of the Late Time Implosion

In our initial presentation of late time 3D RAGE simulation results for our idealized OMEGA implosion, we will focus on the case in which the P30 drive asymmetry has the relatively large amplitude of 50%. We have chosen this large amplitude for the P30 component in order to exaggerate the effect of the asymmetry on the implosion and to more clearly exhibit the mechanism that links the asymmetry to the development of turbulent jets in the fuel. Later we will present results from a 3D RAGE simulation with a P30 drive asymmetry but with only a 5% imposed asymmetry. This later simulation demonstrates quite clearly that even with only a modest 5% asymmetry and a low convergence of 8, turbulent gas jets can still form in a time that is relevant to the "compression burn." We begin with a consideration of the detailed simulation results for the 50% case.

A central property of the implosion that we are not directly modeling is the actual Reynolds number of the flow. Our 3D RAGE simulations are Eulerian and as such the shortest resolved wavelengths are limited by the numerical dissipation on the AMR grid rather than a physical parameter such as the viscosity. The actual Reynolds number of the flow is likely much too large to permit a direct numerical simulation using the Navier–Stokes equations. In addition, detailed modeling of the viscosity would require accurate calculation of the temperatures. The high temperatures in the center may indeed lead to high viscosity but electron conduction in the small regions between the pusher fingers would likely lead to lower temperatures and much lower viscosity making the bubble regions late in time high Reynolds number flow. A qualitative understanding of at least some of the possible aspects of high Reynolds number flow in this system can be obtained by performing a spatial resolution study using 3D RAGE. The RAGE code allows the user to directly control the maximum level of mesh refinement in regions adjacent to material interfaces. This capability is particularly useful in the current problem since much of the vorticity generation and evolution observed in this problem occurs in the spatial regions near the CH/gas interface as a result of the interaction between the reflected gas shocks and the interface. Using 3D RAGE we performed a series of three simulations with increasing AMR spatial resolution near the CH/gas interface, all of which were initialized from the same 2D RAGE implosion at $t = 1.4$ ns. The first of these 3D RAGE simulations was linked from the 2D RAGE simulation of Figures 10.2 and 10.3 at $t = 1.4$ ns, spun into 3D as an octant of the full problem and continued forward with a fixed maximum AMR spatial resolution near the CH/gas interface of 0.20 μm from link time out to a final problem time of $t = 1.938$ ns, a time well past minimum volume. In the second 3D RAGE simulation, the calculation was again linked at $t = 1.4$ ns and run from link time to $t = 1.5$ ns with a maximum AMR resolution of 0.20 μm. Then at $t = 1.5$ ns the maximum AMR resolution was increased to 0.10 μm and the problem was continued out to a final problem time of $t = 1.75$ ns, corresponding to the time of stagnation in the 1D RAGE simulation. In the third 3D RAGE simulation, the problem was again linked at $t = 1.4$ ns and run from t = 1.4 ns to t = 1.5 ns with a maximum AMR resolution of 0.20 μm. From $t = 1.5$ ns to $t = 1.6$ ns the maximum AMR resolution was increased to 0.10 μm. Finally, at $t = 1.6$ ns the AMR resolution was further increased to 0.05 μm and the problem was continued out to a final problem time of $t = 1.71$ ns, corresponding to a time just prior to the time of stagnation in the 1D RAGE simulation. By $t = 1.71$ ns, the final problem time achieved, the total number of AMR grid cells in 3D had grown to nearly 1×10^9. For convenience we refer to these three different 3D RAGE simulations as the 0.20 μm, 0.10 μm, and 0.05 μm simulations respectively.

The procedure followed in all three of our 3D RAGE simulations was to run our P30 implosion simulation of the OMEGA capsule from $t = 0$ out to a link time of $t = 1.4$ ns as an axisymmetric 2D RAGE simulation. Then at the chosen link time of $t = 1.4$ ns we rotated one quadrant of the 2D axisymmetric problem into 3D to create a 3D octant version of the axisymmetric data. Figure 10.13(a) shows a 3D view of the octant at link time. This octant was used to initialize a 3D RAGE problem to continue the simulation to late time. An octant of the full capsule geometry was chosen to maximize the spatial resolution achieved in the 3D simulations.

The chosen link time of $t = 1.4$ ns corresponds to a time just after the first collision of the outgoing reflected gas shock with the incoming CH/gas interface, a time just after the beginning of significant enstrophy production in the 2D 50% asymmetry simulation as Figure 10.8 illustrates. And since we needed to maintain high spatial resolution in the DT gas, linking at a significantly earlier time would have required running a much larger 3D simulation since the DT gas volume decreases rapidly during the implosion. Thus, the choice of $t = 1.4$ ns for the link time represents a balance between physics fidelity and computational practicality for the 3D problem.

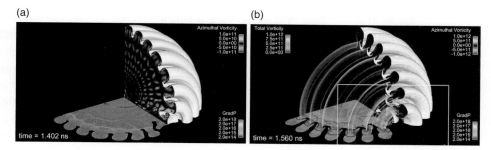

Figure 10.13 Snapshots at (a) link time $t = 1.40$ ns and (b) at a later time t = 1.56 ns, the time of the second collision of the reflected gas shock with the incoming CH capsule surface, for the linked 0.05 μm 3D RAGE simulation of the idealized OMEGA capsule with $a_{30} = 0.50$. In panel (b) the reflected gas shock can be seen traveling up the developing gas bubbles depositing sheets of vorticity which are visible in the volume rendered representation of the total vorticity.

Figure 10.13(b) shows a view of the capsule octant at a slightly later time $t = 1.56$ ns, the time of the second collision of the reflected gas shock with the incoming CH shell, in the 0.05 μm 3D RAGE simulation of capsule implosion. In Figure 10.13(b) only the central gas region of the 3D octant is shown with the surrounding CH shell material removed. The white surface represents the CH/gas interface. The vertical face of the gas is colored by the azimuthal vorticity and the horizontal face is colored by the gradient of pressure. The total vorticity in the interior of the gas is displayed in volume rendered representation with a transfer function for color and opacity chosen to visualize regions of the flow with total vorticity above 5×10^{11} sec^{-1}, a technique that is effective in revealing the dynamical evolution of the vortex cores in the interior of the gas bubbles. In Figure 10.13(b) the reflected gas shock can be seen traveling up the gas bubbles between the developing fingers of shell material depositing sheets of vorticity of opposite sign along the opposite sides of the bubbles that are visible in the volume-rendered representation of the total vorticity.

Figure 10.14 focuses in on the spatial region of Figure 10.13(b) near the polar axis of the capsule enclosed by the white rectangle. Figure 10.14 is a sequence of eight time snapshots from the 0.05 μm 3D RAGE simulation that show a close up view of the late time evolution of the vorticity in the gas bubbles nearest the polar axis of the capsule illustrating how the growth of azimuthal instabilities on the strong, counterrotating vortex rings present in the gas bubbles leads to turbulence in the bubbles and the resulting interpenetration of CH and gas.

In Figure 10.14(a) vortex sheets of opposite sign can be seen along opposite sides of the gas bubble nearest the polar axis. These sheets are pushed together by the radial convergence of RM fingers and are quickly rolled up into counterrotating vortex rings that are trapped in the bubble and that immediately begin to undergo azimuthal instability growth in 3D (Fig. 10.14(b)–(c)) with a long wavelength Crow instability[7] clearly evident. This instability first appears in the interior of the gas and it is only when the increasingly turbulent vorticity of the gas flow impacts the CH surface (Figs. 10.14(e)–(h)) that fully 3D interpenetration and mixing occurs. This process may be envisioned as the injection of a gas jet into the bubble though the constriction caused by convergence of the RM fingers and the impact of this turbulent jet on the CH shell wall. A similar process can be seen occurring in the other bubbles as well with azimuthal instabilities of the short wavelength Widnall[6], long wavelength Crow[7], and combined type[8] all contributing to the very rapid evolution of the coherent vortical structures trapped in the bubbles to a fully turbulent state. As time progresses in the sequence of Figure 10.14 the vortical structures in the bubble become increasingly complex and vortical structures of smaller and smaller spatial scale appear. In animations of these vortical structures we observe reconnection of vortex loops and it appears that repeated reconnection of closed vortex loops in 3D is the process that drives the cascade to smaller and smaller length scales observed in the panels of Figure 10.14.

Our linked P30 simulation is intended not to model the results of an actual OMEGA experiment but rather to demonstrate the generic effects of an asymmetry in the pressure drive in a simple capsule geometry of physical interest. For this purpose we have deliberately chosen to impose a perturbation from spherical symmetry in the pressure drive that is simple, axisymmetric, and relatively large and whose symmetry is constant

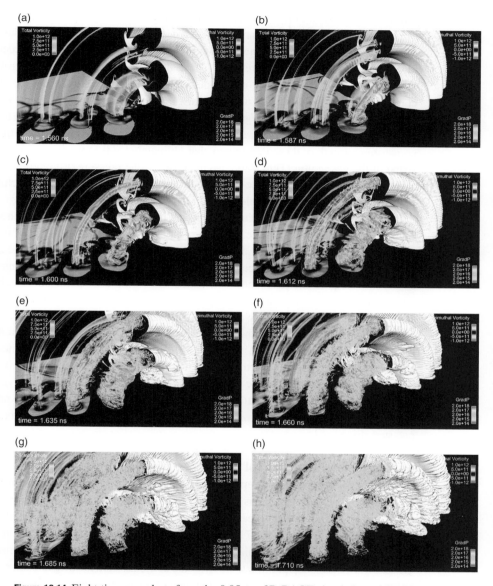

Figure 10.14 Eight time snapshots from the 0.05 μm 3D RAGE simulation of the P30 OMEGA capsule. Close up view of the development of the turbulence in the two bubbles nearest the polar axis. A black and white version of this figure will appear in some formats. For the color version, please refer to the plate section.

in time. As we have seen, this initial drive asymmetry induces nonradial motions of the gas leading to the formation of vortex sheets in the gas. These vortex sheets are then rolled up into counterrotating vortex rings by the outgoing reflected gas shock. The reflected gas shock also deposits intense sheets of azimuthal vorticity of opposite sign on nearby regions of the CH/gas interface. These vortical structures are unstable to azimuthal instabilities in 3D and quickly evolve to a fully turbulent state in which the turbulent vorticity in the gas adjacent to the CH/gas interface drives the interpenetration of gas and shell material. This same drive asymmetry also imprints a perturbation on the plastic shell by nonradial flow away from high pressure regions of the asymmetric drive into regions with low drive pressure. These density enhancements then grow due to convergence (the Bell–Plesset effect). Later in time, depending upon the degree of convergence and the amplitude of the initial asymmetry, these body perturbations induce surface perturbations in the CH/gas interface. These perturbations act as seeds for the RM instability as the outgoing reflected gas shock collides with the interface. It is the radially converging growth of these RM fingers that ultimately leads to the formation of the final shape of the gas region. Thus, large asymmetries in the pressure drive produce both the RM fingers, which determine the shape of the gas region, and the vortical structures in the gas, which lead to the gas turbulence and mix via 3D azimuthal instabilities.

In this simple example we can see, clearly exhibited, a mechanism connecting the asymmetry of the pressure drive on the capsule with the development of turbulence and mix. This mechanism is fully 3D in nature. The asymmetry of the drive leads to the formation of a well-defined pattern of coherent vortical structures, counterrotating sheets and rings in the gas. These structures are unstable to azimuthal instabilities in three dimensions and it is the unstable evolution in 3D of these coherent structures that leads to the development of the gas turbulence and mix.

Further, it can be seen in this simple example that both the number and angular distribution of the fingers and the spatial distribution of the gas turbulence and its associated mix are largely deterministic in nature and are a direct result of the deviations from spherical symmetry in the pressure drive. The unstable growth of the gas turbulence from coherent vortical structures is a stochastic process. But the nature and spatial distribution of these coherent vortical structures is determined by the asymmetry of the pressure drive.

10.2.2.1   Spatial Resolution Study of 3D RAGE Simulations

An overview of the results from a resolution study comparing the 0.20 μm, 0.10 μm, and 0.05 μm 3D simulations is summarized in Figure 10.15, which shows the same close up view of the bubbles nearest the polar axis for each of the three simulations at a fixed simulation time of $t = 1.71$ ns, the last simulation time where data from all three simulation resolutions are available, as the spatial resolution of the 3D simulation is increased. In each snapshot the total vorticity in the interior of the gas is displayed in a volume-rendered representation with the same transfer function for color and opacity chosen to visualize regions of the flow with total vorticity above $5 \times 10^{11} \text{ sec}^{-1}$. The peak vorticity observed at $t = 1.71$ ns in the simulations grows from $9.2 \times 10^{11} \text{ sec}^{-1}$ to $5.3 \times 10^{12} \text{ sec}^{-1}$ as the resolution is increased from 0.20 μm to 0.05 μm. As the resolution progresses from 0.20 μm in Figure 10.15(a) to 0.05 μm in Figure 10.15(c), the degree of turbulent development observed in the gas bubbles nearest the polar axis at a fixed simulation time steadily increases with a corresponding increase in the three dimensional structure of the CH/gas interface as a result of the interaction between the gas turbulence and the adjacent CH/gas interface consistent with the higher resolution calculations representing a higher effective Reynolds number. To the extent that the real

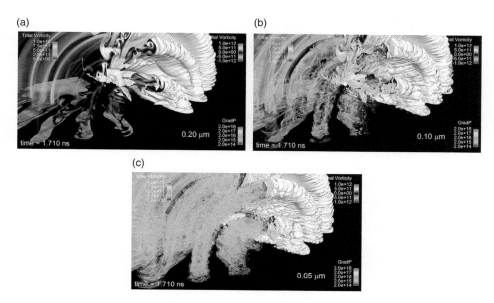

Figure 10.15 Effect of spatial resolution on the development of turbulent structure at a fixed simulation time of $t = 1.71$ ns. The maximum permitted AMR resolution for each 3D RAGE simulation is, from top to bottom, 0.20 μm, 0.10 μm, and 0.05 μm.

Reynolds number is significantly higher than that of the calculation, then the expected level of 3D effects would also be elevated compared with these calculations.

A standard method for demonstrating that turbulence has been observed in a fluid simulation is to compute the power spectrum of the kinetic energy for the simulation and look for the appearance of an inertial subrange in the spectrum. An inertial subrange is a region of the spectrum in k space over which energy cascades from longer to shorter wavelengths before being dissipated as heat either by viscosity in the case of a real fluid or by the effective dissipation terms of the numerical method as in the case of our current inviscid simulations. The kinetic energy spectrum in the inertial subrange typically has a power law dependence of the form $k^{-\alpha}$, where for homogeneous, isotropic turbulence this power takes the Kolmogorov value of $\alpha = 5/3$.

To demonstrate the presence of turbulence in our 3D RAGE simulation, our procedure was to select the fixed square spatial box 33 μm on a side shown in Figure 10.16(a). This box was chosen to completely enclose the two turbulent bubbles nearest the polar axis. Values within the box for the density and the three Cartesian velocity components were resampled from the AMR mesh onto a uniform mesh whose spatial resolution was equal to the finest resolution used in the corresponding AMR mesh. The resampled uniform mesh data was then used to compute the power spectrum by the following method.

By Parseval's theorem the total kinetic energy within the box can be written in the form

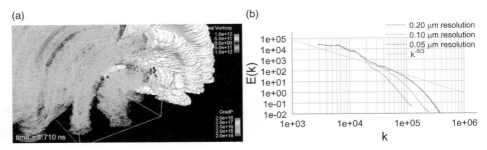

Figure 10.16 (a) Box showing the sampling region selected for Fourier analysis. This box contains the two turbulent rings nearest the polar axis of the capsule. (b) Power spectra of the total kinetic energy at $t = 1.71$ ns for the three spatial resolutions shown in Figs. 10.15. At a spatial resolution of 0.05 μm an inertial subrange is emerging that is consistent with a Kolmogorov type spectrum.

$$\int \frac{1}{2} \rho |\vec{u}|^2 d^3\vec{x} = \int \vec{H}(k) \cdot \vec{H}(k)^* d^3\vec{k} = \int E(k)dk, \tag{10.5}$$

where $k = |\vec{k}|$. $\vec{H}(k)$ is the Fourier transform of the mass-weighted velocity $\sqrt{\frac{\rho}{2}} \, \vec{u}$ and

$$E(k) = 4\pi k^2 \vec{H}(k) \cdot \vec{H}(k)^* \tag{10.6}$$

is the spectral power of the kinetic energy between k and $k + dk$.

Using the resampled data on the uniform mesh, an FFT was utilized to compute the discrete Fourier transform values for $\vec{H}(k)$ on a mesh in 3D k space. $E(k)$ is then computed by averaging over the directions in k space within a shell between k and $k + dk$. Figure 10.16(b) shows the resultant power spectra of the kinetic energy for each of the three simulations at spatial resolutions of 0.20 μm, 0.10 μm, and 0.05 μm at time $t = 1.71$ ns.

The spectrum for the 0.20 μm simulation shows no clear evidence that an inertial subrange has yet developed in the turbulence growing in the bubbles near the polar axis for the 0.20 μm simulation by $t = 1.71$ ns. However, Figure 10.16(b) shows that as the spatial resolution of the simulation is increased, an inertial subrange appears to emerge in the turbulence near the axis at $t = 1.71$ ns. In the spectrum for the 0.05 μm simulation at $t = 1.71$ ns, an inertial subrange is present with a Kolmogorov type $k^{-5/3}$ power law scaling from $k = 25,000$ cm^{-1} out to $k = 80,000$ cm^{-1} followed by a steep dissipation regime at higher values of k. The location of the start of the dissipation region at $k = 80,000$ cm^{-1} corresponds to a length scale of about 16 grid cells in the 0.05 μm RAGE simulation. The spectrum for the 0.10 μm resolution case shows a similar short inertial subrange which again starts at around $k = 25,000$ cm^{-1} and falls off into the dissipation regime at about $k = 40,000$ cm^{-1}, again corresponding to a length scale of 16 grid cells in the 0.10 μm simulation. The 0.20 μm simulation at $t = 1.71$ ns cannot see this inertial subrange at all because the dissipation regime begins at $k = 20,000$ cm^{-1} in the 0.20 μm case. The results of this spectral analysis demonstrate that the turbulence near the polar axis observed at $t = 1.71$ ns in the 0.05 μm 3D RAGE simulation exhibits an inertial subrange with $k^{-5/3}$ power law scaling at a time prior to the stagnation time $t = 1.75$ ns obtained in the 1D RAGE simulation.

The appearance of turbulence with a Kolmogrov inertial subrange in our 0.05 μm 3D RAGE simulation is significant for ICF applications because it demonstrates how asymmetry can lead to hydrodynamic turbulence by way of instability of large coherent features in a time that is short enough to be of interest for degrading the compression burn of an ICF implosion. This result presents a plausible alternative picture to the usual paradigm that degradation in ICF implosions is caused by growth of small scale surface imperfections by acceleration driven instabilities and/or shock induced instabilities. These results may also provide a plausible answer to the question of how the vortices at small scales in turbulent flows begin. One direct answer to this question is to note the appearance in our simulations of the small scale Widnall type waves visible in the volume rendered representation of the total vorticity. These waves have wavelengths of the order of the vortex core size, which is determined by the residual dissipation of the numerical simulation. Thus, from the large scale coherent structures significant wave activity can be generated near the dissipation region of the fluctuations by the appearance of short wavelength instabilities of the Widnall type. A different source of enhanced fluctuations at intermediate scale appears to result from longer wavelength Crow type instabilities of the large coherent structures. Such waves are seen, for example, in the $t = 1.587$ ns panel of Figure 10.14(b). It has been proposed that these Crow instabilities lead to vortex reconnection loops with smaller scales than the original structures and this process then repeats as the energy in longer wavelengths cascades to the dissipation regime (Takakai and Hussain [10] and Kerr and Hussain [11]). We may, in fact, be observing this process in the 0.05 μm simulation discussed earlier. More work is required to clarify the role of repeated reconnection of vortex loops in creating the inertial cascade observed in our 3D ICF simulations. We do note, however, that both processes appear to be fast enough to populate the inertial range with a sizeable amount of fluctuations allowing the generation of the inertial range shown in Figure 10.16(b). Both processes are the result of hydrodynamic instability of coherent large scale hydrodynamic structures created by asymmetry of the drive. Beautiful experimental observations and a comprehensive discussion of the development of both short wavelength and long wavelength azimuthal instabilities in a system of two counterrotating fluid vortices may be found in the paper of Leweke and Williamson [8].

10.2.2.2 Enstrophy Production in the 3D RAGE Simulations

A more quantitative measure of the turbulent state of the gas in our 3D RAGE simulation is the total enstrophy defined as the volume integral,

$$\int \frac{1}{2} |\vec{\omega}|^2 d^3\vec{x} = \int \frac{1}{2} \left\{ \omega_z^2 + \omega_r^2 + \omega_\phi^2 \right\} d^3\vec{x}, \tag{10.7}$$

where

$$\vec{\omega} = \nabla \times \vec{v}$$

is the vorticity of the velocity field $\vec{v}$ in the 3D simulation. $(\omega_z, \omega_r, \omega_\phi)$ are the axial, radial, and azimuthal components of vorticity in cylindrical coordinates (z, r, ϕ) and the integral is extended over the entire 3D simulation volume. Quantitative information on the evolution of the gas turbulence in the 3D RAGE simulations of Figure 10.15 can be obtained by examining the time history of the total enstrophy as well as the separate contributions to the total enstrophy from the axial, radial, and azimuthal components of the vorticity in Eq. (10.7).

In Figure 10.17(a) the total enstrophy and the azimuthal, radial, and axial enstrophy are shown for the 3D RAGE simulation with fixed 0.20 μm maximum AMR resolution. For a fully axisymmetric simulation the total enstrophy would exactly equal the azimuthal enstrophy and the nonazimuthal radial and axial contributions to the total enstrophy would be identically zero. In Figure 10.17(a) we see that the azimuthal enstrophy tracks the total enstrophy very closely until a time just slightly after $t = 1.6$ ns corresponding to the beginning of turbulence growth in the gas bubble nearest the polar axis. By $t = 1.71$ ns significant nonazimuthal radial and axial contributions to the total enstrophy have begun to be apparent and the time period from

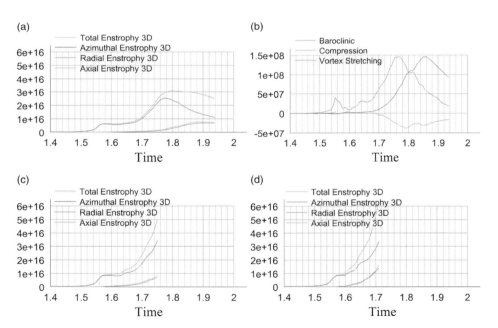

Figure 10.17 (a) Time history for the total enstrophy and for all three components of the enstrophy of the gas for the 0.20 μm simulation in 3D. The axial and radial components track closely and grow as the turbulent rings develop. (b) Time history for the three enstrophy source terms in the total enstrophy production Eq. (10.9). The production of nonazimuthal enstrophy observed in panel (a) corresponds to the vortex stretching term. (c) Time history for the total enstrophy and for all three components of the enstrophy of the gas for the 0.10 μm simulation in 3D. (d) Time history for the total enstrophy and for all three components of the enstrophy of the gas for the 0.05 μm simulation in 3D. Enstrophy is in units of cm³/sec².

$t = 1.71$ ns to $t = 1.938$ ns corresponds to a period of rapid growth in the nonazimuthal components of the total enstrophy. During this same period the azimuthal component reaches a peak at $t = 1.77$ ns and then falls rapidly until, by the end of the simulation at $t = 1.938$ ns, all three components of the total enstrophy have become comparable in magnitude as would be the case in isotropic turbulence. For the maximum AMR resolution 0.20 μm simulation only a small amount of nonazimuthal enstrophy growth is observed by $t = 1.75$ ns, the time of stagnation in the 1D simulation of the implosion, consistent with Figures 10.15(a) and 10.16(b).

More detailed quantitative information can be obtained by considering the Euler equation for the enstrophy production. If we define the quantity Ω as

$$\Omega \equiv \tfrac{1}{2}|\vec{\omega}|^2,$$

then the enstrophy production equation can written in the form

$$\frac{\partial \Omega}{\partial t} + \nabla \cdot \left(\Omega \vec{v}\right) = \frac{1}{\rho^2} \vec{\omega} \cdot (\nabla \rho \times \nabla P) - \Omega \left(\nabla \cdot \vec{v}\right) + \omega^i S_{ij} \omega^j \qquad (10.8)$$

where the quantity

$$S_{ij} = \frac{1}{2}\left(\frac{\partial v^i}{\partial x^j} + \frac{\partial v^j}{\partial x^i}\right)$$

is the strain rate tensor and summation over the repeated indices i and j is understood. No explicit dissipation terms have been included in Eq. (10.8), which corresponds to the enstrophy production by the inviscid Euler equations. Such terms will, of course, arise from the numerical hydro method used in RAGE.

Integrating Eq. (10.8) over the total volume of the simulation and applying the divergence theorem to the second term on the left to convert it to a surface integral over the boundary of the simulation volume yields a simple equation for the time rate of production of the total enstrophy,

$$\frac{\partial}{\partial t} \int \Omega d^3 \vec{x} =$$
$$\int \left[\frac{1}{\rho^2} \vec{\omega} \cdot (\nabla \rho \times \nabla P) - \Omega \left(\nabla \cdot \vec{v}\right) + \omega^i S_{ij} \omega^j \right] d^3 \vec{x}, \qquad (10.9)$$

where the surface terms on the left side are identically zero because of the reflecting boundary conditions at the boundary of the simulation volume. The first term on the right is the baroclinic enstrophy production term, the second is the compression term, and the third is the vortex stretching term. The vortex stretching term exists only in 3D and plays a dominant role in the development of 3D turbulence.

Figure 10.17(b) shows the time histories for the baroclinic, compressible, and vortex stretching enstrophy production terms in Eq. (10.9) obtained from our 3D RAGE simulation with fixed maximum AMR resolution of 0.20 μm. It can be seen that at early times the enstrophy production is predominantly baroclinic whereas at later time the increase in the radial and axial components of the enstrophy observed in

Figure 10.17(a) are associated with a significant enstrophy production contribution from the vortex stretching term. Significant growth of the vortex stretching contribution is seen beginning at roughly $t = 1.68$ ns and reaching a peak value at around $t = 1.855$ ns. Contributions from the compressible term remain relatively small at all times and are negative after minimum volume in the 3D simulation as would be expected.

A more quantitative characterization of the turbulence growth as a function of resolution can again be obtained by considering the time behavior of the nonazimuthal components of the enstrophy for each of the three resolutions considered. Recall that the nonazimuthal enstrophy components, the radial and axial enstrophy components, are initially zero at link time since each of our 3D simulations is initialized from an 2D axisymmetric simulation in which both the radial and axial components of the vorticity are identically zero. The nonazimuthal enstrophy components grow in time as a result of the development of fully 3D turbulence and provide a useful quantitative measure of the degree of turbulence growth. It is therefore of interest to compare the time development of the nonazimuthal enstrophy components for each of the three resolutions.

The time history of total enstrophy and of all three enstrophy components for the 0.20 μm, 0.10 μm, and 0.05 μm simulations are shown in Figure 10.17(a), (c), and (d), respectively. Comparison of the time histories of nonazimuthal enstrophy components in Figure 10.17(a), (c), and (d) show that the nonazimuthal enstrophy grows more rapidly in time as the spatial resolution of the simulation is increased. For example, comparing Figure 10.17(a) and (c) shows that the total enstrophy at $t = 1.71$ ns in the 0.10 μm simulation is 2.1 times greater than the corresponding value in the 0.20 μm simulation, while the sum of nonazimuthal enstrophy components, radial plus axial, at $t = 1.71$ ns in the 0.10 μm simulation is 5.3 times greater than the corresponding value in the 0.20 μm simulation. Similarly, the total enstrophy at $t = 1.71$ ns in the 0.05 μm simulation is 2.22 times greater than the total enstrophy in the 0.10 μm simulation, while the sum of nonazimuthal enstrophy components at $t = 1.71$ ns in the 0.05 μm simulation is 3.83 times greater than the corresponding value in the 0.10 μm simulation at $t = 1.71$ ns. Overall, as the spatial resolution of the simulation is increased from 0.20 μm to 0.05 μm, the total enstrophy at $t = 1.71$ ns increases by a factor of 4.67, while the sum of the nonazimuthal enstrophy components at $t = 1.71$ ns increases by a factor of 20.37. The observed growth of the nonazimuthal enstrophy components further emphasizes the point that for higher resolution 3D RAGE simulations, the turbulence develops more rapidly in time.

10.2.2.3 3D RAGE Simulations of the 5% Amplitude Case

We now consider a 3D RAGE simulation for the case of a 5% imposed asymmetry in our simplified ICF implosion system. This simulation was performed using the same 2D–3D linking procedure previously utilized for the case of a 50% imposed asymmetry. Here the simulation was run with an AMR resolution of 0.05 μm near the CH/gas interface from the chosen link time of $t = 1.6$ ns. This later link time was chosen in order to focus on the late time hydrodynamics occurring near stagnation.

The sequence of eight time snapshots of the full capsule for the case of 5% imposed asymmetry is shown in Figure 10.18. In each snapshot the white surface, as usual,

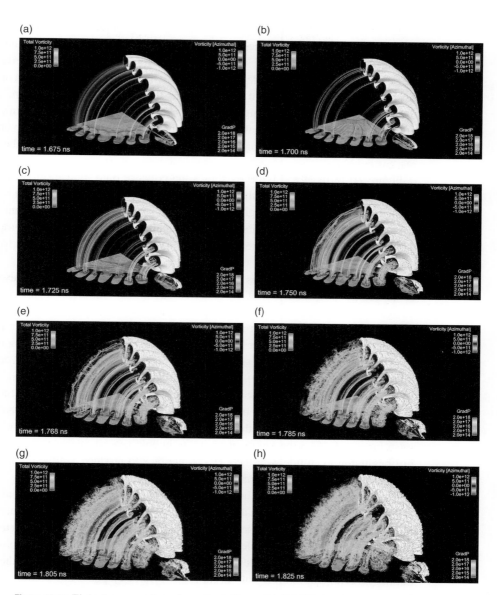

Figure 10.18 Eight time snapshots from the 0.05 μm 3D RAGE simulation of the P30 Omega capsule with a 5% imposed asymmetry $a_{30} = 0.05$. Full view of the capsule showing the turbulent development of the bubbles.

represents the CH/gas interface. The vertical face of the gas is colored by the azimuthal vorticity and the horizontal face is colored by the gradient of pressure. The total vorticity in the interior of the gas is displayed in a volume rendered representation with the same transfer function as that used in Figure 10.14 for the 50% case.

Figure 10.18(a) at $t = 1.675$ ns shows the implosion just after the third shock collision with the CH/gas interface. Note the complex reflections off the perturbed CH/gas interface that can be seen in Figure 10.18(a) traveling inward toward the center of the capsule. Part of the incident shock can also be seen still traveling in the radial direction toward the end of the gas bubbles in Figure 10.18(a). Since the interface is strongly perturbed at this time, the incident shocks will in general be misaligned with the density gradients, producing significant shear layers at the CH/gas interface that are visible as the vortex sheets forming along the edges of the fingers in Figure 10.18(a). As time progresses these sheets are, as we have already pointed out, rolled up into vortex rings of opposite sign in the azimuthal vorticity that are trapped together inside the developing gas bubbles.

Figure 10.18(b) at $t = 1.700$ ns shows the collapse onto the capsule center of this complex reflected gas shock, which bounces off the capsule center at about $t = 1.708$ ns.

Figure 10.18(c) shows a slightly later time $t = 1.725$ ns with the resultant outgoing reflected gas shock again moving radially outward. Note by the time of Figure 10.18(c) that both short and long wavelength unstable azimuthal waves have begun to grow on the vortex rings trapped in the two bubbles nearest the polar axis.

Figure 10.18(d) at $t = 1.75$ ns is approximately at stagnation time and is also the time of the fourth shock collision of the reflected gas shock with the now highly perturbed CH/gas interface. The vortex rings trapped in the two gas bubbles nearest the polar axis have already become turbulent at this time and unstable waves have begun to grow to large amplitude in several other bubbles as well. The fourth gas shock collision can also be seen generating additional strong vortex sheets along the edges of the developing fingers.

The high pressure gas at the center of the capsule is driving high speed jets of gas through the now constricted bubble openings, which can be viewed as "nozzles" forming turbulent jets of gas that traverse the developing gas bubbles. By $t = 1.768$ ns in Figure 10.18(e) the turbulent jet in the bubble nearest the polar axis can be seen just beginning to collide with the CH/gas interface to produce as time progresses fully 3D interpenetration of the gas and CH.

Figure 10.18(f)–(h) shows the appearance and growth of numerous additional long and short wavelength azimuthal instabilities on the vortex rings in the other bubbles, their turbulent evolution, and the development of significant perturbations in the CH/gas interface as the resulting turbulent jets impact the interface.

Figure 10.19 illustrates the development of the turbulent jets observed in the 5% case. Figure 10.19(a) shows a 2D cut on the lower symmetry plane in the simulation at the time $t = 1.768$ ns slightly after stagnation corresponding to Figure 10.18(e). Four panels are shown with the gas colored by the azimuthal vorticity on the top left, by velocity on the top right, by pressure on the bottom left, and by the density on the bottom right. At this time the fingers of shell material converging radially inward have penetrated deeply into the gas constricting the bubble openings. The 6 gigabar pressure peak at the center of the capsule forces a high speed flow of gas radially outward through the constrictions and into the bubbles at speeds of up to 500 km/sec. This flow carries with it the embedded vorticity generated as a result of the gas shock collisions with the incoming perturbed CH/gas interface. As Figure 10.19(b) shows, this embedded vorticity is becoming turbulent on a very short timescale as it is carried across the bubbles by the

(a)

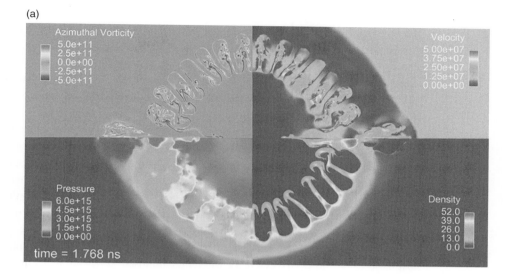

(b)

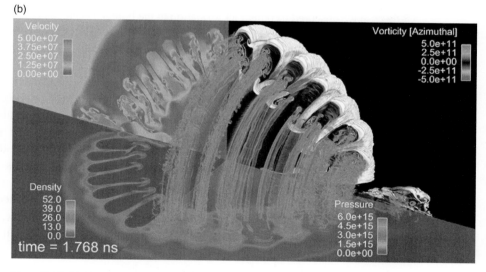

Figure 10.19 (a) Four panels showing azimuthal vorticity, velocity, pressure and density on a symmetry plane of the 5% simulation of Fig. 10.18. Note the 6 gigabar pressure in the capsule center which drives turbulent gas jets outward with velocities of 500 km/sec. (b) Another view of the four hydrodynamic quantities of panel (a) in the 5% 3D RAGE simulation of the P30 capsule with the developing turbulent vorticity shown in a volume rendered representation. The white surface is, as usual, the CH/gas interface. A black and white version of this figure will appear in some formats. For the color version, please refer to the plate section.

bulk gas flow until it impacts the CH/gas interface as a turbulent jet producing 3D interpenetration of gas and CH. At the time pictured the turbulent jet in the bubble on the far right in Figure 10.19(b), the bubble nearest the polar axis, can be seen just beginning to impact the CH/gas interface at the end of the bubble.

This phenomenon is then clearly seen to be a convergence effect since higher convergence leads to more significant penetration of the fingers for a given asymmetry, greater constriction in the bubble openings, and higher relative pressures. Convergence also forces the strong, counterrotating vortex rings trapped inside the bubbles closer together, further increasing the growth rate of the azimuthal instabilities that lead to the turbulent evolution of the rings. All of these effects should work together to create stronger and presumably more unstable jets as convergence is increased. It should be noted that the situation in an ICF implosion is somewhat more complicated than that which occurs for a simple gas jet because many regions near the evolving fingers of shell material have significant embedded vortical flow from previous encounters with the reflected shocks. This vortical flow induced earlier in the implosion may become unstable earlier in time, but in any event may be expected to be carried into the bubbles by the jet flow and further contribute to the turbulence in the bubbles.

Comparing the case with a 50% imposed asymmetry with that of the 5% case leads to several observations. First, more of the cavity is hydrodynamically disrupted for the 50% case as would be expected for the larger imposed asymmetry. Second, the 50% case is unstable in 3D earlier in time than the case with 5% imposed asymmetry. In the 50% case we observed turbulent jet formation in the bubbles nearest the polar axis of the capsule shortly after the second collision of the reflected gas shock with the CH/gas interface. In the 5% case turbulent jet formation was delayed until after the third collision of the reflected gas shock with the CH/gas interface. Again, this is as might be expected since the case with the larger imposed asymmetry generates more enstrophy, as seen from the 2D simulations, but also exhibits stronger convergence effects as the fingers of shell material penetrate earlier and farther into the fuel. This also leads to earlier formation of the gas jets in the more perturbed case. Nonetheless, the behavior of both the 50% and 5% asymmetry cases show turbulence generation and transport occurring in the same generic fashion. More simulations, including ones with fully 3D asymmetries, would help to elucidate just how generic this behavior is. That the processes observed in these 3D simulations are seen at low radial convergences of 8 and drive asymmetries as small as 5% suggests that this mechanism of turbulence generation may be significant for many real high convergence ICF implosions even with quite small drive asymmetries.

10.2.3   Advantages and Limitations of RAGE for ICF Implosions

Traditionally, ICF implosion simulations have been performed using 2D Lagrangian codes with spatial resolution in the gas so coarse that only the main gas shock is visible in the simulation. Low resolution 2D Lagrangian codes are ill suited to the problem of asymmetric implosions because the high shear flows that naturally arise in this case are handled poorly if at all by such codes. In contrast, the high resolution adaptive Eulerian technique used in RAGE robustly handles high shear flows and, as we have seen, our 2D RAGE simulations of asymmetric implosions reveal an enormous wealth of essential detail about the dynamical behavior of shocks and vorticity in the DT gas that remain largely unseen in low resolution 2D Lagrangian calculations.

3D RAGE has an additional important property that makes it well suited for modeling the generation of the gas turbulence in ICF implosions. The numerical hydro scheme being used in RAGE is a monotone integrated large eddy simulation or MILES method [12]. In such methods no explicit subgrid model is required to enforce the physically correct dissipation of energy at small length scales in the modeling of turbulent flows. Rather, the necessary physics of conservation, monotonicity, causality, and locality required to correctly model the dissipation are all built into the numerical hydro method itself. For such methods, it has been shown that if sufficient spatial resolution is utilized in 3D to capture most of the energy containing length scales in the flow, then the turbulent structure of the flow is correctly calculated. Hence, we can have some confidence that the structure of the turbulence being observed in our 3D ICF simulations is being correctly calculated.

RAGE also presents some limitations for ICF applications. One of these is the fact that the RAGE simulations being presented here are inviscid. For many ICF implosions the center of the implosion may in fact have a relatively large Spitzer viscosity and a corresponding low Reynolds number because the temperature in this central region is high. As a result vortical structures generated in this region may not lead to significant turbulence generation. However, near the pusher wall the temperature is likely to be much lower, especially if effects such as electron thermal conduction between the relatively cool pusher and the adjacent DT gas are important. In these regions the viscosity is likely to be much lower and the Reynolds number of the flow much higher than in the center of the DT fuel. As we have seen in our RAGE simulations the majority of the enstrophy generation in our example ICF implosion takes place in these regions adjacent to the CH/gas interface. It is thus reasonable to hope that our inviscid simulation results may accurately reflect how turbulent mixing is really generated in an asymmetrically driven ICF implosion.

It is worth pointing out that some direct experimental evidence exists that changing the viscosity of the DT fuel can have a measureable impact on the performance of the capsule. In experiments on the OMEGA laser, Wilson et al.[13] reported that adding very small admixtures of high Z components to the DT gas had a deleterious effect on the observed performance of the capsule. One effect of these high Z contaminants is to decrease the viscosity of the DT gas and increase its effective Reynolds number. Does the degraded yield observed when high Z contaminants are added to the fuel reflect a decreased viscosity for the gas and an increase in turbulent mixing? This may be an interesting area for future investigations.

An important simplification introduced in the RAGE simulations presented here is that we have considered only the case of an imposed drive asymmetry that is axisymmetric. Our simulations have shown, even in this simple case of an axisymmetric perturbation in the pressure drive, that the resultant turbulent flow has a fully 3D character. While it is almost certainly true that drive asymmetries in real ICF implosions are fully 3D in nature, a fully 3D asymmetry in the pressure drive is unlikely to lead to a significantly altered physical picture for how the turbulence arises in the implosion. A fully 3D asymmetry in the pressure drive will still cause nonradial flow in the drive shell, leading to density enhancements that will be further amplified by Bell–Plesset related convergence effects. These enhancements will act as seeds for the growth of fully 3D RM fingers of shell material that will penetrate the gas. At the same time coherent vortical structures will again be created in the gas by the same 3D asymmetries in the pressure drive. The resultant vortex rings will have a more fully 3D character than the highly axisymmetric rings seen in the case of an axisymmetric perturbation but will undergo the same unstable interactions in 3D nevertheless. Hence, the physical picture of an asymmetric drive leading to coherent vortical structures in 3D, which, in turn lead to turbulence, remains relatively unchanged. And this process should be viewed as essentially deterministic since a well-defined final state is achieved with a given pressure drive.

A significant limitation in the use of RAGE for ICF applications is its inability to simulate the implosion of very high convergence systems without the appearance of significant numerical artifacts. In the aforementioned 2D RAGE simulations we limited ourselves to implosions of the example OMEGA capsule in which the convergence ratio of the initial to final pusher/gas interface radius was about 8. For this case, as we have shown, the implosion of a round capsule with a spherically symmetric pressure drive gives a final configuration just before stagnation with only minor numerical artifacts. However, we have noted empirically that for convergence ratios somewhat greater than 11, RAGE gives increasingly noticeable numerical artifacts for a round capsule that is symmetrically driven. Unfortunately, this means that very high convergence problems of great interest such as the NIF ignition target with a convergence ratio of the order of 30 cannot be directly simulated using RAGE. However, despite this limitation, as we have shown, the 2D and 3D RAGE simulations provide a great deal of insight into the physical processes that lead to turbulent mix in ICF targets.

We also note more generally that the entire problem of simulating high convergence ICF implosions is unresolved. We do not know under what conditions any given type of simulation can correctly calculate the behavior of asymmetrically driven ICF implosions, especially at high convergence.

10.3   Implications for NIF Ignition

As we have already mentioned, a direct RAGE simulation of NIF ignition implosions is not possible because the convergence ratio for the NIF ignition target is of order 30 and this is much too high for credible RAGE calculations of the implosion from $t = 0$. Nevertheless, the physical picture of the relationship between drive asymmetry and turbulence generation in an ICF implosion that emerges from the RAGE simulations presented earlier has potentially important implications for NIF ignition.

The NIF ignition campaign is, of course, an indirect drive experimental program in which a round capsule is enclosed in a cylindrical gold lined hohlraum designed to absorb and reradiate the incoming laser energy in order to produce a symmetric radiation drive on the capsule. A recent point design for the capsule itself consists of a germanium doped CH ablator shell enclosing a cryogenic DT ice layer with a central region containing DT gas. When the ablator absorbs the drive energy and blows off, the capsule is compressed to very high convergence. Most of the DT fuel remains cold and is compressed to high density while the central region of DT gas, the so-called hot spot, heats to thermonuclear temperatures. When the hot spot ignites, the thermonuclear burn propagates out into the surrounding cold compressed fuel producing a high gain yield.

The details of how this compression is achieved are of some interest. The 20 ns long laser drive pulse has four distinct power peaks that launch four separate drive shocks into the capsule, which are designed to coalesce into a single large shock several micrometers inside the inner surface of the DT ice layer. This shock then bounces several times between the capsule center and the DT ice layer in order to build up sufficient temperature and ρR in the hot spot region to achieve ignition.

Our RAGE simulation results raise several physics issues about the basic NIF ignition scheme outlined earlier. In this discussion the central DT gas region of the NIF hot spot plays the role of the DT gas fill in our example RAGE implosion while the DT ice layer plays the role of the cold pusher shell. We have already seen in our 2D RAGE simulations that at a convergence ratio of 8, even a 1% amplitude for the drive asymmetry was sufficient to cause significant penetration of macroscopic RM fingers of pusher material into the DT gas whose structure is determined directly by the asymmetry. In the context of the NIF ignition target with its convergence ratio of 30, this suggests that even a few percent amplitude of low to intermediate ℓ mode perturbations in the drive symmetry may be sufficient to produce macroscopic fingers of DT ice that penetrate the hot spot. And unlike our simplified RAGE implosion simulations using an axisymmetric drive perturbation, the density enhancements in the DT ice that might arise for the case of an asymmetric drive in NIF could have a more intrinsically 3D character. The convergence effects may be more pronounced for fully 3D drive perturbations because there are two directions of convergence for 3D perturbations versus a single direction of convergence for purely 2D axisymmetric perturbations. Thus, Bell–Plesset effects with 3D asymmetries may lead to longer spikes in the gas earlier in time.

In any case such fingers of cold material in the hot spot would be deleterious for several reasons. First, the effective ρR of the hot spot would be reduced because the cold fingers absorb alpha particle energy instead of the hot spot, impeding hot spot ignition. Second, the cold fingers displace the hot spot gas to larger radius, resulting in lower density and lower temperature, again impeding hot spot ignition. Finally, we have seen in our 3D RAGE simulations that the coherent vortical structures deposited in the bubbles between fingers by collisions of the reflected gas shock with the interface lead to turbulence in 3D on a timescale that is short compared to the implosion time. Hence, we might expect that turbulent mixing of the hot spot gas in the bubbles with the fingers of cold DT ice material could provide a significant impediment to hot spot ignition. Each time the reflected gas shock hits the ice/hot spot interface this effect would be exacerbated. For the NIF ignition target several such collisions occur in the course of the target compression to ignition.

The expected experimental signature of this turbulent mixing phenomenon would be similar to the behavior observed in the high Z experiments of Wilson et al.[13] and the intentionally asymmetric implosion experiments of Rygg et al.[1], where significantly reduced thermonuclear burning was noticed after the reflected gas shock hit the pusher (i.e., the DT ice layer) for the first time. Further degradation will occur with subsequent shock collisions. This possibility is in contrast to the expectations of Wilson et al.[14], where an asymmetric hot spot is expected to show significant deviation in burn rate from the ideal 1D situation only late in time.

One additional aspect of the NIF ignition target is the four drive shocks that are designed to coalesce into one as they exit the ice layer. If the asymmetry of the radiation drive is changing over the time period of the laser pulse due to laser plasma interactions or other effects, then the individual shocks may be asymmetric and may, in fact, have different asymmetries. This is a new, more complex possibility than the case of the simple drive pulse with constant drive asymmetry that we considered in our RAGE simulations. In this case, it is possible for additional vorticity production to occur in the gas near the DT ice layer as a result of the interactions of these shocks, which could lead to further turbulent mixing of the hot spot and the DT ice layer.

It has been commonly supposed that mixing in the NIF ignition target is dominated by large ℓ mode RT instabilities. Our 2D and 3D RAGE simulations suggest, however, an additional possibility. For very high convergence targets such as the NIF ignition target, even small asymmetries in the drive can lead to penetration of the hot spot by macroscopic fingers of cold fuel. The same drive asymmetry that leads to these fingers also leads to the coherent vortical structures in the gas whose 3D instability generates turbulent mixing of the hot spot gas with the cold fuel. Experimentally, these two different routes to degraded capsule performance, one due to amplification of small scale perturbations by acceleration induced instabilities, the other due to turbulent evolution of coherent vortical structures in 3D created by large scale drive asymmetries, may not be distinguishable without spatially and temporally resolved measurements of the hot spot and cold ice layer. However, the overall x-ray emission from the assembly may allow inferences about the symmetry of the inside of the implosion. Here the question to ask is whether the overall symmetry is spherical, in which case one might expect the degradation of the burn to result from small scale imperfections in the target. However, for the case in which significant nonspherical features are seen in the x-ray emission images even at large radii, it seems fair to infer the presence of significant asymmetry in the actual implosion of the fuel. In this case the yield degradation might be dominated by effects such as the ones seen in our 3D RAGE simulations.

10.4   Conclusions

In this chapter we have used high resolution 2D and 3D RAGE simulations to examine the effect of imposing a simple P30 drive asymmetry on the implosion of an example ICF capsule similar in design to plastic capsules fielded in actual OMEGA experiments. In the 2D simulations we showed for the case with zero imposed drive asymmetry that 2D RAGE gave a spherically symmetric implosion with only minor numerical artifacts out to the time of stagnation for a convergence ratio of around 8. For this symmetrically driven case, RT growth seeded by numerical perturbations associated with the AMR grid was observed at stagnation as would be physically expected in this case. However, for the cases with a nonzero amplitude for the P30 drive asymmetry the nonradial mass flow induced in the shell by the asymmetric drive leads to a well-defined pattern of density and pressure enhancements in the shell that are further amplified by Bell–Plesset related convergence effects. When the outgoing reflected gas shock collides with the incoming shell, these density enhancements act as seeds for the growth of RM fingers of shell material that penetrate the DT fuel. In our 2D simulation study varying the amplitude of the asymmetry we showed even for a 1% amplitude for the P30 asymmetry and a convergence ratio of 8 that a clear pattern of 15 fingers of shell material was observed to dominate the structure of the CH/gas interface at stagnation as a result of the action of the P30 asymmetry. At the same time the asymmetry of the drive leads directly to the formation of a pattern of counter-rotating vortex rings in the body of the gas, and the multiple collisions of the reflected gas shock with the perturbed shell result in the formation of strong counterrotating vortical structures that are trapped in the DT gas bubbles between the converging RM fingers.

In our linked 3D RAGE simulations with the 50% asymmetry we saw that the counterrotating vortical structures trapped in the bubbles are unstable in 3D to both short and long wavelength azimuthal instabilities that quickly lead to turbulence in the gas bubbles and complex 3D interpenetration of gas and shell material driven by the turbulent vorticity. We showed that the turbulence observed in our 3D RAGE simulations has all of the properties one expects of fully 3D turbulence, including a power spectrum for the kinetic energy with an inertial subrange that is well fit by the Kolmogorov $k^{-5/3}$ law. Finally, we showed that as the spatial resolution of the simulation is increased in 3D, the development of the turbulence occurs more rapidly in time, suggesting, for sufficiently well-resolved 3D simulations, that all of the gas bubbles may become fully turbulent before stagnation.

We also demonstrated that even an implosion with 5% imposed asymmetry and a relatively low convergence ratio of 8 displayed similar phenomena near the time of hydrodynamic stagnation, suggesting that the hydrodynamics seen in these simulations is quite relevant to real ICF implosions. Furthermore, the unifying concept of turbulent jets was introduced as a way to understand the hydrodynamics near stagnation. This process with the turbulent jets has also been demonstrated to be a convergence effect, possibly providing a qualitative explanation for the experimental observation that high convergence ICF implosions are much more degraded than predicted by current calculations. The NIC failure is a good example of this problem.

It is important to emphasize that the existence of these jets and their turbulent evolution is a convergence effect, making the study of them problematic in planar geometry. Further, this process is not a surface instability but rather an azimuthal instability of the coherent vortical structures in the body of the gas. The surface is surely affected, however, by the impact of the turbulent gas jet upon the pusher surface.

Our convergence scaling calculations in 2D demonstrated the profound effect convergence has on the integrity of our simplified ICF implosion. For instance, at the 5% asymmetry level, the capsule cannot withstand a convergence much larger than 10 without complete hydrodynamic disruption. Also, our results have shown that simple scaling for asymmetry growth due to convergence does not hold once the reflected shocks hit the interface. Therefore conclusions for high convergence ICF implosions must depend largely on calculations. This presents unique challenges since the real ICF capsule asymmetries are likely fully 3D and time dependent and the calculations themselves have not been demonstrated to give physical answers at high convergence. In the ICF community one often hears the assertion that mix in ICF implosions arises from the amplification of a spectrum of small initial surface imperfections in the capsule by the action of acceleration induced instabilities such as the RT and RM instabilities. The 3D simulations presented here provide evidence for an additional mechanism for mix in ICF capsule implosions. Asymmetries in the pressure drive on the shell create both fingers of shell material that intrude into the DT fuel as well as the coherent vortical structures in the gas bubbles between the fingers that are unstable in 3D to azimuthal instabilities of the Widnall and Crow type. These azimuthal instabilities quickly lead to the development of turbulence in the DT gas in the form of outward going turbulent jets. The subsequent interaction between this gas turbulence and the CH/gas interface creates a complex 3D structure for this interface, an interpenetration of the gas and metal that represents a type of turbulent mixing in 3D. Thus, in this mechanism the drive asymmetry is responsible for the creation of the coherent vortical structures in the gas and it is the unstable evolution of these structures in 3D that is the source of the gas turbulence. While the detailed turbulent evolution of the vortex rings in 3D is a stochastic process, the nature and distribution of the fingers of shell material and of the coherent vortical structures in the gas are largely deterministic, a result of the detailed asymmetry of the pressure drive.

The 2D and 3D simulation results presented here emphasize the critical role of drive symmetry in ICF ignition attempts such as the ongoing ignition experiments at the NIF. We have already seen in these simulations that asymmetries in the capsule's pressure drive lead to both turbulence and fingers of shell material in the DT fuel. In the aforementioned P30 example we imposed a relatively large drive asymmetry of 50% in an implosion with a relatively low convergence ratio of only around 8 in order to exaggerate these effects for purposes of illustration. But in high convergence systems we might expect to see similar effects for much smaller initial asymmetries in the laser radiation drive. In the NIF ignition target the convergence ratio is around 30, more than three times that of the example OMEGA implosion considered here. In this circumstance it might be expected that only a few percent asymmetry in the laser radiation drive could lead to significant turbulence in the fuel and fingers of DT ice mixing into the central hot spot. An additional complication in the NIF ignition target is the fact that multiple shocks are used to implode the capsule. Any differences in symmetry between these shocks could lead to production of additional vorticity and associated turbulence in the hot spot. Evidence for these phenomena may be seen in the ongoing NIF ignition experiments.

Our RAGE simulations serve to clarify the essential role of convergence and shear in an asymmetrically driven ICF implosion. Pressure drive asymmetries on the capsule lead to nonradial flow in the shell, which, in turn, leads to macroscopic density enhancements in the shell that grow in time as a result of Bell–Plesset related convergence effects. In our 2D RAGE simulations we saw that this Bell–Plesset growth was not just a surface effect but was rather associated with density enhancements throughout the body of the shell. This is in contrast to the typical treatment of the Bell–Plesset effect for incompressible perturbations, which allows for uniform compression of the shell only and does not treat the compressible case with strong density gradients in the body of the shell. In the compressible case considered here the Bell–Plesset growth of the density perturbations leads to radially inward and outward extension of the shell. These macroscopic perturbations act as seeds for the development of RM fingers of shell material in the gas as a result of multiple interactions with the reflected gas shock. This same interaction results in strong shear flow near the fingers in the form of counter-rotating vortical structures in the gas. Radial convergence pushes both the fingers and their associated vortical flow closer together as the implosion progresses. This radial convergence pushing the fingers together creates nozzles by which the high pressure in the center of the gas creates high speed flows into the gas bubbles trapped between the fingers. This radial convergence also guarantees that the counterrotating vortical structures will suffer azimuthal instability growth in three dimensions resulting in rapid development of turbulence in the bubbles trapped between the fingers. It is this turbulent growth that leads to interpenetration of shell material and gas that represents fully 3D mixing. The physical picture described here suggests why it is particularly difficult to achieve the ideal 1D yield in a high convergence ICF implosion.

The issue being addressed here is one of fundamental importance: Is the yield degradation observed in ICF implosions typically determined by small scale surface features or by large scale drive asymmetries? We suggest that the large scale coherent structures associated with asymmetries in the pressure drive may play an important role in determining the yield performance of ICF implosion systems, especially at high convergence.

10.5   Acknowledgments

The authors would like to thank the National Nuclear Security Administration's Advanced Simulation and Computing program for its support of this work and for providing computing resources on the ASC Purple and Cielo supercomputers. Thanks also to the members of the RAGE development team at LANL for their help with the use of the RAGE code. Finally, we would like to add our sincere thanks to Anders Grimsrud and the EnSight development team at Computational Engineering International, Inc., for their help with the use of CEI's EnSight visualization software, which was used to create all of the visualizations that appear in this paper. This work was performed under the auspices of the U. S. Department of Energy by Los Alamos National Security (LANS), LLC under Contract No. DE-AC52-06NA25396.

References

1. J.R. Rygg, J.A. Frenje, C.K. Li, F.H. Seguin, R.D. Petrasso, F.J. Marshall, J.A. Delettrez, J.P. Knauer, D.D. Meyerhofer, and C. Stoeckl, "Observations of the collapse of asymmetrically driven convergent shocks," *Phys. Plasmas* **15**, 034505 (2008).
2. C.K. Li, F.H. Seguin, J.A. Frenje, R.D. Petrasso, J.A. Delettrez, P.W. McKenty, T.C. Sangster, R.L. Keck, J.M. Soures, F.J. Marshall, D.D. Meyerhofer, V.N. Goncharov, J.P. Knauer, P.B. Radha, S.P. Regan, and W. Seka, "Effects of nonuniform illumination on implosion asymmetry in direct-drive inertial confinement fusion", *Phys. Rev. Lett.* **92**, 205001 (2004).
3. M.S. Plesset, "On the Stability of Fluid Flows with Spherical Symmetry," *J. Appl. Phys.* **25**, 96 (1954). G. I. Bell, "Taylor instability of cylinders and spheres in the small amplitude approximation," *Los Alamos National Laboratory Report LA-1321*, (November, 1951).
4. V.A. Thomas and R.J. Kares, "Drive asymmetry and the origin of turbulence in an ICF implosion," *Phys. Rev. Lett.* **109**, 075044 (2012).
5. M. Gittings, R. Weaver, M. Clover, T. Betlach, N. Byrne, R. Coker, E. Dendy, R. Hueckstaedt, K. New, W.R. Oakes, D. Ranta, and R. Stafan, "The RAGE radiation-hydrodynamic code," *Comput. Sci. Disc.* **1**, 015005 (2008).
6. S.E. Widnall, D.B. Bliss, and C.Y. Tsai, "The instability of short waves on a vortex ring," *J. Fluid Mech.* **66**, 33 (1974).
7. S.C. Crow, "Stability theory for a pair of trailing vortices," *AIAA J.* **8**, 2172 (1970).
8. T. Leweke and C.H.K. Williamson, "Cooperative elliptic instability of a vortex pair," *J. Fluid Mech.* **360**, 85 (1998).
9. T. Yabe, "The compression phase in ICF targets," in *Nuclear Fusion by Inertial Confinement: A Comprehensive Treatise*, editors G. Velarde, Y. Ronan, and J. Martinez-Val, CRC Press, Inc., 283 (1993).
10. R. Takaki and A.K.M.F. Hussain, "Recombination of vortex filaments and its role in aerodynamic noise," *Fifth Symp. Turb. Shear Flows, Cornell U.*, 3.19–3.26, (1985).
11. R.M. Kerr and F. Hussain, "Simulation of vortex reconnection," *Physica* **D 37**, 474 (1989).
12. F. F. Grinstein, L.G. Margolin, and W.J. Rider, *Implicit Large Eddy Simulation*, Cambridge University Press (2007).
13. D.C. Wilson, G.A. Kyrala, J.F. Benage Jr., F.J. Wysocki, M.A. Gunderson, W.J. Garbett, V. Yu. Glebov, J. Frenje, B. Yaakobi, H.W. Hermann, J.H. Coley, L. Welser-Sherril, C.J. Horsfield, and S.A. Roberts, "The effects of pre-mix on burn in ICF capsules," *Journal of Physics: Conference Series* **112** (2008).
14. D.C. Wilson, P.A. Bradley, C.J. Cerjan, J.D. Salmonson, B.K. Spears, S.P. Hatchett II, H.W. Hermann, and V. Yu. Glebov, "Diagnosing ignition with DT reaction history," *Review of Scientific Instruments* **79**, 10E525 (2008).

11 Rayleigh–Taylor Driven Turbulence

Nicholas A. Denissen, Jon Reisner, Malcolm J. Andrews, and Bertrand Rollin

11.1 Introduction

Modeling turbulent, multimaterial mixing is important in theoretical and practical applications. Specifically, the behavior of turbulence in the presence of strong density gradients and compressibility is fundamental in applications ranging from inertial confinement fusion (ICF) [1], supernovae [2], and environmental flows. The dominant physical mechanisms at work in variable density turbulence include Kelvin–Helmholtz (KH) [3, 4], Rayleigh–Taylor (RT) [5, 6], and Richtmyer–Meshkov (RM) [7, 8] instabilities. All three mechanisms must be accounted for in a unified way in multiphysics simulations. Even in simplified test problems, such unstably stratified, unsteady, inhomogeneous flows pose a challenge to turbulence modeling.

RT mixing, where heavy fluid is placed (unstably) over a light fluid under the influence of gravity, is the focus of the present work. Experimental work [9–14] provides visualizations of the mixing process as well as limited measurements of the turbulent statistics. Recent work has focused on high resolution numerical simulations of the RT mixing process. These simulations include large eddy simulations (LES) [15–20] and direct numerical simulations (DNS) [21–24] and provide an important level of detailed analysis not possible with current experiments alone.

In addition to LES and DNS, Reynolds-averaged Navier–Stokes (RANS) models have been proposed to model the buoyancy driven mixing process across a range of complexity. At one end of the range, two equation models [25–28], supplemented with additional production terms to account for buoyancy effects, have been used. Single fluid second moment closures [29] add another level of complexity, and multifluid models [16, 30–32] capture additional physics such as demixing, but may be challenging to incorporate into existing single fluid codes, and require additional interspecies closures that are less well studied than traditional RANS approaches.

To address the span of KH, RT, and RM mechanisms, the Besnard-Harlow-Rauenzahn (BHR) family of variable density turbulence models have been developed and implemented on multiple platforms at Los Alamos National Laboratory to model the necessary physics of compressible turbulent mixing [33]. Importantly, the BHR framework allows for basic and elaborate RANS models, and incorporates insight from a multifluid perspective. A three equation implementation has been demonstrated for RT, KH, and RM[34], as has a second moment closure implementation [35]. The present work focuses on a third variant, a four equation turbulence model known as BHR-2 [36, 37].

The effectiveness of the BHR-2 implementation has been demonstrated for variable density mixing in the KH, RT, and RM cases in an Eulerian framework [36, 37]. The present work uses an implementation of BHR-2 in the arbitrary Lagrangian–Eulerian (ALE) hydrodynamics code FLAG [38]. The consistency between the codes has been demonstrated for one-dimensional (1D) test problems [39], here we examine the model for a 2D test problem.

The various models described here have been validated using canonical RT/RM/KH flows where the mixing is occurring in 1D [16, 20, 27, 29, 34, 35]. However, in many problems of interest, the interface is not planar. There can be significant curvature of the interface as well as dynamic interface movement. Understanding how RANS models simulate curved interfaces, and the interaction of turbulence and the mean motion, is essential for complex applications. Validating these models remains an important task. In this paper we demonstrate implicit large eddy simulations (ILES) as an important tool for simulating complex high Reynolds number flows in their own right, and for validating RANS models as well.

Here we focus on the tilted rocket rig experiments that include multifluid mixing and a dynamic interface [11, 40–42]. In these experiments a heavy fluid initially was placed in a tank below a light fluid (a stable configuration). Rocket motors fired, accelerating the tank downward effectively changing the direction of "gravity" so that the system was RT unstable. The acceleration was documented, and photographic images were taken of the growing mixing layer. For the "tilted rig," the initial tank was effectively at an angle relative to gravity, so when the rockets fired the acceleration was not perpendicular to the interface. This resulted in RT mixing and a bulk overturning motion in the fluid that strained the interface. Since the overturning introduces significant, large scale, 2D effects on the mix region, the present work uses the tilted rocket rig to validate the BHR-2 model. This problem has been studied experimentally [40–42], and numerically using ILES [43–45] and DNS [46]. The present work contrasts multiple ILES simulations, with and without explicit viscosity, with state-of-the-art RANS and DNS simulations.

11.2   Modeling and Computational Approach

Despite advances in computational power, RANS models remain the tool of choice for many multiphysics problems such as ICF [28, 47]. However, ILES has become a powerful tool for simulations and validating RANS models and is contributing to our understanding in important ways [48, 49]. The RANS models must give a useful estimate of the amount of turbulent mixing, reproduce canonical KH, RT, and RM behavior, and function in one-, two-, and three-dimensional computations. Further, it should produce grid converged results at manageable resolution, as the turbulent mixing is often a small part of the overall computation. The BHR family of mix models fits these criteria, as they reproduce the physics of variable density turbulent mixing without ad hoc source terms. The computational savings of the RANS model are obvious in the present work, where the RANS solutions use three orders of magnitude fewer computational cells compared with the ILES. Similarly, the computational savings of ILES over the DNS is significant, and they provide a realistic simulation of 3D mixing. Successful validation of these models is needed for future multiphysics simulations.

11.2.1 Details about the Codes Used for the Study

The RANS simulations are performed in the ALE code FLAG [50, 51]. FLAG is a multidimensional (1D, 2D, and 3D), multiphysics code, that solves the bulk hydrodynamic equations in a Lagrangian reference frame on an arbitrary polyhedral mesh. FLAG uses a single bulk velocity, but separate mass and energy equations are solved for each species [38]. This results in conservation equations for species mass fraction, c_i, and specific internal energy, e_i.

The equations are discretized using a compatible, staggered grid formulation [52], and, when called for, the code follows the Lagrangian hydrodynamics step with a mesh relaxation/remap step [53]. Time integration for the hydrodynamics is a second-order, predictor–corrector scheme[51]. The initial BHR framework in FLAG was implemented by Waltz and Gianakon [38]. The present BHR-2 model has been implemented by the authors, and validated against other implementations and canonical RT experiments and DNS [39].

For comparison with the RANS results, ILES computations are performed using RTI3D. RTI3D is an ILES code developed for RT type flows that solves the incompressible, multispecies, Euler, or Navier–Stokes equations [43]. Advection is up to third order, van Leer limited, and the Poisson equation for the pressure update is solved using an algebraic multigrid approach. That is, the implicit solve for the pressure occurs on a succession of changing grid resolutions to increase the rate of convergence. RTI3D has been validated against a large suite of other numerical simulations and experiments [17].

The DNS results are provided by Wei and Livescu [46] using CFDNS, and discussed in [54, 55]. CFDNS is direct numerical simulation code developed for multispecies mixing problems that solves the full Navier–Stokes equations using both spectral and compact finite difference methods. It has provided data for various fundamental studies of RT mixing [22, 23].

11.2.2 BHR-2 Governing Equations and Modeling Rationale

The derivation of the governing equations and closure assumptions in the BHR framework can be found in Ref. [33], and several "flavors" of BHR can be found in Refs. [34–36]. Thus, a detailed derivation will not be presented here. Rather, the relevant terms, governing equations and their model constants are given necessary to highlight characteristics specific to BHR-2 in FLAG [39].

The bulk hydrodynamic equations are the ensemble averaged Navier–Stokes equations in the limit of vanishing molecular viscosity/diffusivity.

$$\frac{D\overline{\rho}}{Dt} = -\overline{\rho}\frac{\partial\widetilde{u}_j}{\partial x_j} \tag{11.1a}$$

$$\overline{\rho}\frac{D\widetilde{c}_k}{Dt} = \frac{\partial}{\partial x_j}\frac{\mu_T}{\sigma_c}\frac{\partial\widetilde{c}_k}{\partial x_j} \tag{11.1b}$$

$$\overline{\rho}\frac{D\widetilde{u}_i}{Dt} = -\frac{\partial}{\partial x_j}\left(\overline{p}\delta_{ij} + R_{ij}\right) + \overline{\rho}g_i \tag{11.1c}$$

$$\overline{\rho}\frac{D\widetilde{e}}{Dt} = -\overline{p}\frac{\partial\widetilde{u}_j}{\partial x_j} + \overline{p}\frac{\partial a_j}{\partial x_j} + \frac{\partial}{\partial x_j}\left(\frac{c_v\mu_T}{Pr_T}\frac{\partial\widetilde{T}}{\partial x_j}\right) + \sum_k\frac{\partial}{\partial x_j}\left(\frac{\mu_T}{\sigma_c}\frac{\partial\widetilde{c}_k}{\partial x_j}\widetilde{e}_k\right) + \overline{\rho}\frac{K^{3/2}}{S} \tag{11.1d}$$

and the necessary equations of state. For the present study ideal gases are used with $\gamma = 1.4$ consistent with the test problem definition. Overbars denote ensemble averages and tildes denote mass weighted ensemble averages. ρ, c_k, u, p, and e are the density, mass fraction of species k, velocity, pressure, and specific internal energy, respectively, and g is gravity. Fluctuating quantities are defined as the differences:

$$\phi = \overline{\phi} + \phi' = \widetilde{\phi} + \phi''$$

with:

$$\widetilde{\phi} = \frac{\overline{\rho\phi}}{\overline{\rho}}.$$

In FLAG these equations are solved in the Lagrange frame (with advection treated separately) so $\frac{D}{Dt}$ is interpreted as either the Lagrangian derivative, or the material derivative if a remap of the mesh takes place. The BHR family of models derives transport equations for the primary quantities of interest, and uses gradient diffusion approximations for remaining terms and higher order moments. BHR-2 employs transport equation for the turbulent kinetic energy ($K = R_{ii}/2\overline{\rho}$), the turbulent length scale ($S = K^{3/2}/\varepsilon$) where ε is the dissipation, the mass-flux velocity ($a_i = \overline{\rho'u'}_i/\overline{\rho}$) and the density specific volume correlation: ($b = -\overline{\rho'v'}$).

In Lagrangian form, the turbulence model equations are given by:

$$\overline{\rho}\frac{DK}{Dt} = a_j\frac{\partial\overline{p}}{\partial x_j} + \frac{\partial}{\partial x_j}\left(\frac{\mu_T}{\sigma_k}\frac{\partial K}{\partial x_j}\right) - R_{ij}\frac{\partial\widetilde{u}_i}{\partial x_j} - \overline{\rho}\frac{K^{3/2}}{S} \tag{11.2a}$$

$$\overline{\rho}\frac{DS}{Dt} = \frac{S}{K}\left[\left(\frac{3}{2} - C_4\right)a_j\frac{\partial\overline{p}}{\partial x_j} - \left(\frac{3}{2} - C_1\right)R_{ij}\frac{\partial\widetilde{u}_i}{\partial x_j}\right] \tag{11.2b}$$

$$+ \frac{\partial}{\partial x_j}\left(\frac{\mu_T}{\sigma_S}\frac{\partial S}{\partial x_j}\right) - \left(\frac{3}{2} - C_2\right)\overline{\rho}K^{1/2}$$

$$\overline{\rho}\frac{Da_i}{Dt} = b\frac{\partial\overline{p}}{\partial x_i} + \overline{\rho}\frac{\partial a_ia_j}{\partial x_j} - \overline{\rho}a_j\frac{\partial\widetilde{u}_i - a_i}{\partial x_j} - \frac{R_{ij}}{\overline{\rho}}\frac{\partial\overline{p}}{\partial x_j} + \frac{\partial}{\partial x_j}\left(\frac{\mu_T}{\sigma_a}\frac{\partial a_i}{\partial x_j}\right) - C_a\overline{\rho}a_i\frac{K^{1/2}}{S} \tag{11.2c}$$

$$\overline{\rho}\frac{Db}{Dt} = 2\overline{\rho}a_j\frac{\partial b}{\partial x_j} - 2(b+1)a_j\frac{\partial\overline{p}}{\partial x_j} + \overline{\rho}^2\frac{\partial}{\partial x_j}\left(\frac{\mu_T}{\overline{\rho}^2\sigma_b}\frac{\partial b}{\partial x_j}\right) - C_b\overline{\rho}\frac{K^{1/2}}{S}b. \tag{11.2d}$$

A turbulent viscosity,

$$\mu_T = C_\mu \overline{\rho} S \sqrt{K},$$

is formed from the model terms and the dissipation ε is recovered using K and S. The Reynolds stress term in the momentum equation is closed using the algebraic relation

$$R_{ij} = 2/3 \overline{\rho} K \delta_{ij} - 2\sigma_1 \mu_T S_{ij} \tag{11.3}$$

with the additional definition

$$S_{ij} = \frac{1}{2} \left[\frac{\partial \widetilde{u}_i}{\partial x_j} + \frac{\partial \widetilde{u}_j}{\partial x_i} - \frac{2}{3} \frac{\partial \widetilde{u}_k}{\partial x_k} \delta_{ij} \right].$$

The coefficient σ_1 is computed to ensure realizability [36]; however, this term does not affect RT results, and $\sigma_1 = 1$ for the cases shown here, which is the standard Boussinesq approximation. With an equation for the turbulent mass-flux velocity, the Reynolds averaged divergence in the internal energy equation can be replaced with a density weighted average and the divergence of a_i.

The additional coupling of the turbulence to the internal energy equation requires some care for compressible mixing [39], but does not affect the nearly incompressible tilted rig test case here. The additional turbulent fluxes are closed with gradient diffusion approximations similar to mass flux. For this nearly incompressible case, no limiting was applied to these fluxes. Thus, the additional terms in the energy equation are given without comment. Each species has an equation of state that gives $\overline{p}_k = \overline{p}_k(\overline{\rho}_k, \widetilde{T}_k)$. Pressure/temperature equilibrium is enforced between species so the total pressure in each zone also depends on the mass fraction $\widetilde{c}_k$.

The coefficients in the BHR model are set from canonical test cases [34, 36]. The default model coefficients that are not set to unity are summarized in Table 11.1. Several of these coefficients differ from the canonical $K - \varepsilon$ values, notably C_μ and σ_S. The canonical values are set by analyzing the Reynolds stress in simple shear flow (C_μ) and the von Kármán constant in the log layer (σ_S) [56]. In the BHR model this coefficients are set by low Atwood number similarity solutions for RT mixing and DNS comparisons [34, 36]. C_μ impacts the growth rate and shape of mixing layer, and low value of σ_S is a result of the flatness of the turbulent length scale across the RT layer. These settings are specific to RT mixing (though they have been validated for RM and KH as well) and would give incorrect behavior for turbulent boundary layer flows.

The BHR model is distinct from other models used in multiphysics applications for multimaterial mixing. In contrast to the K-L model [27, 28], BHR contains no phenomenological (buoyancy drag) production terms, and includes an evolution equation for

Table 11.1 Default Parameter Values for BHR

C_μ	C_1	C_2	C_4	C_a	C_b	σ_c	σ_S	σ_b
28	1.44	1.92	1.05	6.0	2.5	0.6	0.1	3.0

the turbulent mass-flux velocity (a_i), which is the primary turbulent source term for RT mixing. The turbulent mass-flux velocity also enables computing the ensemble averaged velocity ($\overline{u}_i + a_i = \widetilde{u}_i$) if needed. Additionally, the density specific volume correlation (b) provides a measure of molecular mixing that can be included in reacting flow calculations. These two terms require validation that can be provided by the ILES simulations.

In contrast with multifluid models [16, 31], BHR retains the familiarity and ease of implementation that stems from a single bulk velocity. While the higher fidelity of the multifluid models may be necessary in some applications, a successful single fluid model reduces the complexity. Additionally, terms in the BHR equation set have multifluid (volume averaged) interpretations that allow multifluid closures to provide insight into the modeled equations.

For example, in the case of the turbulent mass-flux velocity, the two fluid formula for incompressible, immiscible fluids can be written [33]

$$a_i^{12} = \frac{f^1 f^2 (\rho^1 - \rho^2)(u_i^1 - u_i^2)}{\overline{\rho}}, \tag{11.4}$$

where superscripts indicate species and f^i is the volume fraction of fluid i (i=1,2). Thus, the mass-flux velocity in BHR accounts for the difference in species velocity (while the bulk motion accounts for the average species velocity).

Similarly for incompressible, immiscible fluids, the density specific volume correlation can be computed with

$$b^{12} = \frac{f^1 f^2 (\rho^1 - \rho^2)^2}{\rho^1 \rho^2}, \tag{11.5}$$

which is zero in unmixed regions and maximum when the volume fractions are evenly split. This interpretation of b for immiscible fluids allows an interpretation of the evolved b to represent the level of molecular mixing. The immiscible value of b is the upper bound; departure from this indicates molecular mixing, as a completely mixed fluid will have $b = 0$. To see this we can rewrite b:

$$b = -\overline{\rho' v'} = \overline{\rho v} - 1.$$

If two fluids are molecularly mixed, the volume averaged density is uniform and $\overline{\rho} = 1/\overline{v}$, which gives $b = 0$. See [36] for more discussion of this behavior. These multifluid interpretations will be used later to understand the results.

11.3   Tilted Rig Test Problem

The test problem addressed here is case 110 from [11], and the parameters are summarized in Table 11.2. For this problem, the density of the heavy fluid is $\rho_H = 1.89$ g/cm^3, and the density of the light fluid is $\rho_L = 0.66$ g/cm^3 (the fluids are an NaI solution and hexane, respectively). The associated Atwood number,

Table 11.2 Problem Geometry (cgs units)

ρ_H	ρ_L	A_t	L_x	L_y	L_z	Initial Tilt	$\overline{g}$
89	0.66	0.48	15	2.5	25	5.767°	$35g_0$

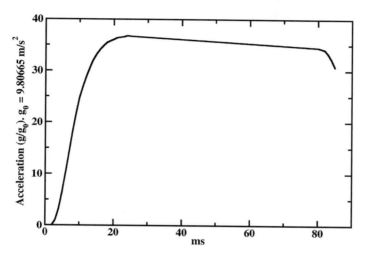

Figure 11.1 Acceleration profile. From [11].

$A_t = (\rho_H - \rho_L)/(\rho_H + \rho_L) = 0.48$. The tank was 15 cm across (L_x), 25 cm tall (L_z), and 2.5 cm deep (L_y). The acceleration from the rocket motors was not constant, but averaged approximately 35 times normal gravity. The RANS and ILES simulations described in the following use the acceleration history measured from the experiment, which is shown in Figure 11.1. Further details of the tilted rig test problem may be found in the test problem definition [54, 55].

The initial tilt was 5.77° which leads to a significant overturning motion in addition to the turbulent mixing. The goal of the present work is to determine how well the turbulence models, which were validated using canonical 1D test cases, perform with the additional complications presented by the 2D overturning interface compared to the ILES. The experimental tank was much narrower than its width ($L_y < L_x$) to force a more 2D flow. The tank height was 25 cm, though the simulations presented here use 24 cm due the presence of a bubble trap in the top of the tank in the experiment.

Figure 11.2 shows a visualization of the mix region for the tilted rig problem from the experiment, and the present simulations. In this paper we examine several features of the interaction of the turbulence and the bulk hydrodynamics. To analyze the effect of the model on the bulk hydrodynamics, the RANS simulations should reproduce the sidewall bubble (right) and spike (left) heights, as well as the tilt angle of the interface. These features are repeatable mean flow phenomena, and should be resolved in the RANS simulations. The large scale bubble and spike associated with the overturning motion should not be confused with individual bubbles and spikes present in the central

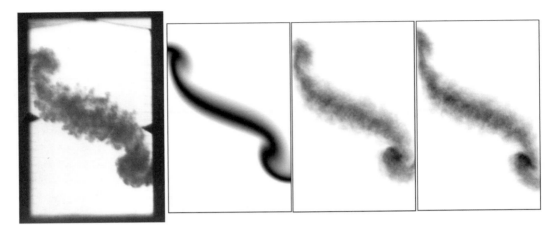

Figure 11.2 Mix region from (left to right) the experiment, RANS, viscous ILES, and inviscid ILES. Simulation mix regions visualized by $4f_1f_2$ at $t = 60$ ms ($\tau = 1.74$). Color range is from $[0 : 1]$. The overall structure of the mixing region in the RANS and viscous ILES is comparable to the experiment. Without the viscosity, the inviscid ILES grows too quickly at the walls. All experimental images © British Crown Owned Copyright 2012/AWE.

mixing layer. The ensemble average of the mixing at the interface is represented by the spread of the density profile in the RANS model, and only the global features should remain. This is compared to the depth averaged ILES results, which approximate an ensemble average.

To analyze the effect of a dynamic interface on turbulence, the RANS simulations should capture the mix layer width, which, as experiments show, grows more slowly than the canonical 1D case [11, 40], and any asymmetry in the turbulence variables arising from the interface motion. Often, experimentally measured mixing widths grow faster than computational mixing widths, owing to long wavelength initial perturbations in the experiment that are difficult to control or other perturbations that are difficult to characterize [17, 19, 57–59]. The BHR-2 model is calibrated to DNS data without any uncontrolled perturbations [36]. So it should agree quantitatively with the ILES results from the RTI3D code [43] and DNS results from CFDNS [22]. Detailed initial conditions of the interface were not available for the small scales in the experiment, only the tilt angle was measured, so discrepancies between the simulation and the experiment are possible.

11.3.1   Initial Conditions and Postprocessing Analysis

Initialization of any turbulence model from a steady initial condition is a challenge as the models assume developed, turbulent flow. The BHR-2 model is initialized with zero mass-flux velocity ($a_i = 0$), and the density specific volume correlation is zero ($b = 0$) everywhere, except cells that are initially mixed. For mixed cells, the two fluid value from Eq. (11.5) is used, consistent with unmixed fluid at t_0. The initial conditions that are uncertain are the initial turbulent length scale and kinetic energy (S_0 and K_0).

Prior work indicates good results are achieved by setting the initial length scale (S_0) to the most amplified wavelength from linear stability analysis. In case 110, a surfactant was used to decrease the surface tension, and the most unstable wavelength is reported as less than 0.5 mm [11]. We find that 0.1 mm gives good results, and is consistent with a mix width smaller than the mesh spacing for all grids. We did not attempt to adjust this value to improve the agreement. Self-similar growth rates for an RT mixing layer give $K_0 = 0.01AgS_0$ as an estimate for the initial turbulent kinetic energy [36] where an average gravity ($\overline{g} = 35g_0$, with g_0 Earth's gravity) has been used. These initialization rules are ad hoc based on prior experience, and work is ongoing to improve the initialization based on interface stability modeling [60].

The LES is initialized with a k^{-2} spectrum ($k = 2\pi/\lambda$), with the wavelength $\lambda_{\min} = 0.2$ cm and $\lambda_{\max} = 7.5$ cm, and a standard deviation of $\sigma = 0.001\lambda_{\max}$. More details can be found in the test problem description [54].

To facilitate comparisons between simulations, it is important to define several quantities [55]. For any quantity, $\phi(x, y, z, t)$, let $\langle \phi \rangle$ be the average in the y direction (depth into the tank):

$$\langle \phi \rangle = \frac{1}{L_y} \int_0^{L_y} \phi(x, y, z, t)dy$$

and $\langle\langle \phi \rangle\rangle_x$ be the average in the x direction (width of the tank):

$$\langle\langle \phi \rangle\rangle_x = \frac{1}{L_x} \int_0^{L_x} \langle \phi \rangle dx.$$

For 2D RANS simulations the computed quantities are all $\langle \phi \rangle$. The DNS/LES apply averaging in y to define the mean quantities.

The integral mix width is defined to give a single scalar value for the total mixing in units of length [42]:

$$W = \int_0^{L_z} \langle\langle f_H \rangle \langle f_L \rangle\rangle_x dz.$$

The mean interface location is computed by integrating the heavy fluid volume fraction:

$$\chi(x, t) = \int_0^{L_z} \langle f_H \rangle dz$$

and the tilt angle θ is found by applying a linear least squares fit of χ over the center third of the domain. The spike height, H_s, is computed by averaging the volume fraction in x and finding the z location where the average goes to 0.001:

$$H_s = \{z : \langle\langle f_H \rangle\rangle_x = 0.001\}.$$

H_b is defined similarly using the light fluid.

The total turbulent kinetic energy in the domain will also be shown, defined as:

$$K_{\text{tot}} = \int \rho K \, dx dz.$$

Since some compressible codes may find it difficult to run the variable gravity computations efficiently, a nondimensional time is defined to account for the differences between constant gravity and variable gravity simulations. This nondimensional time is found using a buoyancy drag model [61]. The nondimensional time is given by

$$\tau = \int \sqrt{\frac{Ag}{L_x}} dt + \delta,$$

where δ is a time offset. $\delta = -0.053$ for the variable gravity cases and zero for constant gravity. See [55] for details. All simulations in the current work use the experimentally measured acceleration profile.

RANS simulations of the tilted rig are fully compressible and performed on square grids in a domain 15 cm × 24 cm. The interface pressure and temperature are set to minimize compressibility effects while maintaining feasible run times. The interface pressure is set at 10,000 Pa, and the temperature set to 300 K. The degree of compressibility was evaluated by examining species density, which was found to vary by less than 1%.

The slanted initial condition results in the interface intersecting several rows of computational cells. This means some cells are initially mixed at $t = 0$, and so there are some numerical perturbations as the interface intersects cells in different places. While it is possible to fit the grid to the interface in FLAG, this was not done so that comparison could be made with other codes that rely on square meshes [44]. For similar reasons, a full Eulerian remap strategy is used to keep the grid fixed throughout the simulation. The initial perturbations are found to have little effect on the results and do not persist in the FLAG calculations.

Three grid resolutions are used for the RANS computations, 152 × 240, 232 × 360, and 304 × 480. Contour plots show results from the highest resolution simulations, and line plots in Figure 11.13 show results from the highest resolution simulation with uncertainty computed using the GCI method of Roache [62] with the recommended safety factor of $F_S = 1.25$. This method uses three grids and Richardson extrapolation to compute the order of convergence and associated grid uncertainty. From the three different grids, the order of convergence p is computed. Then, instead of computing the grid uncertainty as the relative difference between two solutions (ε), the relative uncertainty (E) is reduced to account for numerical convergence,

$$E = F_S \frac{\varepsilon}{r^p - 1},$$

where r is the mesh ratio. The integrated quantities of interest from the RANS simulations all show second-order convergence.

Two sets of ILES results are presented: inviscid and viscous at two grid resolutions. The inviscid simulations use slip walls as specified in [54]. By viscous ILES we mean the material viscosities were included, although material diffusivities were not. The viscosity values are taken from the experiments in [63] (0.31 and 3.3 mNs/m^2 for hexane and NaI, respectively). Unlike for the DNS, not all dissipative length scales were resolved. The results will show that the viscosity has little effect on any of the integral

Table 11.3 Simulation Settings for the Present Work (cgs units)

Simulation	L_x	L_y	L_z	Grid 1	Grid 2	Grid 3
RANS	15	–	24	152×240	232×360	304×480
ILES	15	15	24	$360^2 \times 480$	$512^2 \times 768$	–

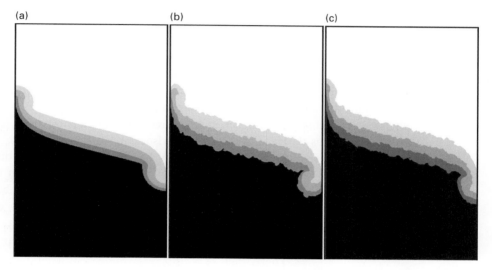

(a)   (b)   (c)

Figure 11.3 Contours of heavy fluid volume fraction (f_1) for RANS (a), viscous ILES (b), and inviscid ILES (c) at $t = 45$ ms ($\tau = 1.26$). Volume fraction contours are [0.025, 0.3, 0.7, 0.975].

quantities; notably, it does not change the total turbulent kinetic energy significantly. This indicates that the numerical dissipation of the ILES approach is still the dominant effect. The inclusion of material viscosity improves the results near the wall by allowing the imposition of no-slip boundary conditions. Settings for the current work are summarized in Table 11.3; details for the other simulations can be found in [54, 55].

It is important to note that this chapter is presenting multiple models of the experiment, none of which is a stand-in for the "truth" of the physical system. They are three methods for modeling the system that have important applications for multiphysics problems of interest. The DNS is also not truth as the current computational resources limit the DNS to using a viscosity ten times larger than that used in the experiment. Thus, as is often the case for complex problems, we must critically evaluate various models to improve our understanding.

The ILES simulations are run in a 3D domain, with the tank depth equal to its width, to improve the convergence of averages in the y (depth) direction. This differs from the experimental tank, which is not as deep as it is wide; the impact of this is discussed in the following. Due to computational constraints only two grids were used for the ILES computations: $320^2 \times 480$ and $512^2 \times 768$. Although capable of third-order advection, the RTI3D advection scheme is run in a second-order mode, so second-order convergence is assumed when computing the grid uncertainty. While this is not a conservative estimate for design purposes ([62] recommends using 3.0 as a safety factor if convergence rates are not calculated), it is a reasonable estimate of the uncertainty. The DNS in [55] suggests these simulations are in a high Reynolds number regime when analyzing the turbulent kinetic energy (see fig. 18 in [55]), so we expect this integrated quantity from the ILES simulations to converge as well. Other quantities, such as the molecular mix parameter (see fig. 20 of [55]), have not reached convergence for early time in the DNS and ILES and thus have greater uncertainty in our comparisons to the model variable b. It is important to note that if the simulations do not ultimately converge with resolution, these metrics are no longer meaningful.

The DNS contours have been reported elsewhere [46, 55] and only the integrated results as a function of time will be used in the present work. Two DNS cases are used, one with a square domain similar to the ILES to increase the statistical convergence, and one with the depth of the experimental tank to investigate these differences. Additionally, the DNS increased the domain in the z direction to avoid effects from the top and bottom walls, thus the validity of the comparisons at late time is limited. The DNS results from [55] are the largest to date, but still use viscosities that are ten times larger than the physical viscosities ($Re_p = 14000$ for the highest case based on maximum wavelength and acceleration, see [55] for definition), and a Schmidt number of one. A true DNS of the experimental conditions remains beyond the limits of current computing platforms.

11.3.2   Simulation Results

This section presents results from the simulations. First, qualitative features of the 2D contours are compared with the limited information available from the experiment. Then, more quantitative contour comparisons are made between the ILES and RANS simulations. All contours shown are at the highest grid resolution at $t = 45$ ms ($\tau = 1.26$) and $t = 60$ ms ($\tau = 1.74$). Finally, quantitative data extracted from the simulations as a function of nondimensional time are presented and discussed, along with their uncertainty, from the RANS, ILES, and DNS.

11.3.2.1   Contour Comparisons

Figure 11.2 shows the mixing region at $t = 60$ ms. All three simulations are able to capture the bubble–spike asymmetry. Specifically, the spike structure (left) is smaller and more confined to the near wall region than the bubble (right). Including viscosity improves agreement with the RANS simulations and experiment near the wall. Without viscosity the near wall structures are smaller and grow more rapidly. The mix width along the center line is slightly smaller in the RANS, and the ILES, than in the experiment. This is expected as simulations generally grow more slowly than do experiments, where uncontrolled and unknown initial perturbations affect late term behavior [17, 19, 59]. In addition, some of this difference is an artifact of the visualization approach. A direct comparison is difficult as the photographic image is a line of sight measurement that does not readily convert to an ensemble average in the depth

(a)   (b)   (c)

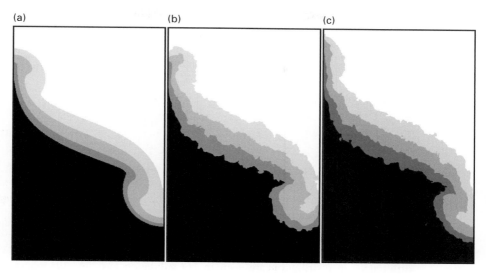

Figure 11.4 Contours of heavy fluid volume fraction (f_1) for RANS (a), viscous ILES (b), and inviscid ILES (c) $t = 60$ ms ($\tau = 1.74$). Volume fraction contours are [0.025, 0.3, 0.7, 0.975]. The RANS contours are closer together on the heavy side, indicating larger asymmetry than the ILES, and inviscid ILES shows significantly faster growth of the side wall bubbles and spikes. [Volume fraction: $t = 45$ ms] [Volume fraction: $t = 60$ ms].

direction. The two ILES simulations agree well along the center line, suggesting the physical viscosity is playing very little role in the mixing region.

The RANS model does well predicting the heights of the evolving spike and bubble as well as the overturning motion of the mixing layer. This figure demonstrates that, at minimum, the BHR model is not interfering with the bulk hydrodynamic evolution in FLAG. The mix layer turns over and creates the mean structures seen in the experiment reasonably well. More importantly, the RANS model is able to capture the mean flow without introducing additional features such as KH rolls along the interface. That is, a sufficient amount of turbulent viscosity is present to eliminate spurious flow features.

We can make more quantitative comparisons between RANS and ILES contours. In all figures, contours are based on the RANS values. The ILES values are often larger in magnitude, and thus the darkest contour region is saturated. The contours are always [0.025, 0.3, 0.7, 0.975] of the peak RANS values. Figures 11.3 and 11.4 show contours of volume fraction (f_1) at $t = 45$ms and $t = 60$ ms. The volume fraction contours use the same contour scale (f_1 is bounded between $[0 : 1]$), and the comparison is good at both mid- and late time. The ILES mix widths are slightly larger at both of these times, which will be seen in the integrated data. Notably, both simulations show asymmetric bubble and spike behavior for the large scale features, as well as some asymmetry in the contour positions across the midregion (smaller contour width on the heavy side).

The asymmetry in the midregion accounts for the difference in penetration distance between small scale bubbles and spikes. The asymmetry in the contour levels is more pronounced in the RANS compared with the ILES. That is, two highest contours are closer together on the heavy side in the RANS. The reason for this is unknown. Some

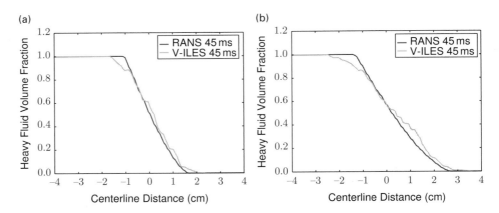

Figure 11.5 Heavy fluid volume fraction (f_1) along the centerline showing the difference in mixture fraction on the heavy side of the interface.

differences in distribution are seen in 1D validation [34, 36, 39] where the BHR-2 model gives density profiles that are not as rounded at the edges of the mix layer. This results in contours closer together near the edges. However, those differences are mostly symmetric across the layer, whereas the current contours seem to only differ on the heavy side. Figure 11.5 shows this more clearly by taking a line out of the volume fraction along the center line of the tank. The edge of the mixing layer appears to agree on the light fluid side, but shows more asymmetry on the heavy fluid side. This behavior has not been observed in 1D tests that compare fully self-similar BHR solutions with DNS results. A potential explanation is the inability to capture all features of the horizontal turbulent mass-flux velocity (a_x), which will be shown later. Difference in production from this term could affect the mass distribution.

The turbulent kinetic energy shown in Figures 11.6 and 11.7 has significant variability along the length of the mix layer, that is, it is strongly affected by the motion of the interface. Several features of the ILES are captured by the RANS, notably the strong area of K in the region of the large scale bubble (right wall) and a somewhat elevated area near the spike (left wall). Further, the thinning and thickening along the mix region can be seen clearly. The dynamic mix model is able to capture these 2D features. Overall, the magnitude of the turbulent kinetic energy is lower in the RANS than the ILES; this will be confirmed and discussed in the following. The viscous ILES has slightly lower turbulent kinetic energy than the invsicid case, especially near the walls, but the overall distribution is similar, suggesting the numerics are providing the majority of the dissipation away from the walls. Similar behavior is seen contrasting the inviscid and viscous solutions for the remaining contours, so we will only show the viscous results for the remaining contours.

The essential modeling term in this class of flows is a_i, the turbulent mass-flux velocity [22]. It is the primary production term for buoyancy driven instabilities and also describes the difference between the mass weighted mean velocity and the non-weighted mean velocity. The main production term, the vertical turbulent mass-flux velocity, a_z, is shown in Figures 11.8 and 11.9. The model does not do as well here as

(a)   (b)   (c)

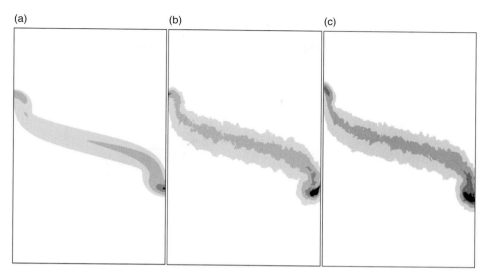

Figure 11.6 Contours of turbulent kinetic energy, K, for RANS (a) and ILES (c) at $t = 45$ ms. Contours are [0.025, 0.3, 0.7, 0.975] of the maximum RANS value; this saturates the ILES. The maximum values are $K = 1.124$ for the RANS, $K = 1.538$ for the viscous ILES, and $K = 1.754$ for the inviscid ILES. Note the elevated levels of K on the spike (left) and bubble (right) side seen in the ILES are captured by the RANS.

(a)   (b)   (c)

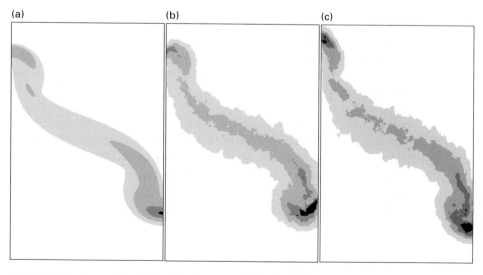

Figure 11.7 Contours of turbulent kinetic energy, K, for RANS (a) and ILES (c) at $t = 60$ ms. Contours are [0.025, 0.3, 0.7, 0.975] of the maximum RANS value; this saturates the ILES. The maximum values are $K = 2.114$ for the RANS, $K = 3.109$ for the viscous ILES, and $K = 3.271$ for the inviscid ILES.

(a)   (b)

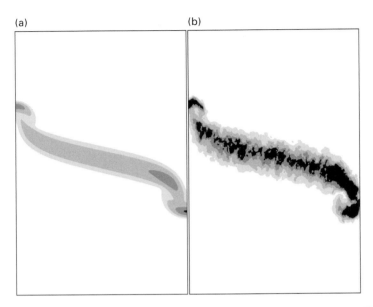

Figure 11.8 Contours of vertical turbulent mass-flux velocity, a_z, for RANS (a) and ILES (b) at $t = 45$ ms. Contours are [0.025, 0.3, 0.7, 0.975] of the maximum RANS value; this saturates the ILES. The maximum values are $a_z = 0.138$ for the RANS and $a_z = 0.5$ for the ILES. The low level contours are comparable, but the RANS underpredicts the center line values. Note also the intermittency of the ILES.

for the turbulent kinetic energy. Large regions of the ILES simulation have values of a_z above the RANS model. However, the shape of the contours at the edges of the mix region is comparable. The larger differences suggest the BHR-2 model may be doing a better job of computing Favre averaged quantities such as the turbulent kinetic energy compared with the terms involving the density fluctuations.

Figures 11.8 and 11.9 also show a_z from the ILES is much more intermittent than the turbulent kinetic energy. Here we mean intermittent in a general sense of the statistics not being as well converged to what one would expect from a true ensemble average. Especially at late time, Figure 11.9 shows there are regions in the center of the mixing layer that are at the same level as the RANS and other regions that show much stronger turbulent mass flux. Thus, the averaged behavior of the LES is more consistent with the RANS prediction than the peak values indicate. Also, it is important to recall that while BHR has an evolution equation for the turbulent mass-flux velocity, species mass fractions are moved using a simple gradient diffusion model. Errors in modeling a_z would not necessarily appear in the volume fraction contours.

The density specific volume correlation, b, is a production mechanism for the turbulent mass-flux velocity and is related to the morphology of the mixing region. Contours of this terms are shown in Figure 11.10 and 11.11. The contour levels are higher for the ILES; however, they are much more intermittent for these correlations. As with the mass-flux velocity, the contours of b are more constant along the mixing layer than the turbulent kinetic energy. This suggests a_z and especially b are more dependent

(a)   (b)

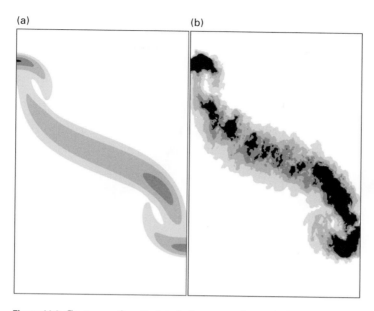

Figure 11.9 Contours of vertical turbulent mass-flux velocity, a_z, for RANS (a) and ILES (b) at $t = 60$ ms. Contours are [0.025, 0.3, 0.7, 0.975] of the maximum RANS value, this saturates the ILES. The maximum values are $a_z = 0.184$ for the RANS and $a_z = 0.66$ for the ILES. The RANS captures the structure of the LES but not the peak values along the center of the mix layer. The ILES shows significant intermittency with regions that are comparable to RANS near regions with much stronger turbulent mass flux.

on the local mixing behavior, while the 2D features affect K. Also, the peak value of b is the same in both figures, suggesting the RANS simulations has reached a steady state value for molecular mixing along the center line while the ILES is still evolving slightly. Overall the results are consistent, with the RANS results that are much more uniform in a range approximately one-third of the peak level of the ILES. The width and shape of the lower valued contours are comparable as well; it is the intermittent center that is not captured.

The two fluid formulas show that the results in Figures 11.10 and 11.11 behave as expected. In particular, Eq. (11.4) shows the vertical turbulent mass-flux velocity should be positive along mix layer as the heavy fluid is rising in the turbulent region. Also, the RANS solution is well below the maximum predicted by Eq. (11.5) ($b_{max} = 0.303$). This is consistent with the high value of molecular mix parameter found in experiments and simulations [55]. The model is respecting the physical limits and shows the state of mixing lies between complete molecularly mixing and immiscible mixing, and no numerical limiting is necessary.

The BHR family of models has an evolution equation for the mass-flux velocity. In contrast, two equation models [27] often use a gradient diffusion model:

$$a_i \sim \frac{\mu_T}{\rho} \frac{\partial \rho}{\partial x_i}. \qquad (11.6)$$

(a)   (b)

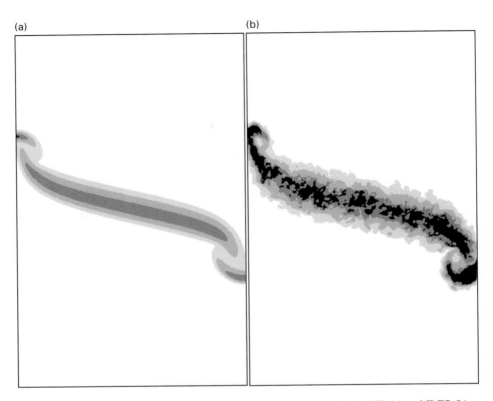

Figure 11.10 Contours of the density specific volume correlation, b, for RANS (a) and ILES (b) at $t = 45$ ms. Contours are [0.025, 0.3, 0.7, 0.975] of the maximum RANS value; this saturates the ILES. The maximum values are $b = 0.064$ for the RANS and $b = 0.182$ for the ILES. The contours of b are much more constant along the mixing layer than other terms.

The gradient diffusion model can be effective modeling 1D RT flows [27] when computing the turbulent mass-flux velocity in the direction of the mean pressure gradient (the z direction). However, the gradient diffusion model cannot capture certain features' turbulent mass-flux velocity in the horizontal direction as shown in Figure 11.12. Specifically, the ILES and the transport model predict a region of negative (countergradient) mass flux in the top half of the mixing layer, which the gradient diffusion approximation cannot capture. Gradient diffusion models will always have the mass flux along the direction of the density gradient and hence across the mixing layer.

In addition to the strong area of countergradient flux, the ILES also shows a smaller and more intermittent region of positive turbulent mass flux in the lower part of the mixing layer (green blobs). The origin of the smaller region of additional positive flux below the mixing layer is unknown at this time. This may account for some of the differences in the volume fraction contours, as the positive a_x would tend to increase production of K and push material to the center of the layer. This would spread the contours on the heavy side of the mix layer, potentially accounting for some of the

(a)   (b)

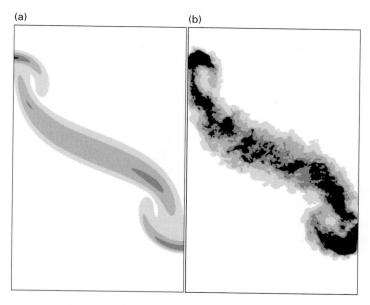

Figure 11.11 Contours of density specific volume correlation, b, for RANS (a) and ILES (b) at $t = 60$ ms. Contours are [0.025, 0.3, 0.7, 0.975] of the maximum RANS value; this saturates the ILES. The peak values are $b = 0.063$ for the RANS and $b = 0.202$ for the ILES. The ILES shows significant intermittency in this quantity as well.

(a)   (b)   (c)

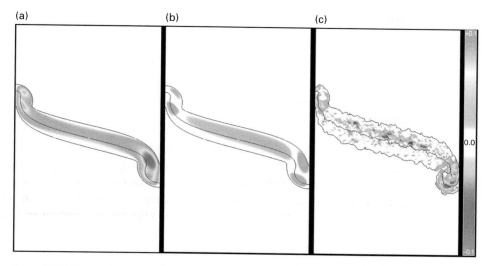

Figure 11.12 Comparison of gradient diffusion model (a) for the horizontal turbulent mass-flux velocity (a_x) and the transport model of BHR (b) and ILES (c) at $t = 45$ ms. Colors are values of a_x and contour lines are [0.025, 0.5, 0.975] of the heavy fluid volume fraction to show the center and edges of the mixing layer. Note the countergradient flux in the top half of the mix layer. A black and white version of this figure will appear in some formats. For the color version, please refer to the plate section.

differences in distribution of volume fractions. While a transport type model for a_i is necessary to capture the countergradient flux, the current model does not capture this additional feature. Further research on the transport equation for a_i should investigate this phenomenon. Note that the gradient diffusion result shown in Figure 11.12 was computed using Equation 11.6 and the mean quantities from the BHR simulations.

11.3.2.2   Time Evolution of Integral Quantities

Greater quantitative detail can be obtained by looking at integrated quantities over time. Figures 11.13(a) and (b) show line plots of what was seen in the contours in Figures 11.2–11.4. In particular, the experiment and RANS simulations show similar large scale bubble and spike growth, while the viscous ILES grows slightly faster. This is partially explained by the DNS data. The ILES has a square domain in the $x - y$ plane, while the experiment was performed on a tank with a depth much less than its width. The DNS ran both cases, and when the DNS was run with the experimental dimensions (labeled DNS-Exp) it resulted in slower growth of the bubbles and spikes, in agreement with the RANS and experiments. When the DNS ran a square platform tank (labeled DNS-TP), increased three-dimensionality was possible and the right side bubble grew faster.

In addition, Figures 11.13(a) and (b) show that the inviscid ILES grows significantly faster at the edges of the domain relative to the viscous case. This is likely due to both viscous drag as well as a regularizing effect at the walls creating a more 2D flow. The DNS incorporated viscous terms, but it used a periodic reflection boundary and thus does not have the same viscous wall effects. The viscous ILES with walls substantially improves the agreement with the experiment and RANS model. The error bars on Figure 11.13(a) and (b) represent the grid uncertainty as computed with the GCI method [62], and are barely visible, indicating the grid uncertainty is negligible for this metric in both codes.

It is important to note that all codes capture the same relative separation between bubble and spike growth, that is, the shift in the bubble height plot is similar to the shift in spike height. Thus, each code is capturing the expected difference in dynamics between spikes and bubbles at this Atwood number. This relative shift is independent of wall treatment.

Figure 11.13(c) shows the integrated mix width, W, for the simulations. The agreement between the simulations is within the mesh uncertainty. Close agreement between the viscous and inviscid ILES shows that the total mix width growth is not dependent on the viscosity, unlike the edge features. Also, differences that occur at early time reflect differences in initialization strategy, and how the initial fluctuations (or K for the RANS models) were imposed.

Error bars shown in Figure 11.13(c) for the mix width are significantly larger than for the bubble/spike heights, indicating more uncertainty in grid convergence. The uncertainty bounds overlap for the ILES and RANS solutions throughout the time evolution. This indicates satisfactory agreement for the RANS initialization strategy chosen. Again, no tuning of initial conditions was done for the RANS study.

The tilt angle, shown in Figure 11.13(d), has similar features and overall agreement with the mix width, again within the grid uncertainty. The DNS begins to overturn faster near the end of its run; however, the DNS simulated a much taller tank and thus does not

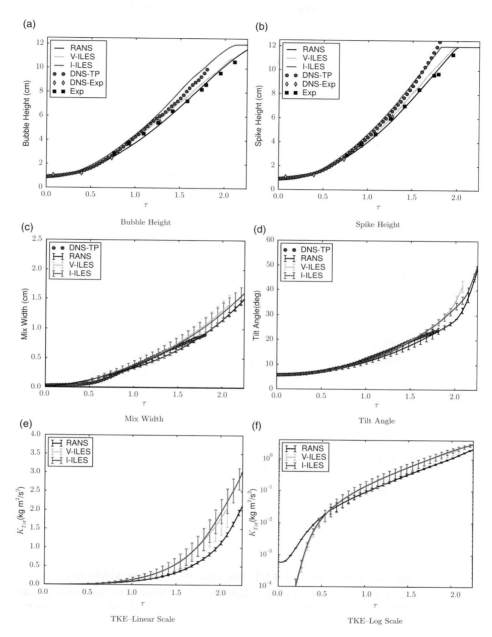

Figure 11.13 Global quantities versus nondimensional time (τ) for FLAG-BHR2, viscous ILES (RTI3D-V), inviscid ILES (RTI3D-I), square domain DNS (DNS-TP), experimental box dimensions DNS (DNS-Exp), and experiment.

experience any blockage from the top of the tank restricting the vertical fluid flow. This accounts for the slightly faster overturning motion. The time where this deviation occurs ($\tau \approx 1.25$) can also be seen in the mix width plot, and indicates where departure from the present work is expected.

The total turbulent kinetic energy in the simulation is shown on a linear scale in Figure 11.13(e) and a log scale in Figure 11.13(f). Integrated K is the only variable shown in the line plots that is computed as part of the turbulence model, as opposed to the bulk hydrodynamics. Inspection of the figure shows results on a linear scale to observe the late time behavior, and a log scale to show the early time behavior. The log scale results for K show the challenges that remain for initializing BHR. The RANS closure models assume fully developed turbulence, which is not true at time zero. The model cannot be initialized the same way the ILES and DNS are initialized. The goal of properly initializing the model is solely to have the right level of K once the high resolution simulations reach self-similarity. It will not get the early time behavior correct.

Figure 11.13(f) does show that the evolution of K has the right power law behavior in the RANS model, which is an essential feature for modeling. The linear scale results show that the initialization prescription given earlier is a reasonable means by which to put the RANS simulation on the right path for the evolution of K. The kinetic energy leaves the error bounds of the ILES around $\tau \approx 1.25$, but follows a similar evolution. While it is possible to achieve a better match to the K evolution by increasing K_0 at the beginning of the simulation, this would result in a reduced agreement at early times. More work is needed to improve the initialization strategy if the RANS simulation is to capture the evolution of K at all times. This initialization problem may be responsible for the systematic underprediction of the higher order flow statistics relative to the ILES. Note that no "tuning" of the RANS initial conditions was done as part of this study.

There is good agreement between the viscous and inviscid ILES simulations. This is evidence that the physical viscosity is not playing a large role dissipating the turbulence, and the numerics are instead controlling the dissipation. However, note that the uncertainty bounds for the viscous simulations are larger, suggesting the convergence is slower for the viscous case. This suggests simulations should proceed cautiously including physical viscosity in LES calculations where the resolution is marginal.

Finally, an important validation issue with ILES is assessing the level of molecular mix that takes place as a result of the numerical treatment. Recall that for both the viscous and inviscid ILES, there is no explicit diffusion term, and all mixing is due to numerics. Figure 11.14 shows the level of molecular mixing given by the global molecular mix percent,

$$\Theta = \iint \frac{\langle f_1 f_2 \rangle}{\langle f_1 \rangle \langle f_2 \rangle} \, dx dz, \tag{11.7}$$

from the two inviscid ILES simulations and the DNS at the highest Reynolds number. The agreement between the two ILES resolutions suggests we are successfully computing the high Reynolds number limit. The DNS, which is still an order of magnitude lower Reynolds number, has approximately the same minimum value, though later in time, and is approaching the final limiting value of the ILES. [55] shows that although the DNS values of turbulent kinetic energy are converged at this Reynolds number, the molecular mix percent is not, and in fact is still moving toward the ILES results. Thus, this metric suggests we are accurately capturing the high Reynolds number mixing behavior and using ILES to validate the RANS mixing behavior is justified.

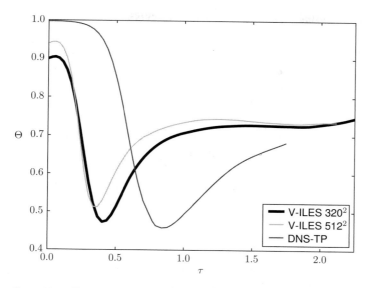

Figure 11.14 Comparison of molecular mix parameter for two different ILES resolutions and the highest Reynolds number DNS. [55] shows the DNS is trending toward the ILES solution but is not yet converged.

11.4   Conclusions

ILES simulations can be a very useful tool for simulating turbulent mixing and validating RANS models. This work has used two ILES simulations to help validate a turbulent mix model for problems with dynamic interfaces. The BHR-2 model can efficiently predict a variety of turbulence quantities, including mixing width, turbulent kinetic energy, and turbulent mass-flux velocity, that are consistent with more computationally expensive ILES models. The higher order terms are consistently below the ILES results, but the total mixing is within the grid uncertainty. This suggests that the modeling errors associated with the higher order terms may be mitigated by the gradient diffusion models that couple the turbulence model to the mean flow. Specifically, the turbulence model couples to the species mass fraction equations via the gradient diffusion approximation and a turbulent Schmidt number (σ_c). This value, which is set from 1D validation, will change the rate of turbulent mass flux for a fixed turbulent viscosity. Decreasing σ_c can compensate for a lower than expected turbulent kinetic energy. Also, different initial conditions could potentially improve agreement here. At the mean flow level the model does not adversely affect the interface motion, and it accurately captures differences in turbulence intensity along the moving interface.

More generally, a four equation, single fluid turbulence model is capable of capturing several distinctive features of the tilted rocket rig. This includes distributions of turbulent kinetic energy and countergradient turbulent mass-flux velocity. This suggests the BHR family of models is a viable option for simulating these flows in addition to more complex multifluid turbulence models. Gradient diffusion models for the turbulent mass-flux velocity that are often used for two equation models cannot capture counter-gradient behavior observed in our simulations of the tilted rig.

In addition, we have demonstrated improved agreement between ILES and experiments by including the physical viscosity in the calculation. The effect of viscosity in the high resolution LES is confined to the near wall region in this test case. While including the viscous terms has a noticeable effect on growth of the near wall features, it has little impact on the mix width or total turbulent kinetic energy. The difference in these integral terms is significantly smaller than the mesh uncertainty for these simulations, so increased resolution would be necessary to further analyze the effect of viscosity. This suggests inviscid ILES simulations can be a useful tool for analyzing the turbulent mixing away from boundaries, but including material viscosity can be an improvement.

Setting appropriate initial conditions for RANS models remains a challenge. The uniform initialization strategy used here is adequate for capturing mean flow behavior such as the mixing width. However, the current initialization of the model may be responsible for underprediction of the higher order statistics relative to the ILES. Future progress will depend on more sophisticated analysis of the interface at early time.

Finally, for the RANS cases studied here, the mixing is dominated by the turbulence model and not by numerical diffusion. This is reflected in the numerical convergence of the results, as well as by explicit calculations of the mix mass. These metrics are important in demonstrating grid independence, and are a necessary part of model validation.

This work was performed for the U.S. Department of Energy by Los Alamos National Laboratory under Contract No. DEAC52-06NA2-5396. The authors thank Tie Wei and Daniel Livescu for access to the DNS data.

References

[1] J.D. Lindl, *Inertial Confinement Fusion: The Quest for Ignition and Energy Gain Using Indirect Drive*, Vol. 2998 (Springer, New York, 1998).

[2] S. Gull, *Royal Astronomical Society, Monthly Notices* 171, 263 (1975).

[3] W. Thomson, *Proceedings of the Royal Society* 7, 63 (1871).

[4] H. von Helmholtz, Monatsberichte der Königlichen Preussische Akademie der Wissenschaften zu Berlin 23, 215 (1868).

[5] Rayleigh, *Proceedings of the London Mathematical Society* 14, 170 (1883).

[6] G. Taylor, *Proceedings of the Royal Society of London. Series A. Mathematical and Physical Sciences* 201, 192 (1950).

[7] R.D. Richtmyer, *Communications on Pure and Applied Mathematics* 13, 297 (1960).

[8] E. Meshkov, *Soviet Fluid Dynamics* 4, 101 (1969).

[9] Y.A. Kucherenko, S. Balabin, R. Cherret, and J. Haas, *Laser and Particle Beams* 15, 25 (1997).

[10] S. Dalziel, P. Linden, and D. Youngs, *Journal of Fluid Mechanics* 399, 1 (1999).

[11] V.S. Smeeton and D.L. Youngs, *AWE Report No. O* 35/87 (1987).

[12] G. Dimonte and M. Schneider, *Physics of Fluids* 12, 304 (2000).

[13] D. Olson and J. Jacobs, *Physics of Fluids* 21, 034103 (2009).

[14] A. Banerjee, W.N. Kraft, and M.J. Andrews, *Journal of Fluid Mechanics* 659, 127 (2010).

[15] D.L. Youngs, *Physics of Fluids A: Fluid Dynamics* 3, 1312 (1991).

[16] D.L. Youngs, *Laser and particle beams* 12, 725 (1994).

[17] G. Dimonte, D. Youngs, A. Dimits, S. Weber, M. Marinak, S. Wunsch, C. Garasi, A. Robinson, M. Andrews, P. Ramaprabhu, et al., *Physics of Fluids* 16, 1668 (2004).

[18] A.W. Cook, W. Cabot, and P.L. Miller, *Journal of Fluid Mechanics* 511, 333 (2004).

[19] D.L. Youngs, *Philosophical Transactions of the Royal Society A: Mathematical, Physical and Engineering Sciences* 367, 2971 (2009).

[20] N.J. Mueschke and O. Schilling, *Physics of Fluids* 21, 014106 (2009).

[21] J. Ristorcelli and T. Clark, *Journal of Fluid Mechanics* 507, 213 (2004).

[22] D. Livescu and J. Ristorcelli, *Journal of Fluid Mechanics* 591, 43 (2007).

[23] D. Livescu and J. Ristorcelli, *Journal of Fluid Mechanics* 605, 145 (2008).

[24] D. Livescu, J. Ristorcelli, R. Gore, S. Dean, W. Cabot, and A. Cook, *Journal of Turbulence* (2009).

[25] V.A. Andronov, S.M. Bakhrakh, E.E. Meshkov, V.N. Mokhov, V.V. Nikiforov, A. V. Pevnitskii, and A.I. Tolshmyakhov, *Sov. Phys. JETP* 44 (1976).

[26] S. Gauthier and M. Bonnet, *Physics of Fluids A: Fluid Dynamics* 2, 1685 (1990).

[27] G. Dimonte and R. Tipton, *Physics of Fluids* 18, 085101 (2006).

[28] V.P. Chiravalle, *Laser and Particle Beams* 24, 381 (2006).

[29] O. Grégoire, D. Souffland, and S. Gauthier, *Journal of Turbulence* (2005).

[30] C. W. Cranfill, Los Alamos National Laboratory Report LA-UR–92-2484 (1992).

[31] A.J. Scannapieco and B. Cheng, *Physics Letters* A 299, 49 (2002).

[32] A. Llor, *Laser and Particle Beams* 21, 305 (2003).

[33] D. Besnard, F.H. Harlow, R.M. Rauenzahn, and C. Zemach, Los Alamos National Laboratory Report LA-12303-MS (1992).

[34] A. Banerjee, R.A. Gore, and M.J. Andrews, *Physical Review E* 82, 046309 (2010).

[35] J.D. Schwarzkopf, D. Livescu, R.A. Gore, R.M. Rauenzahn and J. R. Ristorcelli, *Journal of Turbulence* (2011).

[36] K. Stalsberg-Zarling and R.A. Gore, Los Alamos National Laboratory Report LA-UR–1104773 (2011).

[37] B.M. Haines, F.F. Grinstein, and J.D. Schwarzkopf, *Journal of Turbulence* 14, 46 (2013).

[38] J. Waltz and T. Gianakon, *Computer Physics Communications* 183, 70 (2012).

[39] N. Denissen, J. Fung, J. Reisner, and M. Andrews, Los Alamos National Laboratory Report LA-UR–12–24386 (2012).

[40] N. Ptitzyna, Y. Kucherenko, V. Chitaikin, and A. Pylaev, Proceedings of the 4th International Workshop on the Physics of Compressible Turbulent Mixing (1993).

[41] M. Andrews, Ph.D. Thesis, London University (1986).

[42] M.J. Andrews and D.B. Spalding, *Physics of Fluids A: Fluid Dynamics* 2, 922 (1990).

[43] M. Andrews, *International Journal for Numerical Methods in Fluids* 21, 205 (1995).

[44] B. Rollin and M. Andrews, *Proceedings of the ASME Fluids Engineering Summer Meeting* (2012).

[45] J.M. Holford, S.B. Dalziel, and D. Youngs, *Laser and Particle Beams* 21, 419 (2003).

[46] T. Wei and D. Livescu, International Conference on Numerical Methods in Multiphase Flow, Penn State, PA (2012).

[47] V. Smalyuk, R. Tipton, J. Pino, D. Casey, G. Grim, B. Remington, D. Rowley, S. Weber, M. Barrios, L. Benedetti, et al., *Physical Review Letters* 112, 025002 (2014).

[48] M. Marinak, G. Kerbel, N. Gentile, O. Jones, D. Munro, S. Pollaine, T. Dittrich, and S. Haan, *Physics of Plasmas* (1994–present) 8, 2275 (2001).

[49] C.R. Weber, D.S. Clark, A.W. Cook, L.E. Busby, and H.F. Robey, *Phys. Rev. E* 89, 053106 (2014).

[50] D.E. Burton, Lawrence Livermore National Laboratory Report UCRL–JC–110555 (1992).

[51] D.E. Burton, Lawrence Livermore National Laboratory Report UCRL–JC–118788 (1994).

[52] E. Caramana, D. Burton, M. Shashkov, and P. Whalen, *Journal of Computational Physics* 146, 227 (1998).

[53] J. Waltz, *Numerical Methods for Multi-Material Fluids and Structures* (Pavia, Italy 2009).

[54] M. Andrews, D. Youngs, and D. Livescu, Los Alamos National Laboratory Report LA-UR–12–24091 (2012).

[55] M. Andrews, D. Youngs, D. Livescu, and T. Wei, ASME J. Fluids Eng. Special Issue: IWPTCM13, Vol. 136, Issue 8 (2014).

[56] S.B. Pope, *Turbulent Flows* (Cambridge University Press, 2000).

[57] P. Linden, J. Redondo, and D. Youngs, *Journal of Fluid Mechanics* 265, 97 (1994).

[58] G. Dimonte, P. Ramaprabhu, D. Youngs, M. Andrews, and R. Rosner, *Physics of Plasmas* 12, 056301 (2005).

[59] J. Glimm, J.W. Grove, X.L. Li, W. Oh, and D.H. Sharps, *Journal of Computational Physics* 169, 652 (2001).

[60] B. Rollin and M. Andrews, *Journal of Turbulence* 14, 77 (2013).

[61] K. Read, *Physica D: Nonlinear Phenomena* 12, 45 (1984).

[62] P.J. Roache, *Verification and Validation in Computational Science and Engineering* (Hermosa, Albuquerque, 1998).

[63] K.D. Burrows, V.S. Smeeton, and D.L. Youngs, AWE Report No. O 22/84 (1984).

12 Spray Combustion in Swirling Flow

Suresh Menon and Reetesh Ranjan

12.1 Introduction

Swirl is ubiquitous in nature, appearing in several kinds of flows, from flow in pipes to tornados and hurricanes, and even in the Great Red Spot in Jupiter. Swirling flow can be defined in many ways, but in the present context involves axial or streamwise motion with a strong rotational or azimuthal component. Swirl has to be generated by impacting the mean flow so that it drives primarily the larger scales of motion. As a result, swirl can affect all the downstream processes in a direct manner, thus providing an aerodynamic mechanism for flow manipulation or control.

In engineering applications swirl also appears in many propulsion devices such as internal combustion (IC), gas turbines (GT), and liquid rocket (LR) engines. In all these cases, swirl is deliberately generated in order to achieve some specific design goal. Typically swirl is created by upstream manipulation of the incoming air or oxidizer (and in some cases the fuel) flow. Examples of the means of swirl generation are tangential injection and angled vanes [1, 2]. The amount and the quality of swirl so generated is determined by the design of these devices since in almost all systems, swirl generation is by a passive or a fixed design. In IC and GT combustors swirl enables two critical design goals: fuel–air mixing and flame stabilization [3, 4]. Typically, the liquid fuel is injected into these combustors, therefore, the swirling air flow enables the fuel spray to break up and undergo rapid dispersion, which in turn allows mixing of the vaporized fuel with the incoming hot swirling air. In some cases, air is swirled upstream of the fuel injector and the fuel is injected radially into the airstream to directly entrain into the swirling air flow. This enables formation of very lean mixtures, for example, as observed in the lean direct injection (LDI) combustor [5–9].

Swirling flow also creates flow patterns that can be exploited for flame stabilization. If the swirl intensity is high (i.e., beyond a critical value, see following discussion), the swirling air flow creates a recirculating region downstream and this region, often called the vortex breakdown bubble (VBB), acts as an aerodynamic flame holder. Depending on the swirl intensity, the VBB can be quite compact and therefore the flame is stabilized within a compact region. The reduction in the flame extent with an increase in the swirl intensity enhances level of mixing and entrainment near the nozzle exit and at the boundaries of the VBB. When combined with the geometrical flame holders such as rearward facing steps and center bluff body [10–14], multiple regions of flame stabilization can be created in the combustor. In addition, increased intensity of the swirl leads to a compact flame anchored by the VBB such that the flame does not impinge on the combustor surface, thus reducing maintenance cost and extending life of the combustion system.

The nature of these swirling flows makes both measurements and computations very difficult and, hence, there are still many issues that are still poorly understood even though such flows are employed in most engine designs. For example, since spray injection occurs into a hot, three-dimensional (3D), turbulent flow field in the combustor, the measurements of spray breakup and dense regimes are currently impossible. Therefore, most quantitative measurements are made farther downstream of the injector where the spray has become dilute. However, this implies serious problems for simulations since the breakup process and the initial spray formation are critical to the simulation setup. Some approximations are needed and they can impact the accuracy of the predictions on top of all the other unknowns of such two phase mixing and combustion.

In this chapter we focus primarily on modeling and simulation of swirling flows in gas turbine types of combustors since this is an area where there has been some progress in the recent past. IC engine and LR engine applications are not directly discussed although some references are noted where appropriate. Also, since we are interested in gas turbine combustors, we primarily focus on liquid fueled systems, although swirl is used extensively in premixed and non-premixed (gaseous) systems as well [15–17]. There have been several investigations of such systems using the Reynolds-averaged Navier–Stokes (RANS) modeling approach, see, for example, [1, 18]. Such studies are not discussed here. Recently, some direct numerical simulation (DNS) studies of swirling spray combustion have been performed [8, 19], but such studies are typically relegated to lower Reynolds number due to a huge computational cost and memory requirements. Therefore, we mainly focus on modeling, simulation challenges, and key results obtained using large eddy simulation (LES) studies of systems that are more realistic to lab or flight conditions. However, as noted in the following, LES too has some limitations to address all the issues of interest.

The chapter is organized as follows. Section 12.2 describes general features of swirling flows with a focus on mechanisms used for generating swirl and instantaneous and mean features of such flows. Spray combustion in swirling flow and typical experimental and numerical approaches to study such systems are described in Section 12.3. In Section 12.4 some key LES results are described with a goal to highlight key features of swirl spray systems. Finally, a summary and future prospects are detailed in Section 12.5.

12.2   General Features of Swirling Flows

In this section we discuss some of the key features of swirling spray systems and briefly describe parameters that characterize such systems.

12.2.1 Generation Mechanisms

Swirl is typically generated using tangential injection of the air into an axial flow or by using inclined vanes. Many variants of this exist in the literature and some typical GT combustors are summarized in Table 12.1. The table lists some of the combustors that have been studied both experimentally and numerically in the past. We will be referring to some of these combustors in the following discussion. A general feature of such systems is that although flight systems operate under high pressure conditions, these devices when studied in the laboratory conditions are (typically) operating at 1 atm. Thus, detailed studies of the effect of high pressure operation of such systems is an area with limited progress. Another inherent limitation of laboratory scale combustors is that in most (but not all) cases, only a single injector combustor is investigated, whereas in actual hardware many injectors in multiple combustors are present. Multiple injector studies (e.g., in an annular assembly [26]) are very rare and only addressed in a qualitative sense so far but will be required to investigate complex physics such as ignition cycle, injector to injector interactions, and combustion dynamics in realistic systems.

Figure 12.1 shows some typical swirler assemblies used within some of the aforementioned GT combustors. These assemblies include helical vane (LDI), radial (GE-CFM56), and tangential swirlers. In a typical swirl injector, an array of axially or radially positioned vanes deflects the flow, leading to a desired level of mixing of fuel and oxidizer to have an efficient combustion. In an axial swirl injector, the fuel is delivered to the swirling airflow immediately downstream of the swirler vanes, followed by mixing that occurs within a premixing duct before finally entering the combustor. In a radial entry swirler, air is injected tangentially and fuel is injected from the nozzle placed on the center body of the injector, which eventually disperses into the swirling air. Swirling flows can also be generated without use of swirling vanes as being achieved by a tangential entry swirl injector shown in Figure 12.1(c). Air is delivered into the premixing chamber through tangentially oriented air slots that span the entire axial length of the premixing chamber.

12.2.2 Definitions and Features

Since swirling flow has been extensively studied in nonreacting and reacting gas phase flows [3, 4] the observations from them are not repeated here other than to note that in

Table 12.1 Some Examples of Experimentally and Numerically Studied Swirl Spray Combustors

Combustor	Injector Type	Swirler Type	References
GE-DACRS	Dual annular	Counterswirl	[20]
LDI	Single/Multi	Helical	[7, 21]
GE-TAPS	Twin annular	Cyclone	[22, 23]
GE-CFM56	Single	Dual-swirl radial	[24, 25]
VESTA	Multi	Dual stage	[26]
MERCATO	Single	Single stage	[27]

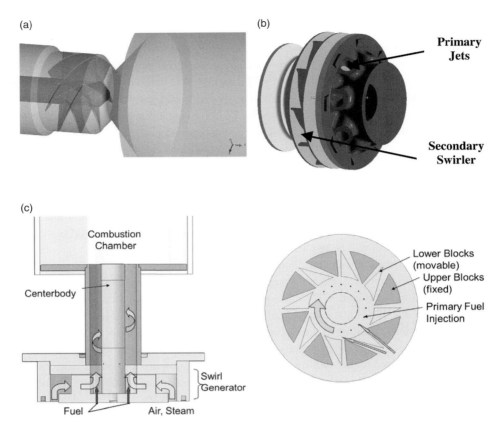

Figure 12.1 Some typical experimentally and numerically studied swirler assemblies. (a) Helical vane swirler (LDI) [7]; (b) radial swirler (CFM56) [24]; (c) tangential swirler [28].

general swirl is known to enhance jet spreading and to increase entrainment and mixing in nonreacting flows. The degree of mixing depends on the intensity of the swirl, defined using a swirl number S_w. In addition, there are several other parameters that characterize such flows and are summarized in Table 12.2. These parameters include Reynolds number (Re), Strouhal number (S_f), momentum parameter, and Stokes number (St). Along with S_w, Re is a parameter that determines whether VBB occurs within such flows or not. A key feature of such flows is presence of precession vortex core (discussed in the following), which can be characterized in terms of its precession frequency through the nondimensional Strouhal number. The momentum parameter is used to characterize flux of the inlet angular momentum. Stokes number is mainly used to characterize the two phase spray flow system and is determined by the ratio of characteristic time scales associated with the disperse (liquid) and the carrier (gas) phases.

The swirl number S_w is an important quantity and is typically defined as a ratio of axial flux of the tangential momentum and the product of the axial flux of the axial momentum with a characteristic diameter D, and is given by [3]

Table 12.2 Typical Nondimensional Parameters Characterizing Swirling Spray Combustion*

Parameter	Definition
Reynolds number (Re)	$Re = \rho U D / \mu$
Swirl number (S_w)	$S_w = 2 G_\theta / G_x D$
Strouhal number (S_f)	$f D^3 / Q$
Stokes number (St)	t_p / t_f
Momentum parameter	$\Omega D / \rho Q^2$

* Here, U, D, μ, G_θ, G_x, f, Q, t_p, t_f, Ω, and ρ denote axial velocity, characteristic diameter, dynamic viscosity, axial flux of the tangential momentum, axial flux of the axial momentum, frequency of precession, flow rate, particle time constant, characteristic flow time, rate of change of angular momentum, and density at the swirler exit, respectively.

$$S_w = 2 \frac{G_\theta}{G_x D} \qquad (12.2.1)$$

where D for a combustor is typically the combustor diameter, and G_θ and G_x are the axial flux of the tangential momentum and axial momentum, respectively, which are expressed as

$$G_\theta = \int_0^R (U_\theta r) \rho U_x \, 2\pi r dr, \qquad (12.2.2a)$$

$$G_x = \int_0^R (U_x) \rho U_x \, 2\pi r dr + \int P \, 2\pi r dr = \int_0^R \rho \left[U_x^2 - \frac{1}{2} U_\theta^2 \right] 2\pi r dr. \qquad (12.2.2b)$$

In Eq. (12.2.2), U_θ and U_x denote azimuthal and axial components of the velocity, P is the pressure, and $R = D/2$ is the radius of the combustor. With some approximations [2], the swirl number can be calculated using the swirl vane angle ϕ through

$$S_w = \frac{2}{3} \tan \phi. \qquad (12.2.3)$$

However, in most practical cases it is necessary to obtain S_w from the actual measurements for accurate prediction. Note that S_w is typically increased by increasing the tangential momentum in the inflow rather than by a reduction of the axial flow since the latter is typically determined by the operational margins of the engine.

Many features of the swirling flow are well understood. Figure 12.2 shows two typical characteristic flow features that are typically observed in a gas turbine swirling flow combustor. These include VBB induced central toroidal recirculation zone (CTRZ), corner recirculation zones (CRZ), precessing vortex core (PVC), and shear layers originating at the inlet of the combustor. Depending on the design of the combustor, the VBB induced recirculation zone can be toroidal (as shown here), or in simpler configurations, as in a swirling pipe flow into a dump combustor [7], a single VBB is seen in the center line region. Regardless, these flow features are unsteady, asymmetric, and 3D, and are affected by the swirl strength and the unsteady heat release

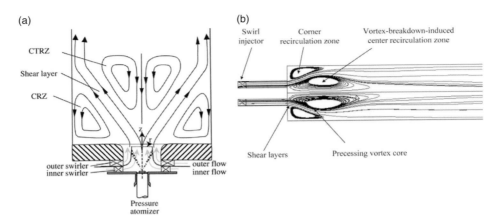

Figure 12.2 Characteristic flow features within a combustion chamber of (a) a dual swirl single injector [29] and (b) an annular swirl injector with a center body [2].

in the flame (that typically resides in the shear layer and around the VBB). These flow features also play a significant role in spray dispersion in axial and lateral directions, which directly affects level of mixing of fuel droplets with the oxidizer. We describe in the following the two important flow features, namely VBB and PVC, and discuss the role of swirl and heat-release on spatial evolution of these features.

VBB is an important feature of a swirl injector and can be described as the formation of a free stagnation point and a recirculation zone (RZ) in the core flow [3, 4]. VBB formation leads to an abrupt change in the shape of the columnar vortex and assists in formation of a recirculation bubble with a surrounding 3D spiral flow pattern. The RZ provides flame stabilization due to the presence of hot burnt products within the recirculation region and the upstream stagnation region. As the flow swirls, it also creates an adverse pressure gradient in the axial direction. When the intensity of swirl is strong enough, the adverse axial pressure gradient can lead to formation of an internal recirculation zone even in a pipe [30]. If the swirling flow enters a dump combustor (with a sudden increase in area/volume) the adverse pressure gradient effect is even more pronounced [31].

Figure 12.3 shows the effect of swirl number on the mean streamline patterns and the temperature field in a coannular swirl combustor. We can observe the presence of three distinct recirculation zones in the low swirl number case ($S_w = 0.44$), including a wake recirculation zone (WRZ) behind the center body, a CRZ, and a CTRZ resulting from the vortex breakdown. However, at the high swirl number ($S_w = 1.1$), WRZ disappears. Without swirl in a flow into the dump combustor only the wake and the corner recirculation zones exist. When S_w exceeds a critical value (typically 0.6 in such flows [32]), vortex breakdown occurs, which assists in the formation of the central RZ. With a further increase of the swirl number, the central RZ moves upstream and merges with the WRZ, which reduces the size of the CRZ. The mean temperature fields clearly exhibit enveloped flames anchored at the rim of the center body and at the corner of the dump combustor. The flame is much more compact for the high swirl case, which is due to the enhanced flame speed resulting from the increased turbulence intensity.

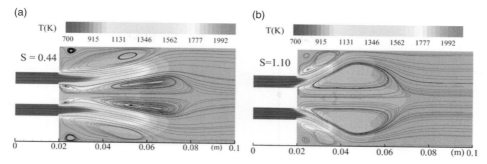

Figure 12.3 Mean temperature field and streamline patterns at (a) low swirl number (S_w=0.44) and (b) high swirl number (S_w=1.10) in a coannular swirl combustor [32].

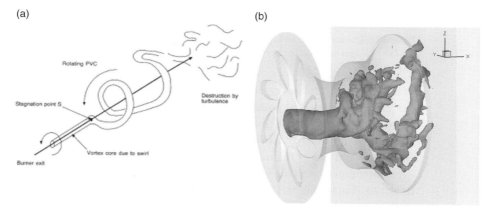

Figure 12.4 (a) A schematic representation of a generic PVC [33] and (b) the PVC identified using iso-surface of pressure from LES of a gas turbine swirl combustor [34].

Swirling flows exhibit a variety of unstable modes and the PVC is one of these modes. The PVC forms due to hydrodynamic instabilities when the central vortex core starts precessing around the axis of symmetry of the combustion device at a well-defined frequency. PVC is a 3D, unsteady, and asymmetric flow feature that is usually linked to the VBB and associated recirculation zones. Figure 12.4(a) shows a schematic representation of the PVC and Figure 12.4(b) shows a PVC extracted from LES [34]. Typically, PVC is located along the outer boundary of the recirculation zone and does not persist far downstream. The frequency of precession depends on the device design and the swirl intensity at the inlet. Further downstream of the injector location, turbulence leads to a breakdown into small scale structures and no coherent PVC is observed.

Identifying the PVC in numerical simulation data is not easy since this feature is embedded inside the reacting flow field. Typically, the unsteady pressure field is used to find the location of this structure and the iso-contour needs to be chosen properly to visualize this feature. For example, Figure 12.4(b) shows the PVC identified using an iso-surface of the unsteady pressure field within an LDI combustor [7]. Near the inlet of the combustor, that is, the dump plane, the PVC tends to align with the central axis

(a)   (b)

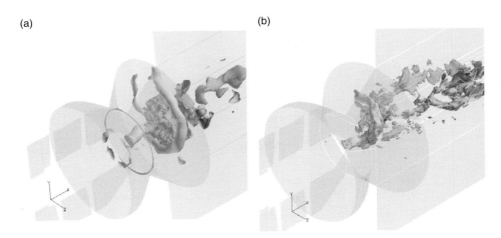

Figure 12.5 Instantaneous visualization of PVC (yellow color) and azimuthal velocity (green color) for (a) non-reacting and (b) reacting cases [7].

of the combustor. However, near the stagnation point, it is driven away from the central axis and forms a spiral pattern further downstream in the axial direction. Due to a centrifugal effect induced by the swirling motion, the large scale, coherent vortical structures are pushed laterally away from the central axis. The PVC rotates in the combustor and peels off structures periodically, which are transported downstream and eventually break up in the turbulent flow field. Such features are seen in many simulations [20, 34–36] but their exact size, shape, and precessing characteristics depend on the device design and the operating conditions.

Compared to reactive flows, the PVC structure is much more coherent and stronger in nonreactive flows [36]. This may be attributed to the volume dilatation by the heat release and an increase in the viscosity of the burnt gases that contribute to the lack of coherence of the PVC. These effects may sometime lead to a disappearance of the PVC. Figure 12.5 shows the PVC identified in nonreactive and reactive cases at a fixed swirl number within the LDI combustor. It can be observed that the PVC structure is altered in the reactive case, leading to quicker breakdown into more small scale vortical motion near the combustor inlet.

The recirculation bubble as shown in Figure 12.3 may appear to be a closed region with a well-defined boundary but in reality, in the instantaneous flow field there is no such thing. In fact, VBB interactions with the injected fuel and the surrounding swirling air are very dynamic and are not very well understood. Figure 12.6(a) and (b) shows typical instantaneous contours of the axial velocity field and Figure 12.6(c) shows the same view of the time averaged field in a high pressure single injector LDI combustor [37]. The location of the zero level of axial velocity is typically used to delineate the RZ from the axial flow and it is clear that there are substantial differences between the instantaneous and time averaged views. The boundary of the RZ is barely seen in the instantaneous flow field, which in fact shows pockets of RZ in the combustor, and is seen to be highly unsteady. However, the time averaged view shows a more benign

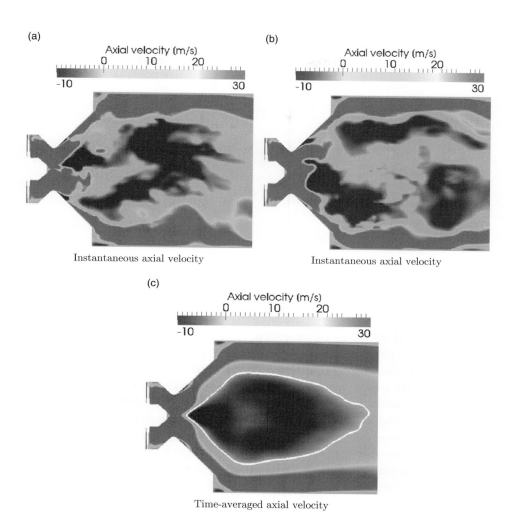

Figure 12.6 Contours of two instantaneous (a, b) and time averaged (c) axial velocity in a high pressure LDI combustor [37].

shape that belies the complex turbulent mixing and interactions that are occurring in this region. Analysis of the flow in the RZ shows that it can contain both entrained fuel and burned products [7, 37]. In case of spray injection, some of the droplets can penetrate into the recirculation regions, vaporize, and mix with the hot products. Two typical spray distributions (droplets colored by temperature and increased in size for visualization) and corresponding to the instantaneous axial velocity fields in Figure 12.6 are shown in Figure 12.7. If there is air still present in the recirculation region (invariably so) then partially premixed combustion can occur in the core. Since there is no fixed boundary for this recirculation in the instantaneous sense, combustion and heat released in the central region can easily interact with the primary flame structure. Swirl flame stabilization is, therefore, quite different in practice from, for example, stabilization due to RZ created by geometrical features such as rearward facing steps and center bodies.

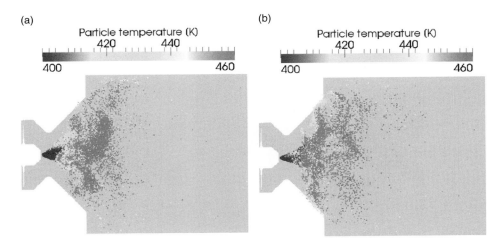

Figure 12.7 Contours of two instantaneous droplet distribution (colored by temperature) in a high pressure LDI combustor [37].

12.3   Swirling Spray Combustion

We focus only on swirling spray combustion and GT engine applications in this chapter. In GT engines (and also in IC engines) liquid fuel is injected and due to the operating conditions, the liquid jet breaks up and forms a spray that is then dispersed into the incoming swirling air. Spray injectors are very complex and many designs are available providing different types of spray patterns (spray angle, size distribution, etc.). Pressure atomizers, air blast atomizers, and hybrid atomizers all are used depending on the design requirements [1, 38]. These devices provide different types of flow patterns but although well known qualitatively, the exact details of the formation of spray are still not very well understood since it is very difficult to capture/resolve all the regimes of the breakup process even in simple canonical problems [39]. Note that this is true for both experiments and simulations.

Some understanding has been obtained using canonical studies in nonreacting sprays but even here there are some unresolved questions. It is typically understood that a liquid jet injection leads to a multitude of breakup processes even without considering combustion [40–43]. Figure 12.8 shows a sketch illustrating the droplet formation process in a combustor. The process can be classified into following regimes [44, 45]: (a) disintegration of liquid sheet into droplets, where the liquid film/jet interacts with the gas phase and disintegrates into filaments due to primary breakup; (b) dense spray regime, where droplets interaction are prevalent and coalescence phenomena are observed; (c) an intermediate regime, where secondary breakup leads to formation of large drops; and finally (d) a dilute regime, where evaporation, fuel vapor mixing, and aerodynamic transport are dominant. The typical length and time scales for these regimes can vary and may depend on the injector design. Many discussions of the breakup process have been reported in the past [46–54] and will not be repeated here.

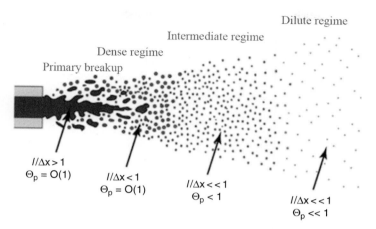

Dilute regime

Intermediate regime

Dense regime

Primary breakup

$l/\Delta x > 1$
$\Theta_p = O(1)$

$l/\Delta x < 1$
$\Theta_p = O(1)$

$l/\Delta x \ll 1$
$\Theta_p < 1$

$l/\Delta x \ll 1$
$\Theta_p \ll 1$

Figure 12.8 A schematic representation of regimes of liquid spray evolution from injectors in a gas turbine engine [45]. Here, Θ_p is the volume fraction of the liquid phase and l and Δx are the characteristic length scales associated with the liquid film and local grid resolution, respectively.

The details of these different regimes are based on physical processes and therefore it means that the simulation model is required to capture all these features. Currently, there is no single model that can handle the entire breakup process. Some past studies [55–58] have attempted to use DNS to understand the breakup process but typically are limited to low *Re* and low speed flows. The challenges are severe from numerical point of view. For example, very close to the injector, the liquid fuel is in the form of a sheet or jet and a continuum formulation is necessary to capture the primary atomization processes [59]. In the dense regime there are no classical droplets and blobs of nonspherical liquid can exist alongside of the liquid core jet. In the intermediate regime, the liquid drops formed during primary atomization undergo further disintegration. Collision, coalescence, and droplet deformation effects are dominant in this region and yet the droplets are distinct entities that may or may not be spherical. In the dilute regime, the droplets are much smaller and the local liquid volume fraction is also very small. In this regime some models exist that have been used in LES and we will discuss some of these results. In reality, there is significant overlap of these regimes and the physical processes of primary/secondary atomization and droplet evaporation may not be distinguishable in a straightforward way. Thus to apply models to LES of spray combustion some modeling of the primary breakup and the dense regime is required.

Figure 12.9 shows instantaneous spray visualization in the LDI combustor, which demonstrates effect of secondary breakup model on the particle dispersion. The droplets are colored by their respective particle radius. Progressive sequence of breakup events can be observed with a $200\,\mu m$ diameter particle/blob injected into the swirling air stream. Effect of PVC is evident from the change in trajectory of the streamwise particle evolution. Within the first 1 mm of the injection location, initial breakup appears to be completed with the formation of several (approximately 70–150 groups) product particles. Furthermore, shear stripping of the parent blob continues, forming more smaller parcels downstream and depending on local conditions and Weber number, both

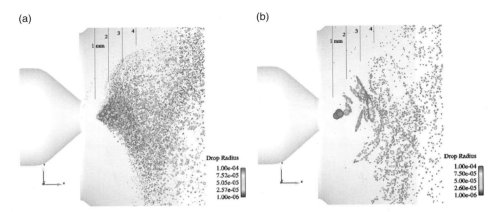

Figure 12.9 Instantaneous visualization of droplets (a) without and (b) with secondary breakup model [7]. All droplets are colored by their particle radius and exaggerated in size for clarity of visualization.

product and parent particles undergo further breakup. All breakup events are completed within the first 4 mm of injection and a finer particle distribution emerges downstream. The secondary breakup affects the fuel evaporation in the region close to the injector, which is different from the case without a secondary breakup model. Since it is very difficult to simulate these processes, the accuracy of the predictions depends strongly on the models used for spray breakup.

12.3.1   Experimental Approaches

Although we do not plan to focus on experimental methods and results, we note briefly some of the challenges for experimental observation of spray breakup, mixing, and combustion in swirling flows. The dense regime is nearly impossible to resolve using nonintrusive methods, although some imaging approaches have been developed in the recent past [60–62]. Most of the techniques have been focused on measurement in the dilute regime where there are some well-known methods that can be employed.

Table 12.3 summarizes some of the experimental techniques typically used to study swirl spray combustion devices. Laser–Doppler velocimetry (LDV) has been used for long time to study the turbulent flow field in a combustor [31, 72]. However, understanding of spray dynamics have been improved greatly by the use of phase Doppler anemometry/analyzer (PDA) technique. This measurement approach has been used to characterize swirling, nonswirling, reactive, and nonreactive sprays. It can be used to obtain droplet characteristics such as mean and fluctuating diameter and velocity. The same method can also be used to investigate statistics of the carrier gas phase by means of a two component PDA [73]. Apart from PDA, there are several other techniques that can be used to measure droplet characteristics. These methods include ensemble light scattering technique, light intensity deconvolution [74], shadowgraph measurements [75], laser induced fluorescence (LIF) [52], and ballistic

Table 12.3 Some Experimental Techniques Used to Study Swirl Spray Combustors

Method	References	Gas Phase	Liquid Phase
LDV	[63]	LDV	LDV
LDV-PDA	[25, 64]	LDV	PDA
PDA	[21, 29, 44, 65–70]	PDA	PDA
CARS-PDA	[71]	CARS	PDA

imaging [60]. Here, we briefly describe the PDA approach, which is commonly used to study swirl spray combustion devices.

The theory behind the PDA approach is described in detail elsewhere [76] and is not repeated here for brevity. In PDA, the droplet velocity is quantified through a standard fringe mode laser anemometry and the droplet size is obtained by measuring the phase shift of the light encoded in the spatial variation of the fringes. The frequency of the Doppler burst is directly proportional to the droplet velocity component perpendicular to the plane of fringes and the measured phase shift is related linearly to the droplet diameter by the phase factor, which is determined by the geometrical arrangement of the PDA system, wavelength of the laser light, and the mechanism of light scattering (reflection or refraction).

A major challenge associated with both LDV and PDA is that it is very difficult to get accurate prediction close to the nozzle exit for both carrier and disperse phases. Additionally, in reacting flows the spray formation, vaporization, mixing, and combustion are all occurring in a small region and in a highly 3D flow field. Furthermore, apart from the gas and droplet velocity field, it is necessary to obtain the flame structure, temperature, and characteristic species (including pollutants, if relevant). Thus, other techniques for such measurement have to be implemented concurrently with the two phase measurements if a full picture is to be obtained. There are several nonintrusive techniques that can accurately measure species concentration, temperature and flame statistics. For example, LIF, shadowgraph, exciplex, and coherent anti-Stokes Raman spectroscopy (CARS) all can be used in gas phase combustion [52, 71] but in the highly unsteady spray combustion field, these methods have inherent limitations. Finally, the time and length scale of interactions in the near field are so small that to resolve these features, very highly resolved and high frequency measurements are needed. Such methods are only now being developed for gas phase flames [77–79] and their application to swirling spray combustion still remains to be addressed.

12.3.2 Computational Approaches

We focus primarily on the modeling aspects for LES of swirling spray combustors. We do not address all the problems and regimes of interest associated with such systems, since there are still numerous uncertainties and ad hoc models are still being used to analyze such systems.

12.3.2.1 Modeling

There are different modeling approaches for simulating two phase flows, which depend upon parameters such as mass and volume loading, Stokes number, spray regime, and the method used for simulating the carrier phase [107–109]. Some of the typical approaches to numerically model practical multiphase flow systems are denoted as Eulerian–Lagrangian (EL), Eulerian–Eulerian (EE), hybrid, and statistical methods (others may exist but are not addressed here). Table 12.4 summarizes some of these models used to study swirling spray combustion (nonswirling spray mixing and combustion are not directly addressed here). A brief description of these methods with their advantages and limitations is provided in the following.

In the EE approach, also referred to as the two fluid method [108–111], both carrier and disperse phases are solved using an Eulerian framework. The two phases are considered to be interpenetrating and require transport equations for the volume fraction, velocity, temperature, and moments of size distribution of the disperse phase, which is obtained after homogenization. The mass, momentum, and energy exchange with the carrier phase occurs through source/sink terms. Since the method utilizes a common Eulerian framework, consistent numerical method can be used for both phases, thus leading to easy implementation and scalable high performance parallel computing. However, the method requires substantial initial modeling effort to obtain the aforementioned transport equations for the disperse phase. Additionally, the method is considered to be expensive for polydisperse systems, although there have been some developments to deal with multisize liquid sprays in a computationally efficient manner [112]. Another limitation of the method is associated with the numerical stability, which requires that the concentration gradient not be very high for the disperse phase within the flow system.

In the EL method, Lagrangian tracking of the disperse phase is performed and the carrier phase is simulated using the conventional Eulerian framework [113–116]. The EL method is the most common approach to simulate flow systems considered here, due to its robustness, accuracy and ability to model complex phenomena such as polydispersity, droplet/wall and droplet–droplet interactions, crossing trajectories, and

Table 12.4 Modeling Approaches for Simulating Typical Swirling Spray Combustors

Method	References	Remarks
Eulerian–Eulerian	[51, 56, 59, 80–85]	Scalable parallelization, consistent numerical method, substantial modeling effort, expensive for polydisperse systems
Eulerian–Lagrangian	[5, 8, 19, 81, 83, 86–99]	Robust for complex systems, slower statistical convergence, inefficient parallelization
Hybrid	[100, 101]	Accurate spray physics, computationally expensive
Statistical	[102–106]	Consistent numerical method, closure is simpler, easy to parallelize, computationally expensive

droplet breakup. However, the method is computationally expensive, as a large number of particles is required within each computational cell used by the carrier phase to allow for a smooth Eulerian reconstruction of the feedback force to the carrier phase. For example, in a typical spray combustor, a sufficient number of Lagrangian droplets is needed at each time step in each computational cell to provide a smooth and accurate continuous Eulerian field of fuel mass fraction, because the fuel vapor distribution, directly produced by the discrete droplet evaporation source terms, controls the propagation of the flame front [117]. Additionally, the method is not easily or perfectly scalable for large scale, high performance parallel computing, which is crucial for simulation of complex combustion systems [118]. A major assumption of these EL methods is that the liquid droplet is a point particle and the gas phase does not directly notice the presence of the particles (although models can be included to account for the gas–particle and particle–particle interactions).

Another major issue with the EL method is related to the accuracy of the disperse phase statistics, which is dependent on the number of particles, particularly in regions of a physically observed sparse distribution of the disperse phase. In general, the convergence rate of the disperse phase statistics by the EL method scales as $N^{-1/2}$, where N is the number of particles in any parcel of the domain. Regions that have low number density tend to be poorly converged because N is small in comparison with regions where N is large. This may or may not be an issue for spray combustion in GT combustors since the spray vaporizes quickly and gas phase combustion occurs very close to the injector. Thus, once droplets become small they are more likely to be consumed.

Regardless, the simulation of flow near the spray injector is a problem in both EE and EL methods due to the complexity of primary atomization. Note that close to the injector, although a Lagrangian approach has been used, dense models and correction for finite loading have to be incorporated. Eulerian methods can be useful in this region, but then methods to track the propagation and breakup of the liquid jet interface also have to be included. Hybrid methods have been developed, which combine the EE approach near the injector with the EL approach further downstream to simulate the full breakup process [100]. However, studies of complex combustion systems by such methods are only now being considered.

The statistical approach is based on transport equation for the particle density function associated with the disperse phase, also referred to as Williams's spray equation [119]. It can be derived from the particle dynamics described in the Lagrangian frame by referring to the Liouville equation [120]. The carrier phase is solved using a conventional Eulerian framework. However, different approaches can be used to solve for the disperse phase. Some of these methods are Lagrangian tracking of "notional" particles [121], sectional methods [122], polynomial chaos expansion [123], method of moments [124], and the Rayleigh–Ritz method [120, 125]. An important feature of the statistical approach is that when LES is performed, then closure in the state space is at a basic level, promising to have a greater chance of success [107]. Furthermore, the number of terms requiring closure is small compared with the deterministic EE, EL, and hybrid methods. It also provides a natural way to formulate correct boundary conditions and it establishes a natural length scale, which can verify validity of the gradient

Table 12.5 Some Typical Droplet Breakup, Collision, Coalescence, and Deformation Models Used for Simulation of Spray Flames

Physical Mechanism	Model
Primary droplet breakup	KH hybrid breakup [126]
Secondary droplet breakup	KH [130], TAB [127], KH–RT hybrid [126], Blob [131], AtoMIST [54]
Collision and coalescence	Stochastic particle method [99, 129, 132]
Droplet deformation	Harmonic oscillator model [133]

diffusion approximation. However, these methods are computationally expensive and require higher order numerical implementation to obtain accurate statistics, and therefore are mainly relegated to investigation of canonical systems.

The breakup is a critical feature to include in all spray calculations. The process can be explained from Kelvin–Helmholtz (KH) instability theory and taking into account of effects of liquid properties, atomization process, and aerodynamics [126]. The secondary breakup models have used variants of the Taylor analogy breakup (TAB) model [127] or a wave model based on KH and Rayleigh–Taylor (RT) stability analysis [128]. Hybrid characterization of primary and secondary breakup [126] utilizes KH and RT stability analysis to account for primary and secondary breakup. Droplet to droplet interaction in terms of collision and coalescence through stochastic particle method have been proposed and employed in a number of studies [129].

Table 12.5 provides some of the models that are typically used to model and couple spray to the gas phase. The spray submodels account for primary breakup, secondary breakup, droplets collision and coalescence and deformation of droplets. In literature [128] various models are reported for these subprocesses and therefore their specific details and their applicability in the current discussion are only highlighted here when reporting relevant results.

In general, so far, for LES of swirling spray combustion modeling, primarily EL and/or EE have been used due to their range of applicability, robustness, accuracy, and computational efficiency. These two methods employ different sets of governing equations based on the assumptions and approximations considered in their derivation. Both these methods have advantages and limitations that are well established and known [107, 109]. A detail hierarchy of the system of equations ranging from RANS to DNS and dilute to dense flow systems can be found elsewhere [109]. Here, we summarize the governing equations for dense two-phase systems for both EE [134–136] and EL [137–139] methods. A comprehensive review of computational modeling approaches for turbulent dilute spray combustion is provided elsewhere [140]. Note that laboratory scale atmospheric swirling spray combustion studies can be performed using low Mach number two phase computational models. However, for the current discussion we focus on the compressible formulation of the governing equations, which is more appropriate for such systems where a nonlinear coupling exists between the unsteady heat release, fluid dynamics, and combustor acoustics. Finally, the generalized form of equations presented here considers coupling of the gas phase, Lagrangian particle (disperse), and/or Eulerian disperse phases, that is, a hybrid formulation based on underlying EE and EL approaches. However, the governing equations for EE or EL formulation can be easily recovered from the general formulation presented as follows.

The density weighted spatially filtered LES equations can be obtained from the compressible form of the multispecies Navier–Stokes equations by employing the well-known Favre filtering approach. The Favre filtered (resolved) quantity corresponding to a field variable $\phi(\mathbf{x}, t)$ is defined as

$$\widetilde{\phi}(\mathbf{x}, t) = \frac{1}{\overline{\rho}(\mathbf{x}, t)} \int_{\Omega} G(\mathbf{x}, \mathbf{x}' - \mathbf{x}) \rho(\mathbf{x}', t) \phi(\mathbf{x}', t) d\mathbf{x}', \qquad (12.3.1)$$

where Ω is the computational domain, G is a spatial filter function, ρ is the density, and $\overline{(.)}$ denotes the conventional spatial filtering, which when applied to the density field $\rho(\mathbf{x}, t)$ leads to the spatially filtered density field $\overline{\rho}(\mathbf{x}, t)$ given by

$$\overline{\rho}(\mathbf{x}, t) = \int_{\Omega} G(\mathbf{x}, \mathbf{x}' - \mathbf{x}) \rho(\mathbf{x}', t) \, d\mathbf{x}'. \qquad (12.3.2)$$

Note that the two filtering approaches are related through $\overline{\rho\phi} = \overline{\rho}\widetilde{\phi}$.

The interphase coupling of the gas phase, the Lagrangian disperse phase, and the Eulerian disperse phase can be tracked in terms of the volume fraction through

$$\overline{\alpha}_g + \overline{\alpha}_p + \overline{\alpha}_d = 1, \qquad (12.3.3)$$

where $\overline{\alpha}_g$, $\overline{\alpha}_p$, and $\overline{\alpha}_d$ denote volume fraction of the gas phase, Lagrangian disperse phase, and Eulerian disperse phase, respectively. The volume fraction of the Lagrangian disperse phase $\overline{\alpha}_p$ is defined as

$$\overline{\alpha}_p = \frac{1}{\Delta V} \sum_{n=1}^{N} n_{p,n} \left[\frac{4}{3} \pi r^3{}_{p,n} \right], \qquad (12.3.4)$$

where ΔV is the volume of the computational cell, n_p is the number of particles in each particle group (referred to as a parcel), and N is the total number of parcels in the computational cell. The volume fraction of the gas phase $\overline{\alpha}_g$ evolves in space and time according to

$$\frac{\partial \overline{\alpha}_g}{\partial t} + \widetilde{u}_{I,i} \frac{\partial \overline{\alpha}_g}{\partial x_i} = \dot{m}_g / \rho_I, \qquad (12.3.5)$$

where $\widetilde{u}_{I,i}$ is the interface velocity field. The term on the right-hand side appears only when there is mass transfer across different phases. From here onward, subscripts g, p, d, and I are used to indicate the gas phase, Lagrangian particle, Eulerian disperse phase, and interface quantity, respectively. Note that the interface terms require additional models [134]. As mentioned before, the governing system of equations for EE and EL formulations can be recovered by setting $\overline{\alpha}_p = 0$ and $\overline{\alpha}_d = 0$, respectively, in the general formulation presented in the following. Note that, in the EL formulation for systems having low mass or volume loading, $\overline{\alpha}_g = 1$ is used.

12.3.2.1.1 Eulerian Gas Phase

Applying a box filter (appropriate for a finite volume based implementation) to the system of transport equations yields the density weighted filtered LES equations for the gas phase, which comprises of transport equations for mass, momentum, energy, and species expressed as

$$\frac{\partial \overline{\alpha}_g \overline{\rho}_g}{\partial t} + \frac{\partial \overline{\alpha}_g \overline{\rho}_g \widetilde{u}_{g,i}}{\partial x_i} = \overline{\dot{\rho}}_D, \tag{12.3.6a}$$

$$\frac{\partial \overline{\alpha}_g \overline{\rho}_g \widetilde{u}_{g,i}}{\partial t} + \frac{\partial}{\partial x_j} \left[\overline{\alpha}_g \left(\overline{\rho}_g \widetilde{u}_{g,i} \widetilde{u}_{g,j} + \overline{P}_g \delta_{ij} + \tau_{g,ij}^{sgs} - \overline{\tau}_{g,ij} \right) \right]$$

$$= \overline{P}_I \frac{\partial \overline{\alpha}_g}{\partial x_j} \delta_{ij} - \overline{\tau}_{I,ij} \frac{\partial \overline{\alpha}_g}{\partial x_j} + \overline{F}_{D,i}, \tag{12.3.6b}$$

$$\frac{\partial \overline{\alpha}_g \overline{\rho}_g \widetilde{E}_g}{\partial t} + \frac{\partial}{\partial x_j} \left[\overline{\alpha}_g \left(\overline{\rho}_g \widetilde{u}_{g,j} \widetilde{E}_g + \widetilde{u}_{g,j} \overline{P}_g + \overline{q}_{g,j} - \widetilde{u}_{g,i} \overline{\tau}_{g,ji} + H_{g,j}^{sgs} + \sigma_{g,j}^{sgs} \right) \right]$$

$$= \overline{P}_I \widetilde{u}_{I,j} \frac{\partial \overline{\alpha}_g}{\partial x_j} - \widetilde{u}_{I,i} \overline{\tau}_{I,ij} \frac{\partial \overline{\alpha}_g}{\partial x_j} + \overline{\dot{Q}}_D + \overline{\dot{W}}_D, \tag{12.3.6c}$$

$$\frac{\partial \overline{\alpha}_g \overline{\rho}_g \widetilde{Y}_{g,k}}{\partial t} + \frac{\partial}{\partial x_i} \left[\overline{\alpha}_g \left(\overline{\rho}_g \left(\widetilde{Y}_{g,k} \widetilde{u}_{g,i} + \widetilde{Y}_{g,k} \widetilde{V}_{g,i,k} \right) + Y_{g,i,k}^{sgs} + \theta_{g,i,k}^{sgs} \right) \right]$$

$$= \overline{\alpha}_g \dot{\omega}_{g,k} + \overline{\dot{S}}_{D,k}, \quad for \quad k = 1, 2, \ldots, N_s, \tag{12.3.6d}$$

where subscript D denotes the source term contribution from Lagrangian and Eulerian disperse phases and N_s is the number of species. In Eq. (12.3.6) u_i is the velocity vector, p is the pressure, E is the total energy, Y_k is the mass fraction of the kth species, τ_{ij} is the viscous stress tensor, q_i is the heat flux vector, and δ_{ij} is the Kronecker delta. The subgrid scale terms are denoted by superscript sgs in Eq. (12.3.6) and they require closure approximation. These terms include subgrid stress tensor τ_{ij}^{sgs}, subgrid enthalpy flux H_i^{sgs}, subgrid viscous work σ_i^{sgs}, subgrid convective species flux $Y_{i,k}^{sgs}$, and subgrid diffusive species flux $\theta_{i,k}^{sgs}$. The closure models for these terms are discussed in Section 12.3.2.3. The source terms $\overline{\dot{\rho}}_D$, $\overline{F}_{D,i}$, $\overline{\dot{W}}_D$, and $\overline{\dot{S}}_{D,k}$, which appear in the transport equations for mass, momentum, energy, and species, respectively, represent the interphase exchange terms. They are obtained in terms of the source terms for Lagrangian and Eulerian disperse phases through

$$\overline{\mathcal{F}}_{D,m} = \overline{\mathcal{F}}_{p,m} + \overline{\mathcal{F}}_{d,m}, \tag{12.3.7}$$

where the vector $\mathcal{F} = \left[\rho, \mathbf{F}^T, Q, W, \mathbf{S}^T \right]^T$ represents source term in different transport equations. Here, $\mathbf{F} = \left[F_1, F_2, \ldots, F_{dim} \right]^T$ and $\mathbf{S} = \left[S_1, S_2, \ldots, S_{N_s} \right]^T$, with dim denoting the dimension of the considered flow system (2D/3D). Note that for mass transfer of single species, we obtain $\overline{\dot{S}}_D = \overline{\dot{\rho}}_D$. The contributions to source terms from the Lagrangian and Eulerian disperse phases are provided in Eqs. (12.3.15) and (12.3.14), respectively.

By assuming a Newtonian fluid with Stokes's hypothesis and Fourier's law of thermal conduction, the filtered viscous stress tensor $\overline{\tau}_{g,ij}$ and the heat flux vector $\overline{q}_{g,j}$ can be approximated as

$$\overline{\tau}_{g,ij} = \mu_g \left(\frac{\partial \widetilde{u}_{g,i}}{\partial x_j} + \frac{\partial \widetilde{u}_{g,j}}{\partial x_i} \right) - \frac{2}{3} \mu_g \frac{\partial \widetilde{u}_{g,k}}{\partial x_k} \delta_{ij}, \qquad (12.3.8a)$$

$$\overline{q}_{g,j} = -\kappa_g \frac{\partial \widetilde{T}_g}{\partial x_j} + \overline{\rho} \sum_1^{N_s} \widetilde{Y}_{g,k} \widetilde{h}_{g,k} \widetilde{V}_{g,j,k} + \sum_1^{N_s} q_{g,i,k}^{sgs}, \qquad (12.3.8b)$$

where $\mu_g \equiv \mu_g(\widetilde{T}_g)$ is the dynamic viscosity of the gas, κ_g is the thermal conductivity of the gas, $\widetilde{h}_{g,k}$ is the resolved specific enthalpy of the kth species, $q_{g,i,k}^{sgs}$ is the subgrid heat flux vector (described in Sec. 12.3.2.3), and $\widetilde{V}_{g,j,k}$ is the resolved species diffusion velocities that can be modeled through a Fickian diffusion approximation through

$$\widetilde{V}_{g,j,k} = -\frac{D_{g,k}}{\widetilde{Y}_{g,k}} \frac{\partial \widetilde{Y}_{g,k}}{\partial x_j}. \qquad (12.3.9)$$

Here, $D_{g,k}$ denotes diffusion coefficient of the kth species and it can be obtained from a constant Lewis number (Le) assumption.

12.3.2.1.2 *Lagrangian Liquid Phase*

The Lagrangian equations for the motion of a single particle within dense spray combustors are obtained with the assumption that the particle density is much greater than the carrier (gas) phase density ($\rho_p/\rho_g \approx 10^3$) and particle diameter is smaller than the Kolmogorov length scale. These equations are given by

$$\frac{dx_{p,i}}{dt} = u_{p,i}, \qquad (12.3.10a)$$

$$\frac{dm_p}{dt} = -\dot{m}_p = \frac{d}{dt} \left(\frac{4}{3} \pi r_p^3 \rho_p \right), \qquad (12.3.10b)$$

$$m_p \frac{du_{p,i}}{dt} = \frac{\pi}{2} r_p^2 C_D \overline{\rho}_g \left| \widetilde{u}_{g,i} + u_{g,i}'' - u_{p,i} \right| \left(\left(\widetilde{u}_{g,i} + u_{g,i}'' - u_{p,i} \right) \right) - \frac{4}{3} \pi r_p^3 \frac{\partial \overline{p}_g}{\partial x_i} + m_p A_{c,i}, \qquad (12.3.10c)$$

$$m_p C_p \frac{dT_p}{dt} = 2\pi r_p \kappa_g Nu (\widetilde{T}_g - T_p) - \dot{m}_p L_v, \qquad (12.3.10d)$$

where x_i, m, r, T denote position vector, mass, radius, and temperature, respectively. Some of the other terms in Eq. (12.3.10) include the drag coefficient C_D, acceleration due to interparticle interactions $A_{c,i}$, the heat capacity C_p, the thermal conductivity of the gas phase κ_g, the Nusselt number Nu, and the latent heat of vaporization L_v. The Nusselt number and the drag coefficient are typically expressed as empirical functions of Re, Prandtl number (Pr), Mach number (M) and volume fraction α_g [138, 139, 141]. The closures for interface quantities can be found elsewhere [138, 139, 142]. The acceleration due interparticle interactions, $A_{c,i}$, is computed as

$$A_{c,i} = -\frac{1}{\alpha_p \rho_p} \frac{\partial \tau}{\partial x_i}, \qquad (12.3.11)$$

where τ is the intergranular stress given by

$$\tau = \frac{P_s \alpha_p^\beta}{\alpha_{cs} - \alpha_p}, \qquad (12.3.12)$$

with α_{cs} being the close packing volume fraction. Also, P_s and β are empirical constants, which are closed based on the nature of the flow being considered [138, 139, 142, 143].

In Eq. (12.3.10c), the sum $(\widetilde{u}_{g,i} + u_{g,i}'')$ represents instantaneous $(u_{g,i})$ gas phase velocity components, consisting of both the LES resolved velocity $\widetilde{u}_{g,i}$ and the unresolved fluctuating velocity $u_{g,i}''$, which can be reconstructed by employing a stochastic model [7, 94, 104]. For example, the unresolved velocity field can be obtained from the subgrid scale turbulent kinetic energy at intervals coincident with the local characteristic eddy lifetime [7] or it can be based on a stochastic Markovian model [144]. In several other numerical implementations in the past, this term has been ignored, however, as mentioned in [94], in poorly resolved regions of the flow, where the subgrid scale turbulent kinetic energy is more than 30% of the resolved turbulent kinetic energy, the effect of unresolved velocity fluctuations on the particle motion becomes important. In a similar way, the gas phase temperature at the particle location in Eq. (12.3.10d) ideally should include the unresolved temperature fluctuation, which in turn can be modeled through stochastic means [104]; however, in the present formulation, we have ignored such contribution.

12.3.2.1.3 Eulerian Liquid Phase

In the limit of a dense mass or volume loading, the disperse phase can be modeled as a continuum fluid. In such cases, the governing equations for the disperse phase can be expressed in the Eulerian framework and are given by

$$\frac{\partial \overline{\alpha}_d \overline{\rho}_d}{\partial t} + \frac{\partial \overline{\alpha}_d \overline{\rho}_d \widetilde{u}_{d,i}}{\partial x_i} = -\overline{\dot{\rho}}_d \qquad (12.3.13a)$$

$$\frac{\partial \overline{\alpha}_d \overline{\rho}_d \widetilde{u}_{d,i}}{\partial t} + \frac{\partial}{\partial x_j} \left[\overline{\alpha}_d \left(\overline{\rho}_d \widetilde{u}_{d,i} \widetilde{u}_{d,j} + \overline{P}_d \delta_{ij} + \tau_{d,ij}^{sgs} - \overline{\tau}_{d,ij} \right) \right]$$

$$= \overline{P}_I \frac{\partial \overline{\alpha}_d}{\partial x_j} \delta_{ij} - \overline{\tau}_{I,ij} \frac{\partial \overline{\alpha}_d}{\partial x_j} - \overline{F}_{d,i}, \qquad (12.3.13b)$$

$$\frac{\partial \overline{\alpha}_d \overline{\rho}_d \widetilde{E}_d}{\partial t} + \frac{\partial}{\partial x_j} \left[\overline{\alpha}_d \left(\overline{\rho}_d \widetilde{u}_{d,j} \widetilde{E}_d + \widetilde{u}_{d,j} \overline{P}_d + \overline{q}_{d,j} - \widetilde{u}_{d,i} \overline{\tau}_{d,ji} + H_{d,j}^{sgs} + \sigma_{d,j}^{sgs} \right) \right]$$

$$= \overline{P}_I \widetilde{u}_{I,j} \frac{\partial \overline{\alpha}_d}{\partial x_j} - \widetilde{u}_{I,i} \overline{\tau}_{I,ij} \frac{\partial \overline{\alpha}_d}{\partial x_j} - \overline{Q}_d - \overline{W}_d, \qquad (12.3.13c)$$

with the source terms in these equations given by

$$\overline{\dot{\rho}}_d = N_d \dot{m}_d = N_d \frac{d}{dt}\left(\frac{4}{3}\pi r_d^3 \overline{\rho}_d\right), \tag{12.3.14a}$$

$$\overline{\dot{F}}_{d,i} = N_d\left[\dot{m}_d u_{d,i} + \frac{\pi}{2}r_d^2 C_D \overline{\rho}_g |\tilde{u}_{d,i} - \tilde{u}_{g,i}|\left(\tilde{u}_{d,i} - \tilde{u}_{g,i}\right) + \frac{4}{3}\pi r_d^3 \frac{\partial \overline{p}_g}{\partial x_i}\right], \tag{12.3.14b}$$

$$\overline{\dot{Q}}_d = N_d\left[\dot{m}_d h_v + 2\pi r_d \kappa_g Nu\left(\tilde{T}_d - \tilde{T}_g\right)\right], \tag{12.3.14c}$$

$$\overline{\dot{W}}_d = N_d\left[\dot{m}_d \tilde{u}_{d,i}\tilde{u}_{d,i} + \frac{\pi}{2}r_d^2 C_D \rho_g |\tilde{u}_{d,i} - \tilde{u}_{g,i}|\left(\tilde{u}_{d,i} - \tilde{u}_{g,i}\right)\tilde{u}_{d,i} + \frac{4}{3}\pi r_d^3 \frac{\partial \overline{p}_g}{\partial x_i}u_{d,i}\right], \tag{12.3.14d}$$

where N_d is the number density of the Eulerian disperse phase.

The source terms from the Lagrangian and Eulerian disperse phases provided through Eqs. (12.3.15) and (12.3.14) can be combined through Eq. (12.3.7) to yield source terms for the gas phase. The governing system of equations given by Eqs. (12.3.6), (12.3.10), and (12.3.13), for the gas phase, Lagrangian disperse phase, and Eulerian disperse phase, respectively, are complete once the subgrid scale models are specified, which are briefly described in the next section.

12.3.2.2 Coupling Between the Phases

As shown in Eq. (12.3.7), the source term needed for the gas phase transport equations given by Eq. (12.3.6) comprises source terms corresponding to the Lagrangian and Eulerian disperse phases. The source term contributions from the Lagrangian disperse phase are obtained through

$$\overline{\dot{\rho}}_p = \frac{1}{\Delta V}\sum_{n=1}^{N} n_{p,n}\dot{m}_{p,n}, \tag{12.3.15a}$$

$$\overline{\dot{F}}_{p,i} = \frac{1}{\Delta V}\sum_{n=1}^{N} n_{p,n}\left[\dot{m}_{p,n}u_{p,i,n} + \frac{4}{3}\pi r_{p,n}^3 \frac{\partial \overline{p}_{g,n}}{\partial x_i}\right. \tag{12.3.15b}$$

$$\left. + \frac{\pi}{2}r_{p,n}^2 C_{D,n}\overline{\rho}_{g,n}|u_{p,i,n} - \tilde{u}_{g,i,n} - u''_{g,i,n}|\left(u_{p,i,n} - \tilde{u}_{g,i,n} - u''_{g,i,n}\right)\right], \tag{12.3.15c}$$

$$\overline{\dot{Q}}_p = \frac{1}{\Delta V}\sum_{n=1}^{N} n_{p,n}\left[\dot{m}_{p,n}h_{v,n} + 2\pi r_{p,n}\kappa_g Nu_n\left(T_{p,n} - \tilde{T}_{g,n}\right)\right], \tag{12.3.15d}$$

$$\overline{\dot{W}}_p = \frac{1}{\Delta V}\sum_{n=1}^{N} n_{p,n}\left[\dot{m}_{p,n}u_{p,i,n}u_{p,i,n} + \frac{4}{3}\pi r_{p,n}^3 \frac{\partial \overline{p}_{g,n}}{\partial x_i}u_{p,i,n}\right. \tag{12.3.15e}$$

$$\left. + \frac{\pi}{2}r_{p,n}^2 C_{D,n}\rho_{g,n}|u_{p,i,n} - \tilde{u}_{g,i,n} - u''_{g,i,n}|\left(u_{p,i,n} - \tilde{u}_{g,i,n} - u''_{g,i,n}\right)u_{p,i,n}\right], \tag{12.3.15f}$$

where h_v is the enthalpy change associated with the mass transfer, and ΔV is the volume of the cell containing N particles.

12.3.2.3 Subgrid Terms and Closure

The density weighted filtering of the governing equations leads to appearance of unclosed terms, also referred to as the subgrid scale terms, which require further closure approximations to obtain a closed system of equations. As mentioned before, these terms include the subgrid stress tensor $\tau_{g,ij}^{sgs}$, subgrid enthalpy flux $H_{g,i}^{sgs}$, subgrid viscous work $\sigma_{g,i}^{sgs}$, subgrid convective species flux $Y_{g,i,k}^{sgs}$, subgrid diffusive species flux $\theta_{g,i,k}^{sgs}$, and the subgrid heat flux $q_{g,i,k}^{sgs}$ [22, 145]. Some of these terms can be closed using the models developed for nonreacting or reacting gas phase problems. Here, we briefly describe some well-known models that are used for closure of subgrid scale terms.

The subgrid stress and heat flux terms are typically closed following the Boussinesq approximation, where an eddy viscosity type gradient closure is employed. Such type of closure is very popular and a model is required for the eddy viscosity. Both algebraic/dynamic Smagorinsky model (ASM/DSM) [146, 147] and the model for subgrid kinetic energy [148, 149] are very popular. For example, when using the subgrid kinetic energy k^{sgs} closure, a transport equation for k^{sgs} needs to be solved, which is given by

$$\frac{\partial}{\partial t}\overline{\alpha}_g\overline{\rho}_g k^{sgs} + \frac{\partial}{\partial x_i}\left(\overline{\alpha}_g\overline{\rho}_g\widetilde{u}_{g,i}k^{sgs}\right) = \overline{\alpha}_g P_{k^{sgs}} + \frac{\partial}{\partial x_i}\left(\overline{\alpha}_g\overline{\rho}_g \nu_t \frac{\partial k^{sgs}}{\partial x_i}\right) - \overline{\alpha}_g D_{k^{sgs}} + \overline{F}_{D,k}^{sgs}.$$

(12.3.16)

Here, $P_{k^{sgs}}$ and $D_{k^{sgs}}$ denote, respectively, the production and dissipation of k^{sgs}, which can be obtained through

$$P_{k^{sgs}} = -\tau_{g,ij}^{sgs}\frac{\partial\widetilde{u}_{g,i}}{\partial x_j}, \qquad D_{k^{sgs}} = C_\varepsilon\overline{\rho}_g\frac{(k^{sgs})^{1.5}}{\Delta},$$

(12.3.17)

where ν_t is the subgrid eddy viscosity, and is modeled as $\nu_t = C_\nu\sqrt{k^{sgs}}\Delta$, where Δ is the local filter width. The constants C_ν and C_ε are obtained theoretically as 0.067 and 0.916, respectively [22], but can be computed using a dynamic procedure, referred to as the locally dynamic kinetic energy model (LDKM) [145, 150] that can be used locally without requiring any ad hoc averaging. With the eddy viscosity model, the subgrid stress tensor $\tau_{g,ij}^{sgs}$ is closed through

$$\tau_{g,ij}^{sgs} = -2\overline{\rho}_g\nu_t\left(S_{g,ij} - \frac{1}{3}S_{g,kk}\delta_{ij}\right) + \frac{2}{3}\overline{\rho}_g k^{sgs}\delta_{ij},$$

(12.3.18)

where $S_{g,ij}$ is the rate of strain tensor given by

$$S_{g,ij} = \frac{1}{2}\left(\frac{\partial\widetilde{u}_{g,i}}{\partial x_j} + \frac{\partial\widetilde{u}_{g,j}}{\partial x_i}\right).$$

(12.3.19)

In Eq. (12.3.16) $\overline{F}_{D,i}^{sgs}$ is the source term due to the Lagrangian and Eulerian disperse phases that can be closed exactly [151, 152].

The subgrid total enthalpy, $H_{g,j}^{sgs}$, can also modeled using the eddy viscosity and gradient diffusion assumption through

$$H_{g,i}^{sgs} = -\overline{\rho}_g\frac{\nu_t}{Pr_t}\frac{\partial\widetilde{H}_g}{\partial x_i},$$

(12.3.20)

where $\widetilde{H}_g$ is the filtered total enthalpy, given by $\widetilde{H}_g = \widetilde{h}_g + \frac{1}{2}\widetilde{u}_{g,i}\widetilde{u}_{g,i} + k^{sgs}$, where $\widetilde{h}_g = \sum_{k=1}^{N_s} \widetilde{h}_{g,k}\widetilde{Y}_{g,k}$. The turbulent Prandtl number Pr_t can be dynamically computed, but is typically assumed to be unity. In a similar manner, the subgrid convective species flux, $Y_{g,i,k}^{sgs}$, is modeled using the gradient diffusion assumption through

$$Y_{g,i,k}^{sgs} = -\frac{\overline{\rho}_g \nu_t}{Sc_t} \frac{\partial \widetilde{Y}_{g,k}}{\partial x_i}, \qquad (12.3.21)$$

where Sc_t is the turbulent Schmidt number. It can also be dynamically computed; however, it is typically assumed to be unity in many studies. The closure for the scalar transport is performed differently in the linear eddy mixing (LEM) model, which employs a subgrid scalar mixing, diffusion, and combustion model and therefore does not close the species equation (Eq. (12.3.6d)) using conventional methods [153–155].

The other subgrid terms, that is, $\theta_{g,i,k}^{sgs}$ and $q_{g,i,k}^{sgs}$, are neglected in most studies [10, 22]. Note that in Eqs. (12.3.18)–(12.3.21), the subscript g can be replaced with d to obtain appropriate closures under the assumption that the Eulerian disperse phase behaves as a pseudogas phase. The so-called pseudofluid assumption is not very easy to justify and therefore the closures for disperse phase, especially in the context of dense flows, are unknown.

A closure model for the reaction rate is also required for the combustion problem. Subgrid turbulence–chemistry interaction models have remained a challenging task even for gas phase combustion and these problems are made even more challenging when spray evaporation, mixing, and combustion have to be included. Due to the discrete nature of the droplets vaporizing and since the mixing is local, therefore, combustion can occur in multitude of manner ranging from non-premixed to fully premixed type. Partially premixing is the norm in spray combustion system and therefore subgrid closures for turbulence–chemistry interaction need to address the complete regime. At this time there are not that many options available for this goal.

Table 12.6 summarizes typical closure models used for the gas phase transport equations and turbulence–chemistry interaction for simulation of swirling spray

Table 12.6 Closure Model Used in Transport Equations for the Gas Phase for Simulation of Typical Swirl Spray Combustors

Model	References	Turbulence Chemistry Interaction	Transport
LEM–LES	[6, 7]	Linear eddy model	LDKM/LEM
PDF–LES	[89, 99, 104]	Statistical	DSM
LFM–LES	[45, 156]	Laminar flamelet model	DSM
FGM–LES	[98]	Flamelet generated manifold	DSM
TF–LES	[80, 84]	Dynamically thickened flame	DSM
EBU–LES	[92, 95]	Eddy breakup	DSM/LDKM
SOM–LES	[96]	Second-order moments SGS	DSM
Laminar–DES	[9]	Laminar chemistry	DES

combustors. Note that all the subgrid scale models have some degree of empiricism, a range of applicability, and differences in computational cost. A review of some of these models, their advantages, and their limitations is provided elsewhere [157].

12.4   Examples

In this section we describe some key results obtained from LES studies of swirling spray combustors, where we focus on instantaneous and mean flow features that can be resolved and uncertainties associated with the predictions. However, the cases included are only representative of such flow systems and are not comprehensive. Further details are provided in the cited references.

12.4.1   Instantaneous Features

Swirling spray dispersion and its interaction with the flow features such as vortex breakdown bubble, precessing vortex core, recirculation zone, and so on, in both reactive and nonreactive systems, have been investigated earlier [7, 20, 45, 63, 99]. Here, we consider studies focused on liquid fuel gas turbine combustors to discuss some of the key features observed in such systems.

Jones et al. [99] studied turbulent mixing, fuel spray dispersion and evaporation and combustion through LES of the DLR Generic Single Sector Combustor (GENRIG). The classical Smagorinsky model was used to obtain a closure for the subgrid scale stress and the subgrid level turbulence–chemistry interaction was modeled through a subgrid scale probability density function approach in conjunction with the stochastic fields solution method. Figure 12.10 shows the time evolution of the instantaneous spray injection process occurring through the swirler. As the droplets are injected in the combustion chamber, they tend to evaporate due to the swirling preheated air. The effect of swirl is apparent from the visualization of the spray dispersion as it evolves in time. We can observe an increased lateral dispersion and the formation of a spiral pattern of the spray.

(a)   (b)   (c)

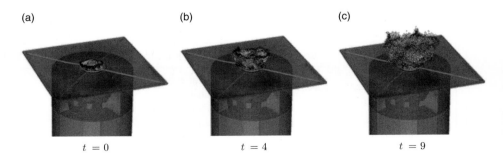

$t = 0$ $t = 4$ $t = 9$

Figure 12.10 Instantaneous spray injection at three different times [99]. Time is nondimensionalized based on the reference time for the first snapshot of the spray injection.

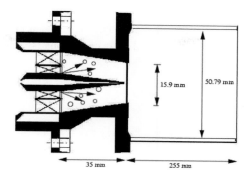

Figure 12.11 A schematic representation of the DACRS gas turbine combustor of GE Aircraft Engines [20].

Figure 12.11 shows a schematic representation of the dual annular counterrotating swirling (DACRS) gas turbine combustor of GE Aircraft Engines (GEAE), which is a cylindrical dump combustor that consists of a converging inlet section with a central injector cone. The inlet pipe corresponds to the region downstream of the swirler. The spray droplets are injected from the lateral sides of the central cone. Sankaran and Menon [20] studied this GT combustor to demonstrate ability to investigate the unsteady interactions between spray dispersion, vaporization, fuel–air mixing, heat release, and effect of swirl intensity.

Figure 12.12 shows iso-surface of the instantaneous azimuthal vorticity and the droplet distribution for non-reacting and reacting cases at two different swirl numbers at the dump plane of the GE-DACRS combustor. The swirl number corresponding to the low swirl case is 0.5, which is smaller than the critical swirl number required for vortex breakdown to occur, whereas the case with higher swirl number has a value of 0.8, which is higher than the critical swirl number. We can observe in Figure 12.12 that in both non-reacting and reacting low swirl cases particle dispersion is rather weak since the transverse motion is not sufficient to overcome the axial momentum. This is due to the large coherence in the shear layer shed from the dump plane and this coherence results in weakly swirling vortex rings (note the iso-surface level is used to enhance the ring coherence and so the helical nature of the shear layer is not fully visible). As a result, droplets enter the combustor modulated by these rings and only begin to disperse once the rings have broken down due to 3D instability. With a high swirl intensity, the coherence of the rings is quickly destroyed and the spray also disperses rapidly into the combustor. The 3D nature of the flow field allows for rapid dispersion and mixing of the two phases. Compared with the non-reacting case, the vortex breakdown is much more vigorous in the reacting case.

The rate of decay of the axial momentum is significantly altered in the reacting case due to the presence of heat release induced thermal expansion. In particular, the decay rate of the axial momentum in the reacting case at low swirl is small in comparison with the non-reacting case. This is attributed to an increase in the axial velocity due to thermal expansion, which leads to a reduction in the axial extent of the CTRZ. Such an increase in the axial velocity in case of low swirl may also lead to difficulty in burning

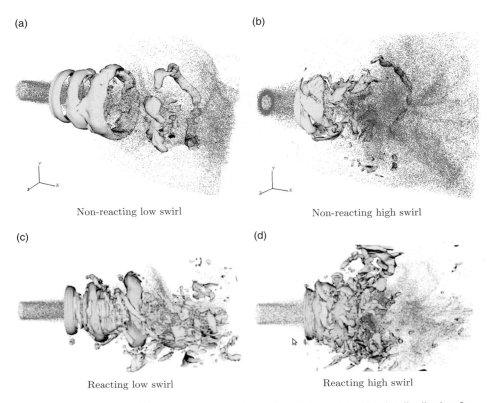

(a)   (b)

Non-reacting low swirl   Non-reacting high swirl

(c)   (d)

Reacting low swirl   Reacting high swirl

Figure 12.12 Iso-surface of the instantaneous azimuthal vorticity and the droplet distribution for nonreacting (a, b) and reacting (c, d) cases at low ($S_w = 0.5$) and high swirl ($S_w = 0.8$) numbers in the GE-DACRS combustor [20]. The yellow and blue iso-surfaces denote positive ($\widetilde{\omega}_\theta = 15000 s^{-1}$) and negative ($\widetilde{\omega}_\theta = -15000 s^{-1}$) values of azimuthal vorticity, respectively. A black and white version of this figure will appear in some formats. For the color version, please refer to the plate section.

and therefore possibly an unstable combustion. However, at higher swirl intensity, presence of the hot central recirculation zone leads to a stable flame holding.

Figure 12.13 shows iso-surfaces of the instantaneous vorticity and temperature fields for the reacting low and high swirl cases, respectively. The flame structure is identified through the iso-surface of the temperature field. We observe that the flame is longer in the low swirl case compared with the high swirl case, where a compact flame occurs close to the dump plane of the combustor, a well-known effect of swirl on the flame length [36]. The flame surface appears to be wrapped around the large scale vortical structures near the dump plane of the combustor. However, further downstream of the dump plane, small scale vortical structures are predominant due to the breakdown of the large scale vortical structures and this induces fine scale wrinkling on the flame surface. The wrinkling is higher in case of a higher swirl intensity. Further description of the effect of swirl on the flame structure is presented elsewhere [2, 7].

The development of PVC has been seen in both low pressure (atmospheric) swirl spray combustion [7] and high pressure combustion [37]. Figure 12.14 shows time evolution of

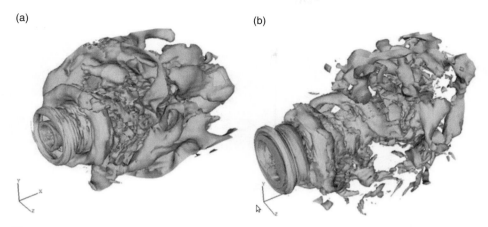

Figure 12.13 Iso-surfaces of instantaneous azimuthal vorticity ($\widetilde{\omega}_\theta = -15000\mathrm{s}^{-1}$) and temperature ($\widetilde{T} = 2200$ K) for the reacting (a) low swirl ($S_\mathrm{w} = 0.5$) and (b) high swirl ($S_\mathrm{w} = 0.8$) cases, respectively [124]. Blue and orange colored surfaces correspond to azimuthal vorticity and temperature iso-surfaces, respectively.

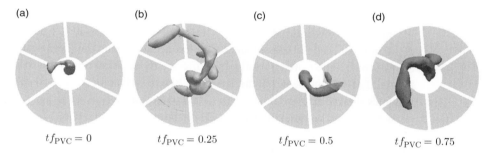

Figure 12.14 Time evolution of the PVC within one cycle of the precession motion identified using iso-surface of the instantaneous pressure $\bar{p} = 1.16$ MPa [37]. Here, f_{PVC} denotes the precession frequency.

the PVC within one cycle of the precession obtained from the LEM–LES study of the LDI combustor at $P_{ref} = 10.3$ MPa [37]. The structure at $tf_{\mathrm{PVC}} = 0$, where f_{PVC} is the precession frequency, is aligned to the central axis, but at $tf_{\mathrm{PVC}} = 0.25$, it is driven away from the core and extends along the axial direction in a spiral manner. At later times, the structure again is driven towards the central axis. Such precessing motion affects the flow and flame evolution in GT combustion systems as it enhances levels of turbulence intensity and mixing. However, they are also undesirable as they may couple with the acoustic pressure oscillations in a resonant manner leading to combustion instability.

The precessing motion plays an effective role in the spray dispersion. In a typical combustor, the flame is confined by the PVC and the reverse flow region created by the VBB. The effect of breakup modeling can impact how the dispersion will occur. Some studies were conducted earlier by comparing identical cases with and without secondary breakup [7]. Figure 12.15 shows two instantaneous snapshots of motion of particles,

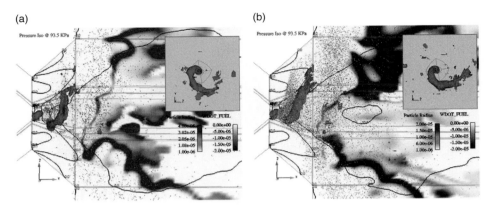

Figure 12.15 Instantaneous droplets, the fuel reaction rate (dark colored), and the PVC (light colored iso-surface and inset) at two instants for simulation with (a) and without (b) secondary breakup model for the disperse phase. The VBB and CRZ extent is identified through a zero value contour of the time averaged streamwise velocity [7].

flame surface, and the PVC for these cases. The rotation of the PVC leads to dispersion of the liquid fuel particles around and into the VBB region. The particles are dispersed in the radial direction right up to the walls, just halfway to the dump plane. A preferential accumulation of small droplets occurs around the PVC, while the heavier droplets move downstream relatively unaffected by the local flow structures due to their inertia. The instantaneous flame structure is relatively thin near the central axis and becomes thicker near the outer part of the combustor.

As mentioned before, both EL and EE methods are typically used to study gas turbine swirling spray combustors. EE and EL methods were also compared earlier in a MERCATO test rig using two different computational codes and closures [27, 83]. Figure 12.16 shows the spray dispersion obtained from EE and EL methods close to the nozzle exit. The PVC is identified using the Q criterion and is seen to strongly influence particle dispersion and in this case to periodically impinge on the walls of the diffuser. Both methods predict spiraling motion of the disperse phase, which is a characteristic feature of swirling flows. Qualitatively, results obtained from EE and EL methods agree reasonably well in terms of prediction of spray–flow interaction, thus demonstrating applicability of these two methods to study realistic GT combustors.

Figure 12.17 shows the instantaneous droplet distribution in the MERCATO combustion chamber obtained using two different computational codes with EE and EL approaches for simulation of the disperse multiphase flow. A preferential concentration of the droplets, which is a main structural feature of the disperse phase is captured by both methods. The central recirculation zone has a low droplet density and dense pockets of droplets can be seen in the shear layer of the swirled air jet. In addition, droplets are also trapped in the recirculation zone in corners of the chamber. Such droplet concentrations in specific regions occur due to their inertia, which is directly associated with the Stokes number and leads to fuel vapor inhomogeneities through the evaporation source terms.

(a)   (b)

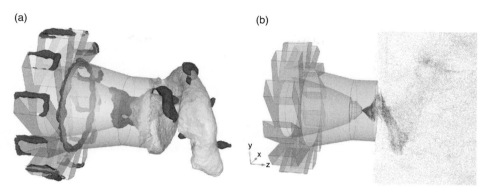

Figure 12.16 Isometric view of the spray close to the injector from EE (a) and EL (b) simulations [27]. In panel (a), iso-surface (white) of liquid volume fraction and iso-surface (black) of Q criterion are shown. In panel (b), black points mark the droplet position.

(a)   (b)

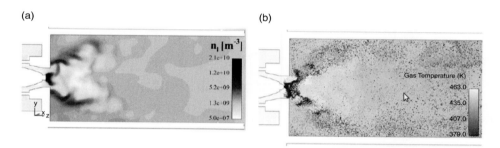

Figure 12.17 Instantaneous droplet distribution in the MERCATO chamber. Comparison between AVBP-EE and CDP-EL research codes. (a) AVBP-EE: droplet density (grayscale); (b) CDP-EL: droplets position (dots); gas temperature in grayscale [83].

12.4.2   Mean Features

Figure 12.18 shows the VBB for non-reacting and reacting cases in an identical setup. The reacting case is simulated using the LEM subgrid model closure. In both cases, VBB is observed. However, in the reacting case, due to the presence of heat release, the flow field is altered significantly. In particular, VBB becomes more compact and the flow motion in the VBB is much stronger compared with that in the non-reacting case. The mean flow streamlines undergo counterclockwise rotation and they move downstream rapidly and at smaller radial distance due to smaller VBB. Although these are qualitative observations, more quantitative comparison with data reported in the cited literature show that these results are in reasonable agreement with the actual flow field seen in these rigs.

Quantitative comparison of the time averaged profiles can be accomplished by using the statistically averaged flow field. Figure 12.19 shows radial profiles of the mean streamwise velocity at different axial locations within the DACRS combustor for non-reacting and reacting cases at two different swirl numbers. The vortex breakdown does not occur at the lower swirl number for both non-reacting and reacting cases. The CRZ

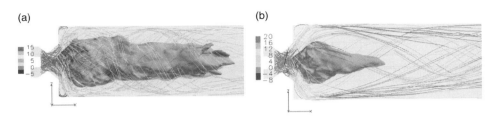

Figure 12.18 Vortex breakdown bubble for nonreactive (a) and reactive (b) cases in the LDI combustor identified using iso-surface of zero mean streamwise velocity [7]. Streamlines colored by the mean streamwise velocity demonstrate spiral pattern of the flow.

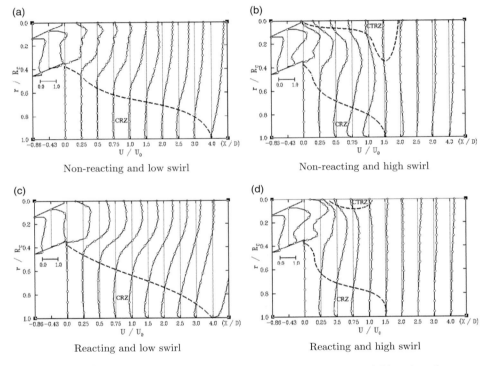

Figure 12.19 Radial profiles of the mean streamwise velocity at different axial locations for nonreacting (a, b) and reacting (c, d) cases, respectively, at low and high swirl numbers [20].

and CTRZ are marked with dashed curves in these figures. Without vortex breakdown at lower swirl number, there is no CTRZ. Additionally, due to appearance of CTRZ at high swirl number, the extent of CRZ reduces, but it is strong compared with the low swirl case. The decrease in the extent of CRZ is also due to a rapid divergence of streamlines in the high swirl case. The size of CTRZ in the reacting case is reduced compared with the non-reacting case because the heat release imposes a favorable pressure gradient in the combustor, which partly compensates the adverse pressure gradient due to the sudden expansion. This demonstrates that the heat release due to

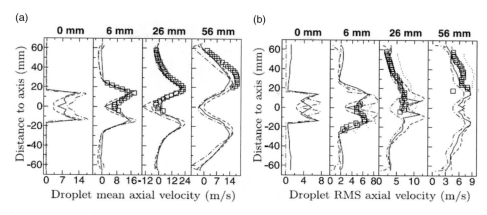

Figure 12.20 Mean (a) and RMS (b) of the droplet axial velocity at different axial location within the MERCATO combustor. Symbol (W) denotes PDA measurement and solid, dashed, and dash-dotted curves denote results from CDP-EL, AVBP-EL, and AVBP-EE codes, respectively [83].

combustion balances the mechanism responsible for the formation of the CTRZ due to dilatation effect and volume expansion, as described earlier.

Figure 12.20 shows results obtained from experiment and three different simulation models (and codes) for droplet mean axial velocity and its RMS at various axial locations for the MERCATO combustor. We can observe a reasonable agreement with some differences in prediction from these three simulations. Close to the axis, where the RZ is located, there is disagreement that may be related to the spatial segregation effect due to varying droplet size. However, the poly-disperse Lagrangian simulations do not provide much more accurate results in this zone than the mono-disperse Eulerian simulation. At other locations, further downstream, the three simulations reproduce similar peak intensity levels, but the shapes of the profiles are not well captured. There are many reasons for these differences and for some of the observed agreement. The current knowledge and modeling strategies are not yet mature enough to resolve these differences and this goal remains a future effort.

12.4.3   Multiple Injectors

Ability to carry out LES studies of multiple injectors is still in its infancy although codes and grid generation capability are probably ready given the current modeling capability and availability of high performance computational resources. However, the computational cost and the lack of detailed data hinder such model studies. Some attempts are underway nevertheless. For example, recent studies of ignition sequence in a realistic combustor with 18 injectors have been reported using the EE method [17]. Here, we briefly describe two representative studies of multiinjector configuration that have been studied in the past using EE and EL methods.

Boileau et al. [80] performed LES investigation of the ignition sequence within the VESTA combustor shown in Figure 12.21 using the EE formulation. The annular combustion chamber comprises 18 sectors through which kerosene spray is provided,

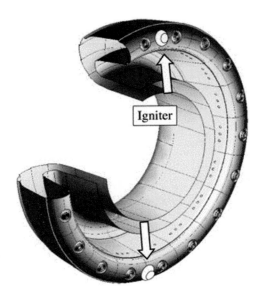

Figure 12.21 Schematic of multiinjector configuration of the VESTA combustor featuring 14 of the 18 sectors [26].

which in turn produces a swirled jet. Such a swirled flow leads to a classical vortex breakdown forming a strong central recirculation zone, which affects the flame propagation and stabilization during ignition. In addition, the combustor utilizes two ignition devices, which can be viewed as jets injecting hot burnt gases. The study analyzed the ignition sequence and flame propagation from one burner/injector to the other.

Figure 12.22 shows a 3D view of the ignition sequence at four successive instants. The turquoise blue colored iso-surfaces represent different cold air inlets and red surfaces correspond to two hot igniter jets. The light blue regions show presence of backflow zones generated by each main injector. The color map shows that the LES grid is able to resolve many small scale turbulent structures. The time instant $t = 14$ ms corresponds to the ignition of the burners surrounding each igniter. Afterward, from $t = 19.2$ to 46 ms, the flame front progresses from the first ignited sectors into the four quadrants of the annulus. The backflow zone makes it possible to stabilize the flame close to the main injector. It can be observed that the thermal expansion of the burnt gases produces a strong acceleration of the flow toward the outlet.

Although study of multiple injectors in a combustor using the EL method was reported earlier [22], detailed assessment of injector to injector interactions using EL methods is still in its infancy due to the computational cost and other modeling challenges. This is a rich area for research and understanding of the dynamics of these interactions. The flow features predictions clearly show that a single injector study cannot be scaled up to a multi-injector case even under identical operating conditions. For example, Figure 12.23 shows an instantaneous LES visualization of the PVC structure in a 9-injector LDI setup that is being studied experimentally [158]. For the LES the same grid and inflow conditions were employed for all 9 injectors. No direct

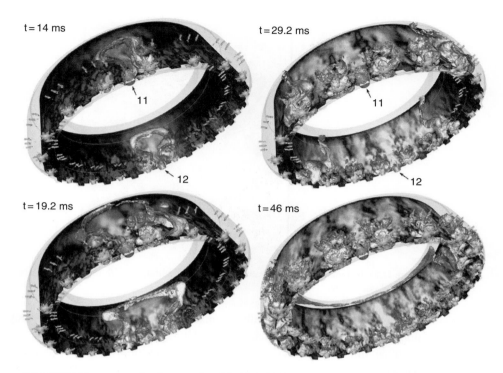

Figure 12.22 Four successive instants (t = 14, 19.2, 29.2, and 46 ms) of the ignition sequence. The cutting surface C is colored by axial velocity (light blue: -20 m/s $\rightarrow$ 20 m/s, iso-surface of the axial velocity $U = 38$ m/s colored by temperature (turquoise blue: 273 K $\rightarrow$ red: 2400 K), and iso-surface of progress rate $Q = 200$ mol/s (shiny light blue) representing the flame front. The two high speed hot jets used for ignition appear as red zones in the pictures (marked I1 and I2) [80]. A black and white version of this figure will appear in some formats. For the color version, please refer to the plate section.

comparison with data has been attempted so far but it can be seen that the PVC originating at each injector is different at the same instant of evolution. Although there could be many reasons for these differences it is apparent that such multiinjector flow fields demonstrate significant 3D interactions and therefore need to be addressed in future numerical studies.

12.5   Summary and Future Prospects

Large eddy simulation of swirling spray combustion in GT combustors is extremely challenging from the modeling and computational perspective and so far only limited studies have been attempted, usually for atmospheric laboratory combustors. Unsteady features such as the efficient dispersion of the spray by the rotating PVC structure and the flame stabilization by the VBB are key features in such combustors and LES has shown an ability to capture these physics. Flame structure shows a wide range of

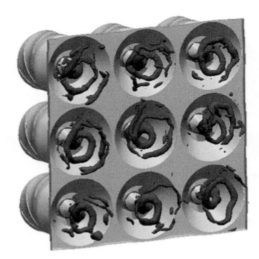

Figure 12.23 Instantaneous PVC features in an LDI combustor with nine injectors operating under identical conditions.

burning modes including partially premixed combustion. Simulation of spray combustion in realistic systems still requires development of subgrid models and also a robust treatment of the breakup process. Finally, multiinjector problems can be attempted now but with limited data and large computational cost of these simulations, such simulations are limited to demonstration purpose at this time.

12.6   Acknowledgments

The results reported here from the Computational Combustion Laboratory, Georgia Institute of Technology have been funded in part by NASA/GRC, AFOSR, and GE Aviation. Computational resources provided by NASA Advanced Supercomputing (NAS), DOD HPC, and CCL are greatly appreciated.

References

[1] A.H. Lefebvre, *Gas Turbine Combustion*, 2nd Edition, Taylor and Francis, 1999.

[2] Y. Huang, V. Yang, "Dynamics and stability of lean-premixed swirl-stabilized combustion," *Prog. Energy. Comb. Sci.* 35 (2009) 293–364.

[3] N. Syred, J. M. Beér, "Combustion in swirling flows: A review," *Combust. Flame* 23 (1974) 143–201.

[4] D. Lilley, "Swirl flows in combustion: A review," *AIAA Journal* 15 (8) (1977) 1063–1078.

[5] M. Kirtas, N. Patel, V. Sankaran, S. Menon, "Large-Eddy Simulation of a Swirl-Stabilized, Lean-Direct Injection Spray Combustor," Proceedings of ASME GT2006 (Barcelona, Spain) GT 2006–91310.

[6] N. Patel, S. Menon, "Simulation of spray combustion in a lean-direct injection combustor," *Thirty-First Symposium (International) on Combustion* 31 (2) (2007) 2327–2334.

[7] N. Patel, S. Menon, "Simulation of spray-turbulence-flame interactions in a lean direct injection combustor," *Combust. Flame* 153 (2008) 228–257.

[8] K. Luo, H. Pitsch, M. G. Pai, O. Desjardins, "Direct numerical simulations and analysis of three-dimensional *n*-heptane spray flames in a model swirl combustor," *Proc. Combust. Inst.* 33 (2011) 2143–2152.

[9] C. Yoon, R. Gejji, W. E. Anderson, V. Sankaran, "Computational investigation of combustion dynamics in a lean direct injection gas turbine combustor," AIAA 2013–0166 (2013) 1–20.

[10] C. Fureby, S.-I. Möller, "Large eddy simulation of reacting flows applied to bluff body stabilized flames," *AIAA J.* 33 (12) (1995) 2339–2347.

[11] B. Dally, A. Masri, R. Barlow, G. Fiechtner, "Instantaneous and mean compositional structure of bluff-body stabilized nonpremixed flames," *Combust. Flame* 114 (1998) 119–148.

[12] A. Ben-Yakar, R. K. Hanson, "Cavity flame-holders for ignition and flame stabilization in scramjets: an overview," *J. Propul. Power* 17 (2001) 869–877.

[13] A.F. Ghoniem, S. Park, A. Wachsman, A. Annaswamy, D. Wee, H.M. Altay, "Mechanism of combustion dynamics in a backward-facing step stabilized premixed flame," *Proc. Combust. Inst.* 30 (2) (2005) 1783–1790.

[14] J. Wan, A. Fan, K. Maruta, H. Yao, W. Liu, "Experimental and numerical investigation on combustion characteristics of premixed hydrogen/air flame in a micro-combustor with a bluff body," *International Journal of Hydrogen Energy* 37 (24) (2012) 19190–19197.

[15] T. Terasaki, S. Hayashi, "The effects of fuel-air mixing on *no_x* formation in non-premixed swirl burners," *Symposium (International) on Combustion*, Vol. 26, 1996, pp. 2733–2739.

[16] A. Masri, S. Pope, B. Dally, "Probability density function computations of a strongly swirling nonpremixed flame stabilized on a new burner," *Proc. Combust. Inst.* 28 (2000) 123–131.

[17] M. Johnson, D. Littlejohn, W. Nazeer, K. Smith, R. Cheng, "A comparison of the flowfields and emissions of high-swirl injectors and low-swirl injectors for lean premixed gas turbines," *Proc. Combust. Inst.* 30 (2005) 2867–2874.

[18] A.D. Gosman, E. Ioannides, "Aspects of computer simulation of liquid-fueled combustors," *Journal of Energy* 7 (1983) 482–490.

[19] K. Luo, H. Pitsch, M.G. Pai, "Direct numerical simulation of three-dimensional swirling *n*-heptane spray flames," *Center of Turbulence Research Annual Research Briefs*, 2009, pp. 171–183.

[20] V. Sankaran, S. Menon, "LES of spray combustion in swirling flows," *Journal of Turbulence* 3 (2002) 011.

[21] J. Cai, S.-M. Jeng, R. Tacina, "The structure of a swirl–stabilized reacting spray issued from an axial swirler," *AIAA Paper* 2005-1424 (2005).

[22] S. Menon, N. Patel, "Subgrid modeling for LES of spray combustion in large-scale combustors," *AIAA Journal* 44 (4) (2006) 709–723.

[23] H.C. Mongia, "Taps: A fourth generation propulsion combustor technology for low emissions," AIAA-03-2657.

[24] M.G. Giridharan, H.C. Mongia, S.M. Jeng, "Swirl cup modelling, part 8: Spray combustion in CFM56 single cup flame tube," *AIAA Paper* 2003–0319.

[25] J.A. Colby, S. Menon, J. Jagoda, "Spray and emission characteristics near lean blow out in a counter-swirl stabilized gas turbine combustor," Proceedings of the ASME Turbo Exposition GT2006-90974 (2006) 1–10.

[26] M. Boileau, G. Staffelbach, B. Cuenot, T. Poinsot, "Bérat, LES of an ignition sequence in a gas turbine engine," *Combust. Flame* 154 (2008) 2–22.

[27] M. Sanjosé, J.M. Senoner, F. Jaegle, B. Cuenot, S. Moreau, T. Poinsot, "Fuel injection model for Euler–Euler and Euler–Lagrange large-eddy simulations of an evaporating spray inside an aeronautical combustor," *Int. J. of Multiphase Flow* 37 (2011) 514–529.

[28] S. Terhaar, B.C. Bobusch, C.O. Paschereit, "Effects of outlet boundary conditions on the reacting flow field in a swirl-stabilized burner at dry and humid conditions," *Journal of Engineering for Gas Turbines and Power* 134 (2012) 111501.

[29] R. Hadef, B. Lenze, "Measurements of droplets characteristics in a swirl-stabilized spray flame," *Exp. Fluid Thermal Sci.* 30 (2005) 117–130.

[30] F. White, *Viscous Fluid Flow*, McGraw-Hill Series in Mechanical Engineering, McGraw-Hill Higher Education, 2006.

[31] R.-H. Chen, J.F. Driscoll, "The role of the recirculation vortex in improving fuel-air mixing within swirling flames," *Symposium (International) on Combustion* 22 (1988) 531–540.

[32] Y. Huang, V. Yang, "Effect of swirl on combustion dynamics in a lean-premixed swirl-stabilized combustor," *Proc. Combust. Inst.* 30 (2005) 1775–1782.

[33] L. Selle, L. Benoit, T. Poinsot, F. Nicoud, W. Krebs, "Joint use of compressible large-eddy simulation and helmholtz solvers for the analysis of rotating modes in an industrial swirled burner," *Combust. Flame* 145 (2006) 194–205.

[34] S. Wang, V. Yang, G. Hsiao, S.-Y. Hsieh, H.C. Mongia, "Large-eddy simulations of gas-turbine swirl injector flow dynamics," *J. Fluid Mech.* 583 (2007) 99–122.

[35] Y. Huang, H.-G. Sung, S.-Y. Hsieh, V. Yang, "Large-eddy simulation of combustion dynamics of lean-premixed swirl-stabilized combustor," *J. Propul. Power* 19 (2003) 782–794.

[36] S. Roux, G. Lartigue, T. Poinsot, U. Meier, C. Bérat, "Studies of mean and unsteady flow in a swirled combustor using experiments, acoustic analysis, and large eddy simulations," *Combust. Flame* 141 (2005) 40–54.

[37] S. Kim, S. Menon, "Large-eddy simulation of a high-pressure, single-element lean direct-injected gas-turbine combustor," AIAA 2014–0131.

[38] A. B. Liu, R.D. Reitz, "Mechanisms of air-assisted liquid atomization," *Atomization and Sprays* 3 (1993) 55–75.

[39] X. Li, M.C. Soteriou, "High-fidelity simulation of fuel atomization in a realistic swirling flow injector," *Atomization and Sprays* 23 (2013) 1049–1078.

[40] G.M. Faeth, L.-P. Hsiang, P.-K. Wu, "Structure and Breakup Properties of Sprays," *Int. J. Multiphase Flow* 21 (1995) 99–127.

[41] P.-K. Wu, K.A. Kirkendall, R.P. Fuller, A.S. Nejad, "Breakup processes of liquid jets in subsonic crossflows," *J. Propul. Power* 13 (1997) 64–73.

[42] P.-K. Wu, K.A. Kirkendall, R.P. Fuller, A.S. Nejad, "Spray structures of liquid jets atomized in subsonic crossflows," *J. Propul. Power* 14 (1998) 173–182.

[43] K.A. Sallam, G.M. Faeth, "Surface properties during primary breakup of turbulent round liquid jets in still air," *AIAA Journal* 41 (2003) 1514–1524.

[44] M. Sommerfeld, H.-H. Qiu, "Experimental studies of spray evaporation in turbulent flow," *Int. J. of Heat and Fluid Flow* 19 (1998) 10–22.

[45] S.V. Apte, P. Moin, "Spray modeling and predictive simulations in realistic gas-turbine engines," in *Handbook of Atomization and Sprays*, Springer, 2011, pp. 811–835.

[46] R.D. Reitz, F.V. Bracco, "Mechanisms of atomization of a liquid jet," *Physics of Fluids* 25 (10) (1982) 1730–1742.

[47] N. Chigier, R. D. Reitz, "Regimes of jet breakup and breakup mechanisms," *AIAA Progress in Astronautics and Aeronautics, Recent Advances in Spray Combustion*, K. Kuo, Ed. Volume 166, 1995, pp. 109–135.

[48] P.K. Wu, R.F. Miranda, G.M. Faeth, "Effects of initial flow conditions on primary breakup of non-turbulent and turbulent round jets," *Atomization and Sprays* 5 (1995) 175–196.

[49] F.X. Tanner, "Liquid jet atomization and droplet breakup modelling of non-evaporating diesel fuel sprays," *Society of Automotive Engineers*, SAE 97-0050.

[50] H.P. Trinh, C.P. Chen, "Modelling of turbulence effects on liquid jet atomization and breakup," *AIAA* 2005–0154.

[51] Y. Renardy, "Effect of startup conditions on drop breakup under shear with inertia," *Int. J. of Multiphase Flow* 34 (2008) 1185–1189.

[52] T.G. Theofanous, G.J. Li, "On the physics of aerobreakup," *Phys. Fluids* 20 (2008) 052103.

[53] H.P. Gadgil, B.N. Raghunandan, "Some features of spray breakup in effervescent atomizers," *Exp. in Fluids* 50 (2011) 329–338.

[54] M. Arienti, T.A. Shedd, M. Herrmann, L. Wang, M. Corn, X. Li, M.C. Soteriou, "Modeling wall film formation and breakup using an integrated interface-tracking/discrete-phase approach," *Journal of Engineering for Gas Turbines and Power* 133 (3) (2011) 031501.

[55] O. Desjardins, V. Moureau, H. Pitsch, "An accurate conservative level set/ghost fluid method for simulating turbulent atomization," *J. Comp. Phys.* 227 (2008) 8395–8416.

[56] R. Lebas, T. Menard, P.A. Beau, A. Berlemont, F.X. Demoulin, "Numerical simulation of primary break-up and atomization: DNS and modelling study," *Int. J. of Multiphase Flow* 35 (2009) 247–260.

[57] P. Zeng, S. Sarholz, C. Iwainsky, B. Binninger, N. Peters, M. Herrmann, "Simulation of primary breakup for diesel spray with phase transition," in *Recent Advances in Parallel Virtual Machine and Message Passing Interface*, 2009, pp. 313–320.

[58] J. Shinjo, A. Umemura, "Simulation of liquid jet primary breakup: Dynamics of ligament and droplet formation," *Int. J. of Multiphase Flow* 36 (2010) 513–532.

[59] S. Pascaud, M. Boileau, B. Cuenot, T. Poinsot, "Large eddy simulation of turbulent spray combustion in aeronautical gas turbines," in ECCOMAS Thematic Conference on computational combustion, 2005, pp. 149–167.

[60] M. Linne, M. Paciaroni, T. Hall, T. Parker, "Ballistic imaging of the near field in a diesel spray," *Exp Fluids* 40 (2006) 836–846.

[61] J. Desantes, F. Salvador, J. López, J. De la Morena, "Study of mass and momentum transfer in diesel sprays based on x-ray mass distribution measurements and on a theoretical derivation," *Exp Fluids* 50 (2011) 233–246.

[62] M.A. Reddemann, F. Mathieu, R. Kneer, "Transmitted light microscopy for visualizing the turbulent primary breakup of a microscale liquid jet," *Exp Fluids* 54 (2013) 1–10.

[63] C. Presser, A.K. Gupta, H.G. Semerjian, "Aerodynamic characteristics of swirling spray flames: Pressure-jet atomizer," *Combust. Flame* 92 (1993) 25–44.

[64] R.J. Sornek, R. Dobashi, T. Hirano, "Effect of turbulence on vaporization, mixing, and combustion of liquid-fuel sprays," *Combust. Flame* 120 (2000) 479–491.

[65] H.Y. Wang, V.G. McDonell, W.A. Sowa, G.S. Samuelsen, "Scaling of the two-phase flow downstream of a gas turbine combustor swirl cup: Part I- mean quantities," *J. Eng. Gas Turbines Power* 115 (1993) 453–460.

[66] D.L. Bulzan, "Structure of a swirl-stabilized combusting spray," *J. Propul. Power* 11 (1995) 1093–1102.

[67] M.R. Soltani, K. Ghorbanian, M. Ashjaee, M.R. Morad, "Spray characteristics of a liquid–liquid coaxial swirl atomizer at different mass flow rates," *Aero. Sci. Tech.* 9 (2005) 592–604.

[68] J. Cai, S.-M. Jeng, R. Tacina, "The structure of a swirl-stabilized reacting spray issued from an axial swirler," AIAA 2005-1424 (2005) 1–16.

[69] R. Hadef, B. Lenze, "Effects of co- and counter-swirl on the droplet characteristics in a spray flame," *Chem. Engg. Process.* 47 (2008) 2209–2217.

[70] A. Tratnig, G. Brenn, "Drop size spectra in sprays from pressure-swirl atomizers," *Int. J. of Multiphase Flow* 36 (2010) 349–363.

[71] F. Takahashi, W.J. Schmoll, G.L. Switzer, D.T. Shouse, "Structure of a spray flame stabilized on a production engine combustor swirl cup," *Symposium (International) on Combustion* 25 (1994) 183–191.

[72] Y.M. Al-Abdeli, A.R. Masri, "Turbulent swirling natural gas flames: Stability characteristics, unsteady behavior and vortex breakdown," *Combust. Sci. Technol.* 179 (2007) 207–225.

[73] M. Sommerfeld, H.-H. Qiu, "Detailed measurements in a swirling particulate two-phase flow by a phase-doppler anemometer," *Int. J. of Heat and Fluid Flow* 12 (1991) 20–28.

[74] J. Zurlo, C. Presser, H. Semerjian, A. Gupta, "Determination of droplet characteristics in spray flames using three different sizing techniques," in AIAA, SAE, ASME, and ASEE, 27th Joint Propulsion Conference, Vol. 1, 1991.

[75] R.J. Kenny, J.R. Hulka, M.D. Moser, N.O. Rhys, "Effect of chamber backpressure on swirl injector fluid mechanics," *J. Propul. Power* 25 (2009) 902–913.

[76] H.-E. Albrecht, *Laser Doppler and Phase Doppler Measurement Techniques*, Springer, 2003.

[77] J. Stenberg, W. Frederick, S. Boström, R. Hernberg, M. Hupa, "Pyrometric temperature measurement method and apparatus for measuring particle temperatures in hot furnaces: Application to reacting black liquor," *Review of Scientific Instruments* 67 (1996) 1976–1984.

[78] J.A. Sutton, J.F. Driscoll, "A method to simultaneously image two-dimensional mixture fraction, scalar dissipation rate, temperature and fuel consumption rate fields in a turbulent non-premixed jet flame," *Exp Fluids* 41 (2006) 603–627.

[79] S. Roy, J.R. Gord, A.K. Patnaik, "Recent advances in coherent anti-Stokes Raman scattering spectroscopy: Fundamental developments and applications in reacting flows," *Prog. Energy. Comb. Sci.* 36 (2010) 280–306.

[80] M. Boileau, S. Pascaud, E. Riber, B. Cuenot, L. Gicquel, T. Poinsot, "Investigation of two-fluid methods for large eddy simulation of spray combustion in gas turbines," *Flow Turbulence Combust.* 80 (2008) 351–373.

[81] E. Riber, V. Moureau, M. García, T. Poinsot, O. Simonin, "Evaluation of numerical strategies for large eddy simulation of particulate two-phase reacting flows," *J. Comp. Phys.* 228 (2009) 539–564.

[82] J.M. Desantes, J.V. Pastor, J.M. Garciá-Oliver, J.V. Pastor, "A 1D model for the description of mixing-controlled reacting diesel sprays," *Combust. Flame* 156 (2009) 234–249.

[83] J.M. Senoner, M. Sanjosé, T. Lederlin, F. Jaegle, M. Garćia, E. Riber, N. Cuenot, L. Gicquel, H. Pitsch, T. Poinsot, "Eulerian and Lagrangian large-eddy simulations of an evaporating two-phase flow," *C. R. Mecanique* 337 (2009) 458–468.

[84] M. Sanjosé, E. Riber, L. Gicquel, B. Cuenot, T. Poinsot, "Large eddy simulation of a two-phase reacting flow in an experimental burner," in *Direct and Large-Eddy Simulation VII*, Springer, 2010, pp. 345–351.

[85] S. Hank, R. Saurel, O.L. Metayer, "A hyperbolic Eulerian model for dilute two-phase suspensions," *J. Modern Phys.* 2 (2011) 997–1011.

[86] W. Kollmann, I.M. Kennedy, "Les model for the particulate phase in sprays," *AIAA* 97-0369 (1997) 1–11.

[87] D. Caraeni, C. Bergström, L. Fuchs, "Modeling of liquid fuel injection, evaporation and mixing in a gas turbine burner using large eddy simulations," *Flow Turbulence Combust.* 65 (2000) 223–244.

[88] V. Sankaran, S. Menon, "Vorticity-scalar alignments and small-scale structures in swirling spray combustion," *Proc. Combust. Inst.* 29 (2002) 577–584.

[89] S. James, J. Zhu, M.S. Anand, "Large-eddy simulation as a design tool for gas turbine combustion systems," *AIAA J.* 44 (2006) 674–686.

[90] S. Menon, N. Patel, "Subgrid modeling for simulation of spray combustion in large-scale combustors," *AIAA J.* 44 (2006) 709–723.

[91] S.B. Kuang, A.B. Yu, Z.S. Zou, "A new point-locating algorithm under three-dimensional hybrid meshes," *Int. J. of Multiphase Flow* 34 (2008) 1023–1030.

[92] Y. Yan, J. Zhao, J. Zhang, Y. Liu, "Large-eddy simulation of two-phase spray combustion for gas turbine combustors," *Applied Thermal Engineering* 28 (11) (2008) 1365–1374.

[93] T. Lederlin, H. Pitsch, "Large-eddy simulation of an evaporating and reacting spray," in *Center for Turbulence Research: Annual Research Briefs*, 2008, pp. 479–490.

[94] J. Pozorski, S.V. Apte, "Filtered particle tracking in isotropic turbulence and stochastic modeling of subgrid-scale dispersion," *Int. J. of Multiphase Flow* 35 (2009) 118–128.

[95] S. Srinivasan, A.G. Smith, S. Menon, "Accuracy, reliability and performance of spray combustion models in LES," in *Quality and Reliability of Large-Eddy Simulations II*, Springer, 2011, pp. 211–220.

[96] K. Li, L.X. Zhou, "Studies of the effect of spray inlet conditions on the flow and flame structures of ethanol-spray combustion by large-eddy simulation," *Numerical Heat Transfer, Part A: Applications* 62 (1) (2012) 44–59.

[97] B. Franzelli, V.A., B. Fiorina, N. Darabiha, "Large eddy simulation of swirling kerosene/air spray flame using tabulated chemistry," *Proceedings of the ASME Turbo Exposition* GT2006-90974 (2013) 1–10.

[98] M. Chrigui, A.R. Masri, A. Sadiki, J. Janicka, "Large eddy simulation of a polydisperse ethanol spray flame," *Flow Turbulence Combust.* 90 (2013) 813–832.

[99] W.P. Jones, A.J. Marquis, K. Vogiatzaki, "Large-eddy simulation of spray combustion in a gas turbine combustor," *Combust. Flame* 161 (2014) 222–239.

[100] M. Arienti, X. Li, M.C. Soteriou, C.A. Eckett, R. Jensen, "Coupled level-set/volume-of-fluid method for the simulation of liquid atomization in propulsion device injectors," AIAA 2010–7136 (2010) 1–10.

[101] X. Li, M. Arienti, M.C. Soteriou, M.M. Sussman, "Towards an efficient, high-fidelity methodology for liquid jet atomization computations," AIAA 2010–210 (2010) 1–16.

[102] M. Mortensen, R.W. Bilger, "Derivation of the conditional moment closure equations for spray combustion," *Combust. Flame* 156 (2009) 62–72.

[103] C. Laurent, G. Lavergne, P. Villedieu, "Quadrature method of moments for modeling multi-component spray vaporization," *Int. J. of Multiphase Flow* 36 (2010) 51–59.

[104] W.P. Jones, S. Lyra, S. Navarro-Martinez, "Large eddy simulation of a swirl stabilized spray flame," *Proc. Combust. Inst.* 33 (2011) 2153–2160.

[105] W.P. Jones, S. Lyra, S. Navarro-Martinez, "Numerical investigation of swirling kerosene spray flames using large eddy simulation," *Combust. Flame* 159 (2012) 1539–1561.

[106] A. Vié, S. Jay, B. Cuenot, M. Massot, "Accounting for polydispersion in the Eulerian large eddy simulation of the two-phase flow in an aeronautical-type burner," *Flow Turbulence Combust.* 90 (2013) 545–581.

[107] S. Elghobashi, "On predicting particle-laden turbulent flows," *Appl. Sci. Res.* 52 (1994) 309–329.

[108] E. Loth, "Numerical approaches for motion of dispersed particles, droplets and bubbles," *Prog Energy Combust Sci* 26 (2000) 161–223.

[109] S. Balachandar, J.K. Eaton, "Turbulent dispersed multiphase flow," *Annual Review of Fluid Mechanics* 42 (2010) 111–133.

[110] O.A. Druzhinin, S. Elghobashi, "Direct numerical simulations of bubble-laden turbulent flows using the two-fluid formulation," *Phys. Fluids* 10 (1998) 685–697.

[111] O.A. Druzhinin, S. Elghobashi, "On the decay rate of isotropic turbulence laden with microparticles," *Phys. Fluids* 11 (1999) 602–610.

[112] J. Réveillon, M. Massot, C. Péra, "Analysis and modeling of the dispersion of vaporizing polydispersed sprays in turbulent flows," in *Proceedings of the Summer Program*, 2002, pp. 393–404.

[113] J.J. Riley, G.S. Patterson, "Diffusion experiments with numerically integrated isotropic turbulence," *Phys. Fluids* 17 (1974) 292–297.

[114] K. Dukowicz, J., "A particle-fluid numerical model for liquid sprays," *J. Comp. Phys.* 35 (1980) 229–253.

[115] M.R. Maxey, "The gravitational settling of aerosol particles in homogeneous turbulence and random flow fields," *J. Fluid Mech.* 174 (1987) 441–465.

[116] S. Elghobashi, "Particle-laden turbulent flows: Direct simulation and closure models," *Appl. Sci. Res.* 48 (1991) 301–314.

[117] L. Selle, G. Lartigue, T. Poinsot, P. Kaufman, W. Krebs, D. Veynante, "Large-eddy simulation of turbulent combustion for gas turbines with reduced chemistry," in *Proceedings of the Summer Program*, 2002, pp. 333–344.

[118] F. Ham, S. Apte, G. Iaccarino, X. Wu, M. Herrmann, G. Constantinescu, K. Mahesh, P. Moin, "Unstructured LES of reacting multiphase flows in realistic gas turbine combustors," in *Center of Turbulence Research Annual Research Briefs*, 2003, pp. 139–160.

[119] F.A. Williams, "Spray, combustion and atomization," *Physics of Fluids* 1 (1958) 541–545.

[120] R. Ranjan, "A novel state-space based method for direct numerical simulation of particle-laden turbulent flows," Ph.D. thesis, University of Illinois at Urbana-Champaign (2013).

[121] J. Candy, "A numerical method for solution of the generalized liouville equation," *J. Comp. Phys.* 129 (1) (1996) 160–169.

[122] A. Mura, R. Borghi, "Introducing a new partial PDF approach for turbulent combustion modeling," *Combust. Flame* 136 (2004) 377–382.

[123] D.B. Xiu, G.E. Karniadakis, "The wiener-askey polynomial chaos for stochastic differential equations," *SIAM J. Scientific Computing* 24 (2002) 619–644.

[124] D.L. Marchisio, R. Fox, "Solution of population balance equations using the direct quadrature method of moments," *Journal of Aerosol Science* 36 (2005) 43–73.

[125] C. Pantano, B. Shotorban, "Least-squares dynamic approximation method for evolution of uncertainty in initial conditions of dynamical systems," *Physical Review E* 76 (2007) 066705.

[126] J.C. Beale, R.D. Reitz, "Modeling spray atomization with the Kelvin–Helmholtz/Rayleigh–Taylor hybrid model," *Atomization and Sprays* 9 (6).

[127] P.J. O'Rourke, A.A. Amsden, "The TAB method for numerical calculation of spray droplet breakup," *Society of Automotive Engineers*, SAE 87-2089.

[128] Y. Ra, R.D. Reitz, "A vaporization model for discrete multi-component fuel sprays," *Int. J. of Multiphase Flow* 35 (2009) 101–117.

[129] A.A. Amsden, "KIVA-3V: Release 2, improvements to KIVA-3V," Los Alamos Report No. LA-UR-99-915, 1999.

[130] R.D. Reitz, "Modelling atomization processes in high-pressure vaporizing sprays," *Atomization and Spray Technology* 3 (1987) 309–337.

[131] M. Rachner, J. Becker, C. Hassa, T. Doerr, "Modelling of the atomization of a plain liquid fuel jet in crossflow at gas turbine conditions," *Aerospace Science and Technology* 6 (7) (2002) 495–506.

[132] G.M. Faeth, "Spray combustion phenomena," *Proc. Combust. Inst.* 26 (1996) 1593–1612.

[133] A.B. Liu, D. Mather, R.D. Reitz, "Modeling the effects of drop drag and breakup on fuel sprays," NASA STI/Recon Technical Report N 93 (1993) 29388.

[134] R. Abgrall, R. Saurel, "Discrete equations for physical and numerical compressible multiphase mixtures," *J. Comp. Phys.* 186 (2) (2003) 361–396.

[135] A. Chinnayya, E. Daniel, R. Saurel, "Modelling detonation waves in heterogeneous energetic materials," *J. Comp. Phys.* 196 (2) (2004) 490–538.

[136] M.V. Papalexandris, "Numerical simulation of detonations in mixtures of gases and solid particles," *J. Fluid Mech.* 507 (2004) 95–142.

[137] J.C. Oefelein, "Large eddy simulation of turbulent combustion processes in propulsion and power systems," *Progress in Aerospace Sciences* 42 (1) (2006) 2–37.

[138] K. Balakrishnan, D.V. Nance, S. Menon, "Simulation of impulse effects from explosive charges containing metal particles," *Shock Waves* 20 (3) (2010) 217–239.

[139] K. Gottiparthi, S. Menon, "A study of interaction of clouds of inert particles with detonation in gases," *Combust. Sci. Technol.* 184 (3) (2012) 406–433.

[140] P. Jenny, D. Roekaerts, N. Beishuizen, "Modeling of turbulent dilute spray combustion," *Prog. Energy. Comb. Sci.* 38 (2012) 846–887.

[141] C.T. Crowe, J.D. Schwarzkopf, M. Sommerfeld, Y. Tsuji, *Multiphase Flows with Droplets and Particles*, CRC Press, 2011.

[142] K. Balakrishnan, S. Menon, "Characterization of the mixing layer resulting from the detonation of heterogeneous explosive charges," *Flow Turbulence Combust.* 87 (4) (2011) 639–671.

[143] M.R. Baer, J.W. Nunziato, "A two-phase mixture theory for the deflagration-to-detonation transition (ddt) in reactive granular materials," *Int. J. of Multiphase Flow* 12 (6) (1986) 861–889.

[144] M. Bini, W. Jones, "Particle acceleration in turbulent flows: A class of nonlinear stochastic models for intermittency," *Phys. Fluids* 19 (2007) 035104.

[145] F. Génin, S. Menon, "Studies of shock/turbulent shear layer interaction using large-eddy simulation," *Computers & Fluids* 39 (5) (2010) 800–819.

[146] J. Smagorinsky, "General circulation experiments with the primitive equations," *Monthly Weather Review* 91 (3) (1993) 99–164.

[147] M. Germano, U. Piomelli, P. Moin, W.H. Cabot, "A dynamic subgrid-scale eddy viscosity model," *Physics of Fluids A* 3 (11) (1991) 1760–1765.

[148] S. Menon, W.-W. Kim, "High reynolds number flow simulations using the localized dynamic subgrid-scale model," AIAA-96-0425.

[149] W.-W. Kim, S. Menon, A new incompressible solver for large-eddy simulations, *International Journal of Numerical Fluid Mechanics* 31 (1999) 983–1017.

[150] W.-W. Kim, S. Menon, H. C. Mongia, Large-eddy simulation of a gas turbine combustor flow, *Combustion Science and Technology* 143 (1999) 25–62.

[151] G. M. Faeth, Mixing, transport and combustion in sprays, *Progress in Energy and Combustion Science* 13 (1987) 293–345.

[152] S. Menon, S. Pannala, Subgrid modeling of unsteady two-phase turbulent flows, AIAA Paper No. 97-3113.

[153] A. R. Kerstein, Linear-eddy model of turbulent scalar transport and mixing, *Combustion Science and Technology* 60 (1988) 391–421.

[154] S. Menon, P. McMurtry, A. R. Kerstein, A linear eddy mixing model for large eddy simulation of turbulent combustion, in: B. Galperin, S. Orszag (Eds.), *LES of Complex Engineering and Geophysical Flows*, Cambridge University Press, 1993, pp. 287–314.

[155] S. Menon, A. R. Kerstein, The linear-eddy model, *Turbulent Combustion Modeling* 95 (2011) 175–222.

[156] S. V. Apte, P. Moin, Large-eddy simulation of realistic gas turbine combustor, *AIAA J.* 44 (2006) 698–708.

[157] E. Knudsen, H. Pitsch, Large-eddy simulation for combustion systems: Modeling approaches for partially premixed flows, *The Open Thermodynamics Journal* 4 (2010) 76–85.

[158] Y. R. Hicks, R. C. Anderson, R. J. Locke, Optical measurements in a combustor using a 9-point swirl-venturi fuel injector, Isabe 2007-1280 (2007).

13 Combustion in Afterburning Behind Explosive Blasts

Ekaterina Fedina, Kalyana C. Gottiparthi, Christer Fureby, and Suresh Menon

13.1 Introduction

For well over a century various applications in mining, construction, and weapon development have prompted studies of detonations/explosions. Detonation of an explosive charge generates a high energy source, which sustains the propagation of a blast wave. The blast wave is followed by a fireball of hot gases emanating from the explosion. These hot gases often comprise of products of detonation, such as soot and carbon monoxide, which can continue to burn and release energy. The afterburn of the detonation products augments the impact and the energy release of the explosion. Thus, investigations aimed to enhance the afterburn are desired, especially, to enable applications where explosives are used as a source of extreme heat. However, the processes involved in the fireball and detonation product gas propagation are complex, involving hydrodynamic instabilities, mixing, and turbulence. If the explosion produces particulate matter, the mixing and combustion are further complicated by gas–particle and interparticle interactions. Although the experimental studies provide knowledge of large scale quantities such as propagation velocities and dimensions of the fireball [1], due to the destructive nature of the explosion event, measurements to characterize the mixing may never be possible. Thus, numerical simulations play a vital role in exploring the details of the mixing and combustion in the postdetonation flows, where coarse grained simulations (CGS) are a cost effective approach in both gaining knowledge and designing full scale experiments.

Past studies, [2], have shown that the afterburning process during a homogeneous explosion can be divided into four different stages. (1) Strong blast wave: the outgoing blast wave will heat up and accelerate the ambient air as it propagates through it. Meanwhile a rarefaction wave propagates inward, initially boosting the acceleration caused by the blast wave, and later forcing an opposite directed acceleration of the detonation gases. The interface between the detonation products and shock compressed air is impulsively accelerated, inducing Rayleigh–Taylor (RT) instabilities [3] to grow due to large density gradients across the contact surface as well as its impulsive acceleration. A short time after the initial blast the ingoing rarefaction wave will overexpand the flow [4], thereby causing a second shock. Therefore a thin mixing layer is formed between the initial blast wave and the secondary shock. (2) Implosion: the secondary shock will eventually strengthen by means of detonation gases accelerating it. Subsequently the shock will strengthen to the point of implosion. The implosion further entrains the air into the mixing layer, enhancing the combustion. (3) Reshock: when the secondary shock reflects from the origin it is set outward and interacts with the existing RT instabilities, depositing additional vorticity into the mixing layer due to misaligned pressure (baroclinic effects). These effects give rise to Richtmyer–Meshkov (RM) instabilities [5], enhancing the turbulent mixing. (4) Asymptotic mixing: constant pressure mixing layer, during which the remainder of fuel is consumed.

When particles are included into the explosive charges, whether these are inert or reacting, the flow field, combustion, and thus afterburning process becomes even more complicated. Augmentation of impulsive loading is known to occur when solid metal particles are added to an explosive charge, termed heterogeneous explosive, due to the momentum and energy delivered by the particles. When such a charge is detonated, the high pressure combustion products rapidly expand to drive a blast wave, which attenuates due to the effects of spreading [6]. At the same time the metal particles pick up momentum from the gas due to drag, penetrate the contact surface, and generate perturbations, which grow into hydrodynamic instabilities. These instabilities are initially RT instabilities in nature, are essentially three-dimensional (3D), and give rise to enhanced mixing, compared to homogeneous charge, of the hot detonation products and the shock compressed air. If the initial charge contains reactive metal particles, the afterburn can provide a significant contribution to the total impulsive loading of the charge [6]. This afterburn is dependent on magnitude of mixing, which in turn depends on, for example, vorticity structures caused by instability generation.

In this chapter we will present how CGS can be used to examine afterburning properties behind homogeneous and heterogeneous explosive blasts. Investigation of the afterburning in heterogeneous nitromethane charges in a sector domain will show the influence of particles added to an explosive charge on the mixing and afterburning processes. Studies of 2,4,6-trinitrotoluene (TNT) explosion at different Height of Burst (HoB) will provide information on how the CGS can be used to evaluate full scale explosion configurations and thereby contribute to performance evaluation of an explosive. This evaluation often requires careful determination of most appropriate charge positioning to achieve the required pressure blast and/or the desired afterburning effect. In Enhanced Blast eXplosives (EBX) metal particles, usually aluminium, are added to the explosive compound in order to increase the afterburning energy release by allowing the metal particles and detonation products to combust with air. This presents another modeling challenge since the combustion becomes two-phased; to address this, application simulations of pure TNT will be compared with TNT/Al mixture compound in order to demonstrate how the afterburning is affected by the inclusion of the aluminium particles. We hope that the discussion included here will enable better understanding of the effect of particles and the HoB on the afterburning properties of condensed phase charges and facilitate the optimal utilization of an explosive compound.

13.2   CGS Modeling of Heterogeneous Afterburning

13.2.1   Governing Equations

The afterburning process behind the explosive blasts is modeled using the filtered Navier–Stokes equations in a compressible, unsteady, multiphase form by employing a finite-volume method. In the initial stages, the postdetonation flow is dominated by hydrodynamic instabilities comprising of a narrow range of length scales. However, as the instabilities grow and break down, the flow undergoes a transition to turbulence and a wide range of scales are generated. Resolving all the scales in this flow is challenging as small scales are continuously generated. Thus, filtered governing equations are employed that uniquely separate the range of scales into large and small scales. This operation allows the significant range of scales to be resolved with a fine enough grid, leaving the unresolved scales (also called subgrid scales), which are computed using subgrid models. The subgrid terms are dependent on the resolved quantities and affect the flow at the resolved scale. Hence, a coupled system of equations involving both the resolved and subgrid scales is employed and accounts for the wide range of scales encountered in the postdetonation flow. However, note that the problem at hand may or may not comprise ambient turbulence. If initial turbulent scales are absent, the transition and break down of instabilities may not result in significant subgrid scales. In that case, all the subgrid quantities are negligible in comparison to the resolved quantities and the governing equations can be modified/simplified by appropriate assumptions. Since the flow in question sometimes comprises particles, which can occupy significant volume, the gas phase must be adjusted to account for the presence of the particles. For this reason, the conventional governing equations [7, 8] have been extended to handle dense flows. Following the approach provided in [9, 10, 11], the Favre filtered gas phase governing equations under dense loading conditions are summarized as follows:

$$\frac{\partial \overline{\alpha}_g \overline{\rho}_g}{\partial t} + \frac{\partial \overline{\alpha}_g \overline{\rho}_g \widetilde{u}_{g,i}}{\partial x_i} = \overline{\dot{\delta}_p} \tag{13.1}$$

$$\frac{\partial \overline{\alpha}_g \overline{\rho}_g \widetilde{u}_{g,i}}{\partial t} + \frac{\partial}{\partial x_j} \left[\overline{\alpha}_g \overline{\rho}_g \widetilde{u}_{g,i} \widetilde{u}_{g,j} + \overline{\alpha}_g \overline{p}_g \delta_{ij} + \overline{\alpha}_g \tau_{g,ij}^{sgs} - \overline{\alpha}_g \overline{\tau}_{g,ij} \right]$$
$$= \overline{p}_I \frac{\partial \overline{\alpha}_g}{\partial x_j} \delta_{ij} - \overline{\tau}_{I,ij} \frac{\partial \overline{\alpha}_g}{\partial x_j} + \overline{\dot{F}}_{p,i}, \tag{13.2}$$

$$\frac{\partial \overline{\alpha}_g \overline{\rho}_g \widetilde{E}_g}{\partial t} + \frac{\partial}{\partial x_j} \left[\overline{\alpha}_g \overline{\rho}_g \widetilde{u}_{g,j} \widetilde{E}_g + \overline{u}_{g,j} \overline{\alpha}_g \overline{p}_g + \overline{\alpha}_g \overline{q}_{g,j} - \overline{u}_{g,j} \overline{\alpha}_g \overline{\tau}_{g,ji} + \overline{\alpha}_g H_{g,j}^{sgs} + \overline{\alpha}_g \sigma_{g,j}^{sgs} \right]$$
$$= \overline{p}_I \widetilde{u}_{I,j} \frac{\partial \overline{\alpha}_g}{\partial x_j} - \overline{u}_{I,i} \overline{\tau}_{I,ij} \frac{\partial \overline{\alpha}_g}{\partial x_j} + \overline{\dot{Q}}_p + \overline{\dot{W}}_p, \tag{13.3}$$

$$\frac{\partial \overline{\alpha}_g \overline{\rho}_g \widetilde{Y}_{g,k}}{\partial t} + \frac{\partial}{\partial x_i} \left[\overline{\alpha}_g \overline{\rho}_g \left(\widetilde{Y}_{g,k} \widetilde{u}_{g,i} + \widetilde{Y}_{g,k} \widetilde{V}_{g,i,k} \right) + \overline{\alpha}_g Y_{g,i,k}^{sgs} + \overline{\alpha}_g \theta_{g,i,k}^{sgs} \right]$$
$$= \overline{\dot{W}}_{g,k} + \overline{\dot{S}}_{p,k}, \tag{13.4}$$

where $\widetilde{(.)}$ denotes the resolved scale, and $\overline{(.)}$ represents a spatial filtering. Also, $\overline{\alpha}_g$ denotes the filtered gas volume fraction (for single phase flow, $\overline{\alpha}_g = 1$ and all particle contribution terms with subscript p are zero), $\overline{\rho}_g$ is the filtered density of the gas, $\widetilde{u}_{g,i}$ is the ith component of the resolved gas velocity, $\overline{p}_g$ is the filtered gas pressure, δ_{ij} is the Kronecker delta, $\widetilde{Y}_{g,k}$ is the resolved kth species mass fraction, and $\overline{\dot{W}}_{g,k}$ is the filtered reaction rate of the kth species. Subscript I indicates interface values and models are needed to describe these, but as a first approximation, these terms can be modeled using the gas (g) values. The resolved total energy per unit mass, $\widetilde{E}_g$, is computed as $\widetilde{E}_g = \widetilde{e}_g + \frac{1}{2}\widetilde{u}_{g,i}\widetilde{u}_{g,i} + k^{sgs}$, where $\widetilde{e}_g$ represents the resolved specific internal energy of the gas and k^{sgs} is the subgrid turbulent kinetic energy of the gas. Other subgrid terms in Eqs. 13.1-13.4 are the subgrid stress tensor $\tau^{sgs}_{g,ij}$, subgrid enthalpy flux $H^{sgs}_{g,i}$, subggrid viscous work $\Sigma^{sgs}_{g,ij}$, sub-grid convective species flux $Y^{sgs}_{g,i}$, and subgrid diffusive species flux $\theta^{sgs}_{g,i}$. Assuming the fluid to be Newtonian, the filtered viscous stress tensor of the gas is obtained as

$$\overline{\tau}_{g,ij} = 2\mu_g S_{g,ij} + \lambda S_{g,kk}\delta_{ij}, \tag{13.5}$$

where $S_{g,ij}$ is the rate of strain tensor and is expressed as

$$S_{g,ij} = \frac{1}{2}\left(\frac{\partial \widetilde{u}_{g,i}}{\partial x_j} + \frac{\partial \widetilde{u}_{g,j}}{\partial x_i}\right). \tag{13.6}$$

In Eq. (13.5), μ_g is the viscosity coefficient of the gas and is assumed to be only a function of the resolved gas temperature ($\widetilde{T}_g$). Following Stokes's hypothesis, the bulk viscosity [12], λ, is related to μ_g as μ_g as $\lambda = -2/3\mu_g$. The filtered heat flux vector, $\overline{q}_{g,j}$, is obtained using Fourier's law of thermal conduction, summarized as

$$\overline{q}_{g,j} = -\kappa_g \frac{\partial \widetilde{T}_g}{\partial x_j} + \overline{\rho}_g \sum_1^{N_s} \widetilde{Y}_{g,k}\widetilde{h}_{g,k}\widetilde{V}_{g,j,k} + \sum_1^{N_s} q^{sgs}_{g,i,k}, \tag{13.7}$$

where $\widetilde{h}_{g,k}$ denotes the resolved specific enthalpy of the kth species, $q^{sgs}_{g,i,k}$ the subgrid heat flux, and κ_g is the thermal conductivity of the gas, assumed to be a function of $\widetilde{T}_g$ only. The Prandtl number ($Pr = C_{p,g}\mu_g/\kappa_g = 0.72$) is used to obtain κ_g from μ_g. The resolved species diffusion velocities, $\widetilde{V}_{g,j,k}$, are modeled assuming Fickian diffusion, and are obtained as

$$\widetilde{V}_{g,j,k} = -\frac{D_{g,k}}{\widetilde{Y}_{g,k}}\frac{\partial \widetilde{Y}_{g,k}}{\partial x_j}. \tag{13.8}$$

Here, the diffusion coefficient $D_{g,k}$ depends on species k, and is obtained from the assumption that the Lewis number ($Le = Sc/Pr$) is unity.

13.2.2   Subgrid Modeling

The terms that appear with a superscript *sgs* represent the subgrid terms that arise in the filtering operation. Conventional subgrid scale models are either of functional or structural nature depending on whether they are intended to mimic the kinetic energy cascade from large to small scales or whether they are intended to mimic the overall structure of the true subgrid stress tensor [13]. The subgrid terms require appropriate closure models [7, 8], which can be accomplished by using these different classes of models.

13.2.2.1   Functional Models

Functional models are the most widely used class of subgrid models, including the well-known Smagorinsky (SMG) model [14, 15] and the One Equation Eddy Viscosity

Model (OEEVM) [16, 17], in which the subgrid stress tensor is assumed proportional to the deviatoric part of the rate of strain tensor, $S_{g,ij}$, such that

$$\tau_{g,ij}^{sgs} \approx -2\overline{\rho}_g \nu_t \widetilde{S_{g,ij}}, \tag{13.9}$$

with ν_t being subgrid eddy viscosity and Sc_t and Pr_t the turbulent Schmidt and Prandtl numbers. In the SMG model, [14], the subgrid viscosity is estimated based on dimensional analysis and can be expressed as $\nu_t = c_D \Delta^2 |\widetilde{S_{g,ij}}|$, in which c_D is a model coefficient. This model offers the possibility of estimating the subgrid kinetic energy as $k \approx c_I \Delta^2 |\widetilde{S_{g,ij}}|^2$, in which c_I is another model coefficient. The model coefficients, c_D and c_I, can be derived from an inertial range spectrum, $E(\kappa) = C_K \varepsilon^{2/3} \kappa^{-5/3}$, with $c_D = 0.02$ and $c_I = 0.002$, respectively. Alternatively, a more advanced version of the SMG model, using dynamic coefficients [15], can be used. In the OEEVM [17], the subgrid viscosity, ν_t, is obtained from solving the modeled subgrid kinetic energy equation that takes the following form,

$$\frac{\partial}{\partial t}\overline{\alpha}_g \overline{\rho}_g k^{sgs} + \frac{\partial}{\partial x_i}\left(\overline{\alpha}_g \overline{\rho}_g \widetilde{u}_{g,i} k^{sgs}\right) = -\overline{\alpha}_g \tau_{g,ij}^{sgs} \frac{\partial \widetilde{u}_{g,i}}{\partial x_j} + \overline{\alpha}_g T_{k^{sgs}} - \overline{\alpha}_g D_{k^{sgs}}. \tag{13.10}$$

Here, $T_{k^{sgs}}$ and $D_{k^{sgs}}$ denote, respectively, the transport and dissipation of k^{sgs}, obtained as:

$$T_{k^{sgs}} = \frac{\partial}{\partial x_i}\left(\overline{\rho}_g \nu_t \frac{\partial k^{sgs}}{\partial x_i}\right); D_{k^{sgs}} = C_\varepsilon \overline{\rho}_g \frac{(k^{sgs})^{1.5}}{\Delta}. \tag{13.11}$$

Here ν_t is modeled as $\nu_t = C_\nu \sqrt{k^{sgs}}\Delta$, where Δ is the local filter width. The constants C_ν and C_ε are set values of 0.067 and 0.916, respectively [7]. However, these constants can also be computed dynamically using Localized Dynamic Kinetic Energy Model (LDKM) [18].

To close the subgrid stress tensor, $\tau_{g,ij}^{sgs}$, the following expression is used:

$$\tau_{g,ij}^{sgs} = -2\overline{\rho}_g \nu_t \left(S_{g,ij} - \frac{1}{3}S_{g,kk}\delta_{ij}\right) + \frac{2}{3}\overline{\rho}_g k^{sgs}\delta_{ij}. \tag{13.12}$$

The subgrid total enthalpy, $H_{g,j}^{sgs}$, is modeled using the eddy viscosity and gradient diffusion assumption:

$$H_{g,i}^{sgs} = -\overline{\rho}_g \frac{\nu_t}{Pr_t}\frac{\partial \widetilde{H}_g}{\partial x_i}. \tag{13.13}$$

Here, $\widetilde{H}_g$ is the filtered total enthalpy, given by $\widetilde{H}_g = \widetilde{h}_g + \frac{1}{2}\widetilde{u}_{g,i}\widetilde{u}_{g,i} + k^{sgs}$, where $\widetilde{h}_g = \sum_{k=1}^{N_s}\widetilde{h}_{g,k}\widetilde{Y}_{g,k}$. Pr_t can be dynamically computed, but is assumed to be unity here. The subgrid convective species flux, $Y_{g,i,k}^{sgs}$, is modeled using the gradient diffusion assumption as well:

$$Y_{g,i,k}^{sgs} = -\frac{\overline{\rho}_g \nu_t}{Sc_t}\frac{\partial \widetilde{Y}_{g,k}}{\partial x_i}, \tag{13.14}$$

where Sc_t is assumed unity here but can also be dynamically computed. Following the approach provided in [7, 19], the subgrid terms, $\sigma_{g,i}^{sgs}$, $\theta_{g,i,k}^{sgs}$, and $q_{g,i,k}^{sgs}$, are neglected.

13.2.2.2   Hyperviscosity Models

The functional models are known to perform well in many situations [13]. However, most of these models are designed for low Mach number flows and therefore usually fail when strong shocks are present. In regions of shocks the dilatation, $\frac{\partial \widetilde{u}_i}{\partial x_i}$, is significant, and in order to extend the use of the subgrid viscosity models to high Mach number flows and flows with shocks Cook and Cabot [20] developed a subgrid viscosity model for shock–turbulence interactions. This model is based on the shock capturing artificial viscosity approach of Neumann and Richtmyer [21] and the SMG subgrid model [14], which, coincidentally, was developed based on the work of Neumann and Richtmyer. More specifically, the subgrid stress and flux vector terms are modeled as

$$\tau_{g,ij}^{sgs} \approx -\lambda_t \frac{\partial \widetilde{u}_i}{\partial x_i} - 2\overline{\rho}_g \nu_t \widetilde{S}_{g,ij}, \quad H_{g,i}^{sgs} \approx -\overline{\rho}_g \left(\frac{\nu_t}{Pr_t}\right) \frac{\partial \widetilde{H}_g}{\partial x_i}, \quad Y_{g,i,k}^{sgs} \approx -\overline{\rho}_g \left(\frac{\nu_t}{Sc_{t,i}}\right) \frac{\partial \widetilde{Y}_{g,k}}{\partial x_i},$$

(13.15)

in which ν_t and λ_t are the subgrid viscosity and subgrid bulk viscosity to be modelled. Following Cook and Cabot, [20], we here adopt the subgrid viscosity and subgrid bulk viscosity models,

$$\nu_t = C_\nu^r \overline{\rho} \Delta^{(r+2)} \left| \frac{\partial^r}{\partial x_i} \widetilde{S}_{g,ij} \right|, \quad \lambda_t = C_\lambda^r \overline{\rho} \Delta^{(r+2)} \left| \frac{\partial^r}{\partial x_i} \widetilde{S}_{g,ij} \right|, \quad r = 2, 4, 6, ..., \quad (13.16)$$

in which the overbar denotes a truncated Gaussian filter function. Inclusion of the bulk viscosity term is the key to capturing shocks without destroying vorticity, that is, λ_t can be made large (to smooth shocks) without impacting small scale turbulence in regions where $\frac{\partial \widetilde{u}_i}{\partial x_i} \approx 0$. Additionally, by setting $r > 0$, the viscosity keys directly on the ringing, rather than indirectly on gradients. This eliminates the need for ad hoc limiters and switches to turn off λ_t in special cases, expansion, isentropic compression, rigid rotation, etc. [22]. It also removes the need for a dynamic procedure [15] to turn off ν_t in regions of uniform shear. For the simulations reported in Section 13.4, $r = 2$, $C_\nu^r = 0.05$, and $C_\lambda^r = 0.1$.

13.2.2.3   Implicit Large Eddy Simulation Models

Yet another subgrid modeling approach is to use Implicit Large Eddies Simulation (ILES), which emerged during the 1990s as an alternative to conventional LES [23, 24, 25]. The primary reason for the development of ILES was the absence of a universal theory of turbulence, the pragmatic and empirically based development of subgrid models, and the fact that the leading order truncation error of the numerical methods used to solve the LES equation often interacts with the subgrid flow model. More specifically, ILES is based on the original (unfiltered) flow equations instead of the filtered ones, and invoked nonoscillatory constraints via nonlinear limiters in finite volume formulations, to implicitly act as a filtering (and nonlinear adaptive regularization) mechanism for small scales. Using Modified Equations Analysis (MEA), Grinstein and Fureby [26] were able to derive expressions for the implicit (or built-in) subgrid models in a finite volume framework utilizing a hybrid flux formulation. More specifically, for the reacting flow equations (Eqs. 13.1–13.4) the implicit (built-in) subgrid models are

$$\tau_{g,ij}^{sgs} = \rho\left(C\frac{\partial \tilde{u}_{g,j}}{\partial x_i} + \frac{\partial \tilde{u}_{g,i}}{\partial x_j}C^{\mathrm{T}} + \chi^2\left(\frac{\partial \tilde{u}_{g,i}}{\partial x_j}d_i\frac{\partial \tilde{u}_{g,i}}{\partial x_j}d_i\right)\right) + \mathrm{h.o.t.},$$

$$H_{g,i}^{sgs} = \rho\left(C\frac{\partial \tilde{H}_g}{\partial x_i} + \chi^2\left(\frac{\partial \tilde{H}_g}{\partial x_i}d_i\right)\frac{\partial \tilde{u}_{g,i}}{\partial x_j}d_i\right) + \mathrm{h.o.t.}$$

$$Y_{g,i,k}^{sgs} = \rho\left(C\frac{\partial \tilde{Y}_{g,k}}{\partial x_i} + \chi^2\left(\frac{\partial \tilde{Y}_{g,k}}{\partial x_i}d_i\right)\frac{\partial \tilde{u}_{g,i}}{\partial x_j}d_i\right) + \mathrm{h.o.t.}, \tag{13.17}$$

in which $C_{i,j} = \chi\left(\tilde{u}_{g,i}d_j\right)$, χ is a nonlinear function of the flux limiter, Γ, used to switch between the underlying high order and low order numerical flux reconstruction schemes, and d_i is a vector between two neighboring cells. These implicit subgrid models are of the same mathematical character as the subgrid viscosity models, but with the subgrid viscosity being a fourth rank tensor instead of a scalar. This allows these models to better handle simultaneous flow and grid anisotropies and nonuniformities, but makes them dependent of the selection of flux limiters, with monotonicity preserving flux limiters found to perform the best [27].

13.2.3   Thermal Equation of State

In order to solve the governing equations, the thermodynamic variables should be closed using an appropriate Equation of State (EoS). The ideal gas law EoS is expressed as

$$\bar{p}_g = \bar{\rho}_g R_0 \sum_k \left(\tilde{Y}_{g,k}/M_k\right)\tilde{T}_g, \tag{13.18}$$

where M_k is the molar mass of species k and R_0, the universal gas constant, will not accurately predict the postdetonation flow field as the flow is characterized by high pressures and densities. Thus, a real gas model that accounts for the dependence of the internal energy on both pressure and density is essential to model explosives and is typically of the form

$$\bar{p}_g = \bar{p}_g\left(\tilde{e}_g, \bar{\rho}_g\right). \tag{13.19}$$

One of the widely used models to relate thermodynamic variables to simulate explosive charges is the Mie–Gruneisen equation of state, where the pressure is given as

$$\bar{p}_g = \Im\left(\bar{\rho}_g\right) + \Omega\bar{\rho}_g\tilde{e}_g, \tag{13.20}$$

where $\Im\left(\bar{\rho}_g\right)$ is a function of $\bar{\rho}_g$ only, and Ω is a constant. Note that the Mie–Gruneisen EoS readily reduces to the perfect gas EoS when $\Im\left(\bar{\rho}_g\right) = 0$ and $\Omega = \gamma - 1$. Here, γ is the ratio of specific heats of the gas. Several EoS belong to the realm of Mie–Gruneisen EoS and in the context of explosive modeling, Jones–Wilkins–Lee (JWL) EoS for the detonation products [28] is one of the more widely used EoS. JWL EoS can be summarized as

$$\overline{p}_g\left(\overline{\rho}_g, \widetilde{e}_g\right) = A\left[1 - \frac{\omega\overline{\rho}_g}{R_1\rho_o}\right]exp\left(-\frac{R_1\rho_o}{\overline{\rho}_g}\right) + B\left[1 - \frac{\omega\overline{\rho}_g}{R_2\rho_o}\right]exp\left(-\frac{R_2\rho_o}{\overline{\rho}_g}\right) + \omega\overline{\rho}_g\left(\widetilde{e}_g - e_0\right),$$

(13.21)

where A, B, R_1, R_2, ρ_o and ω are explosive specific constants and e_0 is a reference internal energy. For several explosives, these constants are documented elsewhere [28]. By using the constant specific heat assumption, the JWL EoS is further simplified [29] to obtain a thermal form of the equation of state, that is, $\overline{p}_g = \overline{p}_g\left(\overline{\rho}_g\widetilde{T}_g\right)$. This simplified form is expressed as

$$\overline{p}_g = A\,exp\left(\frac{-R_1\rho_o}{\overline{\rho}_g}\right) + B\,exp\left(\frac{-R_2\rho_o}{\overline{\rho}_g}\right) + \omega\overline{\rho}_g C_{v,g}\widetilde{T}_g.$$

(13.22)

At low densities, the JWL equation of state becomes asymptotic to the ideal gas law EoS. Although this simplified approach allows for an easy estimation of thermodynamic variables, the assumption that the specific heats are constant in the postdetonation flow is not accurate. An alternate approach is to use a hybrid EoS, where JWL EoS is used in the immediate vicinity of the blast wave and variable γ ideal gas law EoS is used for the detonation products at low densities [30].

Another widely used EoS for explosive modeling is the Nobel–Able EoS [31, 32]. This slightly more elaborate version of ideal gas law (Eq. (13.18)) is defined as

$$\overline{p}_g = \frac{\overline{\rho}_g R_0 \sum_k \left(\widetilde{Y}_{g,k}/M_k\right)\widetilde{T}_g}{\left(1 - An_g\right)},$$

(13.23)

where A is an empirical constant that accounts for the covolume where reactions take place and n_g is the number of moles per unit volume.

13.2.4   Combustion Modeling

The chemical reactions are usually confined to thin layers at small scales that cannot be resolved on most CGS grids. Hence, most of the Turbulence Chemistry Interaction (TCI) needs to be modeled. Different approaches can be used to access the overall importance of this modeling for afterburning behind explosive blasts. These approaches include both infinite chemistry approximation, where the TCI is not modeled but the the reaction rate is dictated primarily by how soon the fuel and oxidizer mix, and the Finite Rate Chemistry (FRC) models, where the reaction rates are commonly expressed using Arrhenius form:

$$\dot{w}_{g,k} = M_k P_{kl}\dot{w}_l \quad \text{with} \quad \dot{w}_l = A_l e^{\left(-T_{a_l}/T\right)}\prod_{l,k}\left(\rho Y_k/M_k\right)^{\alpha_k},$$

(13.24)

in which is P_{kl} is the stoichiometric coefficients, $\dot{w}_l$ the reaction rate of *lth* reaction step. Here, A_l is the Arrhenius preexponential factor, $T_{a,l}$ the activation temperature, and α_k the reaction orders of reaction step *l*. For this approach the reaction rate $\overline{\dot{w}_{g,k}}$ requires closure and the TCI can be modeled using FRC models such as the Quasi-Laminar (QL) TCI model [33], the eddy dissipation TCI model [36, 37], and the Partially Stirred Reactor (PaSR) TCI model [36, 37], which will be described in the following.

13.2.4.1   Infinite Chemistry Approximation

In most cases the assumption that the combustion process in the postdetonation regime is mixing-controlled rather than chemically controlled is sufficient to determine the chemical source term, $\overline{\dot{w}_{g,k}}$, with reasonable accuracy [6, 38, 39, 40]. At each time step, the concentrations of fuel and oxidizer are compared within a computational cell to determine if the mixture is fuel lean or fuel rich. All the fuel is instantaneously consumed for fuel lean mixtures and all the oxidizer is consumed for fuel rich mixtures. Based on the set of chemical reactions for a given postdetonation combustion, the amount of fuel (in fuel rich mixtures) and the amount of oxidizer (in fuel lean mixtures) required are determined.

13.2.4.2   Partially Stirred Reaction TCI Model

The PaSR is an FRC model [41] based on the conjecture that any turbulent flow can be divided into fine structures (*) and surroundings (^{0}). Since most of the mixing occurs in the fine structures, the reactions also take place there as the reactants are mixed at scales down to the molecular scales. The fine structures topologically form complex regions, composed of a muddle of interacting tube, ribbon, and sheet like structures [80], in which most of the dissipation and mixing takes place. The conditions in the fine structures and surroundings are related by the subgrid balance equations of mass and energy

$$\begin{cases} \rho_g \left(Y_{g,k}^* - Y_{g,k}^0 \right) \big/ \tau^* \approx \dot{w}_{g,k}\left(\rho_g Y_{g,k}^* T_g^* \right), \\ \rho_g \sum_{k=1}^{N} \left(Y_{g,k}^* h_{g,k}^* - Y_{g,k}^0 h_{g,k}^0 \right) \big/ \tau^* \approx \sum_{k=1}^{N} h_{k,f}^\theta \dot{w}_{g,k}\left(\rho_g Y_{g,k}^* T_g^* \right), \end{cases} \tag{13.25}$$

in which τ^* is the subgrid time and $h_{k,f}^\theta$ the enthalpy of formation for *k*th species. By introducing the reacting volume fraction, γ^*, as the ratio of the volume of the fine structures to the LES cell volume, and thereby expressing the species mass fraction and temperature as $\widetilde{Y}_{g,k} = \gamma^* Y_{g,k}^* + (1 - \gamma^*)Y_{g,k}^0$ and $\widetilde{T}_g = \gamma^* T_g^* + (1 - \gamma^*)T_g^0$, respectively, these balance equations may be reformulated as

$$\begin{cases} \rho_g \left(Y_{g,k}^* - \widetilde{Y}_{g,k} \right) = (1 - \gamma^*)\tau^* \dot{w}_{g,k}\left(\rho_g Y_{g,k}^* T_g^* \right), \\ \rho_g \sum_{k=1}^{N} \left(Y_{g,k}^* h_{g,k}^*\left(T_g^* \right) - \widetilde{Y}_{g,k}\widetilde{h}_{g,k}\left(\widetilde{T}_g \right) \right) = (1 - \gamma^*)\tau^* \sum_{k=1}^{N} h_{k,f}^\theta \dot{w}_{g,k}\left(\rho_g Y_{g,k}^* T_g^* \right). \end{cases} \tag{13.26}$$

The reacting volume fraction is defined as $\gamma^* = \Delta V^* / \Delta^3$, in which Δ^3 is the cell volume, in which the CGS variables are constant. This implies that it is possible to lump the fine structure and surrounding fluid components together in different parts of the CGS cell. The lumped fine structures are collectively described by the chemical time scale, τ_c, as the reactions are assumed to take place within these structures. Since the dependent variables are constant in each CGS cell, the fine structure volume can be approximated as $\Delta V^* = \Delta^2 |\widetilde{u}_{g,i}| \tau_c$. Similarly, the cell volume can be estimated as $\Delta^3 = \Delta^2 |\widetilde{u}_{g,i}| (\tau_c + \tau^*)$. The definition of the reacting volume fraction the results in

$$\gamma^* = \tau_c / (\tau_c + \tau^*). \tag{13.27}$$

In Eq. 13.27 the chemical time scale must be representative of the overall combustion reaction, and is here represented by $\tau_c \approx \delta_u / s_u \approx \nu / s_u^2$, in which δ_u and s_u are the laminar flame thickness and speed. The modeling of the fine structure residence time, τ^*, is based on the observation that the fine structure area to volume ratio, $\Delta S^* / \Delta V^*$, is defined by the dissipative length scale $l_D = (\nu \Delta / \nu')^{1/2}$, determined by the molecular viscosity, ν, and the subgrid velocity stretch ν'/Δ, and that the velocity influencing these structures is the Kolmogorov velocity, ν_K, such that $\tau^* = l_D / \nu_K$. Combining the expression for l_D and ν_K, utilizing the Kolmogorov length and time scales, $l_K = (\nu^3 / \varepsilon)^{1/4}$ and $\tau_K = (\nu / \varepsilon)^{1/2}$, in which $\varepsilon = (\nu')^3 / \Delta$ is the dissipation, finally yields in the fine structure residence time being modeled as a geometrical mean of the Kolmogorov time scale and the shear time, τ_Δ,

$$\tau^* = \sqrt{\tau_\Delta \tau_K} = \nu^{1/4} \Delta^{3/4} \nu'^{-5/4}, \tag{13.28}$$

in which $\tau_\Delta = \Delta / \nu'$. This shear time scale was recently found by [43] to accurately model the statistics of dissipation, and therefore also of small scale mixing, which, in turn, is essential to the onset of chemical reactions. The values of γ^* in the afterburning regions vary typically between 0.1 and 0.3. The filtered reaction rates can thus be expressed as

$$\overline{\dot{w}_{g,k}(\rho_g, Y_{g,k}, T_g)} = \gamma^* \dot{w}_{g,k}\left(\rho_g, Y_{g,k}^*, T_g^*\right) + (1 - \gamma^*) \dot{w}_{g,k}\left(\rho_g, Y_{g,k}^0, T_g^0\right). \tag{13.29}$$

13.2.5   Interphase Coupling Terms

The interphase interaction terms in the governing equations represent the mass transfer, $\overline{\dot{\rho}_p}$; ith component momentum transfer, $\overline{\dot{F}_{p,i}}$; heat transfer, $\overline{\dot{Q}_p}$; work transfer, $\overline{\dot{W}_p}$; and production of kth species by evaporation/pyrolysis, $\overline{\dot{S}_{p,k}}$. Due to the constraints set by the computational resource requirements, it is not always possible to solve for every particle in the flow. Hence, particles are grouped and a representative particle from each group, called a parcel, is tracked [6, 38, 39, 44, 45]. The properties for all particles in a group are assumed to be identical. The number of particles per parcel, n_p, is set based on the volume loading, the computational cost, and the required accuracy [46]. As the particle/parcel locations are known at every instant, the local interphase interaction/coupling terms can be obtained by volume averaging over all the particles/parcels in a finite volume. The interphase coupling terms, obtained by averaging over all particles/parcels in a computational cell, are

$$\widetilde{\overline{\rho}}_p = \frac{1}{V} \sum_{n=1}^{N} n_{p,n} \dot{m}_{p,n}, \tag{13.30}$$

$$\overline{F}_{p,i} = \frac{1}{V} \sum_{n=1}^{N} n_{p,n} \left[\dot{m}_{p,n} u_{p,i,n} + \frac{\pi}{2} r^2_{p,n} C_{D,n} \overline{\rho}_{g,n} |u_{p,i,n} - \widetilde{u}_{g,i,n}| \left(u_{p,i,n} - \widetilde{u}_{g,i,n} \right) \right.$$
$$\left. + \frac{4}{3} \pi r^3_{p,n} \frac{\partial \overline{p}_{g,n}}{\partial x_i} \right], \tag{13.31}$$

$$\overline{\dot{Q}}_p = \frac{1}{V} \sum_{n=1}^{N} n_{p,n} \left[\dot{m}_{p,n} h_{v,n} + 2\pi r_{p,n} \kappa_g Nu_n \left(T_{p,n} - \widetilde{T}_{g,n} \right) + 4\pi r_{p,n}{}^2 \varepsilon \sigma \left(T_{p,n}{}^4 - \widetilde{T}_{g,n}{}^4 \right) \right], \tag{13.32}$$

$$\overline{\dot{W}}_p = \frac{1}{V} \sum_{n=1}^{N} n_{p,n} \left[\dot{m}_{p,n} u_{p,i,n} u_{p,i,n} + \frac{\pi}{2} r^2_{p,n} C_{D,n} \rho_{g,n} |u_{p,i,n} - \widetilde{u}_{g,i,n}| \left(u_{p,i,n} - \widetilde{u}_{g,i,n} \right) u_{p,i,n} \right.$$
$$\left. + \frac{4}{3} \pi r^3_{p,n} \frac{\partial \overline{p}_{g,n}}{\partial x_i} u_{p,i,n} \right], \tag{13.33}$$

Where $\dot{m}_p$ is the rate of mass transfer due to evaporation, r_p the particle radius, C_D the drag coefficient, $u_{p,i}$ particle velocity, Nu the Nusselt number, T_p the particle temperature, h_v the enthalpy of mass transfer, ε the emissivity, and σ the Stefan-Boltzmann constant. Note that in the equations, as the interparticle collision force is the force due to collisions between the particles, the effects of interparticle stress on the gas phase are neglected. The properties and quantities of the particle phase are further described in Section 13.2.7.

13.2.6 Dispersed Phase Modeling

The dispersed phase is generally solved using Eulerian–Lagrangian (EL) method or Eulerian–Eulerian (EE) method. The dispersed phase is modeled using Lagrangian approach in EL and Eulerian approach in EE. In both of the methods (EL and EE), the gas phase is modeled using Eulerian approach. Lagrangian tracking can accurately predict the particle dispersal and mixing in complex flows such the one being considered here. Thus, the explosive dispersal is modeled using EL approach. The particle position vector, $x_{p,i}$, is obtained from the particle velocity vector, $u_{p,i}$, where $u_{p,i}$ is calculated based on the forces acting on the particle. The equations used to evaluate $x_{p,i}$ and $u_{p,i}$ using Lagrangian tracking are:

$$\frac{dx_{p,i}}{dt} = u_{p,i}, \tag{13.34}$$

$$m_p \frac{du_{p,i}}{dt} = \frac{\pi}{2} r_p{}^2 C_D \overline{\rho}_g \left| \widetilde{u}_{g,i} - u_{p,i} \right| \left(\widetilde{u}_{g,i} - u_{p,i} \right) - \frac{4}{3} \pi r_p{}^3 \frac{\partial \overline{p}_g}{\partial x_i} + m_p A_{c,i}, \tag{13.35}$$

where m_p is the solid particle mass and $A_{c,i}$ is the i component of net acceleration/deceleration on a particle due to interparticle collisions [44, 45]. The drag coefficient,

C_D, is usually expressed as an empirical function of Reynolds number, Mach number, and solid volume fraction [4, 46, 51].

As the flow can be dense, the collision/contact of particles is inevitable. Thus, the interparticle interaction, $A_{c,i}$, can be very significant [38] and is expressed as an empirical function of particle volume fraction (α_p).

$$A_{c,i} = -\frac{1}{\alpha_p \rho_p} \frac{\partial \tau}{\partial x_i}. \tag{13.36}$$

Here, ρ_p is the particle density. Since the location of each particle is calculated using Lagrangian tracking, α_p is evaluated by volume averaging and the gas phase volume fraction $\overline{\alpha_g}$ is obtained as $\overline{\alpha_g} = 1 - \alpha_p$. The intergranular stress, τ, is obtained as [51, 45]:

$$\tau = \frac{P_s \alpha_p{}^\beta}{\alpha_{cs} - \alpha_p}, \tag{13.37}$$

where P_s and β are empirical constants, and α_{cs} is the close packing volume fraction. Based on past experience, P_s generally takes values in the range of 100–500 MPa for the problem of heterogeneous explosions. Further discussions on the choice of P_s on the blast and particle dispersion characteristics can be found elsewhere [18]. The interparticle acceleration always directs the particle motion away from the regions of high α_p to low, that is, from dense to dilute zones, and the minus sign in Eq. (13.38) ensures this.

The temperature of each particle is obtained by considering the heat transfer between two phases as:

$$m_p C_p \frac{dT_p}{dt} = 2\pi r_p \kappa_g Nu\left(\widetilde{T}_g - T_p\right) - \dot{m}_p L_v + 4\pi r_p{}^2 \varepsilon \sigma\left(\widetilde{T}_g{}^4 - T_p{}^4\right), \tag{13.38}$$

where C_p is the specific heat capacity of the particle, $\dot{m}_p$ is the rate of mass transfer due to evaporation, and L_v is the latent heat of evaporation. The Nusselt number, Nu, is generally represented as an empirical function of Reynolds and Prandtl numbers.

For the reactive particles, the rate of mass transfer, $\dot{m}_p$, is defined as

$$\frac{dm_p}{dt} = -\dot{m}_p = \frac{\pi}{6} r_p{}^2 \rho_p \frac{dr_p}{dt}, \tag{13.39}$$

where the burning rate of a particle, $\frac{dr_p}{dt}$, is expressed as

$$\frac{dr_p}{dt} = -\frac{r_p}{t_b}, \tag{13.40}$$

Here, t_b is the burn time of the particle, which is the most important modeling parameter in particle combustion. Since the application presented in Section 13.5 deals with aluminium particles, several burn time models are available in [52, 53, 54, 55, 56, 57, 58] but no standard model for the burn time of aluminium exists. Here, the burn time model for an aluminium particle is taken from [59], which is also used in [53, 55] and takes into account the surrounding pressure and temperature as well as the oxidizer used in aluminium combustion, according to:

$$t_b = \frac{ar_p^n}{X_{eff}p^{0.1}T^{0.2}},\qquad(13.41)$$

where $a = 1.33 \cdot 10^5$ and $n = 1.8$ are constants [59], and $X_{eff} = C_{O_2} + 0.6C_{H_2O} + 0.22C_{CO_2}$ is the effective concentration of the oxidizers present.

13.2.7 Numerical Methods

The governing equations are solved using a fully explicit finite volume scheme. Employing an unstructured collocated finite volume (FV) method [27], the discretization can be based on Gauss's theorem together with an explicit time integration scheme. Given the vector of unknown variables, $\bar{\mathbf{v}}_i = (\bar{\rho}, \bar{\rho}\tilde{u}_{g,i}, \bar{\rho}\tilde{E}_g, \bar{\rho}\tilde{Y}_{g,k})^T$, the semidiscretized equations are

$$\partial_t(\bar{\mathbf{v}}_i) + \frac{1}{\delta V}\sum_f \left(\mathbf{F}_f^C(\bar{\mathbf{v}}_i) - \mathbf{F}_f^D(\bar{\mathbf{v}}_i) + \mathbf{F}_f^B(\bar{\mathbf{v}}_i)\right) = s(\bar{\mathbf{v}}_i),\qquad(13.42)$$

where $\mathbf{F}_{f,i}^C(\bar{\mathbf{v}}_i)$, $\mathbf{F}_{fi}^D(\bar{\mathbf{v}}_i)$, $\mathbf{F}_{fi}^B(\bar{\mathbf{v}}_i)$ and $s(\bar{\mathbf{v}}_i)$ are the convective, diffusive, subgrid fluxes and the source terms, respectively. The flux reconstruction scheme for the convective fluxes, $\mathbf{F}_{fi}^C(\bar{\mathbf{v}}_i)$, is based on hybridizing a high order linear reconstruction algorithm, $\mathbf{F}_{fi}^{C,H}(\bar{\mathbf{v}}_i)$, with a low order upwind biased reconstruction algorithm, $\mathbf{F}_{fi}^{C,L}(\bar{\mathbf{v}}_i)$, using a nonlinear flux limiter, resulting in a TVD convection scheme. To minimize the non-orthogonality errors in the viscous and subgrid fluxes, $\mathbf{F}_{fi}^D(\bar{\mathbf{v}}_i)$ and $\mathbf{F}_{fi}^B(\bar{\mathbf{v}}_i)$, respectively, these are split into orthogonal and nonorthogonal parts. Central difference approximations are applied to the orthogonal part while face interpolation of the gradients of the variables is used for the nonorthogonal parts. The time integration is performed by a second-order accurate TVD Runge–Kutta scheme [46], such that

$$\left\{\bar{\mathbf{v}}_i^* = \bar{\mathbf{v}}_i^n - \Delta t\left(\frac{1}{\delta V}\sum_f \left(\mathbf{F}_{f,i}^C(\bar{\mathbf{v}}_i^n) - \mathbf{F}_{f,i}^D(\bar{\mathbf{v}}_i^n) + \mathbf{F}_{f,i}^B(\bar{\mathbf{v}}_i^n)\right) - s(\bar{\mathbf{v}}_i^n)\right)\bar{\mathbf{v}}^{n+1}\right.$$

$$= \frac{1}{2}(\bar{\mathbf{v}}_i^n + \bar{\mathbf{v}}_i^*) - \frac{1}{2}\Delta t\left(\frac{1}{\delta V}\sum_f \left(\mathbf{F}_{f,i}^C(\bar{\mathbf{v}}_i^*) - \mathbf{F}_{f,i}^D(\bar{\mathbf{v}}_i^*) + \mathbf{F}_{f,i}^B(\bar{\mathbf{v}}_i^*)\right) - s(\bar{\mathbf{v}}_i^*)\right).\qquad(13.43)$$

An alternative numerical method for dense flows is to use the discrete equations method (DEM) [48, 49], with an Eulerian–Lagrangian reformulation [38]. The problem of heterogeneous explosive dispersal involves dense flow field only at early times and the flow transforms to dilute regime at late times. Thus, it is essential to account for particle volume in the flow at early times. DEM with monotone upstream centered schemes for conservation laws (MUSCL) shock capturing scheme and HLLC Riemann solver [50] can be used for this purpose. At late times, that is, after transition to a dilute flow, a hybrid approach that combines the shock capturing scheme with a central scheme to resolve both shocks and shear flow [8] is used. Further details of this approach can be found elsewhere [11].

13.3   Afterburning behind Heterogeneous Nitromethane Charges

When an explosive charge embedded with solid particles is detonated, the particles augment the impact of the explosive. The momentum and energy imparted by the explosive convert the particles into effective projectiles capable of penetrating target surfaces. Some of these target surfaces withstand the detonation of explosive charges without particles, called homogeneous explosives, and are only susceptible to damage by explosive charges with particles, called heterogeneous explosives. Owing to multiple advantages and applications, heterogeneous explosives have attracted the attention of several researchers. In this section, we aim to address the mixing and the combustion in the postdetonation flow of a heterogeneous explosive. A nitromethane charge of initial radius 5.9 cm is considered for this purpose as these charges have been experimentally studied [60], both with and without particles. The modeling employed here utilizes OEEVM subgrid model (Sec. 13.2.2.1), JWL EoS (Eq. (13.22)), infinitely fast chemistry approximation, (Sec. 13.2.4.1) and numerical methods adopted for dense flows.

13.3.1   Setup, Initial Conditions, and Grid Resolution

Accurate modeling of the initial detonation of the explosive charge is vital for the numerical simulation of the postdetonation flow. Since the primary focus of this chapter is to analyze the afterburning process, the initial detonation profile is simulated independently using a 1D approach outlined in [61]. The initial detonation profiles obtained using this approach for different explosive charges, both homogeneous [61] and heterogeneous [38], are provided elsewhere. The computed profiles are initialized to a 3D grid. In this section, the simulations are performed employing a sector grid [61, 38, 62]. The sector grid consists of only a part of a spherical sector with free slip boundaries along the sector sides and an outflow at the radially outermost plane. The radial extent of the sector is ($r = 2.4$ m), and the extent in the azimuthal (θ) and zenith (ϕ) directions is 45°. Even though the full spherical configuration is not used, the sector grid approach can reliably simulate heterogeneous explosions. To confirm this, numerical simulations of homogeneous nitromethane charges of initial radius 5.9 cm using both the sector grid and a full sphere ($r = 2.4$ m) configurations have been performed. The full sphere configuration could capture the postdetonation flow features, which are qualitatively and quantitatively similar to the sector grid results (see Figs. 13.1 and 13.2). The extent of mixing/combustion using both the configurations is almost identical as demonstrated by computing the degree of mixedness [63] (in Fig. 13.2). Here, the degree of mixedness (DM) is defined as [38, 63]

(a) (b)

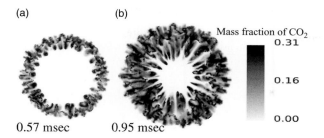

Mass fraction of CO_2
0.31

0.16

0.00

0.57 msec   0.95 msec

Figure 13.1 Formation of CO_2 in the postdetonation flow of homogeneous nitromethane charge during (a) strong blast wave and (b) implosion phases. The time at each phase is indicated.

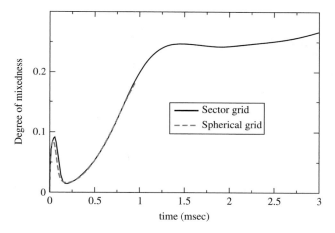

Figure 13.2 Degree of mixedness in the postdetonation flow of homogeneous nitromethane charge evaluated for sector ($1000 \times 60 \times 60$) and spherical ($1000 \times 200 \times 200$) grids.

$$DM = \left[\frac{\int Y_{CO}\left(Y_{N_2} - Y^i_{N_2}\right)dV}{\int dV} \right] \Bigg/ \left[\left(\frac{\int Y_{CO}dV}{\int dV} \right) \left(\frac{\int \left(Y_{N_2} - Y^i_{N_2}\right)dV}{\int dV} \right) \right], \quad (13.44)$$

where $Y^i_{N_2} = 0.2296$ is the mass fraction of nitrogen (N_2) in the initial detonation products and is evaluated from the balanced chemical reaction for decomposition of nitromethane (CH_3NO_2),

$$CH_3NO_2 \rightarrow CO + H_2O + \frac{1}{2}H_2 + \frac{1}{2}N_2. \quad (13.45)$$

For rest of the studies in this section the sector grid approach is used.

Sufficient grid resolution is one of the key requirements to model the afterburning process. The resolution requirement is verified by employing three grids: $1000 \times 45 \times 45$, $1000 \times 60 \times 60$, and $1000 \times 75 \times 75$. To justify the resolution used, the flow quantities for different grids are compared. In order to analyze the flow parameters, it is

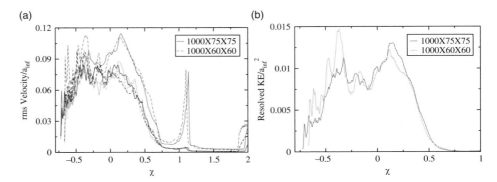

Figure 13.3 Grid independence study: (a) rms of velocity profiles (blue: u_r, red: u_θ, green: u_ϕ); (b) resolved turbulent KE at 4 ms. Note that the quantities are normalized using ambient speed of sound, a_{inf}. A black and white version of this figure will appear in some formats. For the color version, please refer to the plate section.

often convenient to define a scaled distance (χ) [39, 64], defined based on the mixing layer boundaries as:

$$\chi = \frac{r - r_{0.5}}{r_{0.05} - r_{0.95}},$$ (13.46)

where $r_{0.5}$, $r_{0.05}$, and $r_{0.95}$ denote, respectively, the radial distance at which the azimuthally averaged mass fraction of carbon monoxide, Y_{CO}, is 0.5, 0.05, and 0.95 times the Y^i_{CO}. Here, $Y^i_{CO} = 0.459$ is the mass fraction of CO in the initial detonation products computed based on Eq. (13.45). Also, the mixing layer width is defined as

$$\delta = r_{0.05} - r_{0.95},$$ (13.47)

that is, the distance between the locations where the azimuthally averaged mass fraction of CO is 0.05 and 0.95. The grid independence is demonstrated in Figure 13.3, where the root mean square (rms) of the velocity components and the resolved kinetic energy (KE) are shown. Here, u_r, u_θ, and u_ϕ represent the radial (r), the azimuthal (θ), and the zenith (ϕ) velocity components. The velocity components and KE are normalized with a_{inf} and a^2_{inf}, respectively, where a_{inf} is ambient speed of sound. It is evident from Figure 13.3 that the both $1000 \times 60 \times 60$ and $1000 \times 75 \times 75$ grids give reasonably matched profiles of rms velocity components and KE. Minor discrepancies, especially near $\chi \approx 0.4$, are due to the presence of dense particle clusters at the size comparable to the grid. However, these difference quickly disappear due to the transient nature of the flow and the flow is accurately predicted for both grids. In particular, the turbulence decay in the outer regions of the mixing layer, that is, $0.25 \le \chi \le 0.6$ (see Fig. 13.3(b)), is captured almost identically by grids of different resolution. With the confidence in results obtained using different grids, $1000 \times 75 \times 75$ is used for the rest of this section.

Steel (Fe) particles of radius 463 μm are uniformly distributed within the explosive charge to study afterburning in heterogeneous charges. The initial volume fraction of the particles is set to 62.0% based on the experimental investigations [60]. The explosive dispersal generates a particle front that conforms to spherical symmetry due to the large size of the particles (compared to the micron sized particles usually used) and the material density. This has been observed in the experiments. Further, the blast wave trajectory, the particle front trajectory, and the blast wave overpressure computed based on the approach described here are in good agreement with the experimental data [60]. Although the global particle front appears spherically symmetric, the local variations in the flow result in microscale asymmetry. Also, if particle size is in the order of a micron, the hydrodynamic instabilities can lead to localized clusters of particles resulting in preferential combustion in local hot zones. These effects are not discussed here and can be found elsewhere [39].

13.3.2 Mixing Layer Characteristics

13.3.2.1 Chronology of the Postdetonation Flow

The dynamics of the mixing layer ensuing from the detonation of a heterogeneous charge can be described in four stages, which correspond to Figure 13.4. These stages are as follows: (1) The high density product gases formed after the initial detonation expand outward with velocities greater than that of the solid particles and over take the particle front. Due to the momentum and the energy transfer between the particles and the expanding product gases, nonspherical flow is enforced, which results in small azimuthal and zenith gas velocity components. Subsequent interaction of this non-spherically symmetric flow with the decelerating contact surface gives rise to the source of the initial perturbations in the contact surface as shown in Figure 13.4(a). These perturbations develop into the initial hydrodynamic structures due to RT instabilities even before the particles over take the contact surface. This phenomenon is also reported by Frost et al. [65]. (2) The solid particles subsequently over take the contact surface as shown in Figure 13.4(b). By interacting with the gases in the mixing layer through momentum and energy exchange, the solid particles further enhance the growth of the perturbations. (3) The mixing layer widens due to the dragging of its lower boundary by the inward propagating secondary shock (see Fig. 13.4(c)). Subsequent interaction of the secondary shock, which reflects from the origin and propagates outward, with the mixing layer generates further hydrodynamic structures due to RM instabilities. This reshock phase is characterized by deposition of vorticity by baroclinic torque effects, that is, misaligned pressure and density gradients. (4) Finally, the vorticity production distorts and wrinkles the flame, which sustains the mixing process. The hydrodynamic instabilities result in loss of memory of the shape of the structures in the mixing layer at late times (Fig. 13.4(d)).

13.3.2.2 Turbulent Kinetic Energy and Baroclinic Torque

Illustration of the role of the perturbations/instabilities is carried out best by analyzing the resolved and subgrid turbulent KE, as shown in Figure 13.5. As the gas expands, due to the decrease of the velocity magnitudes, the resolved and subgrid KE decay in time. The value of the subgrid KE, about 3–4% of the resolved value, indicates that most of the turbulent KE is resolved with the $1000 \times 75 \times 75$ grid. The contribution of k^{sgs}, as shown in Figure 13.5, to the total KE is minimal. However, it is still essential to

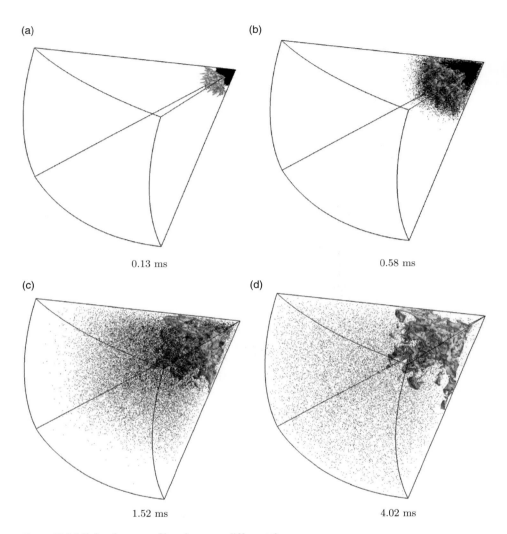

Figure 13.4 Mixing layer profiles shown at different times.

capture the finer scales, which would otherwise not be accounted for. The peak observed near $\chi = -0.6$ at 1.5 ms is a consequence of the secondary shock, which enhances the perturbation intensity locally through the baroclinic mechanism. The same peak is observed near $\chi = 0$ at 2.2 ms, albeit of a lower magnitude due to the decay in the strength of the secondary shock as it propagates outward. The vorticity generation due to misalignment of pressure and density gradients, termed the baroclinic torque effect, explains the mechanism of turbulence production. This is prominent inside the mixing layer during the reshock phase.

The role played by vorticity in the dynamics of the flame that exists in the outer regions of the mixing layer during the reshock phase is of great interest and is analyzed in detail. It is to be noted that, due to the infinite chemistry assumption, the flame location is dictated by where and how fast the CO in the inner detonation products

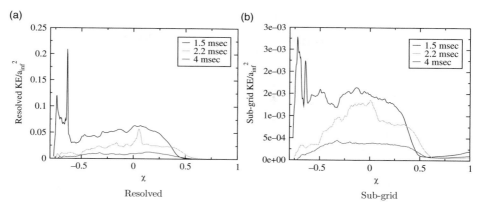

Figure 13.5 Turbulent kinetic energy profiles.

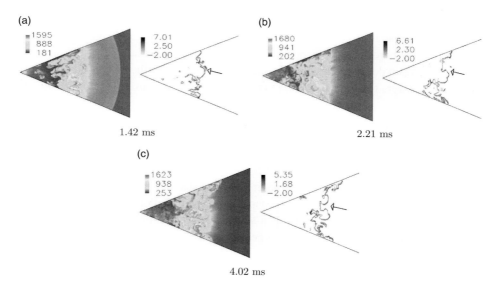

Figure 13.6 Flame dynamics in the mixing layer: Temperature [K] and $\dot{w}_{CO}$ ($\log(\exp(-2) - \dot{w}_{CO})$) profiles at different times. Temperature on the left and $\log(\exp(-2) - \dot{w}_{CO})$ on the right.

mixes with the O_2 in the outer air. Due to the baroclinic effects during the reshock phase, the secondary shock deposits vorticity in the mixing layer [64, 2, 6]. This deposition of vorticity plays a prominent role in the subsequent flame wrinkling. Starting from just before the reshock phase until a short while later, the temperature and reaction rate of CO ($\log(\exp(-2) - \dot{w}_{CO})$), with the temperature in [K] and $\dot{w}_{CO}$ in [Kg/m^3/ sec], at different times is shown in Figure 13.6. The reaction rate here refers to how fast CO and O2 meet/mix so as to react, the reaction rate refers to how fast CO and O_2 meet/mix so as to react. In Figure 13.6(a), the secondary shock is propagating outward after reflection from the origin, but before its interaction with the structures in the mixing layer. Subsequently, at 2.21 ms (Fig. 13.6(b)), the secondary shock interacts with the flame, across which the density gradients exist. This generates vorticity due to baroclinic effects. The arrows in Figure 13.6 indicate a location of interest used to elucidate the role played by vorticity on the flame dynamics. By 4.02 ms (Fig. 13.6(c)), the secondary shock has traversed the entire region of the flame, and is outside the mixing layer. In the vicinity of the arrow in Figure 13.6(c), the flame has started to convolute/wrinkle. This is a consequence of the vorticity deposited by the secondary shock. Due to the enhancement of the flame surface area by vorticity deposition, fresh sources of O_2, which were hitherto unreachable to the inner CO, are now made available for reaction and for sustaining the flame. At the same time, smaller pockets of air that were earlier driven inward during the implosion phase get trapped inside the core detonation products and are being slowly consumed. These pockets of flames are also seen in the inner regions at later times (Fig. 13.6(c)). Subsequently, the pockets of flames get extinguished after all the O_2 in the inner regions of the detonation products is consumed. The peak fireball temperature is 1600 K at 1.42 ms (Fig. 13.6(a)), which is lower than the CO/air flame temperatures. This is a consequence of the CO_2 produced from afterburn acting as a diluent. Later, during the reshock phase, the peak fireball temperature is 1700 K. This slight increase is a consequence of the mixing enhanced by the baroclinically deposited vorticity by the secondary shock. However, the vorticity again decays, subsequently, and the CO_2 that is further produced acts as a blanket. This lowers the peak fireball temperatures, for instance to 1625 K at 4.02 ms (Fig. 13.6(c)).

13.3.3   Comparison to a Homogeneous Explosive Charge

Homogeneous explosive (NM) containing the same amount of the high explosive is compared with the heterogeneous explosive charge (NM/Fe) to illustrate the role played by the particles in the mixing and perturbations/instabilities in the mixing layer. For homogeneous charge, Gaussian random perturbations are added in the outer vicinity of the NM charge to serve as a source of the initial perturbations [6, 38, 64, 2]. As the particles naturally introduce perturbations to the flow, NM/Fe charge does not need an external source of perturbations. For NM, the perturbations introduced are allowed to grow "unforced" whereas for NM/Fe the particles provide perturbations that are constantly "forced." Thus, the mixing layers for NM and NM/Fe charges differ in the underlying physics because of the way the initial perturbations are enforced.

The rms of velocity and concentrations are higher for the NM/Fe charge than the NM charge by about 25–40%, as shown in Figure 13.7, illustrating that the former has higher perturbation levels in the mixing layer. Constantly forced perturbations for NM/Fe charge, owing to the presence of the particles, result in higher perturbation levels and, thus, higher rms of velocity and concentrations. Also, due to the exchange of the momentum and energy between the two phases in NM/Fe charge, the implosion and reshock phases are delayed. Hence, the baroclinic torque, which plays an important role in reenergizing the turbulence levels in the mixing layer, is also delayed. Thus, the perturbation intensity persists longer for NM/Fe, as is confirmed in the turbulent KE profiles (Fig. 13.7(c)).

Comparison of the mixing layer boundaries and the mixing layer width (δ) for NM and NM/Fe, shown in Figure 13.8, indicates that the implosion phase terminates later for the

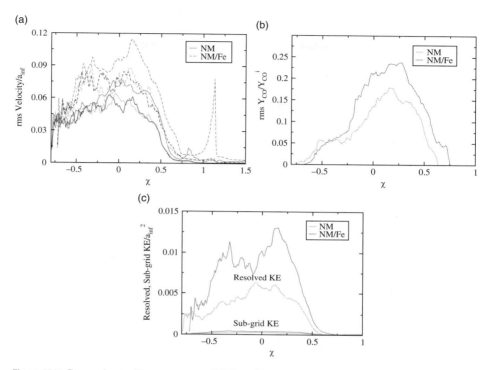

Figure 13.7 Comparison of homogeneous (NM) and heterogeneous (NM/Fe) explosives at 4 ms: (a) rms of velocity profiles (blue: u_r, red: u_θ, green: u_ϕ); (b) rms of Y_{CO}; (c) resolved and subgrid turbulent KE. A black and white version of this figure will appear in some formats. For the color version, please refer to the plate section.

NM/Fe charge (0.75 ms for NM, 1.5 ms for NM/Fe). Since perturbations and mixing levels are higher for the NM/Fe charge, the outer boundary of the mixing layer extends farther by about 2.5 charge radii for the NM/Fe charge. As a result of greater mixing, the afterburn energy release is enhanced, which further drives the outer boundary of the mixing layer. The dependence of the mixing layer widths on the perturbation/instability effects is compared by plotting, δ, normalized with the charge radius (r_o) (see Fig. 13.8 (b)) for the NM and NM/Fe charges. Also, the curve fit expressions for δ/r_o as functions of time (t in ms) are obtained. δ grows linearly with time ($\sim t$), during the early blast wave phase, for NM, but nonlinearly for NM/Fe ($\sim t^{0.75}$). Also, δ is wider for the NM/Fe at the same time instant during these early times (note that for small t, $t^{0.75}$ is a faster growth than t^1). The linear growth for homogeneous explosives have been reported earlier [64, 6], however, the nonlinear growth for NM/Fe is presumed to be due to the continued forced perturbations by the particles. Small perturbations begin to grow linearly at early times as first predicted by Richtmyer [5]. These perturbations later grow to large enough sizes at which time the nonlinear growth ensues. This theory may not be applicable to systems with continuous forcing as in NM/Fe case. Since the implosion phase starts earlier for the NM charge (see Fig. 13.8(a)), δ grows faster for NM and overtakes δ for the NM/Fe charge near 0.35 ms (see Fig. 13.8(b)).

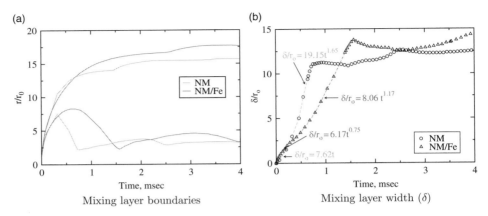

Figure 13.8 Comparison of the homogeneous (NM) and heterogeneous (NM/Fe) explosive charges containing the same amount of the high explosive. t is in ms for the curve fit expressions in panel (b).

13.4   TNT Afterburning at Different Heights of Blast

Afterburning in unconfined domains has been studied by means of combined numerical simulations and experiments in the previous chapter and by Kuhl et al. in [66]. Fully confined afterburning has also been studied by Kuhl et al. [67, 68, 69, 70] and by Kim et al. [32]. The semiconfined case has, however, not been given much attention, with respect to the numerical simulations of mixing and afterburning in the vicinity of a surface and the effect HoB has on these properties. This fairly simple configuration lies between the unconfined and totally confined cases and has its own peculiarities in terms of distinctive shock propagations patterns, shock–mixing layer interactions, and the subsequent afterburning. The main objective of this investigation is to elucidate the physical processes involved in a near ground air blast and what effects the HoB has in this particular configuration. To achieve this we address and compare three HoB cases of 0.15 m, 0.5 m, and 1.0 m above ground.

In this work the reaction rates (from Eq. (13.4)) are low pass filtered and are modeled using the PaSR model (Sec. 13.2.4.2). We employ a reduced TNT mechanism developed by Tran, [71], implemented here as a two step mechanism for the afterburning of TNT detonation products:

$$
\begin{aligned}
\text{(C0)} \quad & C_7H_5N_3O_6 \;\rightarrow\; 1.5N_2 + 2.5H_2O + 3.5CO + 3.5C \\
\text{(C1)} \quad & C + \tfrac{1}{2}O_2 \;\rightarrow\; CO \\
\text{(C2)} \quad & CO + \tfrac{1}{2}O_2 \;\rightarrow\; CO_2
\end{aligned}
\tag{13.48}
$$

where the zero step (C0) represents the initial decomposition of TNT ($C_7H_5N_3O_6$) into N_2, H_2O, CO, and C whereas the subsequent two reaction steps represent the afterburn process, in which C and CO are oxidized into CO and CO_2, respectively. The decomposition of $C_7H_5N_3O_6$ into combustion products is much more complicated than described by the one step mechanism presented earlier, e.g. [72], and includes many other species appearing typically in small concentrations. However, in order to simplify the computational model to something tractable for CGS we adopt the simplification that the combustion products consist of N_2, H_2O, CO, and C.

Current CGS model uses Hyperviscosity subgrid model (Sec. 13.2.2.2) and a Noble–Able EoS (Eq. (13.23)), the single phase CGS equations are solved using a fully explicit finite volume scheme, based on the C++ library OpenFOAM [73] as the computational platform. The code employs an unstructured collocated FV method described in Section 13.2.6.

13.4.1   Effect of HoB on Afterburning

The simulations are initiated with a spherical cloud, with a radius of 6 cm, containing the detonation products in air at the center of three semiconfined domains, composed of $6 \times 6 \times 2.56$ m^3, $5 \times 5 \times 3.56$ m^3, and $6 \times 6 \times 4.56$ m^3 boxes, respectively, where the detonation product cloud was initiated at a HoB of 0.15 m, 0.5 m, and 1.0 m, respectively. The boundary conditions for the ground were modeled by a no slip boundary condition and a constant temperature of 298 K, whereas at the rest of the boundaries the Neumann condition was applied. The initial conditions were obtained from [4] and consisted of 1D profiles of velocity, pressure, density, temperature, and stoichiometric species compositions from an equilibrium decomposition of TNT, spherically mapped onto the 3D grid. This spherically symmetric detonation product cloud was then perturbed by a small random perturbation, varying between 0 and 5×10^{-4} m, to facilitate the development of flow instabilities. The unburned air was set to have atmospheric pressure and temperature and air mass fractions for oxygen and nitrogen of 0.23 and 0.77, respectively.

Figure 13.9 shows the results for HoB of 0.15 m above ground. The primary blast wave expands in all directions and at 0.2 ms the primary shock collides with and reflects from the ground. The ground reflected shock travels upward, interacting with the thin mixing layer and giving rise to RT instabilities. Simultaneously, near ground level, high density regions are formed due to collision. The ground reflected shock collides with a rarefaction wave propagating inward at 1.0 ms and creates two up- and down-going shock waves, which elongate the mixing layer. At 1.5 ms the second collision with the ground occurs, the aftermath of which forms strong vorticity regions near the ground as RM instabilities are developed; that is seen as two rings in the $\widetilde{Y_{CO_2}}$ distribution at 1.5 ms. The new ground reflected shock travels upward again, leaving low pressure regions behind it, which entrain more air into the mixing and afterburning regions. As the previous up- and down-washes of shock system meet again in the center of the domain, another shock gets sent toward the ground and the third ground collision occurs at 3.3 ms. In this system, the mixing layer does not get a chance to develop and thicken as the primary blast wave expands it sideways and the frequent traffic of the vertically moving shocks through it inhibit the growth of the instabilities, which are responsible for the mixing. Only large, energy dense, vortex structures survive the shock passages and hence the afterburning region remains thin.

Figure 13.10 shows the flow at a HoB of 0.5 m above ground. The initial blast wave expands outward in the air as well as parallel to the ground. After the blast wave is reflected by the ground, at 0.5 ms it propagates inward into the mixing layer, initially as

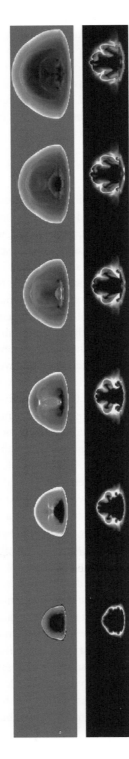

Figure 13.9 TNT afterburning at HoB of 0.15 m. Pressure distribution, $\bar{p}_g$, (top) and mass fraction of CO_2, $\widetilde{Y_{CO_2}}$, (bottom) for times (left to right) 0 ms, 0.5 ms, 1 ms, 1.5 ms, 2 ms, 2.5 ms, and 3.29 ms.

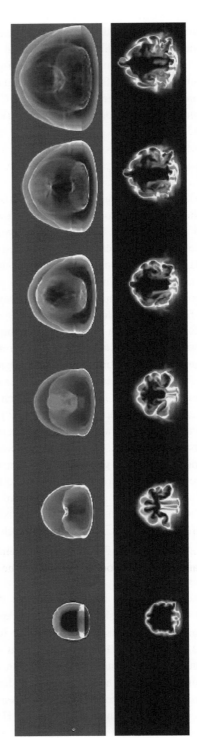

Figure 13.10 TNT afterburning at HoB of 0.5 m. Pressure distribution, $\bar{p}_g$, (top) and mass fraction of CO_2, $\widetilde{Y_{CO_2}}$, (bottom) for times (left to right) 0 ms, 0.5 ms, 1 ms, 1.5 ms, 2 ms, 2.5 ms, and 3.26 ms.

two separate shocks, creating behind it a low pressure region that entrains air into the combustion zone. This effect is seen as a mushroom shape of an initially spherical fireball, visible, for example, in the $\widetilde{Y}_{CO_2}$ distribution at 1.0 ms. The ground reflected shocks collide (at 1.0 ms) with an ingoing rarefaction wave (the origin of which is seen in Fig. 13.10 at 0.5 ms) forming a complex mixing zone around the collision point, seen in $\bar{p}_g$ distribution. Two new shocks are formed, one propagating upward and one downward, towards the ground. These shock systems form strong up- and down-washes of the detonation products that enhance the mixing. Development of these is seen from 1.0 ms to 2.0 ms in the pressure distribution. These shock systems internally expand the afterburning region, primarily in the vertical direction, as can be seen by comparing $\bar{p}_g$ and $\widetilde{Y}_{CO_2}$ distributions at times 1.5–2 ms. Later, at 2.5–3 ms, the shock systems then internally contract both the mixing layer and the afterburning region, as they propagate toward each other and later collide.

In the highest HoB case of 1.0 m above ground, which is presented in Figure 13.11, the primary shock wave expands in all directions and the first ground collision occurs at 0.55 ms. While the ground reflected shock travels upward, the rarefaction wave is propagating inward, toward the origin, resulting in an implosion at 1 ms, seen in pressure distribution. This is the only case that exhibits both ground collisions and an implosion. This is also manifested in the mixing layer evolution as both the ground collision and the implosion entrain plenty of air into the mixing layer; combining this with RT instability growth just before and slightly after the implosion creates a thick and vorticity rich afterburning region. The origin reflected shock and the ground reflected shock collide and also in this case create strong vertically moving shock systems and a complicated shock pattern. One of these shocks collides with the ground, again, at 2.5 ms. The mixing and the afterburning regions become trapped inside the origin reflected shocks and are only influenced by the ground reflected shock propagations, thus imposing the most spherical mixing layer of all three HoB cases, where more instabilities are developed with each shock passage.

13.4.2 Characteristics of the Mixing Layer at Different HoB

To further analyze the effect of the HoB on the mixing and afterburning during an explosive blast, Figure 13.12 presents (a) the mixing layer thickness, δ, (Eq. (13.47)) and (b) the Atwood number, At. The Atwood number is defined as

$$At = \frac{\rho_H - \rho_L}{\rho_H + \rho_L} \tag{13.49}$$

in which ρ_H is the heavy fluid, that is, the detonation products and ρ_L is the light fluid, that is, the air. The Atwood number is presented as an average At of the mixing layer. For $At > 0$ the turbulent mixing is considered unstable, the evolution of perturbations that are initiated as RT instabilities continues [74]. As the At decreases and changes sign the interfaces between ρ_H and ρ_L become stable and the turbulent mixing subsequently stops. In order to prolong the effect of the afterburning, the intention is to sustain the mixing layer, hence the high values of At numbers seems to be desired.

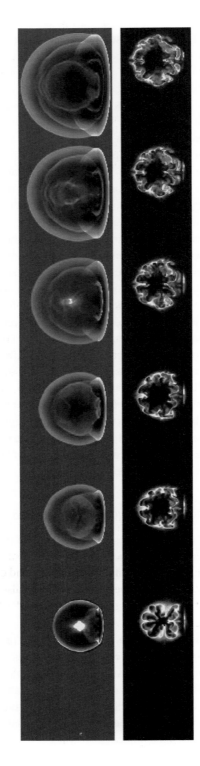

Figure 13.11 TNT afterburning at HoB of 1.0 m. Pressure distribution, $\bar{p}_g$, (top) and mass fraction of CO_2, $\widetilde{Y}_{CO_2}$, (bottom) for times (left to right) 0 ms, 1.25 ms, 2 ms, 2.5 ms, 3.5 ms, 4.0 ms, and 4.8 ms.

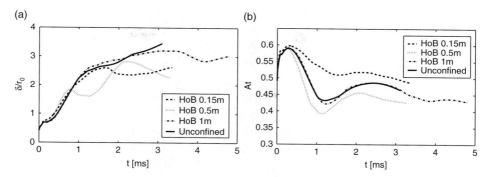

Figure 13.12 Mixing and afterburning characteristics at different values of the HoB. (a) Mixing thickness, δ; (b) Atwood number, At.

Figure 13.12 shows the dominating mechanisms of the different HoB cases, as a reference case, the values from the unconfined case (free air blast) are also included in Figure 13.12. The results and discussion from this case can be found in [75, 76]. In the lowest HoB of 0.15 m, the ground reflected shock passage through the mixing layer is clearly visible in both the δ and At profiles. The HoB 0.15 m profile of δ shows clear peaks at 0.2 ms, 1.7 ms, and 3.3 ms, representing the three ground collisions. In the At profile these collisions are seen as a peak and two following valleys at those times. Between the first and the second collision, the expansion rate of the primary blast wave is faster than the propagation speed of the collision weakened reflected shock, leading to an increase in δ and a decrease in At as the stretching rate of the mixing layer is higher than the vorticity production by means of instabilities (seen also in $\widetilde{Y_{CO_2}}$ distribution, in Fig. 13.9, between 0.5 ms and 1.5 ms). After the second collision, the expansion rate seems to slow down and the upward going reflected shock manages to infuse the mixing layer with RM instabilities, as shown by an increase in At. These instabilities slightly widen the existing vorticity structures, stretching out the already thin mixing layer. This vorticity production and the simultaneous stretching slowly continues as the shock travels toward the ground again, slowly increasing the mixing layer thickness and decreasing the At number as the difference between the ρ_H and ρ_L becomes smaller due to the subsequent afterburning.

In the case of HoB of 0.5 m, the dominating mechanism is also the shock wave propagation; here however, the shock wave system is more complex, as the distance to ground collision is increased. Also in this case the ground collisions are seen as peaks in δ profile, but between the first and the second collision also a shock merge is visible as a valley at 1.25 ms. At the time between the first ground reflection and the internal shock collision, the mixing layer is contracted by the ground reflected shocks propagating through it and the mixing and the afterburning are enhanced, thereby decreasing the density of the detonation products. This is visible as a decrease in At between 0.5 m and 1.25 ms. Just before the collision of the ground reflected shocks, the mixing layer is compressed at its maximum and the combustion is intensified, thereby thinning out the mixing layer. After the center collision at 1.25 ms, seen in the pressure distribution in Figure 13.10 at 1.0 ms and as valleys in both δ and At profiles at 1.25 ms, strong up- and down-washes of shock waves are formed, propagating through the mixing layer and generating RM instabilities. This causes an increase in At and δ as the vorticity production dominates the mixing and the afterburning is debilitated. This continues until the second ground collision at 2.15 ms, after which the entrapment of more oxidizing air, behind the ground reflected shock traveling upward, yet again fires the afterburning, thinning out the mixing layer and decreasing At number. This case has a time averaged lowest At number and mixing layer thickness, suggesting the least unstable mixing layer, the growth of which is only fueled by the passage of shock waves.

The case of a HoB 1.0 m initially exhibits the same behavior as an unconfined case. In the early stages of the explosion the outgoing blast wave heats up and accelerates the ambient air as it propagates through it. RT instabilities occur early as the interface between hot detonation products and the cold ambient air is impulsively accelerated by the passage of a shock wave. A rarefaction wave propagates inward, boosting the acceleration caused by the blast wave and later forcing an outward directed acceleration of the hot detonation gases, a short time after the initial blast the ingoing rarefaction wave will over expand the flow causing a secondary shock resulting in the formation of a thin mixing layer, between the initial blast wave and the secondary shock. The secondary shock will eventually strengthen by means of detonation product gases accelerating it, and subsequently further strengthen to the point of implosion. Here, however, there is also a ground collision at 0.5 ms. Since the flow is dominated by the interaction between the RT instabilities and the ingoing rarefaction wave, this collision does not have a great impact on δ and At as was seen in previous cases. The intensity of the afterburning is at its highest just before the implosion as the mixing layer is vorticity rich and the compression by the rarefaction wave stimulates the reactions between fuel and air. This is seen as a decrease in At and a steady increase in δ between 0.2 ms and 1 ms, and also in $\widetilde{Y_{CO_2}}$ distribution between 0 and 1.25 ms in Figure 13.11. The implosion occurs at 1 ms and captures more air into the mixing layer, nevertheless, as the origin reflected shock travels outward again the vorticity production is initiated by RM instabilities, seen clearly as an increase in At. After the implosion the mixing layer becomes cocooned inside the origin reflected shock system and the afterburning subsides until the second ground collision at $\sim$ 2.5 ms. This further ignites the afterburning, the mixing layer starts to thicken and expand due to volumetric expansion caused by the exothermicity of combustion reactions, and At decreases again. At 3.25 ms the internal, countermoving shocks (one from the ground reflection) collide, slightly compressing the mixing layer and spiking up the combustion. As these shocks travel away from each other, the pattern repeats itself again, generating vorticity and slowing down combustion. The rise and fall of δ and At between 3.25 ms and 4.8 ms show just that, the passage of shocks through the mixing layer that set of combustion and generate vorticity. At this HoB the mixing layer is thickest, which is mostly attributed to the initial stage of the explosion, the long expansion time, and the implosion.

13.5   TNT/Al Afterburning

In EBX charges metal particles, usually aluminium, are added to the explosive compound in order to increase the afterburning energy release by allowing the metal particles and detonation products combust with air. This presents another modeling challenge since the combustion becomes two-phased. This section presents simulations of pure TNT and TNT/Al mixture (80/20% weight) at HoB's of 0.5 m and 1.0 m. The intention is to visualize and elucidate how the afterburning is affected by the inclusion of the reacting aluminium particles.

13.5.1   Aluminium Combustion

When an aluminium (Al) particle is burning, the combustion process that converts the solid Al into the gaseous combustion products is more complex than what is modelled in this work. This combustion process is crudely threefold [77].

1.   Melting and oxidation: when the temperature of the surrounding gas starts to reach the melting temperature of Al the solid Al melts into liquid Al, which at the boiling point temperature (T = 2971 K) undergoes phase change into the gaseous Al. When the gaseous Al is present, it oxidizes with surround gas forming Al suboxides, for example, AlO and AlO_2.

2.   Condensation: aluminium suboxides condense to a liquid aluminium oxide, Al_2O_3. This condensation plays a major role in Al combustion and is the source of considerable amount of heat release during combustion. The condensed aluminium oxide can also deposit back on the particle surface and form an oxide cap, which in its turn changes the Al gasification rate since the cap blocks the vaporization from the region it covers, due to the fact that Al_2O_3 has a higher boiling temperature than solid Al. This in turn alters the temperature and other quantities around the particle.

3.   Dissociation: in the flame zone the heat is sufficient to dissociate the liquid Al_2O_3 back to gaseous aluminium suboxides and oxygen.

It should also be noted that the burn time for Al particles is not modeled according to the standard "D^2 law", which states that the burning time of a droplet/particle is proportional to its area [78], which is common in hydrocarbon combustion. Here, as Eq. (13.43) shows, the diameter has an exponent of 1.8. This is done to model the oxide cap formation, which reduces the exposed area of Al able to combust and therefore the resultant diffusion flame [77].

Following reaction mechanism has been used in the TNT/Al simulations for gaseous aluminium and carbon combustion, reactions C1 and C2 are the same as in Eq. (13.48) for pure TNT cases.

$$
\begin{aligned}
&\text{(C1)} && C + \frac{1}{2}O_2 \rightarrow CO \\
&\text{(C2)} && CO + \frac{1}{2}O_2 \rightarrow CO_2 \\
&\text{(A1)} && Al_{(g)} + O_2 \rightarrow AlO + O \\
&\text{(A2)} && AlO + O_2 \rightarrow AlO_2 + O \\
&\text{(A3)} && AlO_2 \leftrightarrow AlO + O \\
&\text{(A4)} && O + O + M \leftrightarrow O_2 + M \\
&\text{(A5)} && 2AlO + \frac{1}{2}O_2 \rightarrow Al_2O_3.
\end{aligned}
\tag{13.50}
$$

In the model presented here, the processes of condensation and dissociation have been sacrificed to the benefit of cost effectiveness of the code. There are models described in [77, 79, 80, 81, 82] that can be adopted to model these processes; here, however, several approximations and assumptions have been made:

1. An unburned Al particle is always covered by a very thin coating of aluminium oxide, Al_2O_3, which has to melt before the aluminium can be gasified and Al suboxides start to form; in this work, however, this coating is disregarded.

2. The vaporization of Al occurs when $T \geq 2971$ K and the amount of gaseous Al is determined by Eqs. (13.42) and (13.43).

3. The condensation reaction is neglected, and is instead modeled by a global reaction, A5 in Eq. (13.50), to incorporate this exothermic heat release.

4. The dissociation reaction is neglected.

13.5.2   Effect of Aluminium on Afterburning

The following simulations are computed on a finer mesh than simulations presented in Section 13.4, with a slightly lower charge weight and with particle addition, all to facilitate the comparisons with the experimental data [83].

An explosive charge (TNT or TNT/Al) with a 20 cm radius is computationally detonated in air at 0.5 m and 1.0 m HoB. The computational domains have a rectangular shape of sizes $5 \times 5 \times 3$ m^3 and $6 \times 6 \times 4$ m^3 with a spatial resolution of 0.01 m, resulting in domains containing 75 M and 144 M cells. The detonation product cloud is initiated at a given HoB above ground, while the rest of the domain was filled with air at atmospheric conditions, as for cases in previous section. These simulations were conducted using ILES subgrid model (Sec. 13.2.2.3) and a Noble–Able EoS (Eq. (13.23)), the two phase CGS equations are solved using a fully explicit finite volume scheme, based on the C++ library OpenFOAM [73] as the computational platform. The code employs an unstructured collocated FV method described in Section 13.2.6.

Figures 13.13 and 13.14 show simulation results for HoB 0.5 m and HoB 1.0 m, respectively, TNT results on top and TNT/Al on bottom. The results are presented in terms of red–white maps of the λ_2 (defined in Eq. (13.51)) on which the contours of the mass fraction of carbon dioxide, $\widetilde{Y}_{CO_2}$, in copper color at the center plane are superimposed. For TNT/Al cases gray scale contours of the mass fraction of aluminium oxide, $\widetilde{Y}_{Al_2O_3}$ are also included. λ_2 is defined as

$$\lambda_2 = \sqrt{\left| \frac{\partial \widetilde{u_{g,x}}}{\partial x_i} \frac{\partial \widetilde{u_{g,x}}}{\partial x_i} + \frac{\partial \widetilde{u_{g,y}}}{\partial x_i} \frac{\partial \widetilde{u_{g,y}}}{\partial x_i} + \frac{\partial \widetilde{u_{g,z}}}{\partial x_i} \frac{\partial \widetilde{u_{g,z}}}{\partial x_i} \right| - \left| \varepsilon_{ijk} \frac{\partial \widetilde{u_{g,k}}}{\partial x_j} \right|} \qquad (13.51)$$

and is intended to visualize the flow and shock wave propagation along with mixing. The contours of $\widetilde{Y}_{CO_2}$ and $\widetilde{Y}_{Al_2O_3}$ are illustrating the combustion region. Figures 13.13 and 13.14 are presented as a time series for HoB 0.5 m at times 0 ms, 0.5 ms, 1 ms, 1.5 ms, 2 ms, 2.5 ms, 3 ms, and 3.5 ms; and for HoB of 1 m cases at times 0 ms, 0.8 ms, 1.5 ms, 2 ms, 2.5 ms, 3 ms, 3.5 ms, and 4.5 ms.

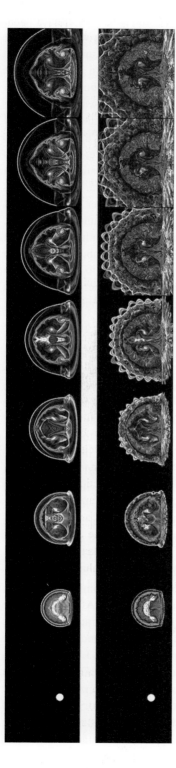

Figure 13.13 Time series of simulation results for HoB 0.5 m cases, TNT (top) and TNT/Al (bottom) in terms of red–white maps of λ_2 on which the contours of $\widetilde{Y_{CO_2}}$ in copper color and contours of $\widetilde{Y_{Al_2O_3}}$ in gray color at the centerplane are superimposed. Times are (left to right) 0 ms, 0.5 ms, 1 ms, 1.5 ms, 2 ms, 2.5 ms, 3 ms, and 3.5 ms.

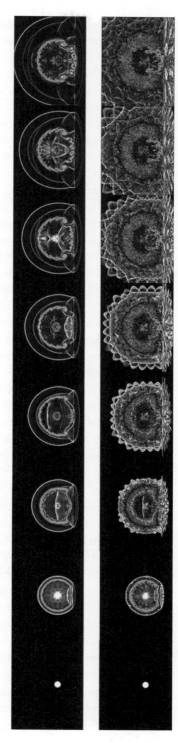

Figure 13.14 Time series of simulation results for HoB 1.0 m cases, TNT (top) and TNT/Al (bottom) in terms of red–white maps of λ_2 on which the contours of $\widetilde{Y_{CO_2}}$ in copper color and contours of $\widetilde{Y_{Al_2O_3}}$ in gray color at the centerplane are superimposed. Times are (left to right) 0 ms, 0.8 ms, 1.5 ms, 2 ms, 2.5 ms, 3 ms, 3.5 ms, and 4.5 ms.

TNT/Al cases exhibit the same major flow features as for the pure TNT cases, that is, the expansion of the primary shock, the ground reflections interacting with the mixing layer and the combustion products, the implosion (in HoB 1m cases), and the continuous shock–mixing layer interactions. Visually, the inclusion of Al has a profound effect on the mixing and the subsequent afterburning. Comparing top and bottom figures in Figure 13.13, in the early times, as the primary shock expands outward, the mixing layer, seen mostly as the expansion of the $\widetilde{Y_{CO_2}}$ contours, is more perturbed in TNT/Al for times 0.5 ms and 1 ms, suggesting an enhanced mixing that enables the afterburning. Between 1 ms and 1.5 ms the particles overtake the primary shock wave, perturbing it as well, the same phenomena as observed in Section 13.3. These perturbations are seen in the λ_2 contours as spikes forming on the perimeter of the shock wave. These perturbations of the shock create local shock merges that are locally stronger than the initial shock. For both TNT/Al cases after 1.5 ms the generation of aluminium oxide, $\widetilde{Y_{Al_2O_3}}$, becomes pronounced and soon dominates the mixing layer. The aluminium combustion seems to occur in fine vorticity structures around the carbon combustion, as both reactions are dependent on oxygen and thus mixing to continue to react. Particle inclusion creates these fine vorticity structures, together with the exothermic aluminium oxide reactions, that enable volumetric expansion, leading to a prolonged afterburning effect.

To further analyze the effects on the afterburning inside mixing layer Figure 13.15 presents profiles of (a) temperature, (b) mass fraction of CO_2, (c) area of the mixing layer, and (d) mass fractions gaseous of Al and Al_2O_3 inside the mixing layer. The extent of the mixing layer is here defined as the volume where $At \leq 0.9$. The profiles in Figure 13.15 are averages of each variable at the center plane over the extent of the mixing layer. For each variable the values are normalized by the maximum value of the variable for all cases. From Figure 13.15(a) one observes that the highest temperature of all cases is found for HoB 0.5 m–TNT/Al, which is in line with the previous studies [84, 83] that concluded that in order to have the best afterburning a continuous shock wave interaction is needed with the mixing layer, which is achieved in HoB 0.5 m cases, and the inclusion of Al further increases the temperature by, among other mechanisms, the exothermic aluminium oxide reactions. However, the duration of the high temperature region is 0.5 ms longer in HoB 1 m cases, which is attributed to the larger mixing layer area (Fig. 13.15(c)), with the TNT/Al case having highest temperature of the two. From Figure 13.15(b) and (d) it is evident that the carbon dioxide (CO_2) generation, that is, carbon combustion, dominates at early times, while the phase change of aluminium occurs, up to ~ 1 ms. After ~ 1 ms for TNT/Al cases the particles overtake the primary shock and the generation of gaseous aluminium ceases. Here, aluminium combustion starts to accelerate and dominates the afterburning at later times, as is also seen from Figures 13.13 and 13.14. This keeps the temperature high in for TNT/Al cases compared with pure TNT cases, where temperature decreases after 1 ms, following the decrease in the $\widetilde{Y_{CO_2}}$ magnitude. The only exemption to this trend is an increase in temperature and $\widetilde{Y_{CO_2}}$ for TNT–HoB 1.0 m at 3.5 ms, which is attributed to a third ground reflection for this case, indicating that the shock–mixing layer interaction is more crucial for sustainment of the

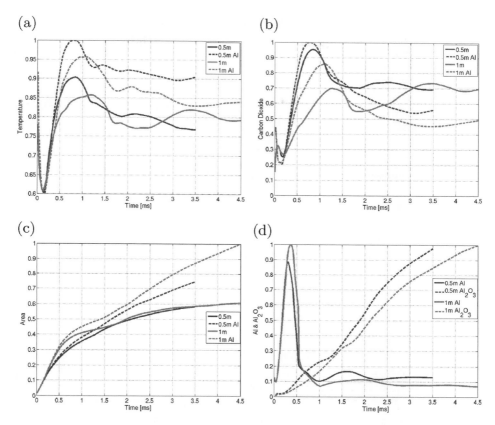

Figure 13.15 Analysis of afterburning inside the mixing layer in terms of (a) temperature; (b) mass fraction of CO_2; (c) area of the mixing layer; and (d) mass fractions gaseous of Al and Al_2O_3.

afterburning for cases without aluminium inclusion. Regarding the area of the mixing layer, Figure 13.15(c), the largest areas are found for cases containing aluminium, with the TNT/Al HoB 1.0 m having an overall largest mixing layer area. While the pure TNT cases stabilize in mixing layer area growth at 3 ms, TNT/Al continue to increase in mixing layer growth with the production of aluminium oxide, sustaining the afterburning effect.

For validation purpose, Figure 13.16 shows pressure traces from the simulations and the experiments [83] for all cases. The experimental pressure traces are shown with small error bars to account for the differences in air pressure, temperature, and humidity between the simulations and the experiments. The experimental data that are presented here are the raw data from the pressure probes, thus the only certain quantitative comparison can be made with the arrival time of the primary shock. The best agreement is achieved for the HoB 0.5 m, with pure TNT case having the best agreement with experiments regarding both the arrival time of the shock wave and its magnitude. Both cases containing Al overpredict the pressure magnitude, and in the case of TNT/Al HoB 1.0 m, quite gravely, with the pressure magnitude

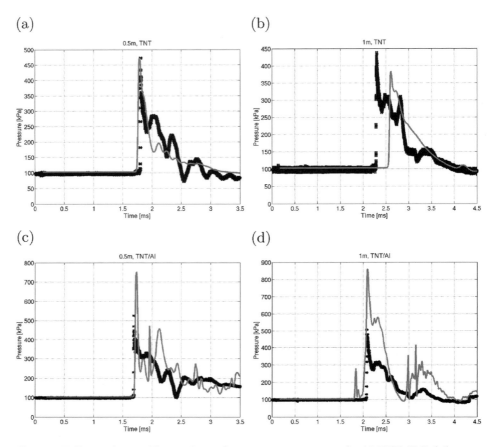

Figure 13.16 Comparisons with experimental pressure measurements for (a) TNT–HoB 0.5 m; (b) TNT–HoB 1 m; (c) TNT/Al–HoB 0.5 m; and (d) TNT/Al–HoB 1.0 m. Thick black lines: simulations; thin gray lines: experiments.

overpredicted by a factor of 5. This overprediction can be attributed to the issues with the unknown initial aluminium mass fractions and modeling choices.

13.6   Concluding Remarks

This chapter has shown how CGS is able to illustrate the complex processes involved in combustion during heterogeneous explosive blasts. The main mechanism responsible for the mixing, and therefore afterburning, is rise of RT instabilities, which trigger the build up of a mixing layer. Shock–mixing layer further create RM instabilities, these arise and generate vorticity through baroclinic effects. The presence of particles increases the vorticity generated by RT and RM instabilities since the particles create perturbations in the detonation product cloud, hence disrupting the alignment of the pressure and density gradients. Burning particles improve the mixing even further through volumetric expansion induced by increased heat release from particle combustion.

In detail, it was observed in Section 13.3 that the shape of the main outer flame at the late time is very different from that at the early time due to the significant wrinkling/convolution of the flame during and after the reshock phase. The peak flame temperatures observed in the fireball decrease in the blast wave and implosion phases marginally increase during the reshock phase, and again decrease in the asymptotic mixing phase at late times. Comparisons between heterogeneous and homogeneous charges showed constantly forced perturbations for NM/Fe charge, owing to the presence of the particles, hence resulting in higher perturbation levels and, thus, higher rms of velocity and concentrations. Also, due to the exchange of the momentum and energy between the two phases in NM/Fe charge, the implosion and reshock phases are delayed. Thereby the baroclinic torque, which plays an important role in reenergizing the turbulence levels in the mixing layer, is also delayed. The mixing layer characteristics also differ between heterogeneous and homogeneous charges, as the homogeneous explosives exhibit a linear growth in mixing layer thickness while heterogeneous charges have a nonlinear mixing layer thickness profile due to the continued forced perturbations by the particles. Small perturbations begin to grow linearly at early times as first; later, these perturbation grow to large enough sizes, at which time nonlinear growth ensues.

Section 13.4 has shown that all three examined HoB exhibited common flow features, such as shock–turbulent mixing layer interaction, the rise of RT and RM instabilities, and the ground reflected shocks affecting the mixing layer dynamics. A further investigation of mixing layer thickness and At number showed that the largest At number was found in HoB of 0.15 m and that this case was dominated by the ground reflected shock propagation. The thinnest mixing layer, with the lowest At number, was the HoB of 0.5m case. The simulation results indicated that in order to achieve maximum effect of the afterburning during an explosive blast, the turbulent mixing layer, which has to contain enough vorticity structures to support semistable mixing, has to be combined with frequent shock propagation through it, thereby sustaining the duration of the afterburning.

Section 13.5 has demonstrated the examined EBX effect of the inclusion of reacting aluminium particles has on the afterburning. The TNT/Al simulation results showed that the model is able to predict the enhanced blast effects and prolonged afterburning associated with aluminium inclusion in an explosive compound. The presence of the particles contributed to an enhanced mixing compared with pure TNT cases and aluminium reactions with air contributed to a more prolonged afterburning.

13.7   Acknowledgments

Work in Section 13.3 is supported by the Defense Threat Reduction Agency (Dr. S. Peiris, Program Manager). The computational resources were provided by DoD HPC Centers at the U.S. Army Research Laboratory DoD Supercomputing Resource Center, Engineer Research and Development Center and the Maui High Performance Computing Center. Work in Sections 13.4 and 13.5 was sponsored by the Swedish Armed Forces.

References

[1] D.L. Frost, Y. Gregoire, O. Petel, S. Goroshin, and F. Zhang. "Particle jet formation during explosive dispersal of solid particles." *Physics of Fluids*, 24:091109, 2012.

[2] A.L. Kuhl, R.E. Ferguson, and A.K. Oppenheim. "Gasdynamic model of turbulent exothermic fields in explosions." *Progress in Astronautics and Aeronautics*, 173:251–261, 1997.

[3] G.I. Taylor. "The instability of liquid surfaces when accelerated in a direction perpendicular to their planes." *Proceedings of Royal Society of London. Series A, Mathematical and Physical Sciences*, 201:192–196, 1950.

[4] K. Balakrishnan. *On the High Fidelity Simulation of Chemical Explosions and their Interaction with Solid Particle Clouds*. PhD thesis, Georgia Institute of Technology, 2010.

[5] R.D. Richtmyer. "Taylor instability in shock acceleration of compressible fluids." *Commun. Pure Appl. Math.*, 13(297), 1960.

[6] K. Balakrishnan and S. Menon. "On the role of ambient reactive particles in the mixing and afterburn behind explosive blast waves." *Combust. Sci. Technol.*, 182(2):186–214, 2010.

[7] S. Menon and N. Patel. "Subgrid modeling for simulation of spray combustion in large-scale combustors." *AIAA Journal*, 44:709–723, 2006.

[8] F. Génin and S. Menon. "Studies of shock / turbulent shear layer interaction using large-eddy simulation." *Computers & Fluids*, 39:800–819, 2010.

[9] J.C. Oefelein. "Large eddy simulation of turbulent combustion processes in propulsion and power systems." *Progress in Aerospace Sciences*, 42:2–37, 2006.

[10] K. Balakrishnan and S. Menon. "On turbulent chemical explosions into dilute aluminum particle clouds." *Combust. Theor. Model.*, 14(4):583–617, 2010.

[11] K. Balakrishnan and S. Menon. "Characterization of the mixing layer resulting from the detonation of heterogeneous explosive charges." *Flow Turbul. Combust.*, 87:639–671, 2011.

[12] F. M. White. *Viscous Fluid Flow*, third edition. McGraw-Hill, 2006.

[13] P. Sagaut. *Large Eddy Simulation for Incompressible Flows*. Springer Verlag, 2001.

[14] J. Smagorinsky. "General circulation experiments with the primitive equations. I: The basic experiment." *Month. Weath. Rev.*, 91:99–165, 1963.

[15] M. Germano, U. Piomelli, P. Moin, and W.H. Cabot. "A dynamic subgrid–scale eddy viscosity model." *Phys. Fluids A*, 3:1760–1765, 1991.

[16] A. Yoshizawa and K. Horiuti. "A statistically–derived subgrid scale kinetic energy model for large eddy simulation of turbulent flows." *J. Phys. Soc. Japan*, 54:2834, 1985.

[17] W.W. Kim and S. Menon. "A new in-compressible solver for large-eddy simulations." *International Journal for Numerical Methods in Fluid Mechanics*, 31:983–1017, 1999.

[18] F. Génin and S. Menon. "Dynamics of sonic jet injection into supersonic crossflow." *J. Turbul.*, 11(4):1–30, 2010.

[19] C. Fureby and S.I. Moller. "Large-eddy simulations of reacting flows applied to bluff-body stabilized flames." *AIAA Journal*, 33:2339, 1995.

[20] A.W. Cook and W.H. Cabot. "Hyperviscosity for shock–turbulence interactions." *J. Comp. Phys.*, 203:379–385, 2005.

[21] J. von Neumann and R.D. Richtmyer. "A method for the numerical calculations of hydro-dynamical shocks." *J. Appl. Phys.*, 21:232–237, 1950.

[22] E.J. Caramana, M.J. Shashkov, and P.P. Whalen. "Formulations of artificial viscosity for multi-dimensional shock wave computations." *J. Comp. Phys.*, 144:70–97, 1998.

[23] J.P. Boris. "Whither turbulence? Turbulence at crossroads," in *On Large Eddy Simulation Using Subgrid Turbulence Models*, 344. Springer, 1990.

[24] J.P. Boris, F.F. Grinstein, E.S. Oran, and R.J. Kolbe. "New insights into large eddy simulation." *Fluid Dynamics Research*, 10:199, 1992.

[25] F.F. Grinstein, L. Margolin, and B. Rider, editors. *Implicit Large Eddy Simulation: Computing Turbulent Fluid Dynamics*. Cambridge University Press, 2007.

[26] F.F. Grinstein and C. Fureby. "On flux–limiting–based implicit large eddy simulation." *ASME J. Fluids Engng.*, 129:1483, 2007.

[27] D. Drikakis, M. Hahn, F.F. Grinstein, C.R. DeVore, C. Fureby, M. Liefvendahl, and D.L. Youngs. *Numerics for ILES: Limiting Algorithms*, chapter 4a. Cambridge University Press, 2007.

[28] J.A. Zukas and W.P. Walters. *Explosive Effects and Applications*. Springer, 1998.

[29] M. Cowperthwaite. "Significance of some equations of state obtained from shock-wave data." *American Journal of Physics*, 34:1025–1030, 1966.

[30] L. Donahue, F. Zhang, and R.C. Ripley. "Numerical models for afterburning of TNT detonation products in air." *Shock Waves*, 23:559–573, 2013.

[31] I.A. Johnston. "The Noble–Able equation of state: Thermodynamic derivations for ballistic modeling." Technical Report DSTO-TN-0670, Australian Government Department of Defence, Defence Science and Technology Organisation, 2005.

[32] C.K. Kim, J.G. Moon, J.S. Hwang, M.C. Lai, and K.S. Im. "Afterburning of TNT explosive products in air with aluminum articles." AIAA paper 2008-1029, 2008.

[33] F.F. Grinstein and K. Kailasanath. "Three-dimensional numerical simulation of unsteady reactive square jets." *Comb. & Flame*, 100:2, 1995.

[34] B.F. Magnussen. "On the structure of turbulence and generalized eddy dissipation concept for chemical reactions in turbulent flow." *19th AIAA Aerospace Meeting*, 1981.

[35] M. Berglund, E. Fedina, C. Fureby, J. Tegner, and V. Sabel'nikov. "Finite rate chemistry large–eddy simulation of self-ignition in a supersonic combustion ramjet." *AIAA Journal*, 48:540–550, 2010.

[36] V. Sabelnikov and C. Fureby. Extended LES-PaSR Model for Simulation of Turbulent Combustion, volume 4, pages 156–169. 2012. In Advances in Aerospace Sciences.

[37] V. Sabelnikov and C. Fureby. "LES combustion modeling for high Re flames using multi-phase analogy." *Comb. Flame*, 160:83, 2013.

[38] K. Balakrishnan, D.V. Nance, and S. Menon. "Simulation of impulse effects from explosive charges containing metal particles." *Shock Waves*, 20:217–239, 2010.

[39] K. Balakrishnan, S. Ukai, and S. Menon. "Clustering and combustion of dilute aluminum particle clouds in a post-detonation flow field." *Proc. Combust. Inst.*, 33:2255–2263, 2011.

[40] D.A. Schwer and K. Kailasanath. "Numerical simulations of the mitigation of unconfined explosions using water mist." *Proceedings of the Combustion Institute*, 31:2361–2369, 2007.

[41] E. Fedina and C. Fureby. "A comparative study of flamelet and finite rate chemistry LES for an axisymmetric dump combustor." *J. Turb.*, 12:1–20, 2010.

[42] M. Tanahashi, M. Fujimura, and T. Miyauchi. "Coherent fine scale eddies in turbulent premixed flames." *Proceedings of the 28th International Symposium on Combustion*, 579–587, 2000.

[43] P.K. Yeung, S.B. Pope, and B.L. Sawford. "Reynolds number dependence of lagrangian statistics in large numerical simulations of isotropic turbulence." *J. Turb.*, 7:N58, 2006.

[44] D.M. Snider. "An incompressible three-dimensional multiphase particle-in-cell model for dense particle flows." *Journal of Computational Physics*, 170:523–549, 2001.

[45] N.A. Patankar and D.D. Joseph. "Modeling and numerical simulation of particulate flows by the Eulerian–Lagrangian approach." *International Journal of Multiphase Flow*, 27:1659–1684, 2001.

[46] K. C. Gottiparthi and S. Menon. "A study of interaction of clouds of inert particles with detonation in gases." *Combustion Science and Technology*, 184(3):406–433, 2012.

[47] S. Gottlieb and C.-W. Shu. "Total variation diminishing Runge–Kutta schemes." *Mathematics of Computation*, 67:73–85, 1998.

[48] R. Abgrall and R. Saurel. "Discrete equations for physical and numerical compressible multiphase mixtures." *Journal of Computational Physics*, 186(2):361–396, 2003.

[49] A. Chinnayya, E. Daniel, and R. Saurel. "Modelling detonation waves in heterogeneous energetic materials." *Journal of Computational Physics*, 196:490–538, 2004.

[50] E.F. Toro. *Riemann Solvers and Numerical Methods for Fluid Dynamics: A Practical Introduction*. Addison-Wesley Publishing Company, 1999.

[51] I.S. Akhatov and P.B. Vainshtein. "Transition of porous explosive combustion into detonation." *Combustion Explosions and Shock Waves*, 20(1):63–69, 1984.

[52] T. Bazyn, H. Krier, and N. Glumac. "Evidence for the transition from the diffusion–limit in aluminum particle combustion." *Proc. Comb. Inst*, 31:2021–2028, 2007.

[53] A.L. Corcoran, V.K. Hoffmann, and E.L. Dreizin. "Aluminum particle combustion in turbulent flames." *Comb. & Flame*, 160:718–724, 2013.

[54] R.A. Yetter, G.A. Risha, and S.F. Son. "Metal particle combustion and nanotechnology." *Proc. Comb. Inst.*, 32:1819–1838, 2009.

[55] J. Servaites, H. Krier, and J.C. Melcher. "Ignition and combustion of aluminum particles in shocked $H_2O/O_2/Ar$ and $CO_2/O_2/ar$ mixtures." *Comb. & Flame*, 125:1040–1054, 2001.

[56] C. Badiola, R.J. Gill, and E.L. Drezin. "Combustion characteristics of micron-sized aluminum particles in oxygenated environments." *Comb. & Flame*, 158:2064–2070, 2011.

[57] P. Lynch, H. Krier, and N. Glumac. "A correlation for burn time of aluminum particles in the transition regime." *Proc. Comb. Inst.*, 32:1887–1893, 2009.

[58] R.J. Gill, C. Badiola, and E.L. Drezin. "Combustion times and emission profiles of micron-sized aluminum particles burning in different environments." *Comb. & Flame*, 157:2015–2023, 2010.

[59] M.W. Beckstead. "Correlating aluminum burning times." *Comb. Explosion and Shock Waves*, 41:533–546, 2005.

[60] F. Zhang, F.D. Frost, P.A. Thibault, and S.B. Murray. "Explosive dispersal of solid particles." *Shock Waves*, 10:431–443, 2001.

[61] K. Balakrishnan, D.V. Nance, and S. Menon. "Numerical study of blast characteristics from detonation of homogeneous explosives." *Shock Waves*, 20:147–162, 2010.

[62] D.L. Youngs and R.J.R. Williams. "Turbulent mixing in spherical implosions." *Intl. J. Numer. Meth. Fluids*, 56:1597–1603, 2008.

[63] K.C. Gottiparthi and S. Menon. "Simulations of heterogeneous detonations and post detonation turbulent mixing and afterburning." *AIP Conference Proceedings*, 1426:1639–1642, 2012.

[64] A.L. Kuhl. "Dynamics of Exothermicity," in *Spherical Mixing Layers in Explosions*. Gordon and Breach Science Publishers SA, 1996.

[65] D.L. Frost, Z. Zarei, and F. Zhang. "Instability of combustion products interface from detonation of heterogeneous explosives." 20th International Colloquium on the Dynamics of Explosions and Reactive Systems, *Montreal, Canada*, 2005.

[66] A.L. Kuhl, A.K. Oppenheim R.E. Ferguson, and M.R. Seizew. "Visualisation of mixing and combustion of TNT explosions." *Extreme States of Substance Detonation Shock Waves, February 26–March 3, Sarov, Nizhni Novgorod Region, Russia*, 2001.

[67] A.L. Kuhl, R.E. Fergusson, and A.K. Oppenheim. "Gasdynamics of combustion of TNT products in air." *Archivum Combustionis*, 19:67–89, 1999.

[68] A.L. Kuhl, M. Howard, and L. Fried. "Thermodynamic model of afterburning in explosions." *34th International ICT Conference: Energetic Materials: Reactions of Propellants, Explosives and Pyrotechnics, June 24–27, Karlsruhe, Germany,* 2003.

[69] J.B. Bell, V.E. Beckner, and A.L. Kuhl. *Simulation of Enhanced–Explosive Devices in Chambers and Tunnels.* HPCMP Users Group Conference, IEEE, 2007.

[70] A.L. Kuhl, J.B. Bell, V.E. Beckner, and B. Khasainov. "Simulation of aluminum combustion and PETN afterburning in confined explosions." *21st International Colloquim on the Dynamics of Explosions and Reactive Systems (ICDERS), July 23–27, Poitiers, France,* 2007.

[71] T.D Tran, R.L. Simpson, J. Maienschein, and C.M. Tarver. "Thermal decomposition of trinitrotoluene (TNT) with a new one-dimensional time to explosion (ODTX) apparatus." *32nd International Conference of Institute of Chemistry Technology, Karlsruhe, Germany,* 2001.

[72] W.J. Pitz and C.K. Westbrook. "A detailed chemical kinetic model for gas phase combustion of TNT." *Proc. Comb. Inst.,* 31:2343–2351, 2007.

[73] H.G. Weller, G. Tabor, H. Jasak, and C. Fureby. "A tensorial approach to CFD using object oriented techniques." *Comp. in Physics,* 12:620–632, 1997.

[74] E.E. Meshkov. "One approach to the experimental study of hydrodynamic instabilities: Creation of a gas–gas interface using the dynamic tecnique." *Proc. 5th International Workshop on Compressible Turbulent Mixing,* 1996.

[75] E. Fedina and C. Fureby. "Numerical simulation of afterburning during explosions." *28th International Symposium on Shock Waves, July 17–22,* 2562, 2012.

[76] E. Fedina and C. Fureby. "Investigating ground effects on mixing and afterburning during a TNT explosion." *Shock Waves,* 23:251–261, 2013.

[77] M.W. Beckstead, Y. Liang, and K.V. Pudduppakkam. "Numerical simulation of single aluminum particle combustion (review)." *Comb. Explosion and Shock Waves,* 41:622–638, 2005.

[78] C. Crowe, M. Sommerfeld, and Y. Tsuji. *Multiphase Flows with Droplets and Particles.* CRC Press, 1998.

[79] S. Gallier, F. Sibe, and O. Orlandi. "Combustion response of an aluminum droplet burning in air." *Proc. Comb. Inst.,* 33:1949–1956, 2011.

[80] Y. Liang and M.W. Beckstead. "Numerical simulation of quasi–steady, single aluminum particle combustion in air." *AIAA* 98–0254, 1998.

[81] O.G. Glotov and V.A. Zhukov. "The evolution of $100 - \mu m$ aluminum agglomerates and initially continuous aluminum particles in the flame of a model solid propellant." *II. results. Comb. Explosion & Shock Waves,* 44:671–680, 2008.

[82] A.L. Kuhl, J.B. Bell, and V.E. Becker. "Heterogeneous continuum model of aluminum particle combustion in explosions." *Comb. Explosion and Shock Waves,* 46:433–448, 2010.

[83] E. Fedina. "TNT/aluminium afterburning in air blasts." Technical Report FOI-R–3913–SE, Swedish Defence Research Agency – FOI, 2014.

[84] E. Fedina and C. Fureby. "Numerical simulations of TNT afterburning at different heights of blast." 22nd International Symposium on Military Aspects on Blast and Shock (MABS22), *November 4–9, Bourges, France,* 2012.

Epilogue
Vision for Coarse Grained Simulation

Fernando F. Grinstein

E.1   Introduction

Accurate predictions with quantifiable uncertainty are essential to practical turbulent flow applications exhibiting extreme geometrical complexity and broad ranges of length and time scales. Underresolved computer studies are unavoidable in such applications. Capturing the dynamics of all relevant scales of motion with direct numerical simulation (DNS) is prohibitively expensive in the foreseeable future for practical flows of interest at moderate to high Reynolds number (*Re*). The Reynolds-averaged Navier–Stokes (RANS) approach – with averaging carried out over time, over homogeneous directions, or across an ensemble of realizations – is employed for turbulent flows of industrial complexity; the RANS focus is on statistical moments and modeling the turbulent effects.

In coarse grained simulation (CGS) macroscale large energy containing portions of the unsteady turbulent motion are computed while the rest, including molecular diffusion and other microscale physics, remains unresolved. The smallest resolved scales are determined by the resolution cutoff prescribed by discretization or by a resolved spatial filtering process. In either case, unresolved structures are eliminated, and their effects on the resolved scales must be accounted for through modeling. CGS includes classical large eddy simulation (LES) [1] focusing on explicit use of SGS models; implicit LES (ILES) [2], relying on SGS modeling and filtering provided by *physics capturing* numerical algorithms; and more general LES combining explicit/implicit SGS modeling.

We revisited the CGS paradigm for dissipative forward cascade dominated flows: small scales are enslaved to the dynamics of the largest, or, put in other words, the spectral cascade rate of energy (the rate limiting step) is determined by the initial and boundary condition constrained large scale dynamics. The physics, computation, and metrics of enslavement and mixing, and the delicate role of scale separation (the spectral gap [3]) between resolved large and modeled small scales, were examined in this context.

We emphasized the inherently intrusive nature of coarse grained observations of computational and laboratory experiments [4], intimately linked to their SGS and initial and boundary condition – supergrid (SPG) specifics. Difficult challenges are then related to characterizing and modeling the unresolved SGS and SPG aspects and assessing uncertainties associated with CGS predictions and laboratory measurements

(Chapter 5). Verification, validation, and uncertainty quantification (VVUQ) provides the rational basis to decide when the CGS modeling is good enough for its intended purpose. The outcome of the VVUQ process (Chapter 7) is a quantitative assessment of uncertainty, which provides the decision making authority with a degree of confidence to place in the modeling activity's contribution.

The present focus has been on variable density turbulent mixing driven by Richtmyer–Meshkov (RM), Kelvin–Helmholtz (KH), Bell–Plesset (BP), and Rayleigh–Taylor (RT) instabilities. For the turbulent flow applications of interest, well-characterized whole scale laboratory studies are impossible or very difficult, and CGS is the simulation strategy of choice. Fundamental predictability challenges for the complex turbulent mixing consequences relate to CGS being critically constrained by characterization and modeling of SGS and SPG specifics. An outlook vision for CGS is presented.

E.2   Fundamental Challenges

E.2.1   Subgrid Modeling

Underresolved simulations of insufficiently characterized laboratory experiments are typically involved in the complex turbulent flow applications. Robustness of CGS predictions is the frequently unsettled issue. The validity of the CGS scale separation assumptions for turbulent material mixing consequences needs to be carefully examined when driven by SGS physics.

The concept of (resolution dependent) effective Re has been extensively invoked in the past ([5–7], Chapter 1); formal framework for analysis and definition of effective Re were proposed, and practical threshold Re values were identified ([8], Chapters 1 and 2), to decide when enough scale separation might be present. Developed turbulence metrics can be used to assess CGS *convergence* once resolution is fine enough to ensure that Re is above the mixing transition Re threshold [9], near which we have a short, barely existing inertial range with minimal associated scale separation. These issues must be directly projected into the process of establishing suitable procedures and metrics for CGS validation.

SGS modeling is primarily an empirical activity within pragmatic CGS practice. Most SGS models are designed to be dissipative, effectively regularizing the equations at the resolved scale level. In this context, we envision building CGS on the basic SGS modeling provided by suitable physics capturing numerics [2] allowing for high Re convection driven dynamics to be captured well and efficiently, and incorporating additional models for nonconvective and SGS driven physics as suitable source terms in the augmented Navier–Stokes (NS) equations. Important challenges demand evaluating the extent to which particular SGS physical effects can be implicitly modeled as turbulent velocity fluctuations and recognizing when additional explicit models and/or numerical treatments are needed within an appropriate VVUQ process. Modified equation analysis [10, 11] provides the mathematical framework for CGS analysis and development, in seeking to reverse-engineer desired SGS physics through numerics or through mixed (implicit/explicit) SGS modeling.

When convective timescales are much smaller than those associated with molecular diffusion, for moderately high Re, the primary concern of the numerical simulation is given to the convectively driven interpenetration mixing processes (entrainment and stirring due to velocity gradient fluctuations), which can be captured once the dissipation is sufficiently resolved with nominally inviscid (Euler) or NS based CGS – Chapters 1 (Fig. 1.1) and 5.

A proof of concept for enslaved turbulent mixing was presented in Chapter 1. Well-designed, sufficiently resolved ILES can capture the fundamental aspects of the mixing transition and characteristics of developed isotropic turbulence for high Re and order-unity Schmidt number (Sc). The detailed analysis presented was based on statistical turbulence metrics and probability distribution functions (PDFs) of velocity and scalar functions, including effects of Mach number and grid resolution. As the effective (grid resolution dependent) Re is increased, the SGS scalar mixing model implicitly provided by a well-designed ILES numerics was found to be adequate to consistently capture well-established mixing characteristics, namely, the Gaussian behavior of fluctuating velocity and scalar PDFs, non-Gaussian (and appropriately biased when applicable) PDF tails of their derivatives, asymptotically constant nondimensional scalar variance, and increasing squared ratio of the Taylor microscales with Re. The results are regarded as a clear demonstration of the feasibility of predictive underresolved simulations of stirring driven high Re turbulent scalar mixing with ILES. They strongly suggest enslavement of the small scale mixing dynamics to that of the larger scales of the flow for large Re.

We examined fundamental ways to ensure that enough scale separation is involved in the simulations. An analysis allowing for large scales of very high Re turbulent flows to be handled by CGS was inspected in Chapter 2, where our current understanding of the energy transfer process of the turbulent flows, which forms the foundation of our scale separation arguments was surveyed. Two distinctive interactions, namely the distance and near grid interactions, were inspected. SGS distant interactions in an inertial range can be effectively modeled by an eddy damping mechanism, but the near grid interactions must be carefully incorporated. A minimum turbulence state was defined as the lowest Re turbulent flow that captures the energy containing scales of prototypical astrophysical problems in a laboratory or simulation setting. Unsteady flow transition criteria were also presented. The spatial and temporal criterions have found applications to a range of high energy density physics experiments, and the procedure for estimating the numerical viscosity for ILES was advanced.

We revisited the discrete finite scale NS (FSNS) framework, as it reframes the computational questions of turbulence. In the continuum, the usual questions relate to the velocity at a point, a concept limited computationally by finite volumes (FVs), mesh spacing dx, and time-step dt, and there are associated convergence issues as well as issues of consistent definition of FV averaged velocities. The question FSNS proposes to answer is different (Chapters 3 and 4): what is the velocity in a FV (L^3) over a finite time T? There are no convergence issues in the FSNS framework, as the true and desired solution is not the NS solution but rather an average over space and time (L and T) of its solution, and the definition of averaged velocity will not change once $dx < L$ and $dt < T$ – that is, there is no need for SGS modeling once the latter resolution limit can be achieved. The motivation for FSNS then relates to directly addressing the nature of laboratory observables: measurement devices always involve finite space/time scales and cannot compute arbitrarily small scales of high Re turbulence. By design, FSNS actually computes what is observable for a measurement device characterized by length and time scales L and T (e.g., thermocouple or hot wire width L, instrumentation inertia T). Additional motivation for such approaches (beyond addressing the noted inherent limitations of the measuring process), come from practical applications with inherently discrete (noncontinuum) requirements: for example, needed contaminant dispersal prediction for urban consequences management is dosages (contaminant in volume L^3 over time T).

The demonstrated connections between ILES and FSNS provide a rationale for ILES (Chapters 3 and 5, and references therein): ILES works because it solves the equations that most accurately represent the dynamics of FVs of fluid governing the behavior of measurable physical quantities on the computational cells [2].

Ensemble moment equation analysis of FSNS in Chapter 4 provided an objective understanding of CGS of high Re NS turbulence. The analysis shows that passive scalar mixing by FSNS turbulence is mathematically consistent with the mixing fluid physics in NS turbulence; specifically, the NS stirring term in the scalar gradient variance equation appears in the scalar variance equation in a form consistent with the material conservation of the original NS equations. A notable result in Chapter 4 was showing that popular eddy viscosity LES does not exhibit this fundamental conserved invariant property present in NS and FSNS. On a similar note, Chapter 1 showed that asymptotic growth with Re of the ratio of eddy turnover rates of turbulent kinetic energy and scalar variance predicted by theory and DNS can be captured with a (numerically) well-designed ILES – whereas well-regarded LES using explicit SGS modeling predicts constancy. These results point at realizability constraints, which are naturally built into ILES/ FSNS but not necessarily in the mainstream explicit SGS modeling strategies.

E.2.2   Supergrid Modeling

Historically, less attention has been devoted to SPG modeling aspects. A special focus of the discussion in Chapter 5 was devoted to turbulent initial (or inflow) condition issues. We illustrated crucial characterization and modeling difficulties encountered when attempting to integrate CGS and laboratory experiments for complex flow problems of interest. The laboratory data is typically insufficiently characterized; different possible initial conditions (IC) and boundary conditions (BC) consistent with the available laboratory information are not equivalent and can lead to significantly different simulated flow solutions. One issue is that of selecting suitable complete datasets to close the IC and BC formulations in the simulations; the other involves appropriate laboratory data acquisition and reduction to capture the relevant initial (or upwind) unsteady multiscale flow physics.

We examined difficulties and challenges introduced by underresolution of initial material interface conditions in the case of underresolved shock driven turbulent material mixing simulations – when resolving all physical space/time scales in numerical simulations is prohibitively expensive. Here, SGS and SPG issues become unavoidably intertwined, and we find that a crucial issue is that of availability of well-characterized, -resolved, and -modeled SPG sets of conditions.

Traditionally, the loss of memory assumption has been made in turbulence research, that is, that IC effects eventually wash out as the turbulence develops [12]. However, a growing body of fundamental research indicates that only very special turbulent flows are truly (universally) self-similar. Far field (late time) sensitivity to ICs has been extensively reported in recent years, and it is now well recognized that transition to different self-similar end results (far field or late time) [13] is possible depending on IC. Robustness of CGS results can thus become an important unsettled issue when results are sensitive to small scale IC content – Chapters 5, 6, and 8.

Though establishing accurate SPG conditions is extremely important for forecasts of cloud systems such as hurricanes, Chapter 6 exemplified how numerical errors and/or underresolution of the IC physics can lead to predictions that do not agree with observations. The track of hurricane Sandy was not accurately forecasted by the Global Forecast System model using its standard operational resolution, with higher resolution being needed to resolve latent heat release and its subsequent impact on track. Fortunately, unlike what was shown in the stratus and hurricane simulations in Chapter 6, resolving evaporative cooling for hurricane Sandy was not as important as resolving condensational heating. And under circumstances in which evaporation does play a significant role in the overall energy balance, Eulerian cloud models may still have difficulty resolving evaporation at the edges of clouds. A possible solution to this problem is the use of a Lagrangian cloud model approach employing particles that can reasonably resolve evaporation occurring at cloud edges and not distort via advection fields of cloud water. But, as shown in Chapter 6 for the ILES LCM simulation of hurricane Guillermo, while the approach can better represent the motion of cloud water and its impact on hurricane intensity, the resolution employed in the simulation is still not sufficient to resolve fields of latent heat release. Thus, outstanding questions remain on whether the impact of this energy release needs to be accurately represented at coarser model resolutions via some type of explicit SGS model or underresolving these fields may still lead to a predictive model solution.

E.3   Complex Mixing Consequences

For the complex flows selected, deterministic simulation studies are expensive and critically constrained by limitations in characterizing, modeling, and validating all the relevant physical subprocesses. We extrapolate from our understanding and established analysis of CGS performance in equilibrium turbulent flows, and rely on simulation model confidence developed from building block VVUQ and testing.

The challenging problem of underresolved mixing of material scalars promoted by under resolved velocity and underresolved IC in shock driven turbulent flows was examined in Chapter 8. The complexity of shock waves and other compressibility effects add to the physics of material mixing and to difficult issues of characterization and modeling of the initial and dynamic material interfaces. Because the RM instability is shock driven, resolving all physical space/time scales in numerical simulations is prohibitively expensive even on the largest supercomputers. By combining shock and turbulence emulation capabilities on a single model ILES provides an effective CGS strategy for the study of RM. The inherent difficulties with the open problem of predictability of material stirring and mixing by underresolved multiscale turbulent velocity fields are now compounded by the inherent sensitivity of turbulent flows to IC. Dedicated diagnostics do not exist for the inhomogeneous transitional and decaying turbulent flow involved here; analysis relies on versions truly designed for steady homogeneous isotropic regimes. The focus was on effects of initial spectral content and interfacial morphology on transitional and late time turbulent mixing in fundamental shock tube experiments, and examined practical VVUQ difficulties encountered in CGS predictability evaluations in complex configurations for which state of the art laboratory data and diagnostics are available. A single parameter characterizing the initial rms material interface slopes was identified as relevant in determining whether shock driven flow is in linear ballistic or nonlinear mode coupling regimes – the bipolar RM behavior.

Chapter 9 reports ILES studies of laser driven reshock, shear, and ICF capsule Omega experiments in the strong shock high energy density regime. Validation studies were based on direct comparisons of simulation and experimental radiographic data supplemented with spectral analysis. ILES results were also compared with comparable results published from DNS and theory of homogeneous isotropic turbulence, seeking to characterize the small scale flow features and their turbulence statistical measures. While the flow is not isotropic or homogeneous, and is statistically unsteady, the flow presents regions exhibiting features of isotropic homogeneous turbulence over finite times. For the reshock experiments, our analysis shows self-similarity characteristics and effective Re assessments suggesting that the mixing transition is achieved, and that the simulations are reasonably converged at the finest resolution after reshock. An improved strategy for performing simulations initialized in two dimensions (2D) and later rotated to 3D was presented, compensating for the reduced turbulence development in the 2D phase of the simulation and the artificial smoothing effects of the 2D–3D spinning. This method was used to simulate the implosion of a simple Omega type ICF capsule, consisting of a polystyrene shell surrounding deuterium gas. The simulations compare well with the available laboratory data, and are used to demonstrate that the dominant mechanism for yield degradation in the capsule compared with lower dimensional simulations is the displacement of fuel from the hot spot by shell material induced by turbulent instability growth generated by long wavelength surface defects. This effect is compounded by drive asymmetry, which breaks up the burn region both spatially and temporally, further reducing the yield and significantly extending the burn width. Thus, discrepancies between the yield predicted by previous simulations of

ICF implosions can be explained by the reduced growth of turbulent instabilities inherent in reduced dimension hydrodynamic simulations and the absence of appropriate modeling of long wavelength asymmetries that are present due to surface defects and drive asymmetries.

ILES studies in Chapter 10 further investigated dominant mechanisms in an asymmetrically driven ICF implosion, seeking to clarify the essential role of convergence and shear. Pressure drive asymmetries on the capsule lead to nonradial flow in the shell, which in turn leads to macroscopic density enhancements in the shell that grow in time as a result of BP related convergence effects. In the compressible case considered here the growth of the density perturbations leads to radially inward and outward extension of the shell. These macroscopic perturbations act as seeds for the development of RM fingers of shell material in the gas as a result of multiple interactions with the reflected gas shock. This same interaction results in strong KH shear flow near the fingers in the form of counterrotating vortices in the gas. Radial convergence pushes both the fingers and associated vortices closer together as the implosion progresses. Radial convergence also guarantees that counterrotating vortices will suffer azimuthal instability 3D growth, resulting in rapid development of stirring and turbulence in the bubbles trapped between the fingers. Stirring promotes interpenetration of shell material and gas that represents fully 3D mixing. The physical picture described earlier suggests why it is particularly difficult to achieve the ideal 1D yield in a high convergence ICF implosion. The issue being addressed is one of fundamental importance: is the yield degradation observed in ICF implosions typically determined by small scale surface features or by large scale drive asymmetries? Chapter 11 suggests that the large scale coherent structures associated with asymmetries in the pressure drive may play an important role determining yield performance of ICF implosion systems, especially at high convergence.

Chapter 11 demonstrated ILES use as a tool for validating RANS models in the context of RT driven turbulent mixing. The RANS model can efficiently predict a variety of turbulence quantities including mixing width, turbulent kinetic energy and turbulent mass–flux velocity that are consistent with more computationally expensive ILES models. The turbulence model couples to the species mass fraction equations via the gradient diffusion approximation and a turbulent Sc. At the mean flow level the model does not adversely affect the interface motion, and it accurately captures differences in turbulence intensity along the moving interface. The four equation, single fluid turbulence model is found capable of capturing several distinctive features of the tilted rocket rig experiment. This includes distributions of turbulent kinetic energy and countergradient turbulent mass–flux velocity. This suggests that this family of models is a viable option for simulating these flows in addition to more complex multifluid turbulence models. Gradient diffusion models for the turbulent mass–flux velocity that are often used for two equation models cannot capture countergradient behavior observed in our simulations of the tilted rocket rig. Euler based ILES can be a useful tool for analyzing the turbulent mixing away from boundaries. Setting appropriate IC for RANS models remains a challenge.

Chapter 12 was devoted to modeling and simulation of swirling flows in gas turbine type of combustors, with primary focus on liquid fueled systems, with swirl being used extensively in premixed and non-premixed (gaseous) systems as well. There have been several investigations of such systems using RANS, and, recently, some DNS of swirling spray combustion have been performed, but such studies are typically relegated to lower *Re* due to a huge computational cost and memory requirements. The main focus of the discussion was on modeling, simulation challenges, and key results obtained using LES of systems that are more realistic to laboratory or flight conditions. LES of swirling spray combustion in gas turbine combustors is extremely challenging from the modeling and computational perspective and so far only limited studies have been attempted, usually for atmospheric laboratory combustors. Unsteady features such as the efficient dispersion of the spray by the rotating precessing vortex core structure and the flame stabilization by the vortex breakdown bubble are key features in such combustors and LES has shown an ability to capture these physics. The flame structure shows a wide range of burning modes including partially premixed combustion. Simulation of spray combustion in realistic systems still requires development of SGS models and also a robust treatment of the breakup process. Finally, multiinjector problems are currently possible with LES, but with limited data and large computational cost such simulations are only limited to demonstration purposes at this time.

Finally, CGS is used in Chapter 13 to exhibit the complex combustion processes involved during heterogeneous explosive blasts. The main mechanism responsible for the mixing, and therefore afterburning, is rise of RT instabilities, which trigger the build up of a mixing layer. Shock–mixing layer further create RM instabilities; these arise and generate vorticity through baroclinic effects. The presence of particles increases the vorticity generated by RT and RM instabilities since the particles create perturbations in the detonation product cloud, hence disrupting the alignment of the pressure and density gradients. Burning particles improve the mixing even further through volumetric expansion induced by increased heat release from particle combustion. Common flow features observed are extensively described, including shock–turbulent mixing layer interaction, the rise of RT and RM instabilities, and the ground reflected shocks affecting the mixing layer dynamics. The simulation results indicated that in order to achieve maximum effect of the afterburning during an explosive blast, the turbulent mixing layer, which has to contain enough vorticity structures to support semistable mixing, has to be combined with frequent shock propagation through it, thereby sustaining the duration of the afterburning.

E.4   Outlook for CGS

As we move away from the CGS realm proper – dissipative forward cascade dominated flows – additional physics based modeling may be needed to address SGS driven flow dynamics. There are potentially more SGS effects than the energy transfer dissipation that numerical regularization or common explicit LES models are designed to represent. This includes modeling material mixing, shock driven physics, and combustion processes in nonconvectively dominated regimes – when significant backscatter effects may be present. Such phenomena act at and affect the small scales of turbulence that are not resolved in CGS – which can invalidate the (dissipative) assumptions on which SGS models (implicit or explicit) are based in current pragmatic CGS practice. For turbulence that is strongly inhomogeneous, anisotropic, out of equilibrium, or has other physical phenomena that act at small scales, interactions between SGS and resolved turbulence – and associated scale separation requirements – can be more complex (and subtle) than those for a simple exchange of energy.

Nondissipative SGS modeling issues are presumably less important for high enough *Re* (large scale separation), but need to be carefully addressed when relatively short scale separations are involved. Largely uncharacterized small scale turbulence processes to be modeled remain in the turbulent variable density context, including effects of SGS perturbations to initial material interfaces; effects of interacting shocks and sharp material interfaces with turbulence; baroclinic production of vorticity and other effects of small scale density variations; and exothermicity effects of chemical (or thermo-nuclear) reactions resulting from molecular scale material mixing.

Robust CGS for dissipative turbulent phenomena exhibiting enslavement of small scale dynamics is in principle achievable with suitable SGS modeling, enough scale separation, and well-resolved IC. However, late-time predictability assessments for high *Re* phenomena can not be robust when inherent CGS grid resolution (or explicit spatial filtering) sensitivities are present: simulations and analysis are constrained by characterization and modeling of (intertwined) SGS and SPG specifics, while nature controls the flow physics independently.

The impossibility of very long range weather forecasting [14] – beyond about two weeks [15] – appropriately comes to mind here. Because of chaotic variability associated with unavoidable small perturbations (uncertainties) of presumed SGS and SPG conditions, it may be impossible – even within a mathematically well-posed dissipative flow simulation framework – to provide realistic late time solutions good enough to address the specific questions of interest (e.g., the very long range predictions for weather forecasting). Chapters 5, 6, and 8 specifically exemplified such inherent difficulties in the context of serious sensitivity of late-time effects to initial prescription of material interface physics and conditions (e.g., see discussion of Figs. 8.5, 8.16, and 8.17 in Chapter 8).

If the IC information contained in the filtered-out smaller and SGS spatial scales can significantly alter the evolution of the larger scales of motion and practical integral measures, then the utility of any CGS for their prediction as currently posed is dubious and not rationally or scientifically justifiable [4].

In practice, mathematical simulation framework, specific questions to be addressed, and predictability metrics in the VVUQ context must be qualified for the particular intended simulation purpose – and only certain statistical predictions and data types will be typically useful in this context. Ensemble averaged CGS over a suitably complete set of realizations covering the relevant IC variability is a simulation strategy of choice – albeit computationally expensive for many 3D turbulent flows of practical interest.

Beyond the complex multiscale resolution issues, me must also address the equally difficult problem of transition prediction. Transition to turbulence involves unsteady large scale 3D vortex dynamics – which can be captured by CGS [13].

By effectively blending computed and modeled dissipation we can also capture relevant aspects of transition physics with a well-designed hybrid RANS/CGS strategy [17, 18] – where CGS ensemble choices correspond to IC and closure choices in the RANS context. Hybrid RANS/CGS thus provide computationally feasible ways of implementing the ensemble averaging goals – while addressing unsteady small scale content and additional associated degrees of freedom. Hybrid RANS/CGS is a well-established industrial standard for aerospace compressible flows [19]. Research extending these hybrid strategies to variable density turbulent flows is clearly warranted.

References

1. Sagaut P. 2006, *Large Eddy Simulation for Incompressible Flows*, 3rd ed., Springer.
2. Grinstein, F.F., Margolin, L.G., and Rider, W.J., eds., 2010, *Implicit Large Eddy Simulation: Computing Turbulent Fluid Dynamics*, 2nd printing, Cambridge University Press.
3. George, W.K., and Tutkun, M., 2009, "Mind the gap: A guideline for large eddy simulation," *Phil. Trans. R. Soc. A*, 367, 2839–47.
4. Grinstein, F.F., 2009, "On integrating large eddy simulation and laboratory turbulent flow experiments," *Phil. Trans. R. Soc. A*, 367, 2931–45
5. Cohen, R.H., Dannevik, W.P., Dimits, A.M., Eliason, D.E., Mirin, A.A., Zhou, Y., Porter, D.H., and Woodward, P.R., 2002, "Three-dimensional simulation of a Ritchmyer-Meshkov instability with a two-scale initial perturbation," *Physics of Fluids*, 14, 3692–709.
6. Drikakis, D., Fureby, C., Grinstein, F.F., and Youngs, D., 2007, "Simulation of transition and turbulence decay in the Taylor–Green Vortex," *Journal of Turbulence*, 8, 020
7. Wachtor, A.J., Grinstein, F.F., Devore, C.R., Ristorcelli, J.R., and Margolin, L.G., 2013, "Implicit large-eddy simulations of passive scalar mixing in statistically stationary isotropic turbulence," *Physics of Fluids*, 25, 025101.
8. Zhou, Y., Grinstein, F.F., Wachtor, A.J., and Haines, B.M., 2014, "Estimating the effective Reynolds number in implicit large eddy simulation," *Phys. Review E*, 89, 013303.
9. Dimotakis, P.E., 2000, "The mixing transition in turbulent flows," *J. Fluid Mech.*, 409, 69–98.
10. Hirt, C.W., 1969, "Computer studies of time-dependent turbulent flows," *Phys. Fluids Suplement II*, 219–227.
11. Ghosal, S., 1996, "An analysis of numerical errors in large-eddy simulations of turbulence," *J. Comp Phys.*, 125, 187–206.
12. Townsend, A.A., 1976, *Structure of Turbulent Shear Flow*, Cambridge University Press.
13. George, W.K., and Davidson, L., 2004, "Role of initial conditions in establishing asymptotic flow behavior," *AIAA Journal*, 42, 438–446.
14. Lorenz, E.N., 1963, "Deterministic nonperiodic flow," *J. of Atmospheric Sciences*, 20, 130–141.
15. Lorenz, E.N., 1972, "Predictability: Does the flap of a butterfly's wings in Brazil set off a tornado in Texas," *American Association for the Advancement of Science Meeting, Washington DC, December*, 29, 1972.
16. Spalart, P.R., 2000, "Strategies for turbulence modeling and simulations," *Heat and Fluid Flow*, 21, 252–263.

17. Frolich, J., and von Terzi, D., 2008, "Hybrid LES/RANS methods for the simulation of turbulent flows," *Prog. Aerosp. Sci.*, 44, 349–77.

18. Leschziner, M., Li, N., and Tessicini, F., 2009, "Simulating flow separation from continuous surfaces: routes to overcoming the Reynolds number barrier," *Phil. Trans. R. Soc. A*, 367, 2885–2903.

19. Fasel, H.F., von Terzi, D.A., and Sandberg, R.D., 2006, "A methodology for simulating compressible turbulent flows," *J. Applied Mechanics*, 73, 405–412.

Index